Springer Series in Statistics

For further volumes:
http://www.springer.com/series/692

Shlomo Yitzhaki • Edna Schechtman

The Gini Methodology

A Primer on a Statistical Methodology

Springer

Shlomo Yitzhaki
Department of Economics
The Hebrew University
Mount Scopus, Jerusalem
Israel

Edna Schechtman
Department of Industrial Engineering
and Management
Ben-Gurion University of the Negev
Beer-Sheva, Israel

ISSN 0172-7397
ISBN 978-1-4614-4719-1 ISBN 978-1-4614-4720-7 (eBook)
DOI 10.1007/978-1-4614-4720-7
Springer New York Heidelberg Dordrecht London

Library of Congress Control Number: 2012946727

Printed on acid-free paper

Springer is part of Springer Science+Business Media (www.springer.com)

To Ruhama, Guy, and Nili who lived under the shadow of Gini, and Ido, Ella, and Roni who are a lovable distraction.

Shlomo

To Gideon

Edna

Acknowledgement

This book covers research that was carried out for about 40 years. However, it is clear to us that the task of investigating and implementing the Gini methodology is far from being completed. Many researchers, some of them anonymous referees, have contributed ideas that shaped the development of the ideas in this book. In some of the cases we can identify and acknowledge their contributions.

The interest of the first author in the Gini coefficient was when trying to convince academic members of the Ben-Shahar committee for tax reform in Israel, 1975, to belittle the recommendation of economic theory which claims that increasing the penalties is a sufficient instrument for decreasing tax evasion. The major argument was that if a crime is committed by many people then it is not considered as a crime. It was Martin Feldstein who suggested searching for an argument that is not related to the classical utility theory, which led to the connection between the Gini and the relative deprivation theory.

Ingram Olkin should be credited for the idea that one should use concentration curves instead of parameters to investigate the relationships between variables, an idea that led to concepts like marginal conditional stochastic dominance and inspecting whether relationships between variables are monotonic, a research agenda which is still in its initial stages.

Collaborations with other researchers brought into this book the specialization and the intimate knowledge of specific areas that helped spread the Gini methodology. The late Joachim Frick, Yoel Finkel, Jan Goebel, Bob Lerman, Joram Mayshar, Branko Milanovic, Joel Slemrod, Dan Slottje, Wayne Thrisk, Gert Wagner, and Quentin Wodon were influential in shaping the implementations in the areas of public finance and income distribution, while Haim Shalit was in charge of the area of finance, Oded Stark and Ed Taylor shaped the implementation in demography, and Manuel Trajtenberg in diffusion processes. Joseph Heller introduced us to the fascinating world of zoology. Yevgeny Artsev, Yolanda Golan, Vika Roshal, Taina Pudalov, and Amit Shelef assisted in implementing the Gini methodology, while Alexandra Katzenelbogen and Tamir Erez programmed the computer programs used in the book. The mathematical skills of Peter Lambert and Gideon Schechtman were crucial in overcoming some obstacles, while Joel Slemrod was the reader and advisor of many papers. Correspondence with Jim Heckman and conversations with Josh Angrist were helpful in understanding the issues in Econometrics.

Over 25 years of cooperation between the two authors has led to the development of the statistical background of the Gini methodology, while Peter Lambert and Gideon Schechtman have helped in overcoming some mathematical difficulties. The reading and advice of Joel Slemrod provided the intellectual support needed to overcome some of the difficulties in the understanding and the exposition of the theory.

The idea of writing the book was initiated by John Kimmel, while Hillel Bar Gera, Michael Beenstock, Ingram Olkin, Peter Lambert, David Johnson, and Christian Toft kindly commented on the final draft. The tedious task of bringing the book into its final form is carried out by Hannah Bracken under the supervision of Marc Strauss but all the mistakes are ours.

Contents

Chapter 1
Introduction

Gini's mean difference (hereafter, GMD) was first introduced by Corrado Gini in 1912 as an alternative measure of variability. GMD and the parameters which are derived from it (such as the Gini coefficient, also referred to as the concentration ratio) have been in use in the area of income distribution for almost a century, and there is evidence that the GMD was introduced even earlier (Harter, 1978). In other areas it seems to make sporadic appearances and to be "rediscovered" again and again under different names. It turns out that GMD has at least 14 different alternative representations. Each representation can be given its own interpretation and naturally leads to a different analytical tool such as L_1 metric, order statistics theory, extreme value theory, concentration curves, and more. Some of the representations hold only for nonnegative variables while others need adjustments for handling discrete distributions. On top of that, the GMD was developed in different areas and in different languages. Corrado Gini himself mentioned this difficulty (Gini, 1921). Therefore in many cases even an experienced expert in the area may fail to identify a Gini when he or she sees one.

Covering all the approaches in detail can become tiresome and possibly uninteresting. Therefore, in order to overcome this "curse of the plenty" we set one target in mind. We shall focus on imitating the analyses that are based on the variance by replacing the variance by the GMD and its variants. We intend to show that almost everything that can be done with the variance as a measure of variability can be replicated by using the Gini. With this target in mind we will mainly focus our attention on one representation—the covariance-based approach—and limit the coverage of other approaches.

The use of GMD as a measure of variability is justified whenever the investigator is not ready to impose, without questioning, the convenient world of normality. When the underlying distribution is univariate and normal, the sample mean and variance are sufficient statistics to describe it and the GMD is redundant. Likewise, when dealing with multivariate distributions, the case of multivariate normality is fully described by the individual means, the individual variances, and Pearson's correlation coefficients. The GMD and the equivalents of the correlation coefficient have nothing to add to the understanding of the data, nor to the analysis. However

S. Yitzhaki and E. Schechtman, *The Gini Methodology: A Primer on a Statistical Methodology*, Springer Series in Statistics 272, DOI 10.1007/978-1-4614-4720-7_1,

when the distribution is not multivariate normal, then, as Lambert and Decoster (2005) put it—the GMD reveals more! As will be demonstrated in this book, it reveals whether the relationships between random variables (as described by the covariance and by the correlation) are symmetric or not, whether the population is stratified and to what extent, whether the assumption of linearity in regression analysis is supported by the data, and more. The use of GMD may add insight and understanding of the data at hand. For example, it can be used whenever one wants to see if the assumption that the underlying distribution is multivariate normal holds, or if the regression model is truly linear. However it comes with a price tag on it. It turns out that using the GMD as a substitute for the variance implies that the number of economic models is doubled because every variance-based model will now have beside it a Gini-based model that may give different results. We will show that many of the properties of the variance-based models are included as special cases of Gini-based models. As a result, we argue that if the estimates of the variance-based and Gini-based models are close to each other then we obtain reassurance that the model is robust in the sense that it is not sensitive to the implicit assumptions imposed on the data by treating it as if the underlying distribution is multivariate normal. On the other hand, if the estimates differ then it is an indication that the implicit assumptions of the variance-based model are responsible for the deviation. As far as we can see, in many cases using the Gini methodology in addition to the variance-based methods will lead to a reduction in the number of possibilities of generating "empirical proofs" that support the researcher's theory but are not supported by the data itself.

This book is a first attempt to present the family of parameters based on GMD and to illustrate its applications in different areas, mainly in the areas of economics and statistics. The main thrust is to "translate" the commonly used analyses based on the variance and the parameters based on it into the Gini world. Parameters such as the covariance and Pearson's correlation coefficient, as well as methodologies such as ANOVA and Ordinary Least Squares (OLS) regressions, are "translated" where the variance is replaced by (the square of) the GMD, the covariance and Pearson's correlation are replaced by Gini covariance and Gini correlation, ANOVA is changed to ANalysis Of GIni (ANOGI), and OLS regression is replaced by Gini regression. As will be shown, the above "translation" gives rise to additional parameters and the alternative approach reveals more when the underlying distribution deviates from the multivariate normal. The slogan of this book is "(almost) everything you can do (with the variance), we can do better (with the Gini)." By "doing better" we mean that the approach offers richer tools for statistical analyses and that the additional parameters that the Gini method offers enable the researcher to adjust the statistical analysis to the needs of the area of research. We argue that the convenience of the assumption of multivariate normality could blur some of the issues that are relevant in several areas of research such as risk analysis, income distribution, economics, and sociology. It should be stated that the task of "translating" the variance world into the Gini regime is not yet completed. In some sense we feel that we are touching the tip of the iceberg and plenty of additional theoretical work as well as user-friendly software are called for to fully utilize the Gini methodology as an analytical tool.

One of the advantages of using the Gini methodology is that it provides a unified system that enables the user to learn about various aspects of the underlying distribution. Almost every property of the underlying distribution that the Gini method enables us to present or test can also be described and tested using other approaches but the advantage here is that we provide a systematic method and a unified terminology.

Let us illustrate this point. Consider the methodology of estimating and verifying a simple linear regression model. The Gini methodology enables the user to estimate the regression coefficient, draw inferences on it, check whether the model is linear, and verify that the residuals are normally distributed—all under one systematic method.

The variance is the most popular measure of variability. There are two properties which seem natural and are implicit when dealing with the variance: the symmetry and the decomposition (to be detailed below). The Gini approach deviates from this conventional (and convenient) approach. Understanding these two points will make the ideas that are stressed in the book easier to follow.

(a) Symmetric relationship: There are two kinds of symmetric relationships that are imposed in the conventional statistical analysis in general but are not followed in this book. The first one is the symmetry of the variability measure with respect to the underlying distribution and the second one is the symmetry in the relationship between variables. The first symmetry can be described as requiring that the variability of X will be equal to the variability of (−X). The justification for not following this kind of symmetry is because some of the subject matters that we are dealing with such as the areas of risk and income distributions are governed by theories that call for asymmetric treatments of the distributions. This issue is handled in Chap. 6 which presents the extended Gini and in Chaps. 13, 17, and 19 which present applications in the areas of welfare economics and finance and in econometrics. The second deviation from the symmetry properties is concerned with the treatment of the relationship between two random variables. Most measures of association are symmetric with respect to the two variables, as is the case in $\text{cov}(X, Y) = \text{cov}(Y, X)$, even if the underlying bivariate distribution is not symmetric. Symmetry between random variables is a convenient property to have, but it comes with a price tag. The price paid is in the value imposed on the correlation. To see this consider two normally distributed random variables X and Y with a Pearson correlation coefficient of (−1). A researcher not knowing what the underlying distribution is may decide to use the exponential transformation to get e^X and e^Y changing the distributions to be lognormal. By doing that, the researcher inadvertently reduces the Pearson correlation coefficient to −0.36 (De Veaux, 1976). We argue that the change of the Pearson correlation from (−1) to (−0.36) should be attributed to the symmetry imposed by the covariance. We can also reverse the example. By taking the natural logarithm of two lognormally distributed random variables with a Pearson correlation coefficient of (−0.36) the researcher changes the Pearson correlation coefficient to (−1). The above example is a bit extreme and only rarely occurs in practice. Consider a more plausible story. Given two normally distributed random variables with a Pearson correlation of 1, a researcher transforms one of them into a binary variable, a procedure intended to describe the participating/nonparticipating dichotomy. This is a common practice when applying Instrumental Variable procedure in regression analysis. In this case the researcher reduces the Pearson correlation from 1 to 0.8. The conclusion from these two examples is that a transformation can change the correlation, enabling the researcher to change the conclusion of the research. The Gini approach offers a remedy to this problem. There are two correlation coefficients defined between each pair of variables. These two correlations are equal if the distributions are exchangeable up to a linear transformation, which we will refer to as symmetric relationships. Applying a transformation to a variable changes only one (Gini) correlation coefficient leaving the other intact. Hence the difference in the correlations enables one to see the vulnerability of the correlation. This issue will be dealt with in Chaps. 3, 4, 8, 18, and 19.

There are at least two other major applications of having two (Gini) correlations between each pair of variables instead of one. First, as will be shown in Chap. 8, every optimization results in first order conditions that can be described as "orthogonality conditions." Those conditions can be interpreted as setting a covariance to zero. Having two correlations (and covariances) between two variables enables one to test whether the other covariance is also equal to zero so that one can have a specification test with respect to the underlying model. The second application is related to the properties of the decomposition of Gini of a linear combination of random variables as is discussed next.

(b) Decompositions: There are two types of decompositions. One is the decomposition of a variability measure of a linear combination of random variables into the contributions of the individual variables and the contributions of the relationships between them. The other decomposition is the one that decomposes the variability of a population that is composed of several subpopulations into the contributions of the subpopulations and some extra terms. In both cases the decomposition of the GMD includes the structure of the decomposition of the variance as a special case. We refer to the assumptions that lead to the structure of the decomposition of the variance as hidden assumptions imposed on the data that lead to the simplicity of the structure of the variance decompositions. We refer to this property as the property of "revealing more." The Gini of a linear combination of random variables does not, in general, decompose into two components as neatly as the variance does. (In the variance decomposition one component is based on the individual variances and the other is based on the correlations among the variables.) Instead, it extracts more information about the underlying distributions, as will be discussed in Chaps. 3 and 23.

The decomposition of the Gini leads, under certain conditions, to a decomposition formula with the same structure as the decomposition of the variance. This fact enables us to test for the hidden (implicit) assumptions that are leading to the simplicity that has made the variance-based analysis so convenient. More specifically, the Gini of a population does not decompose neatly (i.e., additively) into intra- and inter-group Ginis. For this reason it was rejected by several economists who tried to imitate Analysis of Variance. As will be shown in this book, this disadvantage may turn into an advantage. The decomposition of the Gini coefficient of a population extracts more information from the underlying distribution than just the inter- and intra-components. It gives a quantitative measure of the amount of the overlapping between the subgroups which is important whenever one is interested in stratification and/or in evaluating the quality of the classification of a general population into groups. The decomposition will be discussed in Chap. 4 while the empirical applications will be demonstrated in Chaps. 13 and 22.

The usefulness of the GMD and its contribution to our statistical analysis is especially important whenever the concepts that are used are not symmetric by definition. Among those concepts are regression in statistics and elasticity in economics. The properties of the Gini enable one to check the validity of the implicit use of symmetry whenever those concepts are used. In the regression concept the use of the Gini plays an important role in checking whether the assumptions that led to the estimates are supported by the data or not. For example, the Gini methodology can be used to check whether the relationship between Y and X is monotonic over the entire range of X or not by a simple graphical technique. This will be demonstrated in Chaps. 5, 19, and 21.

Having listed these advantages of using the Gini, it is worth mentioning the "cost" of using it. First, its use is cumbersome because sometimes the additional information that the Gini offers may be redundant. Second, in order to use the Gini one has to ignore some of the intuition and conventional wisdom that come with the variance. As will be shown, the Gini describes the variability by two attributes: the variate and its rank. For the economist, this should resemble the intuition that comes with what is known as "the index number problem" that is taught in intermediate economic theory. The index number problem arises whenever one tries to describe a phenomenon by two attributes: the price and the quantity of a commodity. In these cases one attribute is kept unchanged, while the other is allowed to change. Because in real life the two attributes can change simultaneously, the choice of which attribute is held constant and which one is allowed to change may result in some cases in contradicting conclusions. The cases of contradictions are the cases that diverge from the analysis based on the variance, and remembering them may help in understanding the intuition needed for evaluating the results.

An alternative approach to be taken when reading this book is to view the GMD and the parameters that are related to it as representing several theories that originate in the social sciences. Among these theories are (a) the expected utility hypothesis which represents the main paradigm in the area of risk and social welfare, (b) the relative deprivation theory which plays a major role in explaining social unrest, and (c) mobility, horizontal equity, and similar concepts that are used in the social sciences. In this respect the book presents the essence of these theories and advocates the use of the decomposition properties of the Gini so that one can offer statistical tools for understanding, analyzing, and developing these theories. These theories and the relationships with the Gini are presented in the applications part of this book.

To be able to fully utilize the properties of the Gini we will not make any assumptions concerning the distribution of the random variable throughout this book. The only case in which we will assume a particular distribution is to illustrate a point.

Finally, we wish to add an apology. Darling, in his Annals of Mathematical Statistics paper (1957), writes:

> The reader is advised that the relative amount of space and emphasis allotted to the various phases of the subject do not reflect necessarily their intrinsic merit and importance, but rather the author's personal interest and familiarity. Also, for the sake of uniformity the notation of many of the writers quoted has been altered so that when referring to the original papers it will be necessary to check their nomenclature (Darling, 1957, p. 823).

We could not find better words for describing our approach in this book. Also, we apologize in advance for not giving the appropriate credit to the appropriate authors in some occasions. One serious difficulty is to define the meaning of an innovation and to decide to whom to give the credit for it in this area of research. The reason is that on top of independent developments, where researchers could not read the language or were not aware of the developments in their or other areas, there is a difficult issue in this crowded area. Is the person who wrote a formula in passing the one who should be credited for it, or is it the person who correctly interpreted it and developed its implications? In order to illustrate this issue let us investigate the history of expressing the GMD as a covariance. The fact that one can express GMD as a covariance and use the covariance properties to further develop the theoretical aspects is in our opinion a major breakthrough.

As far as we know, the first step in this direction was to write the Gini as a covariance without noticing that it actually *is* a covariance. This was done by Corrado Gini (1914). The next step, some 40 years later, was to realize that Gini can be expressed as a covariance, with no further implications. This fact was realized by Stuart (1954). Fei, Ranis, and Kou (1978) constructed the Gini-covariance, referring to it as pseudo-Gini. Pyatt, Chen, and Fei (1980) used the term covariance in constructing the pseudo-Gini. The final breakthrough was made by Lerman and Yitzhaki (1984) who pointed out that because the Gini can be expressed as a covariance it is possible and helpful to use the properties of the covariance in handling it. This observation opened the way to investigating the Gini covariances and correlations and their properties.

On the anecdotal side, the person who triggered Lerman and Yitzhaki to write the Gini as a covariance was an anonymous referee of Yitzhaki (1982a). He/she argued that the covariance is more important than the variance in the area of finance, and therefore a sentence should be added to say whether it is possible to develop a covariance that is suitable for the Gini or not.

Similar issues arose concerning the development of the extended Gini which was discovered independently and from different angles by Donaldson and Weymark (1980, 1983) and Yitzhaki (1983). Kakwani (1980) mentions the possibility of the extended Gini in passing. Moreover, all the above-mentioned papers and Chakravarty (1983) can be classified as the Gini response to Atkinson (1970) who suggested an inequality measure that depends on a parameter. Clearly, there can be other scenarios for describing the development of the extended Gini and the expression of the Gini as a covariance. In order not to enter into such a debate, we apologize in advance for not taking the appropriate actions to attribute each concept to the original person who developed and coined it. In addition, in order to keep the presentation flowing, and to avoid sidetracking the reader into what we consider as dead end from the point of view of our target, some papers that may be important in the future are only mentioned in passing.

The target audience of this book is mainly applied economists, statisticians, and econometricians who are interested in applications for which the variance is not suitable. These applications arise mostly (but not only) when the underlying distribution deviates from normality. Possible areas of application are welfare economics, finance, and general econometric theory. As will be seen in this book, Gini-based analyses are robust to the asymmetry of the distribution and to the existence of outliers. In addition, the use of the Gini allows one to identify and test the existence of implicit assumptions about the underlying distributions that make the variance-based analyses so simple to apply, yet may not be satisfied by the data, or, alternatively, violate basic principles of economic theory.

The complexity and the different representations and applications of the GMD in different fields forced us to use different notations to represent the GMD in different areas. The reason is that in some areas it is convenient to use GMD/4 as the GMD, and in other areas GMD/2 or simply GMD. This implies the need to carry constants that affect all equations in a specific application and complicate the representation without adding any content. To overcome this problem, we use different representations of the GMD in different chapters and we will state in the introduction of each chapter which definition is used.

The book consists of two main parts. The first part (Chaps. 2–11) contains the theory while the second part (Chaps. 12–22) deals with applications. The applications chapters contain a short review of the needed theory to make them readable on their own. The structure of the book is the following: In Chap. 2 we provide the various definitions of the Gini. The Gini covariance, correlation, and regressions are introduced in Chap. 3. In Chap. 4 we present the decompositions of the Gini while Chap. 5 deals with the relation to the Lorenz and the concentration curves. The extended Gini family of measures is introduced in Chap. 6. Next, two chapters are devoted to Gini regression: the simple regression case is detailed in Chap. 7 while the extension to the multiple regression case is detailed in Chap. 8. The next three chapters are devoted to the statistical inference. Estimation of the Gini-based parameters is the topic of Chap. 9, a selection of formal tests is presented in Chap. 10, while tests that are related to the intersection of concentration curves are the topic of Chap. 11.

The second part of the book contains applications of the Gini methodology in various areas. We start with an introduction to the applications part (Chap. 12). In Chap. 13 we demonstrate the role of the Gini coefficient in two major competing theories that dominate the theoretical considerations in the area of income distribution, namely: the social welfare function approach and the theory of relative deprivation.

In Chap. 14 we illustrate the use of the concentration curves and the Gini methodology in the areas of taxation and progressivity of public expenditure.

Chapter 15 deals with the usefulness of several decompositions of the Gini and the extended Gini in analyzing government policies by non-marginal analyses, while in Chap. 16 the marginal analysis is illustrated. The applications in finance are the topic of Chaps. 17 and 18. These applications are relevant whenever one is interested in decision making under risk. Chapters 19–21 are devoted to applications of the Gini regression: in Chap. 19 we apply the simple Gini and extended Gini regressions, in Chap. 20 the multiple regression is applied, and in Chap. 21 we apply the mixed OLS, Gini, and extended Gini regressions. Chapter 22 deals with one application of the GMD and the Gini coefficient in statistics—an application that replicates the commonly used ANOVA and is denoted by ANOGI (ANalysis Of GIni). The last chapter (Chap. 23) concludes and lists several topics for further research.

Readers who will read the book will find some repetitions between the theoretical and the applications parts of the book. The reason for those repetitions is that each chapter in the applications part is written as a self-contained application. This approach is intended to enable the specialist in a field to read the relevant application chapter without having to read the whole book. Readers who want to see a proof for an argument are referred to the theoretical part.

Part I
Theory

Chapter 2
More Than a Dozen Alternative Ways of Spelling Gini

Introduction

Gini's mean difference (GMD) as a measure of variability has been known for over a century.[1] It has more than 14 alternative representations.[2] Some of them hold only for continuous distributions while others hold only for nonnegative variables. It seems that the richness of alternative representations and the need to distinguish among definitions that hold for different types of distributions are the main causes for its sporadic reappearances in the statistics and economics literature as well as in other areas of research. An exception is the area of income inequality, where it is holding the position as the most popular measure of inequality. GMD was "rediscovered" several times (see, for example, Chambers & Quiggin, 2007; David, 1968; Jaeckel, 1972; Jurečková, 1969; Olkin & Yitzhaki, 1992; Kőszegi & Rabin, 2007; Simpson, 1949) and has been used by investigators who did not know that they were using a statistic which was a version of the GMD. This is unfortunate, because by recognizing the fact that a GMD is being used the researcher could save time and research effort and use the already known properties of GMD.

The aim of this chapter is to survey alternative representations of the GMD. In order to simplify the presentation and to concentrate on the main issues we restrict the main line of the presentation in several ways. First, the survey is restricted to

This chapter is based on Yitzhaki (1998) and Yitzhaki (2003).

[1] For a description of its early development see Dalton (1920), Gini (1921, 1936), David (1981, p. 192), and several entries in Harter (1978). Unfortunately we are unable to survey the Italian literature which includes, among others, several papers by Gini, Galvani, and Castellano. A survey on those contributions can be found in Wold (1935). An additional comprehensive survey of this literature can be found in Giorgi (1990, 1993). See Yntema (1933) on the debate between Dalton and Gini concerning the relevant approach to inequality measurement.

[2] Ceriani and Verme (2012) present several additional forms in Gini's original writing that as observed by Lambert (2011) do not correspond to the presentations used in this book.

S. Yitzhaki and E. Schechtman, *The Gini Methodology: A Primer on a Statistical Methodology*, Springer Series in Statistics 272, DOI 10.1007/978-1-4614-4720-7_2,

quantitative random variables. As a result the literature on diversity which is mainly concerned with categorical data is not covered.[3] Second, the survey is restricted to continuous, bounded from below but not necessarily nonnegative variables. The continuous formulation is more convenient, yielding insights that are not as accessible when the random variable is discrete. In addition, the continuous formulation is preferred because it can be handled using calculus.[4] As will be shown in Sect. 2.4 there is an additional reason for the use of a continuous distribution: there is an inconsistency between the various tools used in defining the GMD when the distribution is discrete. This inconsistency complicates the presentation without adding any insight. To avoid problems of existence, only continuous distributions with finite first moment will be considered. The distinction between discrete and continuous variables will be dealt with in Sect. 2.4, while properties that are restricted to nonnegative variables will be discussed separately whenever they arise. Third, the representations in this chapter are restricted to population parameters. We deal with the estimation issue in Chap. 9.

Finally, as far as we know these alternative representations cover most, if not all, known cases but we would not be surprised if others turn up. The different formulations explain why the GMD can be applied in so many different areas and can be given so many different interpretations. We conclude this chapter with a few thoughts about the reasons why Gini was "rediscovered" again and again and with four examples that illustrate this point.

The structure of this chapter is as follows: Section 2.1 derives the alternative representations of the GMD. Section 2.2 investigates the similarity between GMD and the variance. Section 2.3 deals with the Gini coefficient and presents some of its properties. In sect. 2.4 the adjustments to the discrete case are discussed and Sect. 2.5 gives some examples. Section 2.6 concludes.

2.1 Alternative Representations of GMD

There are four types of formulas for GMD, depending on the elements involved: (a) a formulation that is based on absolute values, which is also known to be based on the L_1 metric; (b) a formulation which relies on integrals of cumulative distribution functions; (c) a formulation that relies on covariances; and (d) a formulation that

[3] For the use of the GMD in categorical data see the bibliography in Dennis, Patil, Rossi, Stehman, and Taille (1979) and Rao (1982) in biology, Lieberson (1969) in sociology, Bachi (1956) in linguistic homogeneity, and Gibbs and Martin (1962) for industry diversification. Burrell (2006) uses it in informetrics, while Druckman and Jackson (2008) use it in resource usage, Puyenbroeck (2008) uses it in political science while Portnov and Felsenstein (2010) in regional diversity.

[4] One way of writing the Gini is based on vectors and matrices. This form is clearly restricted to discrete variables and hence it is not covered in this book. For a description of the method see Silber (1989).

relies on Lorenz curves (or integrals of first moment distributions). The first type is the most convenient one for dealing with conceptual issues, while the covariance presentation is the most convenient whenever one wants to replicate the statistical analyses that rely on the variance such as decompositions, correlation analysis, ANOVA, and Ordinary Least Squares (OLS) regressions.

Let X_1 and X_2 be independent, identically distributed (i.i.d.) continuous random variables with F(x) and f(x) representing their cumulative distribution and the density function, respectively. It is assumed that the expected value μ exists; hence $\lim_{t\to-\infty} tF(t) = \lim_{t\to\infty} t[1 - F(t)] = 0$.

2.1.1 *Formulas Based on Absolute Values*

The original definition of the GMD is the expected absolute difference between two realizations of i.i.d. random variables. That is, the GMD in the population is

$$\Delta = E\ \{|X_1 - X_2|\}, \tag{2.1}$$

which can be given the following interpretation: consider an investigator who is interested in measuring the variability of a certain property in the population. He or she draws a random sample of two observations and records the absolute difference between them.

Repeating the sampling procedure an infinite number of times and averaging the absolute differences yield the GMD.[5] Hence, the GMD can be interpreted as the expected absolute difference between two randomly drawn members from the population. This interpretation explains the fact that for nonnegative variables the GMD is bounded from above by twice the mean because the mean can be viewed as the result of infinite repetitions of drawing a single draw from a distribution and averaging the outcomes, while the GMD is the average of the absolute differences between two random draws. Note, however, that this property does not necessarily hold for random variables that are not restricted to be nonnegative.

Equation (2.1) resembles the variance, which can be presented as

$$\sigma^2 = 0.5E\{(X_1 - X_2)^2\}\ . \tag{2.2}$$

Equation (2.2) shows that the variance can be defined without a reference to a location parameter (the mean) and that the only difference between the definitions of the variance and the GMD is the metrics used for the derivations of the concepts. That is, the GMD is the expected absolute difference between two randomly drawn

[5] See also Pyatt (1976) for an interesting interpretation based on a view of the Gini as the equilibrium of a game.

observations, while the variance is the expected square of the same difference. It is interesting to note that replacing the power 2 by a general power r in (2.2) is referred to as the generalized mean difference (Gini, 1966; Ramasubban, 1958, 1959, 1960). However, as far as we know, they were not aware of the fact that when $r = 2$ it is identical to the variance.

An alternative presentation of the GMD that will be helpful when we describe the properties of the Gini regressions and their resemblance to quantile regressions can be developed in the following way:

Let Q and X be two i.i.d. random variables; then by the law of iterated means the GMD can be presented as the average (over all possible values of Q) of all absolute deviations of X from Q. In other words

$$\Delta = E_Q E_{X|Q}\{|X - Q|\}. \tag{2.3}$$

Next, we note that Q in (2.3) can represent the quantile of the distribution. The reason is that the quantile can be assumed to have the same distribution function as X does, and can be assumed to be independent of X. To see that let $F_X(Q) = P$; then $F_X(Q)$ is uniformly distributed on [0, 1]. It follows that $Q = F_X^{-1}(P)$ is distributed as X,

$$G_Q(t) = P(Q \le t) = P(F_X^{-1}(P) \le t) = P(P \le F_X(t)) = F_X(t),$$

and independent of it. Therefore the term $E_{X|Q}\{|X - Q|\}$ in (2.3) can be viewed as the conditional expectation of the absolute deviation from a given quantile Q of the distribution of X. Hence equation (2.3) presents the GMD as the average absolute deviation from all possible quantiles.

From (2.3) one can see that minimizing the GMD of the residuals in a regression context (to be discussed in Chap. 7) can be interpreted as minimizing an *average* of all possible *absolute deviations* from all possible quantiles of the residual. We note in passing that (2.3) reveals the difference between the GMD and the expected absolute deviation from the mean. The former is the expected absolute difference from every possible value of Q, while the latter is the expected absolute deviation from the mean. We will return to this point in Chap. 23.

A slightly different set of representations relies on the following identities: let X_1 and X_2 be two i.i.d. random variables having mean μ. Then

$$\begin{aligned}|X_1 - X_2| &= (X_1 + X_2) - 2\mathrm{Min}\{X_1, X_2\} = \mathrm{Max}\{X_1, X_2\} - \mathrm{Min}\{X_1, X_2\} \\ &= 2\mathrm{Max}\{X_1, X_2\} - (X_1 + X_2).\end{aligned} \tag{2.4}$$

Using the first equation from the left of (2.4), the GMD can be expressed as

$$\Delta = 2\mu - 2E[\mathrm{Min}\{X_1, X_2\}]. \tag{2.5}$$

That is, the GMD is twice the difference between the expected values of one random draw and the minimum of two random draws from the distribution. Alternatively, we can use the middle part of (2.4) to write

$$\Delta = E[\mathrm{Max}\{X_1, X_2\}] - E[\mathrm{Min}\{X_1, X_2\}]. \tag{2.6}$$

Here, the interpretation of the GMD is as the expected difference between the maximum and the minimum of two random draws. Finally, one can use the right-hand side of (2.4) to write the GMD as twice the expected value of the maximum of two random draws minus twice the expected value of one random draw. These presentations can be easily extended to involve more than two draws, leading to the extended Gini (Yitzhaki, 1983). (This issue will be discussed in Chap. 6.) They can be useful whenever the interpretation of the GMD is related to extreme value theory.

2.1.2 *Formulas Based on Integrals of the Cumulative Distributions*

This section focuses on representations of the GMD that are based on integrals of the cumulative distribution. The basic equation needed in order to develop such representations is an alternative expression for the expected value of a distribution.

Claim Let X be a continuous random variable distributed in the range $[a,\infty)$. Then the expected value of X is given by[6]

$$\mu = a + \int_a^\infty [1 - F(x)]dx. \tag{2.7}$$

Proof The standard definition of the expected value is $\mu = \int_a^\infty xf(x)dx$. Using integration by parts with $u = x$ and $v = -[1 - F(x)]$ yields (2.7).

Using (2.7) and the fact that the cumulative distribution of the minimum of two i.i.d. random variables can be expressed as $\{1 - [1 - F(x)]^2\}$ we can rewrite (2.5) as

$$\Delta = 2\int [1 - F(t)]dt - 2\int [1 - F(t)]^2 dt, \tag{2.8}$$

and by combining the two integrals

[6] The GMD is based on the difference of two such formulae, so this restriction on the range (to be bounded from below) does not affect the GMD. See Dorfman (1979).

$$\Delta = 2\int F(t)[1 - F(t)]dt. \tag{2.9}$$

See Dorfman (1979). Equation (2.9) can be given an interesting interpretation. Let $F_n(x)$ be the empirical cumulative distribution of X based on a sample of n observations. Then for a given x, $F_n(x)$ is the sample mean of n i.i.d. Bernoulli variables with $p = F(x)$. The variance of $F_n(x)$ is equal to

$$\sigma^2_{F_n(x)} = F(x)[1 - F(x)]/n \tag{2.10}$$

(Serfling, 1980, p. 57) and the GMD can be interpreted as $2n\int \sigma^2_{F_n(x)}dx$.
A similar (and older) variant of this formula is

$$\Delta = 2nE\left\{\int [F_n(x) - F(x)]^2 dx\right\}, \tag{2.11}$$

which is the original Cramer–Von Mises–Smirnov criterion for testing goodness of fit of a distribution.[7] In some sense (2.11) can be viewed as a "dual" approach to the central moments of a distribution. Central moments are linear in the probabilities and power functions of deviations of the variate from its expected value. In the GMD, the power function is applied to the deviation of the cumulative distribution from its expected value while the linearity is applied to the variate itself. Hence the "duality."[8] This interpretation also suggests a possible explanation to some robustness properties of the "dual" approach. The range of F(·) is [0, 1] while the range of the variate can be unlimited. Using a power function as is done in the regular moments may lead to unboundedness of the statistics, while all the moments of the dual approach are bounded, provided that the mean is bounded.

---

[7] This formula, which is a special case of the statistic suggested by Cramer, plays an important role in his composition of elementary errors although it seems that he did not identify the implied GMD (see Cramer, 1928, pp. 144–147). Von Mises (1931) made an independent equivalent suggestion and developed additional properties of the statistic. Smirnov (1937) modified the statistic to be

$$w^2 = n\int [F_n(x) - F(x)]^2 dF(x).$$

Changing the integration from dx to dF(x) eliminates the connection to the GMD and creates a distribution-free statistic. The above description of the non-English literature is based on the excellent review in Darling (1957). Further insight about the connection between the Cramér–Von Mises test can be found in Baker (1997) which also corrects for the discrepancy in calculating the GMD in discrete distributions.

[8] This "duality" resembles the alternative approach to the expected utility theory as suggested by Yaari (1988) and others. While expected utility theory is linear in the probabilities and nonlinear in the income, Yaari's approach is linear in the income and nonlinear in the probabilities. In this sense, one can argue that the relationship between "dual" approach and the GMD resembles the relationship between expected utility theory and the variance. Both indices can be used to construct a specific utility function for the appropriate approach (the quadratic utility function is based on the mean and the variance while the mean minus the GMD is a specific utility function of the dual approach).

Finally, we can write (2.9) as

$$\Delta = 2\int_{a}^{\infty}\left[\int_{a}^{x}f(t)dt\int_{x}^{\infty}f(t)dt\right]dx, \tag{2.12}$$

which is the way Wold (1935) presented it.

An additional presentation by Wold (1935, equation 12, p. 47)[9] which is valid for nonnegative variables is

$$\Delta = 2\int_{0}^{\infty}\left[\int_{0}^{t}F(u)du\right]dF(t). \tag{2.13}$$

Equation (2.13) is listed for completeness.

2.1.3 Covariance-Based Formulas

It is well known that the variance is a special case of the covariance, because it can be written as var(X) = cov(X, X). In this section we show that the GMD can be expressed as a covariance as well. Once the GMD is written as a covariance, the properties of the covariance are called for to define the Gini correlation and the decomposition of a GMD of a linear combination of random variables, which naturally leads to Gini regressions, Gini Instrumental Variable, time-series Gini analysis, and numerous other applications. Generally speaking, one can take an econometrics textbook that is based on the variance and rewrite (most of) it in terms of the GMD. Another advantage of the covariance presentation is that the covariance formula opens the way to the decomposition of the Gini coefficient (to be defined later) of an overall population into the contributions of several subgroups. In addition, it opens the way to the extended Gini family of measures of variability (to be discussed in Chap. 6), which means replicating (almost) everything that was developed with the Gini and finding out which properties carry on to an infinite number of measures of variability.

Let us start with presentation (2.9). Applying integration by parts to (2.9), with $v = F(t)\,[1 - F(t)]$ and $u = t$, one gets, after deleting zeros and rearranging terms,

$$\Delta = 2\int F(t)[1 - F(t)]dt = 4\int t[F(t) - 0.5]f(t)dt. \tag{2.14}$$

---

[9] Wold (1935) used a slightly different presentation, based on Stieltjes integrals.

Recall that the expected value of F, which is uniformly distributed on [0, 1], is 0.5. Therefore one can rewrite (2.14) as

$$\Delta = 4E\{X(F(X) - E[F(X)])\} = 4\ \mathrm{cov}[X, F(X)]. \tag{2.15}$$

Equation (2.15) lets us calculate the GMD using a simple regression program as will be shown next.[10] Recall that F(X) is uniformly distributed on [0, 1]. Therefore, cov[F(X), F(X)] = 1/12 (a constant) and we can write the GMD as

$$\Delta = (1/3)\mathrm{cov}[X, F(X)]/\mathrm{cov}[F(X), F(X)]. \tag{2.16}$$

In order to gain some intuition assume that the observations are arrayed in ascending order (say, by height as in the case of soldiers in a parade) with equal distance between each two observations (soldiers). The following proposition summarizes two interpretations of the GMD.

Proposition 2.1

(a) *The GMD is equal to one-third of the slope of the OLS regression curve of the observed variable (height, the dependent variable) as a function of the observation's position in the array (F(X), the explanatory variable).*

(b) *The GMD is a weighted average of the differences in, say, heights between adjacent soldiers (alternatively, it is a weighted average of the slopes defined by each two adjacent heights in the array). The weights are symmetric around the median, with the median having the highest weight.*

Proof of (a) Trivial. Recall that the OLS regression coefficient in a linear regression model is given by

$$\beta = \frac{\mathrm{cov(Y,X)}}{\mathrm{cov(X,X)}}$$

and see (2.16) above.

Proof of (b) Let X(p) be the height of a soldier as a function of his position, p. For example, X(0.5) is the height of the median soldier. That is, $P(X < X(p)) = p = F(X(p))$. Note that X(p) is the inverse of the cumulative

[10] See Lerman and Yitzhaki (1984) for the derivation and interpretation of the formula, see Jenkins (1988) and Milanovic (1997) on actual calculations using available software, and see Lerman and Yitzhaki (1989) on using this equation to calculate the GMD in stratified samples. As far as we know, Stuart (1954) was the first to notice that the GMD can be written as a covariance. However, his findings were confined to normal distributions. Pyatt, Chen and Fei (1980) also write the GMD as a covariance. Sen (1973) uses the covariance formula for the Gini, but without noticing that he is dealing with a covariance. Hart (1975) argues that the moment-generating function was at the heart of the debate between Corrado Gini and the western statisticians. Hence, it is a bit ironic to find that one can write the GMD as some kind of a central moment.

distribution of X at p. Writing explicitly the numerator in (2.16) we get $\text{cov}(X, p) = \int X(p)(p - 0.5)dp$ and by using integration by parts with $u = X(p)$ and $v = (p - 0.5)^2/2$ we get

$$\text{cov}(X,p) = X(p)(p-0.5)^2/2|_0^1 - 0.5\int X'(p)(p-0.5)^2 dp.$$

Substituting $X(1) - X(0) = \int X'(p)dp$, where X' denotes a derivative, we get

$$\text{cov}(X,p) = 0.5\int X'(p)p(1-p)dp. \tag{2.17}$$

Equation (2.17) shows that the GMD is equal to the weighted average of the slopes $X'(p)$ and the weighting scheme $p(1 - p)$ is symmetric in p around the median ($p = 0.5$). The maximum weight is assigned to the median ($p = 0.5$), and the weights decline the farther the rank of the observation gets from the median.

A consequence of (2.17) is that the flatter the density function of X is, the larger the GMD becomes (which is intuitively clear for a measure of spread). To sum up, according to these presentations the GMD is the weighted average change in a random variable as a result of a small change in the ranks. Because X(p) is the inverse of the cumulative distribution it is easy to see that $X'(p) = \frac{1}{f(x)}dx$. That is, the slope is the reciprocal of the density function.

Equation (2.15), the covariance representation of the GMD, can be used to show that R-regressions (Hettmansperger, 1984) are actually based on minimizing the GMD of the residuals in the regression. To see that, note that the target function in R-regression is to minimize $\sum_i e_i R(e_i)$, where e_i is the error term of the i-th observation in the regression while $R(e_i)$ is its rank. Note that the mean of the residuals is equal to zero, and that the rank of the variable represents the cumulative distribution in the sample. Taking into account those facts, it is easy to see that R-regression is actually based on minimizing the GMD of the residuals. Therefore some properties of these regressions can be traced to the properties of the GMD. We will further elaborate on this point in Chap. 7.

We will be using the covariance formula of the GMD extensively in this book. It makes it very natural and convenient to "translate" the variance-based parameters such as the regression and the correlation coefficients into the Gini language. It is interesting to note that for the discrete case these facts were already mentioned in Gini (1914) and were repeated in Wold (1935). For the continuous case one can find the covariance presentation in Stuart (1954). Fei, Ranis, and Kou (1978) constructed the Gini-covariance, referring to it as pseudo-Gini. Pyatt, Chen, and Fei (1980) used the term covariance in constructing the pseudo-Gini. The contribution of Lerman and Yitzhaki (1984) is in recognizing the implications of the term covariance when dealing with the decomposition of the variability measure and in producing the additional parameters which are based on it—a step that opened the way to applying the GMD method in the multivariate case.

2.1.4 Lorenz Curve-Based Formulas

The fourth set of representations of the GMD is based on the Absolute Lorenz Curve (ALC), which is also referred to as the generalized Lorenz curve.[11] The ALC and the concentration curve play important roles in the understanding of the compositions and the contributions of different sections of the distribution to the GMD and other related parameters such as Gini covariance and Gini correlation. Therefore they will be discussed in detail in Chap. 5. In this section we briefly mention the Lorenz curve-based formulas. There are several definitions of the ALC. We follow Gastwirth's (1971, 1972) definition, which is based on the inverse of the cumulative distribution. Let $F(X(p)) = p$, then $X(p) = F^{-1}(p)$. The ALC is plotted as follows: p is plotted on the horizontal axis while the vertical axis represents the cumulative value of the variate, $_{-\infty}\int^{p} X(t)dt$. The familiar (relative) Lorenz curve (LC) is derived from the ALC by dividing the cumulative value of the variate by its expected value. The vertical axis is then $(1/\mu)_{-\infty}\int^{p} X(t)dt$. The ALC has the following properties:

1. The ALC passes through (0, 0) and (1, μ). The LC passes through (0, 0) and (1, 1).
2. The derivative of the curve at p is X(p), which is the inverse of the cumulative distribution function; hence the curve is increasing (decreasing) depending on whether X(p) is positive (negative). Because X(p) is always a nondecreasing function of p the ALC is convex.

Figure 2.1 presents a typical ALC, the curve OAB. The slope of the line connecting the two extremes of the curve is μ. We refer to this line as the Line of Equality (LOE), because when all observations are equal, the curve coincides with the line. The line OEGB in Fig. 2.1 represents the LOE. It can be shown (details will be given in Chap. 5) that the area between LOE and the ALC, OAB is cov(X, F(X)) (that is, one-fourth of the GMD). We will return to this topic when dealing with the properties of ALC in Chap. 5.

As far as we know, we have covered all known interesting presentations of the GMD. The rest of this chapter is intended to supply some intuition and to compare the GMD with the variance.

[11] The term "generalized Lorenz curve" (GLC) was coined by Shorrocks (1983). Lambert and Aronson (1993) give an excellent description of the properties of GLC. However, it seems that the term "absolute" is more intuitive because it distinguishes the absolute curve from the relative one. Hart (1975) presents inequality indices in terms of the distribution of first moments, which is related to the GLC.

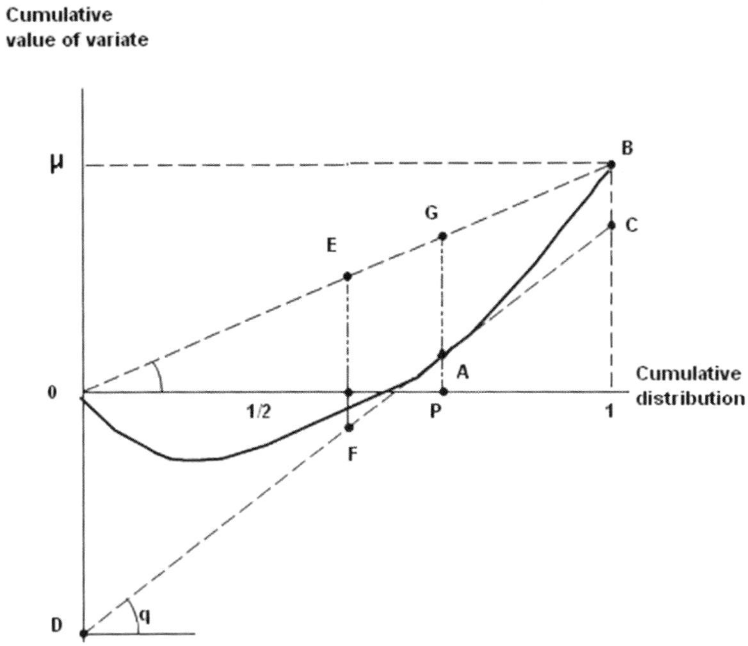

Fig. 2.1 The absolute Lorenz curve. *Source*: Yitzhaki, 1998, p. 21. Reprinted with permission by Physica Verlag, Heidelberg

2.2 The GMD and the Variance

In this section we investigate the similarities and the differences between the GMD and the variance. As will be seen, on one hand they share many properties, but on the other hand there are some fundamental differences.

2.2.1 The Similarities Between GMD and the Variance

The first similarity between the GMD and the variance is the fact that both can be written as covariances. The variance of X is cov(X, X), while the GMD of X is cov(X, F(X)). This similarity serves as the basis for the ability to "translate" the variance world into the Gini world.

The second similarity is the fact that the decomposition of the variance of a linear combination of random variables is a special case of the decomposition of the GMD of the same combination. The decomposition of the GMD includes some extra parameters that provide additional information about the underlying distribution, as will be developed in Chap. 4. If these additional parameters are equal to zero then the decompositions of the GMD and the variance have identical structures. This property makes the GMD suitable for testing implicit assumptions that lead to the convenience of using the variance. This property is also the base for the claim that was put forward by Lambert and Decoster (2005) that "the Gini reveals more."

The third similarity is the fact that both the variance and the GMD are based on averaging the distances between all pairs of observations (see (2.1), (2.2), and Daniels, 1944, 1948) or, alternatively, averaging the distances between random draws of two i.i.d. random variables. However, the difference between them is in the distance function used. The effects of the distance functions on the properties of the indices will be illustrated when we deal with the properties of OLS and the Gini regression coefficients in Chap. 7. The source of this difference will be discussed in the next section.

2.2.2 *The Differences Between the GMD and the Variance: City Block vs. Euclidean*

Let Δx_k denote the difference between adjacent observations. That is, $\Delta x_k = X_{k+1} - X_k$, where the observations are arranged in an increasing order. Then for any two ordered observations $X_i > X_j$

$$X_i - X_j = \sum_{k=j}^{i-1} \Delta x_k. \tag{2.18}$$

The GMD and the variance can be presented as weighted averages of these distances between adjacent observations.[12] In both cases the weighting scheme attaches the highest weight to the mid-rank observation (i.e., the median), and the weights decline symmetrically the farther the rank of the observation is from the mid-rank. The fundamental difference between the two measures of variability is attributed to the distance function they rely on. The GMD's distance function is referred to as the "city block" distance (or L_1 metric), while the variance's distance is Euclidean. It is interesting to note that other measures of variability (e.g., the mean deviation) also rely on the L_1 metric, but they do not share the weighting scheme caused by the averaging of differences between all pairs of observations. To shed some light on the difference between the distance functions, note that the most basic measure of variability is the range, which is equivalent to the simple

[12] In the case of the GMD, the weights are not functions of Δx_k so that it is reasonable to refer to them as weights. In the case of the variance, the "weights" are also functions of Δx_k which makes the reference to them as weights to be incorrect. We refer to them as weights in order to compare with the GMD. See Yitzhaki (1996).

sum of the distances between adjacent observations, so that we end up with the difference between the most extreme parts of the distribution. If the distributions are restricted to have only two observations then the variance and the GMD (and all other measures of variability) will order all distributions in accordance with the ordering of the range. However, the range suffers from two major deficiencies: (1) it is not sensitive to the distribution of the non-extreme observations and (2) there are many important distributions with an infinite range.

In order to illustrate the difference between the distance functions embodied in the GMD and the variance one can ask, for a given range, what characterizes the distribution with the smallest/largest variance (GMD). Alternatively, for a given variance (GMD) one can ask what characterizes the distribution with the smallest/largest range. Presumably by answering those questions we will be able to form an opinion as to which distance function is more appropriate for a given situation and which one reflects our intuition better. To illustrate, let us restrict the distributions to have only three possible values and assume a given (normalized) range (equals to 1 in our example) so that the discussion is restricted to distributions of the type: [0, δ, 1]. Which δ will maximize or minimize each variability index? Ignoring constants, the GMD is

$$\Delta(\delta) = \sum\sum |X_i - X_j| = 1 + \delta + |1 - \delta|, \tag{2.19}$$

and it equals 2 regardless of the value of δ. (More generally, it is equal to twice the range). Thus the position of the middle observation does not change the GMD. Repeating the same exercise with the variance yields (again, ignoring constants)

$$\sigma^2(\delta) = 1 + \delta^2 + (1 - \delta)^2 \tag{2.20}$$

and the variance is maximized for $\delta = 0$ or 1 and minimized for $\delta = 0.5$. That is, for a given range, the more equal the distances defined by adjacent observations are, the smaller is the variance. The conclusion is that the variance is more sensitive to the variability in Δx_i than the GMD. In other words, if one of the differences will be extremely large the variance will be affected by it more than the GMD. This fact is responsible for the sensitivity of the variance to extreme observations.

An alternative way to illustrate the difference between the variance and the GMD can be presented geometrically. Let δ_1 be the difference between the second and first observations and let δ_2 be the difference between the third and second observations. Figure 2.2 presents an example of equal GMD and equal variance curves. That is, we allow the range to vary and instead, we are asking what should δ_1 and δ_2 be so that we get equal Ginis or equal variances. The horizontal axis represents δ_1 while the vertical axis represents δ_2. The range, of course, is equal to $\delta_1 + \delta_2$. Gini (2.19) becomes $1 + \delta_1 + \delta_2$ while the variance (2.20) is $1 + \delta_1^2 + \delta_2^2$. By changing the value of the GMD or the variance, one gets parallel curves of different distances from the origin, which can be referred to as equi-variance (GMD) curves. The point A is on the equi-GMD curve, so that all the points on the

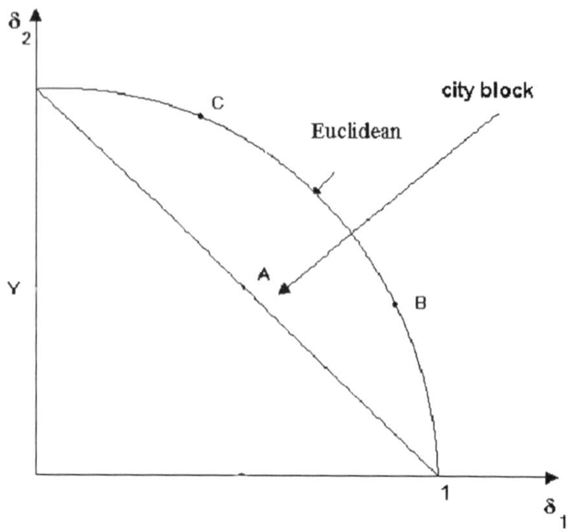

Fig. 2.2 Equal GMD and variance curves. *Source*: Yitzhaki, 2003, p. 292. Reprinted with permission by Metron International Journal of Statistics

graph have a value of GMD that is equal to the GMD at A. The points B and C are on the equi-variance curve. (In Fig. 2.2 the chosen value is 2).

As can be seen, the two measures represent different types of curves of equal distances. Imagine you are in a city. If you are allowed to only move in the east/west or north/south directions then you are in a GMD (city block) world. If, on the other hand, you are allowed to move in any direction you want, and you are Pythagorean, then you are in a variance world. It is hard to determine in general which distance function should be preferred. If one is traveling on the sea then the variance metric makes sense. However, the money metric which is extensively used by economists resembles the city block metric because the distance function embodied in the budget constraint is identical to the distance function of the GMD. One does not get a discount for spending equally on two commodities, as is the case of the variance (see Deaton, 1979; Jorgenson & Slesnick, 1984; McKenzie & Pearce, 1982 on uses of the money metric in economics). Hence when it comes to choosing a metric, the natural choice for the economists should be the GMD-type metric because spending money and the budget constraint follow the money metric rules.

The implication of the difference in metrics can also be seen from the following question which should be answered intuitively, without calculations. Consider the following distributions: [0, 0, 1] vs. [0, 0.575, 1.15]. Which distribution portrays a higher variability? If your intuitive answer points to the former (latter) distribution then you want to be in a variance (GMD) world (the variances are 0.222 and 0.220, respectively, while the Ginis are 0.667 and 0.767, respectively). The extension to more than three observations is straightforward.

This difference is responsible for the sensitivity of OLS regression to extreme observations and for the robustness properties of the GMD regressions as will be discussed in Chap. 7. Another implication of the difference in metrics is that the

GMD exists whenever the expected value exists while the existence of the variance requires the existence of a second moment.

It is interesting to note that if the underlying distribution is normal, then the increase in the distance between adjacent observations when moving from the middle to the extremes is identical to the decrease in the weight due to being farther away from the median, so that each observation gets an equal weight (Yitzhaki, 1996). Our conjecture is that this property leads to the statistical efficiency of variance-based statistics in cases of normality: the weights are distributed equally among observations. The main conclusion from the above discussion is that it is not obvious which metric is the preferred one, and the subject matter one is dealing with should also be taken into account when considering the appropriate metric.

Finally, an important property of the GMD is that it is bounded from above by $\frac{2}{\sqrt{3}}\,\sigma_X$ (as is shown below). The advantage of having the bound from above by a function of the standard deviation is that whenever the standard deviation converges to zero, so does the GMD. Note that

$$1 \geq \rho(X,F(X)) = \frac{\mathrm{cov}(X,F(X))}{\sigma_X\ \sigma_F} = \frac{\Delta_X}{4\ \sigma_X\ \sigma_F}.$$

Recall that F is uniformly distributed. Hence its standard deviation is equal to $\frac{1}{\sqrt{12}}$. Therefore we get the bound

$$\Delta_X \leq \frac{2}{\sqrt{3}}\ \sigma_X. \tag{2.21}$$

Note that if X is uniformly distributed on [0, 1] then $\rho(X,F(X)) = \rho(X,X) = 1$ and (2.21) holds as equality. This implies that (a) it is impossible to improve the bound and (b) for the uniform distribution the GMD is a constant multiplied by the standard deviation. (A similar case occurs under the normality, where the GMD $= 2\sigma/\sqrt{\pi}$.)

Our main purpose in this book is to imitate the applications of variance-based methods. Equation (2.21) enables us to simplify the analysis in this book by restricting the distributions to those with a finite lower bound. The reason for the above statement is that any convergence property that can be attributed to the variance can also be proved for the GMD. For additional bounds on the GMD see Cerone and Dragomir (2005, 2006) and Dragomir (2010).

We mention, for completeness, that there is also a bound which is related to the mean absolute deviation (MAD). The bound can be derived from

$$\frac{1}{2}\frac{1}{n}\sum_{i=1}^{n}|x_i - \bar{x}| \leq \frac{1}{n^2}\sum_{i<j}|x_i - x_j| \leq \frac{1}{n}\sum_{i=1}^{n}|x_i - \bar{x}|. \tag{2.22}$$

For details see Cerone and Dragomir (2006). Further discussion on the relationship between MAD and GMD will be given in Chap. 23.

2.3 The Gini Coefficient

The most well-known member of the Gini family is the Gini coefficient. It is mainly used to measure income inequality. The Gini coefficient can be defined in two alternative ways:

(a) The Gini coefficient is the GMD divided by twice the mean. For this definition to hold, the mean must be positive.
(b) The Gini coefficient, also known as the concentration ratio, is the area enclosed between the 45° line and the actual Lorenz curve divided by the area between the 45° line and the Lorenz curve that yields the maximum possible value that the index can have. This definition which is based on the areas enclosed by the actual and potential Lorenz curves holds for nonnegative variables only. (Zenga (1987) describes the historical development of the connection between the concentration ratio and the GMD.)

There are two differences between the two alternative definitions. The first difference is that the first definition applies only when the expected value of the variable is positive while the second imposes the restriction that the variable is bounded to be nonnegative (otherwise the maximum inequality may be unbounded). The second difference is that the first definition is valid only for continuous distributions while the second definition has a built-in correction for discrete distributions with finite number of observations. To see that assume that the distribution is composed of n observations. Then the upper bound of the Gini coefficient is (n − 1)/n. (It is attained when all observations except one are equal to zero). As a result, the area enclosed between the 45° line and the Lorenz curve is divided by (n − 1)/n. This correction plays a similar role as the correction for degrees of freedom.

The Gini coefficient was developed independently of the GMD, directly from the Lorenz curve and for a while it was called "the concentration ratio." Gini (1914) has shown the connection between the GMD and the concentration ratio. Ignoring the differences in definitions, the relationship between the GMD and the Gini coefficient is similar to the one between the variance and the coefficient of variation, $CV = \frac{\sigma}{\mu}$, a property that was already known in 1914. That is, the Gini coefficient is a normalized version of the GMD and it is unit-free (measured in percent). In order to calculate it one only needs to derive the GMD, and then easily convert the representation into a Gini coefficient by dividing by twice the mean.

The best known version of the Gini coefficient is as twice the area between the 45° line and the Lorenz curve (definition (b) above). For this definition the range of the coefficient is [0, 1], with 0 representing perfect equality while 1 is reached when one observation is positive and all other observations are zero. Similar to the coefficient of variation, the Gini coefficient can be defined for distributions with negative lower bound, provided that the expected value is positive (definition (a) above). However in this case the upper bound for the Gini coefficient can be greater than one. Also similar to the coefficient of variation, the Gini coefficient is not defined for distributions with expected value of zero. Being a unit-free index, the Gini coefficient is unaffected by multiplication of the variable by a constant.

Although the normalization seems innocent—normalization of the units in which the Gini is measured—it may have implications on the notion of inequality and variability. It is worth mentioning that reference to "variability" or "risk" (most common among statisticians and finance specialists) implies the use of the GMD, whereas reference to "inequality" (usually in the context of income distribution) implies the use of the Gini coefficient. To see the implication of the normalization, assume a distribution that is bounded in the range [**a**, **b**]. Try to answer the following question: What characterizes a distribution that is the most unequal according to a relative measure (either the Gini coefficient or the coefficient of variation), and what characterizes a distribution that is most unequal according to an absolute measure like the GMD or the variance? It is easy to see that when an absolute measure is used to rank inequality or variability, then the most unequal distribution is the one with half of the population at **a** and the other half at **b**. On the other hand, the answer according to a relative measure will be that the most unequal distribution is the one with almost all the population at **a** and only a tiny fraction at **b**. Therefore when dealing with issues of justice, a minor unnoticeable change may reflect a major change of opinion. By a seemingly innocent division by (twice) the mean one can switch between what Kolm (1976) refers to as "lcftist" and "rightist" measures of inequality, a point that we will discuss at length in Chap. 13 when we deal with applications of the Gini methodology.

2.4 Adjustments Needed for Discrete Distributions

The discussion so far was limited to continuous distributions. When dealing with discrete distributions, or with empirical distributions that are discrete in nature, or even when the distributions are continuous while one is interested to do a decomposition of the variability according to population subgroups, one encounters a problem of inconsistent definitions of the basic concepts which imply a serious problem of incompatible calculations. For a survey of some of the problems arising in discrete distributions see Niewiadomska-Bugaj and Kowalczyk (2005).

In what follows we point out two inconsistencies. The first is that the definitions of the absolute and relative Lorenz curves and the cumulative distribution are incompatible (to be detailed below) and the second is similar to the issue of degrees of freedom.

2.4.1 Inconsistencies in the Definitions of Lorenz Curves and Cumulative Distributions

Assume the following pairs of observed values and their probabilities of occurrence: $(x_1, p_1), \ldots, (x_k, p_k), \ldots, (x_n, p_n)$. For simplicity of exposition we assume that the observations are ordered in a nondecreasing order. That is, if $i < j$

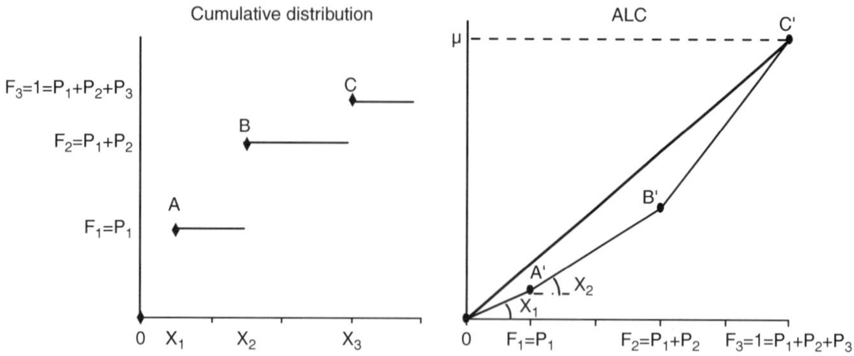

Fig. 2.3 The cumulative distribution and the ALC for discrete distributions

then $x_i \leq x_j$. This means that we observe at most n points on the cumulative distribution and on the absolute (or relative) Lorenz curve. We also assume that at least one p_k is not equal to the others. The cumulative distribution at x_k is

$$F_k = F(x_k) = \sum_{i=1}^{k} p_i. \tag{2.23}$$

The commonly used definition of a cumulative distribution function (cdf) is as a step function, continuous from the right, as shown on the left side of Fig. 2.3 for the case n = 3.

That is, the horizontal axis representing the variate is assumed to be continuous almost everywhere, while the vertical axis representing the cumulative distribution is discontinuous and jumps between the points. The right-hand side of Fig. 2.3 presents the ALC. In the absolute (or relative) Lorenz curve the horizontal axis represents the cdf while the vertical axis represents the cumulative value of the weighted mean. Formally, the vertical axis of the absolute Lorenz curve, q_k, is

$$q_k = q(F_k) = \sum\nolimits_{i=1}^{k} p_i x_i. \tag{2.24}$$

When plotting a Lorenz curve, the usual procedure is to connect the points (F_k, q_k) by linear segments, making the curve continuous in F, and discontinuous in the variate (which is the slope of the Lorenz curve). These definitions of the cumulative distribution and the Lorenz curve are not compatible with each other because if we use definition (2.23) in plotting the Lorenz curve, we should plot the Lorenz curve as a step function and this would change the value of the Gini coefficient. This complicates the curves, the convexity property is lost, and even if this treatment solves the problem for the Gini, it is not clear how to solve this issue when dealing with other parameters of the Gini method such as the Gini covariance, to be presented in Chap. 3. Whenever the number of points is small one should expect large discrepancies between the different methods of calculations.

Several solutions are available in the literature to handle those problems.

The inconsistency between the definitions of the Lorenz and the cumulative distribution and the effect on the Gini as defined in (2.1) have already been mentioned in the original paper by Gini who suggested to plot the Lorenz curve as a step function (see the translation in Metron, 2005, p. 25). But this solution has several disadvantages: (1) the convexity property of the Lorenz curve disappears and (2) to be consistent, the 45° line has to be changed into a step function as well.

Another solution was suggested by Lerman and Yitzhaki (1989). They suggested using a mid-point approximation of the cumulative distribution in the covariance formula. That is,

$$\Delta = 4\sum_{i=1}^{n} p_i(x_i - \mu)\left(\frac{F(x_i) + F(x_{i-1})}{2}\right), \quad \text{with } F(x_0) = 0. \tag{2.25}$$

This solution overcomes the problem of inequality between the Lorenz and the covariance formulas. However, it may raise difficulties in interpretation because the mid-point cumulative distribution is not formally a cumulative distribution. Also, this solution is useful with respect to the Gini but it does not solve the problem in an extended Gini context. See Chap. 6.

2.4.2 Adjustment for a Small Number of Observations

When dealing with discrete distributions two additional problems arise. The first is that the upper bound of the Gini coefficient ceases to be one; instead, it is equal to (n − 1)/n. Therefore, the number of observations affects the estimated inequality. In some sense this is similar to the correction for degrees of freedom. The other problem arises when trying to implement (2.1). The issue is how to handle ties, whether to add them with zero value to the numerator which reduces the average absolute deviation or to omit them both from the numerator and the denominator.

2.5 Gini Rediscovered: Examples

The GMD is an intuitive measure of variability. Therefore it can be easily conceived. It turns out that (as was shown above) there is a large number of seemingly unrelated presentations of the GMD (and other parameters that are derived from it), making it hard to identify that one is dealing with a GMD and also to identify which version of the GMD one is dealing with. The GMD takes different forms for discrete vs. continuous distributions and for nonnegative vs.

bounded-from-below random variables. The fact that one has to differentiate between discrete and continuous distributions and between nonnegative and bounded-from-below variables, together with the wealth of alternative presentations, makes it hard to identify a GMD. In addition, as will be seen throughout the book, the GMD falls in between parametric and nonparametric statistics. Some of the properties resemble nonparametric while the others resemble parametric statistics. This can make it complicated for the "purists" to grasp. And finally, the alternative representations are scattered throughout many papers and languages, spread over a long period of time and in many areas of interest, and not all are readily accessible.[13]

The advantage of being able to identify a GMD is that it enables the investigator to use the existing literature in order to derive additional properties of the parameter at hand and rewrite it in an alternative, more user-friendly way. It also enables the investigator to find new interpretations of the GMD and of Gini-related parameters as well as draw inference about them. In order to illustrate this point we present four cases in details. We start with a recent reinvention of GMD that appeared in 2007 in a paper published in the *American Economic Review* by Kőszegi and Rabin, entitled: "Reference-Dependent Risk Attitudes." In Appendix A of the paper, entitled "Further Definitions and Results" on p. 1063, the authors write: "In this appendix we present an array of concepts and results that may be of practical use in applying our model but that are not key to any of the main points of the paper." Definition 5 is the following:

"DEFINITION 5: *The average self-distance of a lottery F is*

$$S(F) = \int\int |x - y| dF(x) dF(y).$$

The average self-distance of a lottery is the average distance between two independent draws from the lottery. A lower average self-distance is a necessary but not sufficient condition for one lottery to be unambiguously less risky than another." p. 1063. Anyone who is familiar with GMD will recognize the index (see Yitzhaki, 1982a). The Kőszegi and Rabin article may be the starting point of a new branch in the literature that will not be recognized as related to the GMD, and therefore will not rely on the already proven properties, and maybe several years down the road a future author will notice that the properties of the GMD were investigated again, using new terminology.

The second example is what is referred to as R-regression (Hettmansperger, 1984). As will be clear in Chap. 7, R-regression is actually a regression technique based on minimizing the GMD of the residuals of the regression. Knowing this fact can simplify many of the proofs of the properties of R-regression.

---

[13] This phenomenon seems to be a characteristic of the literature on the GMD from its early development. Gini (1921) argues: "probably these papers have escaped Mr. Dalton's attention owing to the difficulty of access to the publications in which they appeared." (Gini, 1921, p. 124).

A third example is the debate between Corrado Gini and the Anglo-Saxon statisticians. The most popular presentation of the variance is as a second central moment of the distribution. The most popular presentation of the GMD is as the expected absolute difference between two i.i.d. variables. See Giorgi (1990) for a bibliographical portrait. Using the expected absolute difference between two i.i.d. variables in order to measure variability characterized the Italian school, led by Corrado Gini, while reliance on moments of the distribution characterized the Anglo-Saxon school. However, as shown by Hart (1975) and the covariance presentation, and as will be shown in Chap. 6, the GMD can also be defined as a central moment. Had both sides known about the alternative presentations of the GMD, this debate which was a source of confrontation between the Italian school and what Gini viewed as the Western schools could be avoided (see Gini, 1965, 1966, p. 199; Hart, 1975).

A fourth example is the presentation of the GMD as four times the covariance between the variate and its cumulative distribution (Lerman & Yitzhaki, 1984). This formula can be seen in Wold (1935, p. 43) except that it was not referred to as covariance. Understanding that the GMD is actually a covariance enables the imitation of the decomposition properties of the variance. This property turns Gini into an analytical tool and enables replicating ANOVA, regression, and correlations—which in some sense doubles and triples the possibilities of modeling in economics and statistics. The result is that almost every model that can be constructed using the variance can be replicated using the Gini.

2.6 Summary

This chapter surveys all (known to us and relevant to the purpose of the book) alternative representations of the GMD and the Gini coefficient. While it is hard to make an accurate count of how many independent alternative definitions exist, there are clearly more than a dozen of them. Each representation is naturally related to a specific area of application. For example, the covariance formulation is natural when one is interested in regression analysis or in the decomposition of a Gini of a population into the contributions of the subpopulations.

The fact that the GMD is an intuitive measure and the need to distinguish between discrete and continuous and between negative and nonnegative variables may explain why the Gini has been "reinvented" so often. It also explains why it is harder to work with the Gini than with the variance.

The Gini is an alternative measure of variability. Therefore it is only natural that it shares some properties with the variance on one hand, and exhibits some differences on the other hand. These similarities and differences are discussed in this chapter. The main difference between the two measures lies in the distance function used. While the Gini uses the absolute value as the distance, the variance uses the square. This difference has practical implications that will be discussed later.

Chapter 3
The Gini Equivalents of the Covariance, the Correlation, and the Regression Coefficient

Introduction

Given two random variables, one may be interested in the correlation or association or concordance between them (Gili & Bettuzzi 1987). This purpose can be generalized by following Daniels who stated the target as "the degree of agreement" (Daniels, 1950, p. 171) between the order and the rank-order of two variables.

Measures of association are treated in the literature for the parametric and the nonparametric settings. In the parametric case the widely known and used measure is Pearson's product-moment correlation coefficient, ρ, which is based on the normalized covariance between the two variables. In the nonparametric case the most commonly used measure is Spearman's correlation coefficient which is based on the normalized sample covariance between the ranks (i.e., cumulative distributions) of the variables. In this chapter we define the Gini correlation which is, in a way, in between the two types of correlations. It is based on the normalized covariance between one variable and the rank of the other variable. We note in passing that there are two Gini correlations between each pair of random variables, depending on which one is taken in its variate value and which one is ranked.

In order to be able to compare the properties of the measures of association we first list the properties that are desired from a good measure of association.

1. Nonparametric: A desirable measure of association should measure a meaningful concept of association regardless of the underlying distributions.
2. Known bounds: In order to have some idea on whether the association is strong or weak, one needs common reference points. The usual reference points are the upper bound (for positive association), the lower bound (for negative association), and the mid-point (no association, or statistical independence between the variables).
3. A desirable measure should be able to detect monotonic and non-monotonic relationships as well as the turning points in relationships if they exist. (By monotonic relationship it is meant that the relationship is monotone over the entire ranges of the variables.) Some relationships tend to be non-monotonic.

S. Yitzhaki and E. Schechtman, *The Gini Methodology: A Primer on a Statistical Methodology*, Springer Series in Statistics 272, DOI 10.1007/978-1-4614-4720-7_3,

The example that comes to mind is related to age. There are several properties of the human body that increase with age and then reach a peak and start to decline (for example, the number of teeth or hair, physical strength, etc.). Also, this kind of relationship typically exists in time series analysis where the main goal is to search for turning points, as pointed out by Raveh (1989). However, as far as we can see, monotonicity cannot be detected by a coefficient but it can be detected by a curve. This subject will be discussed in Chap. 5.

4. Intuitive explanation: It is desirable to have an intuitive explanation to enable the researcher to explain the measure to the nonexpert.
5. Computational simplicity: This property is less important nowadays with the spread of computers, but it is still a desirable property because it is associated with the ability to find simple explanations.

In addition to the above-mentioned properties there are two issues that are of concern.

The first issue is the symmetry. Most measures of association are symmetric.

This property is reasonable for symmetric distributions. We argue that it is not a reasonable property to look for whenever the distributions are nonsymmetric or whenever the relationship we are looking for is nonsymmetric such as in the case of regression analysis where the roles of the dependent and explanatory variables are not symmetric. This can happen if the explanatory variable is assumed to be measured more precisely than the dependent variable or if one believes that there is a causal relationship from the explanatory to the dependent variable.

Actually, one of the major issues that are stressed in this book is the usefulness and importance of the ability to test whether the relationship between two variables is symmetric or not. By imposing symmetry on the correlation coefficient, the ability to use the correlation to test for symmetry is lost. In our view, asymmetry is an advantage of a measure because then it "reveals more" (Lambert & Decoster, 2005). However the need of symmetry goes back to Kendall (1948) who considered the Gini correlation but rejected it on lack-of-symmetry grounds.

The second issue is related to the decomposition of the measure of variability of a linear combination of random variables. The measure of variability of a linear combination of random variables can be decomposed into individual contributions (that are caused by each individual variable) and contributions of several variables simultaneously, which are measured by measures of association. In these cases a good measure of association will enable the decomposition. For the purpose of the analysis it seems that the last issue is the most important one. The decomposition enables one to avoid double counting by classifying the contributions to the variability into individual contributions and those that cannot be associated with one particular random variable.

The structure of the chapter is as follows: Sect. 3.1 provides some preliminaries and terminology. In Sect. 3.2 we review the familiar measures of association: Pearson, Spearman, and Kendall. The Gini correlation coefficient and its properties are discussed in Sect. 3.3. Section 3.4 deals with the similarity between the two Gini correlations between a pair of random variables while Sect. 3.5 introduces the Gini regression coefficient. Section 3.6 concludes.

3.1 Preliminaries and Terminology

There are two standard approaches to analyzing the relationship between two variables X and Y. Both are based on the properties of the covariance. In one approach the variates themselves are used whereas in the other the cumulative distributions are used. If one uses the variates then the key parameter is cov(X, Y). This implies that the natural measure of variability of the individual variable is cov(X, X), which is its variance. Pearson's correlation coefficient emerges as the standardized covariance between X and Y. (The standardization factor is the product of the individual standard deviations.) This method will be referred to as the *variate* method. According to Stigler (1989) the idea of this correlation can be traced to Francis Galton (1888) who was interested in studying heredity. The correlation helped in reconciling a dilemma: according to the central limit theorem father's height and son's height tend to be normally distributed. The role of the correlation was to show that "a normal mixture of normal distributions is itself normally distributed" (Stigler, 1989, p. 75).

The second approach is to consider the covariance between the cumulative distributions, cov(F(X), G(Y)), where F(X) and G(Y) are the cumulative distributions of X and Y, respectively. This leads to Spearman's correlation coefficient as the standardized covariance between F(X) and G(Y). Here the standardization factor is a constant. Note, however, that this approach does not lead naturally to a measure of variability of an individual variable (because cov(F(X),F(X)) is a constant).

Some statisticians (e.g., Barnett, Green, and Robinson, 1976; Daniels, 1944; Kendall, 1955; Stuart, 1954) use a third method which is a mixture of those basic methods, namely, cov(X, G(Y)) and cov(F(X),Y), that is, the covariance between a variate and the cumulative distribution of another variable. This third method is the base for constructing several Gini-based correlations, as was explained in Schechtman and Yitzhaki (1987) and in Yitzhaki and Olkin (1991). In this book we focus on those that enable the decomposition of the GMD of a linear combination of random variables into the basic components, as will be detailed in Chap. 4. We turn now to the definitions and the main properties of the alternative measures of association.

3.2 Measures of Association

3.2.1 *Pearson's Correlation Coefficient*

The most widely used measure of correlation is Pearson's correlation coefficient which is defined as

$$\rho_{X,Y} = \frac{\text{cov}(X,Y)}{\sigma_X \, \sigma_Y}.$$

This measure has the following widely known properties:

(a) If X and Y are statistically independent then $\rho_{X,Y} = 0$.
(b) It is symmetric and bounded $-1 \leq \rho_{X,Y} = \rho_{Y,X} \leq 1$.
(c) It is equal to 1(−1) when Y = a + bX with b > 0 (b < 0).
(d) It is not sensitive to linear nondecreasing monotonic transformations in either of the variables.
(e) Pearson's correlation coefficient is a parameter in the multivariate normal distribution.
(f) It is a parameter in the decomposition of the variance of a linear combination of random variables into the contributions of the individual components.
(g) Pearson's correlation coefficient is the easiest to calculate.
(h) It has an intuitive explanation.

There are two rarely mentioned problems in the interpretation of the Pearson correlation coefficient. The first problem is that in practice researchers tend to compare its value to the boundaries (± 1), where large deviation of its absolute value from 1 is viewed as a weak statistical association between the variables. The second problem is that an absolute value close to zero is viewed as no association. In fact, both assertions may be misleading. Shih and Huang (1992) show that unless the marginal distributions of the two random variables can differ only in their location and/or scale parameters, the range of Pearson's ρ is narrower than [−1, 1] and depends on the marginal distributions F and G. For example, if (X, Y) have a bivariate standard lognormal distribution (that is, their natural logs have a standard bivariate normal distribution), then the range of Pearson's ρ is [−0.368, 1] (De Veaux, 1976). Other examples include a particularly simple class of multivariate distributions where the given marginals are the Eyraud–Farlie–Gumbel–Morgenstern (EFGM) distributions. A bivariate EFGM distribution H(X, Y) with univariate continuous marginal distributions F(X) and G(Y) is a distribution of the form

$$H(X,Y) = F(X)G(Y)(1 + \alpha(1 - F(X))(1 - G(Y)),$$

where α lies in [−1, 1]. It has been shown (Cambanis, 1991; Kotz & Seeger, 1991) that the range of Pearson's ρ for this family is [−1/3, 1/3], and the maximum is achieved when both marginals are uniform. All other marginals will result in a correlation coefficient smaller than 1/3 in absolute value. For example, for EFGM with standard exponential marginals the range is [−0.25, 0.25] (Johnson & Kotz, 1977). Denuit and Dhaene (2003) present an example with Pearson's correlation coefficient converging to zero, while the variables are connected by a monotonic transformation. Their example is based on the random couple (X_1, X_2) where $\ln(X_1)$ is normally distribution with mean zero and a unit standard deviation and $\ln(X_2)$ is normally distributed with mean zero and a standard deviation σ. The extreme value of the correlation is achieved when X_1 and X_2 are connected by a monotonic functional relationship:

1. If $X_2 = X_1^\sigma$ then the maximal value of the correlation coefficient for these marginals is attained and equals

$$\rho_{max}(\sigma) = \frac{e^{\sigma} - 1}{[e^{\sigma^2} - 1]^{0.5}[e - 1]^{0.5}}.$$

2. If $X_2 = X_1^{-\sigma}$ then the minimal value of the correlation coefficient is attained and equals

$$\rho_{min}(\sigma) = \frac{e^{-\sigma} - 1}{[e^{\sigma^2} - 1]^{0.5}[e - 1]^{0.5}}.$$

Now, provided that σ tends to infinity we get

$$\lim_{\sigma\to\infty} \rho_{max}(\sigma) = \lim_{\sigma\to\infty} \rho_{min}(\sigma) = 0.$$

Denuit and Dhaene (2003) conclude: "it is possible to have a random couple where the correlation is almost zero even though the components exhibit the strongest kind of dependence possible for this pair of marginals" (p. 3).

The effect of these properties on econometric analysis should not be underestimated. Note that econometricians tend to almost freely apply monotonic transformations to the variables. By doing so they may affect the correlation between the variables, which in turn may affect the decision on which variable has a higher explanatory power. The conclusion is that one should be careful when applying seemingly innocent monotonic transformations to variables when using Pearson's correlation coefficient.

3.2.2 Spearman Correlation Coefficient

Spearman's rank correlation coefficient or Spearman's ρ was first introduced by Charles Spearman (1904). It is often denoted by r_s and it is a nonparametric measure of correlation. It measures a monotonic association between two variables and inference can be drawn without making any assumptions about the underlying distributions of these variables.

In principle, r_s is simply a special case of Pearson's correlation coefficient in which the two variables X and Y are converted to their rankings R(X) and R(Y) before calculating the coefficient. Formally, the Spearman's correlation coefficient is defined by

$$r_s = \frac{\sum\left[R(X_i) - \frac{n+1}{2}\right]\left[R(Y_i) - \frac{n+1}{2}\right]}{\frac{n(n^2-1)}{12}}. \tag{3.1}$$

We note that when there are ties an alternative formula exists (see, for example, Conover, 1980, p. 252).

The properties of Spearman's correlation coefficient are identical to the properties of Pearson's correlation coefficient with three exceptions. The first is that Spearman coefficient will be equal to 1 (-1) whenever Y is *any* monotonically increasing (decreasing) function of X, not necessarily linear. The second exception—it is *not* a parameter of any distribution, and the third exception—it is not a parameter in the decomposition of a measure of variability of a linear combination of random variables.

3.2.3 Kendall's τ

The Kendall tau (τ) rank correlation coefficient (or simply Kendall's τ) is a nonparametric statistic used to measure the degree of correspondence between two rankings and to assess the significance of this correspondence. It was developed by Maurice Kendall in 1938. The measure resembles Spearman's coefficient in that it is based on ranks rather than on the original data and its distribution does not depend on the underlying distribution of the data (that is, it is nonparametric distribution-free). Kendall's τ is based on concordant and discordant pairs of observations. A pair of observations is said to be concordant if both members of one observation are larger than their respective members of the other observation (for example: (1, 2) and (3, 4)). Otherwise, the pair is discordant.

Kendall's τ coefficient is defined by

$$\tau = \frac{n_c - n_d}{0.5\, n\, (n-1)}, \tag{3.2}$$

where n_c is the number of concordant pairs and n_d is the number of discordant pairs in the data set of size n.

The denominator in the definition of τ can be interpreted as the total number of pairs of observations. Hence a high value in the numerator means that most pairs are concordant, indicating that the two rankings are consistent. Note that a tied pair is not regarded as concordant or discordant. If there exists a large number of ties then the total number of pairs (in the denominator of the expression of τ) should be adjusted accordingly.

The Kendall tau coefficient (τ) has the following properties:

- If the agreement between the two rankings is perfect (i.e., the two rankings are the same) the coefficient is equal to 1.
- If the disagreement between the two rankings is perfect (i.e., one ranking is the reverse of the other) the coefficient is equal to (-1).
- For all other arrangements the value lies between (-1) and 1, and increasing values imply increasing agreements between the rankings. If the rankings are

completely independent, the coefficient will be equal to 0 (up to a random variation).
- Kendall's τ is more difficult to compute than Pearson or Spearman.
- It is intuitively clear.
- It is not associated with the decomposition of any variability measure of a linear combination of random variables.
- It is not a parameter of any distribution.

3.3 Gini Correlations

As was shown in Chap. 2 there are more than a dozen alternative representations of the GMD. Therefore it is only natural that there is more than one way to represent the Gini correlation. The advantage of having several representations is because some properties are easier to introduce when using a specific representation while others require an alternative representation. The two main representations that will be used in this book are the covariance-based (to be detailed in this chapter) and the one based on concentration curves (to be detailed in Chap. 5). While the covariance representation of the Gini correlation is convenient for comparing the properties of Gini's correlation with Pearson's and Spearman's correlation coefficients, an alternative representation of the Gini correlations which is based on concentration curves enables the user to learn more about the behavior of the association along the distribution of one of the random variables involved. However, to fully study the properties and usefulness of the alternative representation we have to start with the properties of the absolute concentration curve (ACC). This will be done in Chap. 5. We note in passing that this alternative representation preceded the covariance representation of the Gini correlation (Blitz & Brittain, 1964).

In general the Gini correlation is based on a mixture of the variate and the cumulative distribution. Daniels (1944), Stuart (1954), Kendall (1948, 1955), and Barnett et al. (1976) gave examples of measures that are based on such a mixture. Blitz and Brittain (1964) introduced a definition, but they left it as based on areas of concentration curves (see Chap. 5). As far as we know, Shalit and Yitzhaki (1984) and Lerman and Yitzhaki (1985) were the first to realize that it is the measure of association that appears in the decomposition of the GMD of a linear combination of random variables. Schechtman and Yitzhaki (1987) investigated its properties, while Raveh (1989) used it to detect a turning point in time-series data.

We start with the most natural definition, the one that is based on the covariances. In order to prevent confusion we note that the Gini correlation used here is not related to the statistic that is referred to by Gini and others by the same name (Gini, 1936).

In order to define the Gini correlation we start with the definition of the equivalent of the covariance. There are two Gini covariances between each pair of random variables. We call them co-Ginis. The co-Ginis are defined as

$$\mathrm{Gcov}(X, Y) = \mathrm{cov}(X, G(Y)); \qquad \mathrm{Gcov}(Y, X) = \mathrm{cov}(Y, F(X)). \tag{3.3}$$

The correlations which are the normalized co-Ginis are written as

$$\Gamma_{X,Y} = \frac{\mathrm{cov}(X,G(Y))}{\mathrm{cov}(X,F(X))}; \qquad \Gamma_{Y,X} = \frac{\mathrm{cov}(Y,F(X))}{\mathrm{cov}(Y,G(Y))}. \tag{3.4}$$

In general the Gini covariances and correlations are not symmetric in X and Y. Moreover, Gcov(X, Y) and Gcov(Y, X) may have different signs. This property may be viewed as a deficiency of the method. On the other hand, there are important instances of asymmetric concepts such as, for example, regression analysis and the concept of elasticity in economics where the asymmetric property may come as an advantage, as will be shown in Chaps. 6 and 7 and in the applications part of the book.

Before we proceed to describe the properties of the Gini correlation we express the three correlation coefficients in a unified way. Let K(X, Y) denote the joint distribution of X and Y; then Pearson's ρ, Spearman's r_s, and Gini's Γ correlation coefficients can be written as

$$\rho_{X,Y} = \frac{\iint (K(x,y) - F(x)G(y))dxdy}{\sigma_X \, \sigma_Y},$$

$$r_{s,X,Y} = 12 \iint (K(x,y) - F(x)G(y))dF(x)dG(y),$$

and

$$\Gamma_{X,Y} = \frac{\iint (K(x,y) - F(x)G(y))dxdG(y)}{\mathrm{cov}(X,F(X))}.$$

Details can be found in Hoeffding (1948), Schweizer and Wolff (1981), and Schechtman and Yitzhaki (1999). These representations hint that the properties of the Gini correlation are a mixture of the properties of Pearson and Spearman correlations: it is similar to Pearson in X (the variable which is taken in its variate values) and it is similar to Spearman in Y (the variable that is taken in its ranks).

We now list the properties of the Gini correlation coefficient. Proofs of the nontrivial properties follow.[1]

(1) $-1 \leq \Gamma_{X,Y} \leq 1$.
(2) If Y is a monotonic increasing (decreasing) function of X, then both $\Gamma_{X,Y}$ and $\Gamma_{Y,X}$ equal +1 (−1).

[1] The complete proofs can be found in Schechtman and Yitzhaki (1987, 1999), Yitzhaki (2003), and Serfling and Xiao (2007).

(3) If X and Y are statistically independent then $\Gamma_{X,Y} = \Gamma_{Y,X} = 0$.
(4) $\Gamma_{X,Y} = -\Gamma_{X,-Y} = -\Gamma_{-X,Y} = \Gamma_{-X,-Y}$.
(5) $\Gamma_{X,Y}$ is invariant under all strictly monotonic transformations of Y.
(6) $\Gamma_{X,Y}$ is invariant under changes of scale and location in X.
(7) $\Gamma_{X,Y}$ is symmetric in (X, Y) if (aX + b, cY + d) is exchangeable for some constants a, b, c, and d with a and c > 0.
(8) If (X, Y) follow a bivariate normal distribution with parameters (μ_X, μ_Y, σ_X^2, σ_Y^2, ρ) then $\Gamma_{X,Y} = \Gamma_{Y,X} = \rho$, where ρ is Pearson's correlation coefficient.

Proofs of properties 1, 2, 7, and 8:

Proof of property (1): Because cov(X, F(X)) is nonnegative, it is enough to show that cov(X, F(X)) $\geq$ cov(X, G(Y)) $\geq$ −cov(X, F(X)).

The proof is based on the following claim:

Claim Given the marginal distribution functions of X and Y, and assuming that the densities exist and are positive everywhere, cov(X, Y) is maximal when E{Y|X = x} is an increasing function of X, and minimal when Y is a decreasing function of X.

Proof of the claim cov(X, Y) = E(XY) − E(X)E(Y). We need to show that E(XY) is maximal when Y is an increasing function of X.

Let (x_1, y_1) and (x_2, y_2) be two pairs of numbers. It is easy to see that if $x_1 > x_2$ and $y_1 > y_2$ then the following relationships hold:

$$x_1y_1 + x_2y_2 \geq x_1y_2 + x_2y_1 \Leftrightarrow x_1(y_1 - y_2) + x_2(y_2 - y_1) \geq 0$$
$$\Leftrightarrow (x_1 - x_2)(y_1 - y_2) \geq 0.$$

In other words, the maximum is achieved when the smaller (larger) number of one pair is multiplied by the smaller (larger) number of the second pair, that is, if Y is a monotonic increasing function of X. The proof of the other part is similar.

We now turn to the proof of property 1.

We need to show that

$$\text{cov}(X,G(Y)) \leq \text{cov}(X,F(X)).$$

Note that F(X) and G(Y) have uniform distributions over [0, 1] (that is, they are U(0, 1) random variables); therefore

$$E[F(X)] = E[G(Y)] = \frac{1}{2}.$$

Note that X and F(X) are nondecreasing functions of X. By the claim, cov(X,G(Y)) achieves its maximal value when G(Y) is an increasing function of X, which implies that F(X) = G(Y). This means that the maximum is achieved at cov(X,F(X)), which completes the proof. The other direction is similar.

Proof of property (2): Let Y = t(X) where t is a monotonic increasing function, then $G(y) = P(Y \le y) = P(t(X) \le y) = P(X \le t^{-1}(y)) = P(X \le x) = F(x)$. Hence $\Gamma_{X,Y}$ and $\Gamma_{Y,X}$ will equal +1.

The proof for a monotone decreasing transformation is similar.

Proof of property (7): Without loss of generality assume that d = b = 0 and a = c = 1. That is, assume that $(X, Y)\underline{\underline{d}}(Y, X)$ and denote the joint distribution by h(X, Y) = h(Y, X). (This can be assumed because $\Gamma_{X,Y} = \Gamma_{aX+b,cY+d}$.)

$$\Gamma_{X,Y} = \frac{\iint xG(y)h(x,y)dxdy - 0.5E(X)}{\int xF(x)f(x)dx - 0.5E(X)}$$

$$\Gamma_{Y,X} = \frac{\iint yF(x)h(y,x)dydx - 0.5E(Y)}{\int yG(y)g(y)dy - 0.5E(Y)}.$$

Because (X, Y) is exchangeable, the denominators are equal. Also,

$$\iint xG(y)h(x,y)dxdy = \iint xF(y)h(y,x)dxdy = \iint yF(x)h(y,x)dydx,$$

which completes the proof.

Proof of property (8): It is sufficient to show that $\Gamma_{Y,X} = \rho$.

The conditional expectation E(Y|X) is given by

$$E(Y|X) = \mu_Y + \rho\,\sigma_Y \frac{(X - \mu_X)}{\sigma_X}.$$

Using this property we can write the numerator as

$$\mathrm{cov}(Y, F(X)) = E_X[(E_{Y|X}(Y|X) - \mu_Y)(F(X) - 0.5)] = \rho\,\sigma_Y E(Z(F(X) - 0.5)),$$

where $Z = \frac{(X - \mu_X)}{\sigma_X}$ is a standard normal random variable. Let $\Phi(Z)$ denote the cumulative distribution of a standard normal random variable. Then $\Phi(Z) = F(X)$ and $\mathrm{cov}(Y,F(X)) = \rho\,\sigma_Y \mathrm{cov}(Z, \Phi(Z))$.

Hence by similar considerations we can express the denominator as $\mathrm{cov}(Y, G(Y)) = \sigma_Y \mathrm{cov}(Z, \Phi(Z))$, which completes the proof.

Property (7) above states that the Gini correlation is symmetric for exchangeable variables. Hence it can be used to test for exchangeability in general and to construct tests for asymmetry (Boos, 1982). We view the ability to test for exchangeability of random variables as an important contribution of this book. Therefore exchangeability and its implications will be described in a separate section, Sect. 3.4.

An additional alternative representation that will be useful later is based on a combination of earlier presentations: Let (X_1, Y_1) and (X_2, Y_2) be two i.i.d. pairs of random variables. Then

$$\Gamma_{X,Y} = \frac{\text{cov}[(X_1 - X_2)(G(Y_1) - G(Y_2))]}{\text{cov}[(X_1 - X_2)(F(X_1) - F(X_2))]}. \tag{3.5}$$

Equation (3.5) presents the Gini correlation as based on the concordance between the variate values and their ranks (denominator) as well as the ranks of the other variable (numerator).

To summarize: The properties of $\Gamma_{X,Y}$ are a mixture of Pearson's and Spearman's correlation coefficients. It behaves like Pearson's coefficient in X, like Spearman's coefficient in Y, and it is equal to Pearson's ρ when the distribution follows a bivariate normal distribution.

3.4 The Similarity Between the Two Gini Correlations of a Pair of Variables

The variability measure (variance or Gini) of a linear combination of random variables can be decomposed into the contributions of the individual variables and the correlations between them. The two decompositions will have the same structure when the two Gini correlations within some of the pairs of variables will be equal (details will be given in Chap. 4). The aim of this section is to investigate the properties of the underlying distribution that will cause the two Gini correlation coefficients between a pair of variables to be equal. It should be pointed out that the discussion is not complete and a full characterization needs further research.

A sufficient condition for the equality of the two Gini correlations between X and Y is that the underlying distributions are exchangeable up to a linear transformation (see property 7 in Sect. 3.3). Therefore, exchangeability up to a linear transformation is the feature of the underlying distribution that distinguishes between the decomposition properties of the GMD vs. the variance. We note in passing that it has not been investigated yet whether the condition is a necessary condition as well.

3.4.1 Formal Definitions of Exchangeability

Exchangeability is a fundamental concept that was skipped upon by most economists but may have major implications on economic theory. A review of this argument and its implications can be found in McCall (1991).

We start with two definitions. First we define exchangeability and then we define "exchangeability up to a linear transformation."

Definition of exchangeability: The random variables X and Y are said to be exchangeable if F(X, Y) = F(Y, X).

Definition of exchangeability up to a linear transformation: X and Y are said to be exchangeable up to a linear transformation if there exist **a**, **b**, **c**, and **d** (**a**, **c** > 0) such that (X, Y) and (**a**Y + **b**, cX + d) are equally distributed. Obviously exchangeability implies exchangeability up to a linear transformation, but not the other way around.

For example, (X,Y)~BVN(0,0,1,1, ρ) are exchangeable. However, (X,Y)~BVN $(2, 3, 1, 4, \rho)$ are only exchangeable up to a linear transformation.

Note that a necessary condition for exchangeability is that the marginal distributions are identical. In addition it was proved that if (X, Y) are exchangeable up to a linear transformation, then

$$\Gamma_{X,Y} = \Gamma_{Y,X}.$$

This means that the equality of the two Gini correlations is a necessary condition for exchangeability up to a linear transformation, and therefore can be used to test for its existence. The statistical test of the equality of two Gini correlations is proposed in Chap. 10.

3.4.2 *The Implications and Applications of Exchangeability*

The main implications of exchangeability are related to the decomposition of the Gini of a linear combination of random variables. Those will be detailed in Chap. 4. Additional applications are the following:

1. An application for testing convergence: If the underlying distributions are not exchangeable up to a linear transformation then the results of the decompositions of the variance and the Gini do not follow the same structure. (The decomposition will be discussed in detail in Chap. 4.) This fact is useful—it allows the user to test for the convergence of the distribution of an average of i.i.d. random variables to the normal distribution. Let $Y_{mn} = a_1X_{m1} + a_2X_{m2} + \ldots + a_nX_{mn}$, where each X_{mi} is an average of m i.i.d. random variables. That is, Y_{mn} is a linear combination of n random variables, each of which is an average of m i.i.d. random variables. According to the central limit theorem for a large enough m all the Xs are approximately normally distributed and so is Y. If this is the case, then the two correlations between every pair of variables should be equal (which implies that the structures of the decompositions of the variance and the Gini are similar). Developing such a test is beyond the scope of this book. We comment on this issue in Chap. 23.
2. Directional movements: In time-series analysis the implication of exchangeability up to a linear transformation is that moving forward in time is the reverse of moving backward in time. If the variables in a time series are not exchangeable, then the trend that we observe when we move forward in time need not be the reverse of the trend observed when we move backward. The literature on

convergence in growth economics includes Barro (1991), Barro and Sala-i-Martin (1992), and Sala-i-Martin (1996) while recent research has put these initial findings in doubt (e.g., Friedman, 1992; Quah, 1993, 1996; Bliss, 1999). For an application of the Gini methodology in this field see Wodon and Yitzhaki (2006) who presented an example where one can find convergence both when moving forward and moving backward in time using the same data. We will elaborate on this issue in Chap. 23, when we discuss the application of the Gini methodology in time-series analysis.

3. Additional applications of exchangeability rely on the suggestions made in an important paper by McCall (1991). McCall translates the concept to mean symmetric dependence, interchangeability, and fairness. He differentiates between local and global exchangeability and applies it to wider situations in terms of both infinite series and economic settings. The importance of global exchangeability is emphasized by: "Each society member is exchangeable with respect to these fundamental traits in the following sense: given that *m* is a member of society *s*, his basic behavior is the same as that exhibited by a randomly drawn individual from *s*." p. 554.
4. McCall (p. 557) also interprets exchangeability as fairness because it means equal treatment. There is a need for a lot of work to translate those deep ideas into testable and workable hypotheses. All we do in this subsection is point out that by using the variance, symmetry is imposed on the data even when the distributions are not symmetric, and therefore we lose the ability to test for symmetry.

3.5 The Gini Regression Coefficient

The literature dealing with Gini regression coefficients is much more developed than the literature dealing with the Gini correlation coefficient. Therefore we devote several chapters to the alternative concepts and presentations of the Gini regression coefficients. However, a complete presentation has to be postponed until we introduce additional tools such as concentration curves (Chap. 5). In this section we briefly introduce the simple Gini regression coefficients.

There are two types of regression coefficients that can be attributed to the GMD. One is based on the minimization of the GMD of the residuals, while the other is based on substituting the variance-based expressions in the Ordinary Least Squares (OLS) regression by the equivalent GMD terms. The former has the advantage of being "optimal" because it is derived by an optimization process while the latter is based on imitation of the OLS procedure, and hence enables to replicate the concepts and intuition and even allows to mix Gini regression with the popular OLS regression (as will be shown in Chap. 8 and illustrated in Chap. 21). In this section we introduce the second type of the Gini regression coefficient. We refer to it as "covariance based" because it is based on the covariance presentation of the GMD and the properties of the covariance are used in order to develop its properties. As will be seen later, it is semi-parametric because it can be interpreted as a weighted average of slopes defined between adjacent observations of the explanatory variable. Hence no model is assumed in order to estimate the regression coefficient.

The OLS regression coefficient can be expressed as

$$\beta^{\mathrm{O}} = \frac{\mathrm{cov(Y,X)}}{\mathrm{cov(X,X)}}. \tag{3.6}$$

Replacing each covariance by the corresponding Gini covariance, we get

$$\beta^{\mathrm{N}} = \frac{\mathrm{cov(Y,F(X))}}{\mathrm{cov(X,F(X))}}, \tag{3.7}$$

where N indicates that we are dealing with the semi-parametric version of the GMD regression coefficient. Note that in the numerator we have a co-Gini, while the denominator is the GMD of the explanatory variable.

It seems natural to view presentation (3.7) as an OLS instrumental variable (IV) method, with F(X) serving as the instrument. We argue that this is not the appropriate interpretation because the Gini regression does not have to fulfill the requirements needed to be qualified as IV method. (The use of the empirical cumulative distribution as an IV can be traced to Durbin (1954) who suggested using the rank of a variable as an IV to overcome the bias caused by errors in the measurement of the explanatory variable.)

As will be seen in Chap. 7, most of the parameters and the concepts in the Gini regression framework are parallel in structure to the OLS concepts. However they are different in their properties and interpretations. For example, one can define an IV in the framework of the Gini regression, but it will be different from the OLS IV. The Gini IV can be referred to in an OLS framework as the "double IV," because the concept of IV is applied twice: first by using the ranks (the sample's empirical cumulative distribution) instead of the variable without questioning the validity of the rank to serve as an IV, and in the second stage one uses another variable which is required to obey all the requirements from an IV. That is, one moves from relating Y to X to relating Y to F(X) and then to relating Y to F(Z) (see Yitzhaki and Schechtman (2004)).

The advantage of the similarity between the Gini and the OLS regression coefficients is that one can use the OLS software to estimate the parameters of the Gini regression, but the similarity is misleading and requires caution. As will be shown in Chap. 9 (estimation), while the OLS software can be used in order to obtain the point estimates of the coefficients, it cannot be used in order to estimate the standard errors of the Gini regression coefficients. The distinction between OLS and Gini regressions should always be kept in mind because the similarity and the proximity of the concepts may mislead the intuition.

Formally, let (Y, X) be a bivariate random variable that follows a continuous distribution with finite first and second moments. Similar to the OLS regression coefficient, a normal equation can be derived from β^{N}.

To see that, assume that the following model is given:

$$Y = \alpha + \beta X + \varepsilon. \tag{3.8}$$

One of the normal equations in OLS is cov(X, ε) = 0. In order to obtain the normal equation for the Gini coefficient we use the covariance properties and get

$$\text{cov}(Y, F(X)) \equiv \beta \ \text{cov}(X, F(X)) + \text{cov}(\varepsilon, F(X)). \tag{3.9}$$

Substituting β in (3.9) by β^N from (3.7) yields a normal equation that is similar to the normal equation in OLS. That is:

$$\text{cov}(\varepsilon_N, F(X)) = 0. \tag{3.10}$$

Note that both regression coefficients imply orthogonality between the explanatory variable and the resultant residual, but the notion of orthogonality is different.

An additional interpretation of the OLS and Gini regression coefficients is that both are weighted averages of slopes defined between all possible pairs of observations chosen from (Y, X) as we show next.

Proposition 3.1 *Let* (Y_1, X_1) *and* (Y_2, X_2) *be two independent draws from the bivariate distribution F(Y, X). Let the slopes be*

$$\beta_{21}(x_2, x_1) = \begin{pmatrix} \frac{Y_2 - Y_1}{X_2 - X_1} & \text{if } X_2 \neq X_1 \\ 0 & \text{if } X_2 = X_1 \end{pmatrix}; \tag{3.11}$$

then

$$\beta = \iint w_{21}(x_2, x_1) \ \beta_{21}(x_2, x_1) f(x_1) f(x_2) dx_1 dx_2, \tag{3.12}$$

where w() represents the weighting scheme. Equation (3.12) can be used to describe the structures of both OLS and Gini regression coefficients. The difference is in the weighting schemes used. In the case of the OLS the weighting scheme is

$$w_{21}(x_2, x_1) = w_{21}^o = \frac{(X_2 - X_1)^2}{E\{(X_2 - X_1)^2\}}.$$

The weighting scheme for the Gini regression coefficient is

$$w_{21}(x_2, x_1) = w_{21}^N = \frac{(X_2 - X_1)(F(X_2) - F(X_1))}{E\{(X_2 - X_1)(F(X_2) - F(X_1))\}}.$$

Proof The proof will be given for OLS. The adjustment to the Gini regression is straightforward. The proof is based on the following property of the covariance:

$$\mathrm{cov}(Y, X) = 0.5E\{(Y_2 - Y_1)(X_2 - X_1)\}.$$

$$\beta^o = \frac{\mathrm{cov}(Y,X)}{\mathrm{cov}(X,X)} = \frac{0.5E\{\ (Y_2-Y_1)(X_2-X_1)\}}{0.5E\{\ (X_2-X_1)^2\}}$$

$$= \frac{1}{E\{(X_2-X_1)^2\}} E\left[\left(\frac{(Y_2-Y_1)}{(X_2-X_1)}\right)(X_2 - X_1)^2\right] = E\left\{\frac{(X_2-X_1)^2}{E\{\ (X_2-X_1)^2\}}\left(\frac{Y_2-Y_1}{X_2-X_1}\right)\right\},$$

$$= \iint w^o{}_{21}(x_2,x_1)\ \ \beta_{21}(x_2,x_1)f(x_1)f(x_2)dx_1dx_2$$

where

$$w^o{}_{21} = \frac{(X_2 - X_1)^2}{E\{(X_2 - X_1)^2\}}.$$

The proof for the GMD is similar, and it is based on

$$\mathrm{cov}(Y, F(X)) = 0.5E\{(Y_2 - Y_1)(F(X_2) - F(X_1))\}.$$

The corresponding weights are

$$w^N_{21} = \frac{(X_2 - X_1)(F(X_2) - F(X_1))}{E\{(X_2 - X_1)(F(X_2) - F(X_1))\}}.$$

Proposition 3.1 stresses the nonparametric nature of the regression coefficients. No assumption of a linear model is required in order to produce the regression coefficients. One can simply interpret them as weighted averages of slopes. This property of the regression coefficients is used when one is interested in a summary statistic of average slopes, as we will show when dealing with the use in public finance and analysis of risk. In these cases assuming and testing for linearity are not needed. The importance of the linearity assumption is clear when one wants to predict, in which case it is essential to be able to test whether the assumptions of linearity and of distributional properties of the residuals are supported by the data. This point will be discussed in Chap. 7 where we offer additional presentations of the Gini regression coefficient.

3.6 Summary

The covariance-based presentation of the GMD opens a wide area of research. In this chapter we have introduced the Gini covariance, the Gini correlation, and the Gini regression coefficient. It was shown that the Gini covariance and correlation are asymmetric measures of association, as opposed to the ordinary covariance and to Pearson's correlation coefficient. The Gini regression coefficient is similar in structure to the OLS regression coefficient. The covariance representation enables the user to replicate almost every model that is based on the variance by using the equivalent parameters of the Gini method. In some cases the two representations are similar in structure. In others the Gini representation contains extra terms. In this latter case we claim that "the Gini reveals more." Whenever the estimates of the variance-based parameters differ from those obtained for the Gini-based equivalent parameters, one can attribute the difference to the different metrics used. Because under the normal distribution the expected values of the estimates are equal, the difference in the estimates may indicate a violation of an assumption. This issue will be further investigated in the chapters that deal with the comparisons between the specific parameters. In Chaps. 4–6 we compare them to the familiar measures that are used in the variance world and discuss some of the differences.

Chapter 4
Decompositions of the GMD

Introduction

Several basic methods of statistical analysis such as regression and analysis of variance are based on the properties of the decomposition of the measure of variability. In the decomposition of a measure of variability we differentiate between two kinds of decompositions:

(a) Decomposition of the variability of a linear combination of random variables into the contributions of the components: The most popular example is the Ordinary Least Squares regression in which the mean of the dependent variable is a linear combination of the explanatory variables and the objective is to find the values of the coefficients that maximize the portion of the variance of the dependent variable that can be explained by the linear combination of the explanatory variables. A variant of this procedure is the decomposition of the covariance between two linear combinations of random variables (instead of a variability measure) into the contributions of the different variables.
(b) Decomposition of the variability of a population composed of several groups (subpopulations) into the contributions of the groups to the overall variability: An example is ANOVA (ANalysis Of VAriance) where the variance of an overall population is decomposed into the intra- and inter-group variances.

The decomposition can be performed on an absolute measure of variability such as the variance or the GMD, on a relative measure such as the coefficient of variation or the Gini coefficient, or on a covariance between two random variables instead of the measure of variability itself. In this chapter we introduce the two decompositions of the GMD and compare them to the respective decompositions of the variance. At a later stage we will analyze the implications of those decompositions. Most of the applications covered in this book in different fields of research rely on the decompositions of the GMD, but they will be referred to by using different names, such as inequality or progressivity or risk, depending on the field.

S. Yitzhaki and E. Schechtman, *The Gini Methodology: A Primer on a Statistical Methodology*, Springer Series in Statistics 272, DOI 10.1007/978-1-4614-4720-7_4,

Hence, presenting the decomposition in advance enables us to transfer additional properties from one field into the other because the models are similar, only the subject matters differ.

It is important to emphasize that both decompositions of the GMD share similar structures to the respective decompositions of the variance. However the decompositions of the GMD contain additional parameters. In general these extra parameters do not vanish and they contribute to our understanding of the phenomena that we want to analyze as will be seen below. This is the origin of the slogan put forward by Lambert and Decoster (2005) "The Gini reveals more." However when these additional parameters are equal to zero then the structures of the Gini decompositions are identical to the structures of the decompositions of the variance. In these cases the components of the decomposition of the variance are sufficient statistics in describing the phenomena under study. Note, however, that even when the additional parameters are equal to zero, they may still add to the analyses because the number of models that exist for the analysis is doubled. For each variance-based model there will be an identical Gini-based model which may result in totally different numerical values. We will return to this subject after describing the additional parameters that appear in the decompositions of the GMD.

The structure of the chapter is as follows: In Sect. 4.1 we present the decomposition of the GMD of a linear combination of random variables. Section 4.2 is devoted to the decomposition of the variability of a population by subpopulations. In Sect. 4.3 we show how the Gini covariance is decomposed and Sect. 4.4 concludes.

4.1 The Decomposition of the GMD of a Linear Combination of Variables

The variability index of a linear combination of variables can be decomposed into two types of components: individual components which represent the contribution of each variable individually and components that are shared by pairs of variables. This fact enables the researcher to evaluate the impact of each component separately. By changing the mixture of the components the user can evaluate the effectiveness of different policies. An additional advantage of the decompositions is that the terms involved in the decomposition must add up to the total variability. Therefore there is no double counting of the same property. For example, if the final distribution does not change, it is impossible that rearrangements of the elements will increase one or several terms without causing some other terms to decline in exactly the same total amount. The downside of this approach is that when dealing with decompositions one cannot choose the correlation coefficient to be used. It is determined by the equation because the elements have to add up to a given total. As far as we know the only measure of variability that allows the decomposition of the variability measure of a linear combination of random variables into its components is the variance through the Pearson's correlation coefficient.[1] Therefore we will be interested mainly in comparing the decomposition of the GMD with the decomposition of the variance. There are two types of terms that cause the decomposition of the Gini of a linear combination to be different in structure from the decomposition of the variance of the same combination. Although their roles in the decomposition are different, they are caused by the same phenomenon: the non-symmetrical nature of the Gini correlation. This same feature distinguishes the Gini correlations from Pearson's and Spearman's correlation coefficients. As is shown below, this difference plays a major role in making the decomposition of the GMD more revealing, and at the same time more complicated than the decomposition of the variance.

Let

$$Y = \sum\nolimits_{k=0}^{K} \beta_k X_k, \tag{4.1}$$

where β_k, k = 0, 1, ..., K are given constants while X_k, k = 1, ..., K are random variables and X_0 is a constant that takes the value of 1 for all realizations. The variance of Y can be decomposed into

$$\sigma_Y^2 = \sum\nolimits_{k=1}^{K} \beta_k^2\, \sigma_k^2 + \sum\nolimits_{k\neq j} \beta_k\, \beta_j\, \sigma_k\, \sigma_j\, \rho_{jk}. \tag{4.2}$$

The decomposition of the (square of the) coefficient of variation is similar in structure to the decomposition of the variance except that the constant β_k is replaced by δ_k which is the share of variable k in the expected value of Y.

$$\frac{\sigma_Y^2}{\mu_Y^2} = \sum\nolimits_{k=1}^{K} \delta_k^2 \frac{\sigma_k^2}{\mu_k^2} + \sum\nolimits_{k=1}^{K} \sum\nolimits_{j\neq k} \delta_k\, \delta_j \frac{\sigma_k\, \sigma_j}{\mu_k \mu_j}\, \rho_{kj},$$
$$\text{where } \delta_k = \frac{\beta_k\, \mu_k}{\mu_Y}. \tag{4.3}$$

The most convenient presentation to use for decomposing the GMD of Y is (4.4) below (same as equation (2.15)), which allows to utilize the properties of the covariance:

$$\Delta_Y = 4\,\mathrm{cov}(Y, F(Y)). \tag{4.4}$$

The covariance is a function of Y and F(Y). Therefore there are two decompositions that we will refer to as one- and two-step decompositions, respectively.

---

[1] Additional potential measures of variability that can be decomposed will be discussed in Chap. 23.

4.1.1 One-Step Decomposition (Marginal Decomposition)

The one-step decomposition is useful in the analysis of policy in the areas of finance and income distributions. Basically it is based on replacing Y in (4.4) by (4.1) and leaving F(Y) untouched.

Proposition 4.1 *One-step decomposition of GMD. Given (4.1) then*

$$\Delta_Y = \sum_{k=1}^{K} \beta_k \, \Gamma_{kY} \, \Delta_k \quad \text{where} \quad \Gamma_{kY} = \frac{\text{cov}(X_k, F(Y))}{\text{cov}(X_k, F(X_k))}. \tag{4.5}$$

Proof The proof follows from the properties of the covariance and it is straightforward:

$$\text{cov}(Y, F(Y)) = \sum_{k=0}^{K} \beta_k \text{cov}(X_k, F(Y)) = \sum_{k=0}^{K} \beta_k \frac{\text{cov}(X_k, F(Y))}{\text{cov}(X_k, F(X_k))} \text{cov}(X_k, F(X_k)).$$

This decomposition plays a crucial role in applications in the areas of income distribution and taxation which are intended to analyze changes in the income distribution due to changes in one of its components. Note that the variable Y still appears on the right-hand side of (4.5) because the rank of Y (i.e., the cumulative distribution) does not vanish. This in turn limits its applications to situations where the rank does not change in a significant way. Therefore it is useful for analyzing marginal changes in the income distribution. We will refer to it as a marginal decomposition. See Chap. 14.

4.1.2 Two-Step Decomposition

The two-step decomposition is intended to fully replicate the decomposition of the variance. It is more complicated than the one-step decomposition. For the sake of simplicity we first assume that there are K = 2 variables and then we extend the decomposition to the general case with K variables.

Proposition 4.2 *Let* $Y = \beta_0 + \beta_1 X_1 + \beta_2 X_2$. *Then the following identities hold:*

(a)

$$\Delta_Y^2 - [\,\beta_1 D_{1Y}\,\Delta_1 + \beta_2 D_{2Y}\,\Delta_2]\,\Delta_Y = \beta_1^2\,\Delta_1^2 + \beta_2^2\,\Delta_2^2 + \beta_1\,\beta_2\,\Delta_1\,\Delta_2(\Gamma_{12} + \Gamma_{21}), \tag{4.6}$$

where Γ_{ij} *is Gini's correlation between* X_i *and* X_j *and* $D_{iY} = \Gamma_{iY} - \Gamma_{Yi}, i = 1,2$.

(b) Provided that $D_{iY} = 0$, *for* $i = 1, 2$, *and* $\Gamma_{12} = \Gamma_{21} = \Gamma$ *decomposition (4.6) can be simplified into*

$$\Delta_Y^2 = \beta_1{}^2\Delta_1{}^2 + \beta_2{}^2\Delta_2{}^2 + 2\,\beta_1\,\beta_2\Delta_1\Delta_2\Gamma. \tag{4.7}$$

Note that (4.7) is identical in structure to the decomposition of the variance (4.2).

Proof (Wodon & Yitzhaki, 2003a)
Proof of (a). Using the one-step decomposition above we get

$$\mathrm{cov}(Y, F(Y)) = \beta_1 \Gamma_{1Y} \mathrm{cov}(X_1, F(X_1)) + \beta_2 \Gamma_{2Y} \mathrm{cov}(X_2, F(X_2)). \quad (4.8)$$

Recall that

$$\Gamma_{iY} = \Gamma_{Yi} + D_{iY} \quad \text{for } i = 1, 2. \quad (4.9)$$

That is, D_{iY} is the difference between the two Gini correlations defined between Y and X_i. Using (4.8) and (4.9), we get

$$\Delta_Y = \beta_1(\Gamma_{Y1} + D_{1Y})\Delta_1 + \beta_2(\Gamma_{Y2} + D_{2Y})\Delta_2.$$

Rearranging terms

$$\Delta_Y - \beta_1 D_{1Y}\Delta_1 - \beta_2 D_{2Y}\Delta_2 = \beta_1 \Gamma_{Y1}\Delta_1 + \beta_2 \Gamma_{Y2}\Delta_2.$$

Using the properties of the covariance

$$\Gamma_{Y1} = \frac{\mathrm{cov}(Y,F(X_1))}{\mathrm{cov}(Y,F(Y))} = \frac{1}{\mathrm{cov}(Y,F(Y))}\{\beta_1 \mathrm{cov}(X_1,F(X_1)) + \beta_2 \mathrm{cov}(X_2,F(X_1))\}$$

$$= \frac{\beta_1 \mathrm{cov}(X_1,F(X_1)) + \beta_2 \mathrm{cov}(X_2,F(X_2))\ \Gamma_{21}}{\mathrm{cov}(Y,F(Y))} = \frac{\beta_1\ \Delta_1 + \beta_2\ \Delta_2\ \Gamma_{21}}{\Delta_Y}.$$

Writing Γ_{Y2} in a similar manner, we get (4.6)

$$\begin{aligned} \Delta_Y^2 - [\beta_1 D_{1Y}\ \Delta_1 + \beta_2 D_{2Y}\ \Delta_2]\ \Delta_Y = \beta_1\ \Delta_1(\beta_1\ \Delta_1 + \beta_2\ \Delta_2\ \Gamma_{21}) \\ + \beta_2\ \Delta_2(\beta_1\ \Gamma_{12}\ \Delta_1 + \beta_2\ \Delta_2) \\ - \beta_1^2\ \Delta_1^2 - \beta_2^2\ \Delta_2^2 + \beta_1\beta_2\ \Delta_1\ \Delta_2(\Gamma_{12} + \Gamma_{21}). \end{aligned}$$

Proof of (b). In order to prove (4.7) we assume equality of the two Gini correlation coefficients between Y and X_i, which means that $D_{iY} = 0$ for $i = 1, 2$. The assumption $\Gamma = \Gamma_{12} = \Gamma_{21}$ completes the proof of (4.7).

The extensions of (4.6) and (4.7) to K variables and to the decomposition of the Gini coefficient are trivial. The decomposition of the Gini coefficient differs from the decomposition of the GMD in one respect. Each β_k is substituted by $\delta_k = \beta_k\mu_k/\mu_Y$. The decomposition of the Gini coefficient G_Y is presented next:

Let $Y = \sum_{k=0}^{K} \beta_k X_k$, and let $\delta_k = \beta_k\mu_k/\mu_Y$, where μ represents the expected value. Then

$$G_Y^2 - G_Y \sum_{k=1}^{K} \delta_k D_{kY} G_k = \sum_{k=1}^{K} \delta_k^2 G_k^2 + \sum_{k=1}^{K} \sum_{k \neq j} \delta_k\ \delta_j G_k G_j\ \Gamma_{kj}. \tag{4.10}$$

If $D_{kY} = 0$ for $k = 1, \ldots, K$ and $\Gamma_{kj} = \Gamma_{jk}$ for $(k, j = 1, \ldots, K)$ then

$$G_Y^2 = \sum_{k=1}^{k} \delta_k^2 G_k^2 + \sum_{k=1}^{K} \sum_{k \neq j} \delta_k\ \delta_j G_k G_j\ \Gamma_{kj}. \tag{4.11}$$

Equation (4.10) is similar in its structure to the decomposition of the coefficient of variation. In order for it to be identical in structure, the two Gini correlations between each pair of variables $Y, X_1, \ldots, X_K$ must be equal. The resultant presentation is given in (4.11). Schechtman and Yitzhaki (1987) show that a sufficient condition for $\Gamma_{kj} = \Gamma_{jk}$ is that the variables are exchangeable up to a linear transformation. (See property (7) in Sect. 3.3). The meaning and implication of exchangeability are discussed in the next section (Sect. 4.3). Examples of such distributions are the multivariate normal and the multivariate lognormal, provided that $\sigma_k = \sigma_j$, where σ is the logarithmic standard deviation. If the Gini correlations between pairs of variables are not equal, one needs to use (4.10), where each "violation" of the equality of the Gini correlations is captured by an additional term in the decomposition (hence, we can treat each violation separately and evaluate its effect on the decomposition; in particular we can see whether the violation tends to increase or decrease the overall variability or inequality).

We conclude the theoretical part with an example, showing that a linear combination of two statistically independent random variables does not necessarily lead to the similarity between the decompositions of the variance and of the Gini.

In other words, statistical independence is not a sufficient condition for similarity in the structures of the decompositions of the variance and the Gini of a linear combination of the variables. More specifically, the example shows that while the variance of a sum of i.i.d. random variables is the sum of the variances, this is not necessarily true for the Gini of the sum. It will be true if the combination and each of the individual components are exchangeable up to a linear transformation. This is an additional insight that the Gini reveals more.

Let X_1, X_2 be i.i.d. random variables having a uniform distribution on [0, 1] and let $Y = X_1 + X_2$. We show that $\Gamma_{X_1,Y} \neq \Gamma_{Y,X_1}$, i.e., the condition $D_{kY} = 0$ of equation (4.6) does *not* hold.

The cumulative distribution of Y is given by

$$F(y) = \begin{pmatrix} 0 & y < 0 \\ \frac{y^2}{2} & 0 < y < 1 \\ 1 - (2-y)^2/2 & 1 < y < 2 \end{pmatrix}$$

and the density function is given by

$$f(y) = \begin{pmatrix} y & 0 < y < 1 \\ 2 - y & 1 < y < 2 \end{pmatrix}.$$

The joint density of X_1 and Y is given by

$$f(x_1,y) = \begin{pmatrix} 1 & 0 < x_1 < 1, x_1 < y < x_1 + 1 \\ 0 & \text{otherwise} \end{pmatrix}.$$

The calculations give

$$\begin{aligned} \text{cov}(X_1, F(Y)) &= \int_0^1 \int_x^1 (x - 0.5)\frac{y^2}{2}dydx + \int_0^1 \int_1^{x+1} (x - 0.5)\left(1 - \frac{(2-y)^2}{2}\right)dydx \\ &= 0.05833 \end{aligned}$$

$$\text{cov}(X_1, F(X_1)) = \int_0^1 (x - 0.5)xdx = \frac{1}{12}.$$

Therefore

$$\Gamma_{X_1,Y} = \frac{0.05833}{1/12} = 0.7.$$

On the other hand,

$$\text{cov}(Y, F(X_1)) = \int_0^1 \int_x^{x+1} (y - 1)xdydx = 0.08333$$

$$\begin{aligned} \text{cov}(Y, F(Y)) &= \int_0^1 (y - 1)\frac{y^2}{2}ydy \\ &+ \int_1^2 (y - 1)\left(1 - \frac{(2-y)^2}{2}\right)(2 - y)dy = 0.116667. \end{aligned}$$

Therefore

$$\Gamma_{Y,X_1} = \frac{0.08333}{0.16667} = 0.714.$$

Because the correlations are not equal, the term D of (4.6) does not vanish. The intuitive reason for the result is that the distribution of the sum of the variables is not identical, up to a linear transformation, to the distribution of the components. This is also an illustration that "the Gini reveals more."

The following conclusions can be drawn:

(a) If the underlying distribution is multivariate normal then the decompositions of the Gini and of the variance of a linear combination of random variables produce identical results (identical in structure, but not necessarily in their numerical results), with the decomposition of the Gini being less efficient than the decomposition of the variance, due to the fact that the estimates of the variance and the Pearson correlation are sufficient statistics for the parameters in the normal distribution.

(b) In general, if the underlying distributions of Y, $X_1, \ldots, X_K$ are exchangeable up to linear transformations then the decompositions of the Gini and of the variance of a linear combination of random variables produce identical *structure*, but with different numerical and expected values. This means that the number of potential models that can be used is doubled because for every model based on the variance there is a parallel model based on the Gini. By parallel it is meant that one can take the solution of a variance-based model, substitute every variance by the square of the GMD and every Pearson correlation coefficient by Gini correlation, and obtain a new set of results. For example, assume that we are interested in minimizing the measure of variability of a linear combination of random variables subject to a given expected value as is the case of constructing portfolios in finance (see Chaps. 17 and 18). Then there are two potential solutions—one for each metric chosen. Another example: When dealing with models in finance and policy design the question of interest is how to change the constants β_k or δ_k ($k = 1, \ldots, K$) in order to minimize the GMD or the variance of the linear combination, Y, subject to additional constraints. In these cases we can take the solution that was derived for the variance-based problem, substitute every variance by the square of the GMD, substitute every Pearson correlation coefficient by the Gini correlation coefficient, and get a Gini-based solution. The numerical results may, of course, differ. Actually, the semi-parametric Gini regression and the model of selecting a portfolio (Chap. 18) are examples of this imitation.

(c) If only some (or all) of the $X_1, \ldots, X_K$ are *not* exchangeable up to linear transformations among themselves (but all are exchangeable with Y), then one can define a symmetric Gini correlation as follows:

$$S_{XY} = S_{YX} = 0.5\ (\ \Gamma_{XY} + \ \Gamma_{YX})$$

and we are back in case (b). (See Yitzhaki & Olkin, 1991.) Yitzhaki and Wodon (2004) suggest an alternative symmetric measure which will be discussed in Chap. 15 when dealing with the concept of mobility.

(d) If $D_{kY} \neq 0$ for some k's then the linear combination Y ceases to belong to the family of distributions of the X_k's. In this case the decompositions of the variance and the Gini do not have the same structure. The Gini decomposition includes some extra terms. This case opens the way to developing tests of convergence, as will be shown in Chap. 23.

We have described the differences in the structures of the decompositions of the GMD and the variance of a linear combination of random variables as depending on whether the distributions are exchangeable up to linear transformations or not. We will define and illustrate the importance of exchangeability at the end of the chapter. We now move to the second kind of decomposition—the decomposition of the variability of a population which is composed of several groups (subpopulations) into the contributions of the individual groups to the overall variability.

4.2 The Decomposition of the Variability of a Population by Subpopulations

The decomposition of the variability of a population which is composed of several groups (subpopulations) into the contributions of the groups to the overall variability leads to similar results as in the decomposition of the variability of a linear combination of random variables: it adds extra terms. If these terms are relevant to the area of investigation then the use of the GMD or the Gini coefficient "reveals more." If, on the other hand, one is not interested in these extra terms, then the use of the Gini is cumbersome and complicated.

Intuitively, the methodology presented below can be referred to as ANOGI (*AN*alysis *O*f *GI*ni)—the equivalent of ANOVA performed with the Gini coefficient (the relative measure). Because in practice most of the decompositions deal with the Gini coefficient (for example, in the area of income distributions) we will present the decomposition of the Gini coefficient rather than that of the GMD. In order to obtain the decomposition of the GMD one only has to substitute the constant terms, similar to the decomposition of a linear combination of random variables.

The decomposition we follow is the one presented in Yitzhaki (1994a). There are other versions of the same decomposition such as the one presented in Yitzhaki and Lerman (1991). The difference between the two is that while Yitzhaki and Lerman decompose the Gini coefficient according to the contribution of one group vs. all the others combined, Yitzhaki (1994a) decomposes the Gini coefficient into the contribution of each group vis-à-vis the entire population, including the group itself. As a result, one can perform a second-stage decomposition that consists of the contribution of each pair of groups. This latter decomposition enables symmetry in the decomposition, and therefore it is the preferred one in our view. It should be mentioned that the decomposition we present is not the only decomposition of the Gini. See, for example, Dagum (1997) and Deutsch and Silber (1999). However, they are not covered because they do not imitate ANOVA.

Let Y_i, $F_i(Y)$, $f_i(Y)$, μ_i, and p_i represent the income (variate), the cumulative distribution, the density function, the expected value, and the share of subpopulation *i* in the overall population, respectively ($i = 1, \ldots, n$). Let $s_i = p_i\mu_i/\mu_u$ denote the share of group *i* in the overall income, where subscript *u* denotes the union of the subpopulations. In other words the income of the overall population is composed of the union of the incomes of the subpopulations, namely, $Y_u = Y_1 \cup Y_2 \cup \cdots \cup Y_n$.

Note that

$$F_u(Y) = \sum_i p_i F_i(Y). \tag{4.12}$$

That is, the cumulative distribution of the overall population is the weighted average of the cumulative distributions of the subpopulations, weighted by the relative sizes of the subpopulations. Two representations of the Gini coefficient of group i are used in this chapter. The first presentation is the covariance-based formula (2.15):

$$G_i = \frac{2\mathrm{cov}_i(Y,F_i(Y))}{\mu_i}, \tag{4.13}$$

which is twice the covariance between the income Y and the rank $F_i(Y)$, standardized by the mean income μ_i, and $\mathrm{cov}_i()$ is the covariance calculated under the distribution F_i. The second presentation is

$$G_i = \frac{E_i(|Y_1 - Y_2|)}{2\,\mu_i}, \tag{4.14}$$

where Y_1 and Y_2 are i.i.d. coming from the ith group and E_i is the expected value under distribution F_i. Using these presentations we can write the Gini mean difference of the overall population as

$$E(|Y_1 - Y_2|) = E_u(|Y_1 - Y_2|) = \sum_{i=1}^{n}\sum_{j=1}^{n} p_i p_j E(|Y_i - Y_j|), \tag{4.15}$$

where Y_1 and Y_2 are i.i.d. with cdf $F_u()$ (i.e., the entire population) and $E = E_u$ is the expected value calculated under the distribution F_u. Next, the expectation can be written as

$$E(|Y_i - Y_j|) = 2\{\mathrm{cov}_i[Y,F_j(Y)] + \mathrm{cov}_j[Y,F_i(Y)] + \mu_i(F_{ji} - 0.5) + \mu_j(F_{ij} - 0.5)\}, \tag{4.16}$$

where

$$F_{ji} = \int F_j(t)dF_i(t)$$

is the expected rank of observations of group i had they been ranked according to the ranking of group j. (See Appendix 4.1 and Yitzhaki (1994a) for details.)

Substituting (4.16) into (4.15) we get

$$E(|Y_1 - Y_2|) = 2\sum_{i,j} p_i p_j\{\, cov_i[Y,F_j(Y)] + cov_j[Y,F_i(Y)]\}$$
$$+ 2\sum_{i,j} p_i p_j\{\mu_i(F_{ji} - 0.5) + \ \mu_j(F_{ij} - 0.5)\}\ . \qquad (4.17)$$

Note that

$$\sum_{i,j} p_i p_j\ \mu_i(F_{ji} - 0.5) = \sum_i p_i\ \mu_i \sum_j p_j(F_{ji} - 0.5) = \sum_i p_i\ \mu_i(F_{ui} - 0.5) = cov_B(\ \mu\ ,\bar{F}_u),$$

where $cov_B()$ is the between-groups covariance, F_{ui} is the expected rank of observations of group i had they been ranked according to the ranking of the entire population, $\bar{F}_u$ is the vector of the expected ranks, and μ is the vector of means of the vector Y of the subpopulations. Therefore (4.17) can be written as

$$E(|Y_1 - Y_2|) = 4\sum_{i,j} p_i p_j O_{ji} cov_i[Y,F_i(Y)] + 4cov_B(\ \mu\ ,\bar{F}_u),$$

where

$$O_{ji} = \frac{cov_i(Y,F_j(Y))}{cov_i(Y,F_i(Y))} \qquad (4.18)$$

is the overlapping index of group j by group i (to be detailed below). Finally, using the fact that the overlapping of group i by the entire population can be expressed as

$$O_i = \sum_i p_j O_{ji}$$

we get that the Gini coefficient of the entire population, G_u, can be decomposed as

$$G_u = \sum_{i=1}^{n} s_i G_i O_i + G_B, \qquad (4.19)$$

where s_i is the share of subpopulation *i* in the total income, G_i is the Gini coefficient of subpopulation *i*, O_i is the overlapping index of subpopulation *i* with the entire population, and G_B measures the between-groups inequality (the terms are defined below). Equation (4.19) decomposes the Gini coefficient of the union into two related components: intra- and inter-group components, connected in a way which is relatively complicated.

The decomposition of the GMD follows a similar pattern and is given by

$$\Delta_u = \sum_{i=1}^{n} p_i \; \Delta_i O_i + \Delta_B. \tag{4.20}$$

Equation (4.20) can be compared to the equivalent equation in decomposing the variance, which is

$$\sigma_u^2 = \sum_{i=1}^{n} p_i \; \sigma_i^2 + \sigma_B^2. \tag{4.21}$$

Equation (4.21) is the theoretical basis for ANOVA. It partitions the total variability into two components—a "within (intra)" component and a "between (inter)" component. Note that while in ANOVA the total variability is partitioned into inter- and intra-variances, in ANOGI there are inter- and intra-Gini's, but in addition there is an extra parameter O_i which is called the overlapping index. We will return to this implication following the explanation of the individual components.

4.2.1 *The Overlapping Parameter*

Overlapping can be interpreted as the inverse of stratification. Stratification is a concept used by sociologists. We follow Lasswell's (1965, p. 10) definition as: "in its general meaning, a stratum is a horizontal layer, usually thought of as between, above or below other such layers or strata. Stratification is the process of forming observable layers, or the state of being comprised of layers. Social stratification suggests a model in which the mass of society is constructed of layer upon layer of congealed population qualities."

According to Lasswell, perfect stratification occurs when the observations of each subpopulation are confined to a specific range of income, and the ranges of the subpopulations do not overlap. An example of a perfect stratification is the division of the society into deciles. Stratification plays an important role in the theory of relative deprivation (Runciman, 1966), which argues that stratified societies can tolerate greater inequalities than non-stratified ones (Yitzhaki, 1982b). The relationship between the Gini and social stratification will be discussed in Chap. 13. Stratification is also important in cases where the interest is in the quality of the classification into groups. The more stratified the classification is, the higher is its quality. For example, Heller and Yitzhaki (2006) used the overlapping index to evaluate the quality of the classification of families of prehistoric snails according to observable characteristics of the shells. Stratification also plays a role in regression trees but this topic is beyond the scope of this book. (See for example Lewis, 2000.)

One can rarely find a perfect stratification. Therefore an index which will quantify the degree of stratification is called for. The index of overlapping (to be defined below) quantifies the extent to which the different subpopulations are stratified.

Formally, overlapping of the overall population by subpopulation *i* is defined as

$$O_i = O_{ui} = \frac{\mathrm{cov}_i(Y,F_u(Y))}{\mathrm{cov}_i(Y,F_i(Y))}, \tag{4.22}$$

where cov_i is the covariance according to distribution *i*. (For convenience, the index u is sometimes omitted.) That is,

$$\mathrm{cov}_i(Y,F_u(Y)) = \int (y - \mu_i)\,(F_u(y) - \overline{F}_{ui})\, f_i(y)dy, \tag{4.23}$$

where $\overline{F}_{ui}$ is the expected rank of subpopulation *i* in the union (all observations of subpopulation *i* are assigned their ranks in the union $F_u(y)$, and $\overline{F}_{ui}$ represents the expected value of those ranks). Here it is worth pointing out two issues: (a) ranking observations according to a different distribution is a rare concept in statistics. However, it is common in sports where each athlete is frequently ranked in his or her country as well as according to other scales (world, continent, gender, age group, etc.). (b) The overlapping index resembles the Gini correlation. Its numerator involves a covariance between a cumulative distribution and a variate, but it does not include all the observations of the cumulative distribution, only the ones belonging to subpopulation i.

The overlapping index (4.22) can be further decomposed to identify the overlapping of subpopulation *i* with all other subpopulations that comprise the union. In other words, the total overlapping of subpopulation *i*, O_i, is composed of overlapping of group *i* with all the subpopulations, including group *i* itself. This further decomposition of O_i is

$$O_i = \sum_j p_j O_{ji} = p_i O_{ii} + \sum_{j \neq i} p_j O_{ji} = p_i + \sum_{j \neq i} p_j O_{ji}, \tag{4.24}$$

where p_j is the share of subpopulation j in the union and $O_{ji} = \frac{\mathrm{cov}_i(Y,F_j(Y))}{\mathrm{cov}_i(Y,F_i(Y))}$ is the overlapping of group *j* by group *i*.

The properties of the overlapping index O_{ji} are the following:

(a) $O_{ji} \geq 0$. The index is equal to zero if no member of the *j*th distribution lies in the range of distribution *i* (i.e., group *i* is a perfect stratum).
(b) O_{ji} is an increasing function of the fraction of population *j* that is located in the range of population *i*.

(c) For a given fraction of distribution *j* that is in the range of distribution *i*, the closer the observations belonging to *j* are to the expected value of distribution *i*, the higher O_{ji} is.
(d) If the distribution of group *j* is identical to the distribution of group *i*, then $O_{ji} = 1$. Note that by definition $O_{ii} = 1$. This result explains the second equality in (4.24). Using (4.24) it is easy to see that $O_i \geq p_i$, which is a result to be borne in mind when comparing different overlapping indices of groups with different sizes.
(e) $O_{ji} \leq 2$. That is, O_{ji} is bounded from above by 2. This maximum value is reached if all observations belonging to distribution *j* that are located in the range of distribution i are concentrated at the mean of distribution *i*. Note, however, that if distribution *i* is known then it may be that the upper bound is lower than 2 (see Schechtman, 2005). That is, if we confine distribution *i* to be of a specific type then it may be that the upper bound will be lower than 2, depending on the assumptions of the distribution. For example, in the exponential distribution the overlapping index is bounded from above by 1! The upper bound of 2 is the best bound only when no other assumptions can be made on the distributions.
(f) In general, the higher the overlapping index O_{ji} is, the lower will O_{ij} be. That is, the more group *j* is included in the range of group *i*, the less group *i* is expected to be included in the range of group *j*.

Properties (a) to (f) show that O_{ji} is an index that measures the extent to which subpopulation *j* is included (overlapped) in the range of subpopulation *i*. Note that the indices O_{ji} and O_{ij} are not interrelated by a simple relationship. However it is clear that the two indices of overlapping are not independent.

4.2.2 Between-Groups Component G_B and Its Properties

As will be seen later, we are interested in two alternative parameters representing the between-group inequality. We start with the one appearing in equation (4.19). The between-groups inequality G_B is defined in Yitzhaki and Lerman (1991) as

$$G_B = \frac{2\text{cov}_B(\mu, \bar{F}_u)}{\mu_u}. \tag{4.25}$$

G_B is twice the covariance between the mean incomes of the subpopulations and the subpopulations' mean ranks in the overall population, divided by overall mean income. That is, each subpopulation is represented by its mean income and by the mean rank of its members in the overall distribution. The term G_B equals zero if either the mean incomes or the mean ranks are equal for all subpopulations. In extreme cases G_B can be negative, which occurs when the mean incomes are negatively correlated with the mean ranks of the members of the subpopulations.

For example, imagine a case where in one subpopulation there is one extremely rich person while the others are extremely poor. In such a case, the average income of the group will be relatively large while the average rank of its members will be relatively small.

G_B is not a pure Gini coefficient because $\overline{F}_u$ is not the cumulative distribution of the variable μ_i. An alternative between-groups Gini (G_{BP}) was defined by Pyatt (1976). (Cowell, 1980; Mookherjee & Shorrocks, 1982; Shorrocks, 1984; Silber, 1989 also follow Pyatt.) In this definition, the between-groups Gini is based on the covariance between the mean income in each subpopulation and its rank among the mean incomes of the subpopulations. This between-groups component **is** a pure Gini (of the vector of means). The difference between the two definitions is in the rank that is used to represent the group: under Pyatt's approach it is the rank of the mean income of the subpopulation, while under Yitzhaki–Lerman it is the mean rank of all members belonging to the subpopulation.

Generally, it can be shown (Frick, Goebel, Schechtman, Wagner, and Yitzhaki, 2006) that

$$G_B \leq G_{BP}. \tag{4.26}$$

The upper limit of G_B is reached and (4.26) holds as an equality when the groups occupy nonoverlapping ranges of incomes (i.e., perfect stratification) because in that case the ranking of the means of the groups is equivalent to taking the average of the rankings of the individual members of the groups. Therefore the difference between the two can supply an indication of the quality of the overlapping.

Having explained the different components we now present an alternative version of (4.19) that will be used throughout this book.

$$G_u = \sum_{i=1}^{n} s_i G_i + \sum_{i=1}^{n} s_i G_i (O_i - 1) + G_{BP} + (G_B - G_{BP}). \tag{4.27}$$

Presentation (4.27) is obtained from (4.19) by adding and subtracting several terms. Equation (4.27) presents the decomposition of the Gini coefficient of the entire population into four components. Two of the components are similar in nature to the components of ANOVA: inter (the first component) and intra (the third component) variabilities. The other two components are specific to this decomposition and reflect the overlapping, as is explained below. In addition, note that the first two components are related to the "intra" group variability and the last two are related to the "inter" group variability.

Intuitively the overlapping affects both the "intra" and the "inter" parts of the decomposition: as the amount of overlapping increases we would expect the "intra" part to become larger while the "inter" part should become smaller. We now turn to (4.27). We start with the effect on the "intra" part: if there is complete overlapping, then $O_i = 1$ for all i, and the second term vanishes. Any deviation from this perfect (and unrealistic) case will imply a positive contribution to the "intra" term. We now

Table 4.1 A summary of ANOGI components in comparison to ANOVA

Components parallel to ANOVA	Formula	Range
Intra-group	$IG = \sum_{i=1}^{n} s_i G_i$	$0 \leq IG \leq G_u$
Between-groups-Pyatt	$BG_p = G_{BP}$	$0 \leq BG_p \leq G_u$
Additional information		
Overlapping effect on intra-group	$IGO = \sum_{i=1}^{n} s_i G_i (O_i - 1)$	
Overlapping effect on between-groups	$BGO = G_B - G_{BP}$	$- BG_p - IGO - IG \leq BGO \leq 0$

Source: Frick et al. (2006), p. 439, Table 3
Reprinted with permission from SAGE Journals

turn to the "inter" part. If the groups are perfectly stratified then the between-groups component reaches its highest value G_B (in which case $G_B = G_{BP}$). Overlapping reduces the quality of the classification and as a result decreases the between-groups part (in which case the "between" component is reduced by G_{BP}-G_B which is positive). Heller and Yitzhaki (2006) used this property to evaluate the classification of families of prehistoric snails according to observable characteristics of the shells. Frick et al. (2006) used it to compare the distributions of different independent samples taken from the same population at different points in time. If the samples are drawn from the same population (that is, there was no change in the population over time), then the means and the Ginis should be equal, no stratification should be observed (i.e., perfect overlapping), and the between-groups component should be equal to zero. In Chap. 13 we analyze the implication of stratification on social unrest. Wodon (1999) suggests additional results concerning sequential decomposition in a multidimensional context. Monti and Santoro (2011) restrict the population to be based on only two groups, which enables them to interpret the between-groups as representing the probability that a random member of the lower group (on average) to be higher than a random member drawn from the (on average) higher group. Mussard and Richard (2012) suggest a connection between the Gini decomposition according to subgroups and the decomposition of a linear combination of random variables. However, the method is applicable to nonoverlapping distributions.

4.2.3 *ANOGI vs. ANOVA: A Summary Table*

We conclude this chapter with a comparison between ANOGI and ANOVA as shown in the summary table (Table 4.1). The four components that comprise the decomposition of the Gini can be divided into two types: those which carry equivalent information as in ANOVA (when using Gini instead of the variance as a measure of variability), and those which carry an additional information. We note in passing that the measure of variability in ANOGI is the Gini coefficient (i.e., a relative measure), while in ANOVA the commonly used way is to decompose the variance (i.e., an absolute measure). This distinction does not interfere with the comparison below.

(a) Components Which Are Identical in Nature to ANOVA

For a given overall inequality, G_u:

Intra-Group component (*IG*). A weighted average of groups' Ginis. It reaches the lower limit if all intra-group Ginis are equal to zero. It reaches the upper limit if the distributions of all groups are identical (identical in nature to MSE in ANOVA).

Between-Groups component, based on Pyatt (BG_P). It reaches the upper limit if all groups are concentrated at their means. It reaches the lower limit, zero, if the means of all groups are equal (identical in nature to MSB in ANOVA). It measures between-groups inequality, assuming a perfect stratification.

(b) Additional Components

The two additional components are related to the overlapping among the groups.

The effect of overlapping on the intra-group component (*IGO*). This term "revises" the contribution of each subpopulation to the intra-group variability, provided that the inequality in the group (as measured by the group's Gini) is greater than zero. If the subpopulation and the overall population are equally distributed, then there is no revision to its contribution ($O_i = 1$). However, if a subpopulation forms a strata in the population ($O_i < 1$), then its contribution to the intra-group component is reduced, while its contribution to the between-groups component is increased. On the other hand, if the scatter of the ranks of the group's members is larger than that of the population ($O_i > 1$), the contribution of the group to the intra-group component is increased, while its contribution to the between-groups component is decreased.

The effect of overlapping on the between-groups component (*BGO*). The effect of overlapping on the between-groups component occurs only if the expected values of the subpopulations are not all equal. It is always non-positive, because the overlapping reduces the ability to distinguish between the groups. It reaches the upper limit (zero) if the ranges occupied by the different groups do not overlap. Note, however, that the combined effect of the between-groups inequality and the impact of overlapping on it can be negative if the means of the groups are negatively correlated with the means of the ranks of the members of the subpopulations. This possibility occurs if, for example, the population is composed of two groups, with one group composed of a majority of poor people and a few very rich people, while the second group is composed of the middle class. In this case the expected income of the first group is high (because of the few rich) while its expected rank (that is, the mean of the ranks of its members) is low (because of the majority of poor people), causing the correlation to be negative.

Finally, an alternative and technical interpretation of equation (4.27) is as follows: the first term represents the variability of the variate within each group, the second term represents the variability of the ranks in each group in the overall population, the third term represents the variability of the expected values among groups, while the fourth term represents the variability of the expected ranks.

Having explained the individual components of the decomposition, we now discuss the meaning of their convergence to zero.

We can interpret the convergence to zero as follows:

(a) $G_{BP} = 0$ implies that all the mean values are equal.
(b) $G_{BP} > 0$ and $G_B = 0$ imply that the mean ranks of the subpopulations in the overall population are equal.
(c) $\sum_{i=1}^{n} s_i G_i (O_i - 1) = 0$ implies that it is most likely that each subpopulation fully overlaps with the entire population.

Clearly, we are using terms that are connected. However each parameter adds insight and there is no redundancy or double counting because the sum of all of them adds up to the overall Gini, and one can produce examples where one term is equal to zero and the others are not. The advantage of ANOGI over ANOVA is that the decomposition of the Gini coefficient adds a new parameter to the existing inter and intra terms, namely, the overlapping index. Hence, not only are the equivalents of the first and second moments examined, but the extent of subpopulations intertwining is also considered.

4.3 The Decomposition of Gini Covariance

Similar to the decomposition of the GMD into the contributions of different groups, one can also decompose the Gini covariance into the contributions of different groups. This decomposition is especially useful when one wants to see the contributions of different groups to the Gini correlation and the Gini regression coefficients. The concepts are general and can be applied to Pearson's correlation coefficient and the OLS regression coefficient. Therefore we start with a general claim about the decomposition of a covariance. For simplicity, we treat the distribution as discrete.

Assume that the observations of (X, Y) are partitioned into M disjoint groups, denoted by m = 1, …, M.

Claim

$$\mathrm{cov}(X,Y) = \sum_{m=1}^{M} p_m \mathrm{cov}_m(X,Y) + \mathrm{cov}_B(\overline{X}_m, \overline{Y}_m), \tag{4.28}$$

where $p_m = n_m/n$ is the relative size of group m, $\overline{X}_m$ and $\overline{Y}_m$ are the vectors of the group means, and

$$\mathrm{cov_B}(\overline{\mathrm{X}}_\mathrm{m},\overline{\mathrm{Y}}_\mathrm{m}) = \sum_{\mathrm{m}=1}^{\mathrm{M}} \mathrm{p_m}(\overline{\mathrm{X}}_\mathrm{m} - \overline{\mathrm{X}}_{..})(\overline{\mathrm{Y}}_\mathrm{m} - \overline{\mathrm{Y}}_{..}), \tag{4.29}$$

where $\overline{\mathrm{X}}_{..}$ is the overall mean. Equation (4.29) holds in the population and in the sample analogues. Although trivial, a proof (in the sample) is added because we will be repeating its use in several ways.

Proof of the claim Let $(\mathrm{X_{mj}}, \mathrm{Y_{mj}})$, $\mathrm{m} = 1, \ldots, \mathrm{M}$, $\mathrm{j} = 1, \ldots, \mathrm{n_m}$ *be the n observations* $(\mathrm{n} = \sum_{\mathrm{m}=1}^{\mathrm{M}} \mathrm{n_m})$; *then*

$$\mathrm{n^*cov(X,Y)} = \sum_{\mathrm{m}=1}^{\mathrm{M}} \sum_{\mathrm{j}=1}^{\mathrm{n_m}} (\mathrm{X_{mj}} - \overline{\mathrm{X}}_{..})(\mathrm{Y_{mj}} - \overline{\mathrm{Y}}_{..})$$

$$= \sum_{\mathrm{m}=1}^{\mathrm{M}} \sum_{\mathrm{j}=1}^{\mathrm{n_m}} (\mathrm{X_{mj}} - \overline{\mathrm{X}}_{m.} + \overline{\mathrm{X}}_{m.} - \overline{\mathrm{X}}_{..})\mathrm{Y_{mj}}$$

$$= \sum_{\mathrm{m}=1}^{\mathrm{M}} \sum_{\mathrm{j}=1}^{\mathrm{n_m}} (\mathrm{X_{mj}} - \overline{\mathrm{X}}_{m.})\mathrm{Y_{mj}} + \sum_{\mathrm{m}=1}^{\mathrm{M}} \sum_{\mathrm{j}=1}^{\mathrm{n_m}} (\overline{\mathrm{X}}_{\mathrm{m}.} - \overline{\mathrm{X}}_{..})\mathrm{Y_{mj}}$$

$$= \sum_{\mathrm{m}=1}^{\mathrm{M}} \mathrm{n_m cov_m(X,Y)} + \mathrm{n} * \mathrm{cov_B}(\overline{\mathrm{X}}_\mathrm{m},\overline{\mathrm{Y}}_\mathrm{m}).$$

Using the claim, it is easy to see how the regression coefficient can be decomposed.

4.3.1 Decomposing the OLS Regression Coefficient

The OLS regression coefficient is defined by

$$\beta^{\mathrm{OLS}} = \frac{\mathrm{cov(Y,X)}}{\mathrm{cov(X,X)}}. \tag{4.30}$$

Using the decomposition above, it can be presented as

$$\beta^{\mathrm{OLS}} = \frac{\mathrm{cov(Y,X)}}{\mathrm{cov(X,X)}} = \frac{1}{\mathrm{cov(X,X)}} \left\{ \sum_{\mathrm{m}=1}^{\mathrm{M}} \mathrm{p_m}\ \sigma_\mathrm{m}^2 \beta_\mathrm{m}^{\mathrm{OLS}} + \sigma_\mathrm{B}^2 \beta_\mathrm{B}^{\mathrm{OLS}} \right\},$$

where σ^2 stands for the variance of the explanatory variable. That is, the regression coefficient is a weighted average of the intra- and between-groups regression coefficients, with the share of the variance of the explanatory variable in the group $(\mathrm{p_m}\frac{\sigma_\mathrm{m}^2}{\sigma_\mathrm{X}^2})$ used as the weighting scheme.

The decomposition of the Gini regression coefficient is more complicated as is shown below.

4.3.2 Decomposing the Gini Regression Coefficient

The Gini regression coefficient is defined by

$$\beta^{N} = \frac{\mathrm{cov}(Y,F(X))}{\mathrm{cov}(X,F(X))}. \tag{4.31}$$

Our starting point is the decomposition of the numerator:

$$\mathrm{cov}(Y,F(X)) = \sum_{m=1}^{M} p_m \mathrm{cov}_m(Y,F_m^u(X)) + \mathrm{cov}_B(\overline{Y}.,\overline{F}^u.(X)), \tag{4.32}$$

where $F_m^u(X)$ is the vector of ranks of the members of group m (in the overall distribution), $\overline{Y}.$ is the vector of group means, and $\overline{F}^u.$ is the vector of average ranks of members of the groups in the overall distribution.

Using (4.32) we get

$$\mathrm{cov}(Y,F^u(X)) = \sum_{m=1}^{M} p_m \frac{\mathrm{cov}_m(X,F_m^u(X))}{\mathrm{cov}_m(X,F_m(X))} \mathrm{cov}_m(X,F_m(X)) \frac{\mathrm{cov}_m(Y,F_m^u(X))}{\mathrm{cov}_m(X,F_m^u(X))},$$
$$+ \mathrm{cov}_B(\overline{Y}.,\overline{F}^u.(X))$$

where F^u indicates the cumulative distribution in the overall distribution, while F_m is the cumulative distribution of group m.

We now interpret the terms:

$$\beta_m^{Nu} = \frac{\mathrm{cov}_m(Y,F^u(X))}{\mathrm{cov}_m(X,F^u(X))} \tag{4.33}$$

is the regression coefficient of group m when the X′s are ranked in the overall distribution, which is different from the regression coefficient of group m, would it be handled separately.

$$O_m = \frac{\mathrm{cov}_m(X,F^u(X))}{\mathrm{cov}_m(X,F_m(X))}$$

is the overlapping of group m with the overall population. Let

$$S_m = p_m \frac{\mathrm{cov}_m(X,F_m(X))}{\mathrm{cov}(X,F^u(X))}$$

be the share of the GMD of group m in the overall GMD.

Using the notation above and replicating the exercise with respect to the between-groups component, we get

$$\beta^{N} = \frac{\mathrm{cov}(Y,F(X))}{\mathrm{cov}(X,F(X))} = \sum_{m=1}^{M} S_m O_m \beta_m^{Nu} + S_B \beta_B^{Nu}, \tag{4.34}$$

where $S_B = \frac{\mathrm{cov}(\overline{X},\overline{F}^u(X))}{\mathrm{cov}(X,F(X))}$ is the share of between-groups Gini in the overall Gini, while β_B^{Nu} is the between-groups regression coefficient. In the case where the groups do not overlap, that is when the range of X is divided into nonoverlapping segments, the decomposition of the Gini regression coefficient will be identical to the decomposition of the OLS regression coefficient. To see that, write the left term on the right side of (4.34) as

$$\sum_{m=1}^{M} S_m O_m \beta_m^{Nu} = \sum_{m=1}^{M} S_m \beta_m^{Nu} + \sum_{m=1}^{M} S_m (O_m - 1) \beta_m^{Nu}, \tag{4.35}$$

and note that in the case with no overlapping the ranks of a group within the overall distribution are the same as within the group itself (up to a constant), and hence the value of O_m is 1 and the second term in (4.35) vanishes. Since there is no overlapping, the average ranking of group's members in the overall population will be identical to the ranking of the average of X. Hence, the decompositions of the Gini and OLS regression coefficients are identical in structure.

4.4 Summary

Two types of decompositions are introduced in this chapter: the decomposition of the GMD of a linear combination of random variables and the decomposition of a Gini coefficient (and GMD) of a population into the contributions of the subpopulations. The decomposition properties of the GMD and the Gini coefficient are a vivid proof to the claim that the Gini reveals more. The decomposition of the GMD of a linear combination of random variables reveals whether the individual variables and the linear combination of them are exchangeable up to a linear transformation or not. The decomposition of the Gini coefficient of an overall population to the contributions of several subpopulations reveals the degree of stratification among the subpopulations. Whether these new parameters contribute to our understanding and to the analysis of the data or not depends on the subject matter. One cannot claim that they are always useful or always redundant. It depends on the application in mind. It is clear, though, that using the Gini instead of the variance complicates the analysis. Note that even when the distributions are well behaved in the sense of being exchangeable up to linear transformations (so that the decompositions of the GMD and the variance are identical in structure) the use of the GMD still adds information (unless the distribution is multivariate normal). It doubles the number of models used in economics because for every variance-based model there is a Gini-based model. It is important to know whether the two methods give the same numerical results or not, and when they do not, one should investigate the reason that causes the different outcomes.

Appendix 4.1

Claim

$$E(|Y_i - Y_j|) = 2\{cov_i[Y,F_j(Y)] + cov_j[Y,F_i(Y)] + \mu_i(F_{ji} - 0.5) + \mu_j(F_{ij} - 0.5)\}, \tag{A4.1}$$

where

$$\Gamma_{ji} = \int F_j(t)dF_i(t) \tag{A4.2}$$

is the expected rank of observations of group i had they been ranked according to the ranking of group j.

Proof (Yitzhaki, 1994a)

$$\begin{aligned}
E(|Y_i - Y_j|) &= \int_0^\infty\int_0^\infty |y_1 - y_2| dF_1(y_1)dF_2(y_2) \\
&= \int_0^\infty\int_0^{y_2} (y_2 - y_1)dF_1(y_1)dF_2(y_2) + \int_0^\infty\int_{y_2}^\infty (y_1 - y_2)dF_1(y_1)dF_2(y_2) \\
&= \int_0^\infty\int_0^{y_2} y_2 dF_1(y_1)dF_2(y_2) - \int_0^\infty\int_0^{y_2} y_1 dF_1(y_1)dF_2(y_2) \\
&\quad + \int_0^\infty\int_{y_2}^\infty y_1 dF_1(y_1)dF_2(y_2) - \int_0^\infty y_2(1 - F_1(y_2))dF_2(y_2) \\
&= 2\int_0^\infty y_2 F_1(y_2)dF_2(y_2) - \mu_2 - \int_0^\infty\int_0^{y_2} y_1 dF_1(y_1)dF_2(y_2) \\
&\quad + \int_0^\infty\int_{y_2}^\infty y_1 dF_1(y_1)dF_2(y_2)\,.
\end{aligned} \tag{A4.3}$$

The covariance between Y_2 and $F_1(Y_2)$ is given by

$$\mathrm{cov}(Y_2,F_1(Y_2)) = \int_0^{\infty} y_2F_1(y_2)dF_2(y_2) - F_{12}\ \mu_2. \tag{A4.4}$$

Therefore using (A4.4) we can write the first two terms of (A4.3) as

$$2\int_0^{\infty} y_2F_1(y_2)dF_2(y_2) - \mu_2 = 2\mathrm{cov}(Y_2,F_1(Y_2)) + 2F_{12}\ \mu_2 - \mu_2$$

$$= 2\mathrm{cov}(Y_2,F_1(Y_2)) + 2\ \mu_2(F_{12} - 0.5)\ .$$

The third and the fourth terms in (A4.33) can be presented in a similar way. That is, using

$$\mu_1 - \int_0^{y_2} y_1dF_1(y_1) = \int_{y_2}^{\infty} y_1dF_1(y_1)$$

and integration by parts with

$$u = \int_0^{y_2} y_1dF_1(y_1),\quad u' = y_2dF_1(y_2)$$

$$v' = dF_2(y_2)\ \text{and}\ v = F_2(y_2),$$

we get

$$-\int_0^{\infty}\int_{y_2}^{\infty} y_1dF_1(y_1)dF_2(y_2) + \int_0^{\infty}\int_{y_2}^{\infty} y_1dF_1(y_1)dF_2(y_2) = \mu_1 - 2\int_0^{\infty}\int_0^{y_2} y_1dF_1(y_1)dF_2(y_2)$$

$$= \mu_1 - 2F_2(y_2)\int_0^{y_2} y_1dF_1(y_1)|_0^{\infty} + 2\int_0^{\infty} y_2F_2(y_2)dF_1(y_2) = 2\int_0^{\infty} y_2F_2(y_2)dF_1(y_2) - \mu_1$$

$$= 2\mathrm{cov}(Y_1,F_2(Y_1)) - \mu_1 + 2F_{21}\ \mu_1 = 2\mathrm{cov}(Y_1,F_2(Y_1)) + 2\ \mu_1[F_{12} - 0.5]\ .$$

Chapter 5
The Lorenz Curve and the Concentration Curve

Introduction

The Lorenz and the concentration curves play important roles in the areas of GMD and the related measures such as Gini covariance, Gini correlation, Gini regression, and more. In this chapter we introduce the curves, discuss their properties, and show their connections to the Gini world. In addition, in order to be able to analyze the parallel concepts that are common in the variance world we investigate the equivalents of the Lorenz and the concentration curves that are relevant to the variance and the covariance, respectively. Those parallel curves share some properties among themselves. Therefore one can deduct from the concentration curve about some properties of the covariance and not only about the Gini covariance. In addition we present the relationships between the concepts of second-degree stochastic dominance and welfare dominance on one hand and the concentration and Lorenz curves on the other hand. These relationships enable the Gini methodology to serve as an analytical tool for statistical analyses and to be compatible with economic theory, a property that holds for the variance as well, but only for specific distributions.

Our approach in this chapter deviates from the main body of the literature on Lorenz curves. Historically, Lorenz (1905) presented the Lorenz curve as based on the relationship between the cumulative distribution of the variable (the horizontal axis) and the cumulative value of the *percentage* of the variate (the vertical axis). We have no qualm with this presentation which is useful in the area of income distribution. However our main use of the curve is to investigate the properties of the GMD and to carry out statistical analyses. For these purposes it is useful to use the variate values rather than to normalize the Lorenz curve by dividing the cumulative value of the variate by its mean because by looking at the percentages it implies that we actually investigate the properties of the variance from the properties of the coefficient of variation. Therefore we start with the simplest curve, the Absolute Lorenz Curve (ALC) which is different from the Lorenz curve in the quantity that is accumulated along the vertical axis: in the ALC it is the cumulative value of the

S. Yitzhaki and E. Schechtman, *The Gini Methodology: A Primer on a Statistical Methodology*, Springer Series in Statistics 272, DOI 10.1007/978-1-4614-4720-7_5,

variate itself while in the Lorenz curve it is the cumulative percentage value of the variate. The properties of the Lorenz curve can be derived later from the properties of the ALC in a way which is similar to the derivation of the properties of the coefficient of variation from the properties of the variance.

The structure of the chapter is as follows: In Sect. 5.1 we introduce the ALC. Section 5.2 is devoted to the Lorenz curve of the coefficient of variation. Next, in Sect. 5.3 the absolute concentration curve (ACC) is introduced. Section 5.4 deals with the relationship between the ALC and second-degree stochastic dominance. The connection between the ACC and marginal conditional stochastic dominance is discussed in Sect. 5.5. Section 5.6 deals with the use of the ACC to check for the monotonicity of the correlations and the regression coefficients. Section 5.7 concludes.

5.1 The Absolute Lorenz Curve[1]

There are several possible definitions of the ALC. We follow Gastwirth's (1971, 1972) definition which is based on the inverse of the cumulative distribution. Let $F(X(p)) = p$ and $X(p) = F^{-1}(p)(X(p)$ is the pth quantile of F. In Gastwirth's definition p is plotted on the horizontal axis while the vertical axis represents the cumulative value of the variate, $\int_{-\infty}^{p} X(t)dt$. The familiar Lorenz curve (LC) is derived from the ALC by dividing the cumulative value of the variate by its mean. The vertical axis is then $(1/\mu)$, $\int_{-\infty}^{p} X(t)dt$.

The ALC was called the "Generalized Lorenz Curve (GLC)" by Shorrocks (1983). However, we find the term "absolute" to be more informative because it distinguishes between the absolute curve and the relative one. Therefore it will be used throughout the book.

Note that

1. The ALC passes through (0, 0) and (1, μ). The Lorenz curve passes through (0, 0) and (1, 1).
2. The derivative of the curve at p is X(p); hence the curve is increasing (decreasing) when X(p) is positive (negative).

Lambert (2001) gives an excellent description of the properties of ALC. Hart (1975) presents inequality indices in terms of the distributions of first moments, which are related to the ALC.

Figure 5.1 presents a typical ALC, the curve OAB. Before proceeding with the relationship between the ALC and the GMD some geometrical properties of the curve are worth mentioning. The slope of the line connecting the two extremes of the curve (line OEGB in Fig. 5.1) is μ. We refer to this line as the Line of Equality

[1] This section is based on Yitzhaki and Olkin (1988).

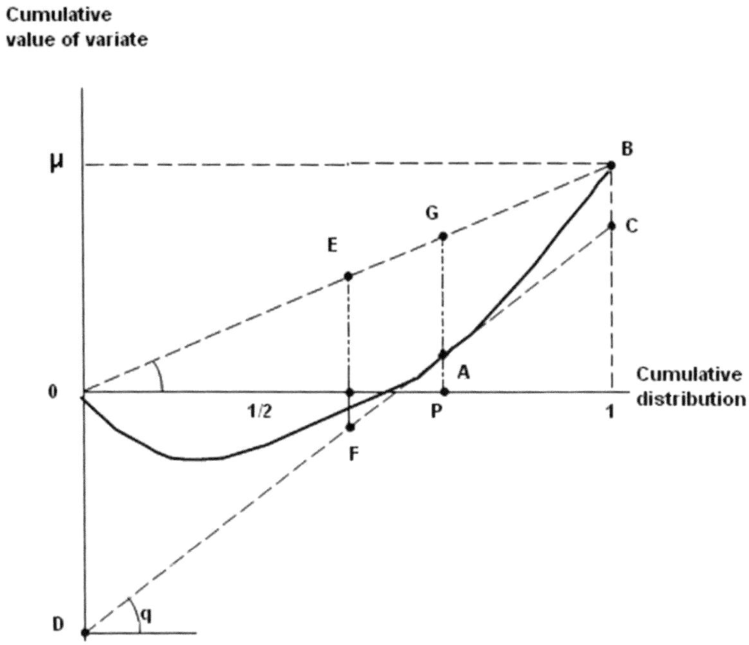

Fig. 5.1 The absolute Lorenz curve. *Source*: This figure is identical to Fig. 2.1

(LOE) because when all the observations are equal the ALC coincides with this line. (When dealing with financial applications we will refer to it as the Line of Safe Asset (LSA).) There are two additional important elements in Fig. 5.1. The first element is the line DFAC which is tangent to the curve at A, where the cumulative distribution is equal to p. Its slope is $q = X(p)$, where q is the pth quantile of the distribution of X. The second element is the vertical segment EF, which passes through $p = 1/2$.

We now turn to the connection between the ALC and the GMD. For any constant q, the expected absolute deviation $E\{|X - q|\}$ of X from q can be divided into two components: a lower absolute deviation LAD(q) and an upper absolute deviation HAD(q). Formally,

$$\mathrm{LAD}(q) = \int_{-\infty}^{q} (q - x)dF(x) = E[(q - X)I(X<q)] = F(q)E\{(q - X)|X<q\},$$

$$\mathrm{HAD}(q) = \int_{q}^{\infty} (x - q)dF(x) = E[(X - q)I(X>q)] = (1 - F(q))E\{(X - q)|X>q\},$$

where I() is the indicator function.

Clearly,

$$E\{|X - q|\} = LAD(q) + HAD(q). \tag{5.1}$$

Let q be a random variable identically distributed as X and independent of it. For convenience, let us replace (X, q) by (X_1, X_2), where X_1 and X_2 are i.i.d. Then,

$$\begin{aligned} E_q\{LAD(q)\} &= E_q\{E_X[(q - X)\ I(X < q)]\} = E_{X2}\{E_{X1}[(X_2 - X_1)\ I(X_1 < X_2)]\} \\ &= E\{|X_2 - X_1|\ I(X_1 < X_2)\} = 0.5\ E\{|X_2 - X_1|\}. \end{aligned}$$

Therefore,

$$\Delta = E\{|X_1 - X_2|\} = 2E_q\{LAD(q)\}.$$

Similarly,

$$\Delta = E\{|X_1 - X_2|\} = 2E_q\{HAD(q)\}.$$

Hence, by (2.1) and (2.15), we get the following connections:

$$\Delta = E\{|X_1 - X_2|\} = 2E_q\{LAD(q)\} = 2E_q\{HAD(q)\} = 4\ \mathrm{cov}(X, F(X)). \tag{5.2}$$

Let q be the tangent of the angle at D, as shown in Fig. 5.1. The following geometrical results can be obtained from Fig. 5.1 and the previous definitions.

(a) The slope of DC is q (q = X(p) is the inverse of the cumulative distribution at p).

Proof: Trivial. DC is the tangent to ALC at p. X(p) is the derivative of the ALC at p.

(b) The segment $\overline{OD}$ (i.e., its length) is equal to LAD(q) and is a nondecreasing function of q.

Proof: Using the figure it can be seen that

$$\overline{AP} = \int_{-\infty}^{p} X(t)dt = \sum_{i=1}^{p} X(i)I(X(i){<}X(p)).$$

Also, from the geometrical presentation it follows that q is given by

$\frac{\overline{AP}+\overline{OD}}{p} = q$ which implies that $(\overline{AP} + \overline{OD}) = pq = pX(p) = \sum_{i=1}^{p} X(i)$.

Using both equations we get (recall that q = X(p))

$$\overline{OD} = \frac{1}{n}\sum_{i=1}^{p} (X(p) - X(i))I(X(i){<}X(p)) = \frac{1}{n}\sum_{i=1}^{p} |X(i) - q|I(X(i){<}q) = LAD(q).$$

(c) The segment $\overline{BC}$ is equal to HAD(q) and is a nonincreasing function of q.

Proof: The proof is similar to the proof of (b).

Properties (b) and (c) imply that $(\overline{OD} + \overline{BC}) = E\{|X - q|\}$.

(d) The segment $\overline{EF}$ which connects the LOE ($\overline{OEB}$) and $\overline{DAC}$ and passes through $p = 1/2$ is equal to $\overline{EF} = 0.5\ E\{|X - X(0.5)|\}$, which is one-half of the expected absolute deviation from the median.

Proof: The segment $\overline{EF}$ is in the middle of the trapezoid OBCD; therefore it is equal to one-half of the sum of the two bases. By (b) and (c) above, plus (5.1), we get $\overline{EF} = 0.5E\{|X - X(0.5)|\}$.

(e) Summation of the segments mentioned above over the entire range of p yields several presentations of the GMD.

(e.1) The area between the LOE and the ALC is equal to cov[X, F(X)]. That is, the summation of all segments $\overline{GA}$ is equal to cov[X, F(X)].

Proof: The area between the LOE and ALC is equal to

$$\int_{-\infty}^{\infty}\int_{-\infty}^{q} (\mu - x)\, dF(x)dF(q) = \int_{-\infty}^{\infty} F(q)(q - \mu)dF(q) = \mathrm{cov}[X, F(X)],$$

which is obtained by integration by parts with $u = \int_{-\infty}^{q} (\mu - x)dF(x)$ and $dv = dF(q)$.

(e.2) Summation of all segments $\overline{OD}$, that is $E_q\{LAD(q)\}$, is equal to $2\mathrm{cov}[X, F(X)]$.

Proof: To obtain this result first note that

$$\begin{aligned} E_q\{LAD(q)\} &= \int_{-\infty}^{\infty}\int_{-\infty}^{q} (q - x)dF(x)dF(q) \\ &= \int_{-\infty}^{\infty} q\, F(q)dF(q) - \int_{-\infty}^{\infty}\int_{-\infty}^{q} xdF(x)\, dF(q). \end{aligned}$$

Using integration by parts in the second integral with $u = \int_{-\infty}^{q} x\, dF(x)$ and $dv = dF(q)$ yields $\int_{-\infty}^{\infty}\int_{-\infty}^{q} x\, dF(x)\, dF(q) = -[1-F(q)] \int_{-\infty}^{q} x\, dF(x) \Big|_{-\infty}^{\infty} + \int_{-\infty}^{\infty} q[1-F(q)]\, dF(q) = \int_{-\infty}^{\infty} q[1-F(q)]\, dF(q)$. Hence $E_q\{LAD(q)\} = \int_{-\infty}^{\infty} q\, F(q)\, dF(q) - \int_{-\infty}^{\infty} q\, [1 - F(q)]\, dF(q) = 2\int_{-\infty}^{\infty} q\, [F(q) - 1/2]\, dF(q) = 2\mathrm{cov}[X, F(X)] = 1/2\ \Delta$, as was shown in (5.2).

(e.3) Summation of all segments $\overline{BC}$ over the entire range of q yields $E_q\{LAD(q)\} = E_q\{HAD(q)\} = 2\ \mathrm{cov}[X, F(X)] = 0.5 \times \Delta$, as was shown in (5.2).
Proof: The proof is similar to the proof of (e.2).

(e.4) The sum of all segments $\overline{EF}$, that is summation of $E\{|X - q|\}$, over all q equals 2cov[X, F(X)], which means one-half of GMD.

Proof: Using (b) and (c) and (5.2)

$$\int \overline{EF}(q)dq = \int 0.5(\overline{OD}(q) + \overline{BC}(q))dq = \int 0.5(LAD(q) + HAD(q))dq$$
$$= 2\mathrm{cov}(X,F(X)).$$

5.2 The Lorenz Curve of the Coefficient of Variation

The aim of this section is to demonstrate that knowledge of the similarity between the GMD and the variance enables us to find additional graphical connections. For example, one of the advantages of the Gini coefficient mentioned in the literature is its graphical representation based on the Lorenz curve (see Sect. 3.3). It turns out that one can imitate the derivation of the Gini coefficient in order to write the coefficient of variation as the area defined by a (transformation of the) Lorenz curve and the (transformed) 45° line (Yitzhaki, 1998).

As explained earlier, the Gini coefficient is twice the area defined between p and LC(p), where LC(p) is the Lorenz curve at p. It turns out that the coefficient of variation can be defined by a curve which is equivalent to the Lorenz curve (to be denoted by LCV) as we show next.

Proposition 5.1 *The square of the coefficient of variation is the area defined between Y = F(X) and Y = LC[F(X)] where LC[F(X)] is the Lorenz curve defined as a function of X. Namely,*

$$LC(F(X)) = LCV(X) = \frac{1}{\mu}\int_{-\infty}^{X} tf(t)dt, \tag{5.3}$$

where LCV is the Lorenz curve which corresponds to the variance (that is, instead of being a function of p = F(X), it is now a function of X).

Proposition 5.1 implies that if one takes the Lorenz curve and applies a monotonic transformation to the horizontal axis (i.e., portraying the curve as a function of X rather than of p), then the square of the coefficient of variation has a geometrical representation that resembles the one for the Gini coefficient.

Proof To simplify the proof, assume that the range of the random variable is bounded in [a, b]. Note that $\{F(X) - LCV(X)\}/\mu$ is the vertical difference between the 45° line and the Lorenz curve, as a function of X. We need to show that

$$(\mathrm{CV})^2 = \frac{\sigma^2}{\mu^2} = \frac{1}{\mu}\int_a^b [\mathrm{F(x)} - \mathrm{LCV(x)}]\mathrm{dx}. \tag{5.4}$$

To prove (5.4), use integration by parts with u = F(x) − LCV(x) and dv = dx:

$$\int_a^b [\mathrm{F(x)} - \mathrm{LCV(x)}]\mathrm{dx} = [\mathrm{F(x)} - \mathrm{LCV(x)}]\ \mathrm{x}\Big|_a^b - \int_a^b \mathrm{x}\left[\mathrm{f(x)} - \frac{\mathrm{xf(x)}}{\mu}\right]\mathrm{dx}$$

$$= \frac{1}{\mu}\int_a^b [\mathrm{x} - \mu]\mathrm{xf(x)dx} = \sigma^2/\mu.$$

By dividing both sides by μ the proof is complete.

The square of the coefficient of variation can thus be presented in a manner that resembles the representation of the Gini coefficient. However the graphical representation of the Gini coefficient has an advantage. The summation of the area for the Gini is limited to the range of [0, 1], whereas the summation of the area for the coefficient of variation extends over the range of the random variable. Hence for variables with infinite range, the geometrical interpretation of the coefficient of variation is problematic. Another difference between the two geometrical presentations is that while the Gini relies on an LOE, which is a straight line, the coefficient of variation relies on the cumulative distribution which is harder to visualize. This difference gives us intuitive rules to distinguish between results that are shared by the Gini methodology and the variance, and those that are limited to the Gini: all properties that are based on intersections of curves hold for both approaches. All properties that are based on convexity/concavity properties hold only for the Gini.

5.3 The Absolute Concentration Curve

We now move to the concentration curve. The ACC which is the extension of the ALC into the two-variable case was used sporadically as a descriptive tool. The idea behind it is similar to the idea behind the ALC, except that the horizontal axis and the vertical axis now represent two different variables. Because of this difference, some of the properties of the ALC (such as the convexity property) disappear while others remain. On top of its connection with the Gini correlation, as discussed in Blitz and Brittain (1964) and in Chap. 3, the ACC has two additional properties that make it an important analytical tool. The first property is that it enables the researcher to form necessary and sufficient conditions for

Marginal Conditional Stochastic Dominance (MCSD, see Chap. 17) and for welfare dominance (see Chap. 14). The second property is that it enables the researcher to investigate the monotonicity of the regression coefficient and of the Gini correlation (Chap. 20). This property of the ACC is useful whenever the association between two variables might change along the range of the distribution of the explanatory variable. However, in order to be able to examine those features it is important not to assume a specific underlying distribution because almost all simple and easy-to-handle distributions tend to have simple relationships among the variables, and therefore impose a specific type of curvature on the curve that describes the relationship between them. For example, the sign of the correlations between two normal, lognormal, or uniform random variables can be either positive *or* negative. The correlation coefficient cannot change its sign along the distributions of the variables.

The concentration curve (CC) is mainly used in the field of income distributions to portray the impact of taxes on income distributions (Kakwani, 1977, 1980; Lambert, 2001; Suits, 1977; Yitzhaki & Semrod 1991).[2] Normally, the horizontal axis would portray the poorest p percent of the population while the vertical axis would present the share of total expenditure on a consumption item spent by the poorest p percent. The ACC differs from the concentration curve by presenting the *cumulative consumption* (rather than the cumulative share of consumption) of the poorest p percent on the vertical axis. The definitions below follow the terminology in Yitzhaki and Olkin (1988, 1991).

Let μ_X and μ_Y denote the means of X and Y, respectively, and let $f_{Y|X}$ denote the conditional density function of Y given X. The conditional expectation is $g(x) = \mu_{Y.X} \equiv E\{Y|X = x\}$. It is assumed that all densities are continuous and differentiable, and all second moments exist.

Definition 5.1 The ACC of Y with respect to X, $ACC_{Y.X}(p)$, is implicitly defined by the relationship

$$ACC_{Y.X}(p) = \int_{-\infty}^{X(p)} g(t)dF_X(t), \tag{5.5}$$

where X(p) is defined by

$$F(X(p)) = p = \int_{-\infty}^{X(p)} dF_X(t). \tag{5.6}$$

In words, X(p) is the pth percentile of the distribution of X. The special case $ACC_{X.X}(p)$ is referred to as the ALC (see Sect. 5.1).

[2] See also Iyengar (1960) who uses it for estimating income elasticities.

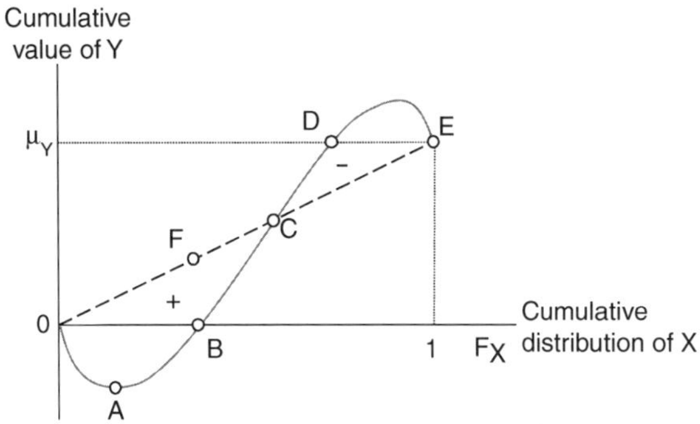

Fig. 5.2 The absolute concentration curve. *Source*: Yitzhaki and Schechtman (2004), Fig. 1, p. 290. Reprinted with permission by Metron International Journal of Statistics

Definition 5.2 The Line of Independence (LOI) is the line connecting (0, 0) with $(1, \mu_Y)$. Let $LOI_{Y.X}(p) = \mu_Y\, p$ denote the LOI of Y with respect to X. Note that the LOI is also a concentration curve of a special case—the case of independent variables. (LOI is the equivalent of LOE in ALC).

Figure 5.2 presents hypothetical ACC and LOI curves. The solid curve is the ACC of Y with respect to X and the dashed line is LOI.

Note that while the ALC is always convex, the ACC is not. To apply the concentration curve to variance-based parameters, it is convenient to redefine the concentration curve and the LOI as functions of the variate, X, rather than its cumulative distribution function. In this case, we denote the ACC by ACCV and define it as

$$ACCV_{Y.X}(x) = \int_{-\infty}^{x} g(t)dF_X(t), \tag{5.7}$$

where $g(x) = \mu_{Y.X} \equiv E\{Y|X = x\}$. The LOI simply changes to $LOI_{Y.X}(x) = \mu_Y F_X(x)$. Note, however, that it is no longer a straight line. In terms of Fig. 5.2, the only difference between (5.5) and (5.7) is that the horizontal axis has changed from $p = F_X(X(p))$ to X. The curvature is changed, but every point of intersection in the (ACC, F(X)) plane would have a parallel point in the (ACCV, X) plane. However convexity/concavity properties may differ.

The ACC has the following properties (proofs of the nontrivial claims are given below). The theorems are scattered in Yitzhaki and Olkin (1991) and in Yitzhaki (1990, 1996, 1998):

(a) The ACC passes through the points (0, 0) and $(1, \mu_Y)$.
(b) The derivative of the ACC with respect to p is $g(x(p)) = E\{Y|X = x(p)\}$. Consequently, $ACC_{Y.X}(p)$ is increasing if and only if $g(x(p)) = E\{Y|X = x(p)\} > 0$ (hereafter we will write g(x) instead of g(x(p))).
(c) The ACC is convex (concave, straight line) if and only if $\partial g(x)/\partial x > 0$ ($\partial g(x)/\partial x < 0$, $\partial g(x)/\partial x = 0$).
$ACC_{X.X}(p)$ is always convex.

Proof The first derivative of the ACC with respect to p is $g(x(p)) = E\{Y|X = x(p)\}$.
The second derivative is

$$\frac{\partial^2 \mathrm{ACC_{Y.X}(p)}}{\partial \mathrm{p}^2} = \frac{\partial g}{\partial \mathrm{X(p)}} \frac{\partial \mathrm{X(p)}}{\partial \mathrm{p}} = \frac{\partial g}{\partial \mathrm{X}} \frac{1}{\mathrm{f(X)}}.$$

Because $f(X) > 0$, the sign of the second derivative is determined by the sign of $\frac{\partial g}{\partial X}$.

(d) If Y and X are independent then ACC coincides with the LOI.

(e) e.1. The area between the LOI and the ACC is equal to $\mathrm{cov}(Y, F_X(X))$. That is,

$$\mathrm{cov}(Y, F_X(X)) = \int_0^1 \{\mu_Y \, p - \mathrm{ACC_{Y.X}(p)}\} \, dp. \tag{5.8}$$

e.2. The area between the shifted LOI and the shifted ACC is equal to $\mathrm{cov}(Y, X)$. That is,

$$\mathrm{cov(Y,X)} = \int_{-\infty}^{\infty} \left\{ \mu_Y \, F_X(t) - \int_{-\infty}^{x} g(t) f_X(t) \right\} dtdx = \int_{-\infty}^{\infty} \{ \mu_Y \, F_X(t) - \mathrm{ACCV_{Y.X}}(t)\} \, dt. \tag{5.9}$$

Note that $\mathrm{ACCV_{Y.X}}(x)$ is the transformed ACC, while $\mu_Y F_X(x)$ is the transformed LOI. The variance of X is the area between the (transformed) LOI and the (transformed) ACC, denoted by $\mathrm{ACCV_{X.X}}$ (Yitzhaki, 1998).The proofs of properties (e.1) and (e.2) are almost identical to the proofs dealing with similar properties in the case of ALC. They can be found in Yitzhaki and Olkin (1988).

(f) The ACC is above the LOI for all p if and only if $\mathrm{cov}(Y, T(X)) < 0$ for all continuous differentiable monotonically increasing functions T(X). (The ACC is below the LOI if and only if the covariance is positive.)

Proof A point on LOI has coordinates $(p, p\,\mu_Y)$. The condition that ACC is above the LOI for all p is

$$\mathrm{ACC_{Y.X}(p)} - p\,\mu_Y \geq 0 \quad \text{for all p.} \tag{5.10}$$

Equivalently, it is

$$\int_{-\infty}^{X(p)} (g(x) - \mu_Y) dF(x) \geq 0 \quad \text{for all p.} \tag{5.11}$$

Note that

$$\int_{X(p)}^{\infty} (g(x) - \mu_Y) dF(x) = \int_{-\infty}^{\infty} g(x) dF(x) - \int_{-\infty}^{X(p)} g(x) dF(x) - \mu_Y(1-p)$$
$$= \mu_Y - \mu_Y(1-p) - \mathrm{ACC_{Y.X}(p)} = \mu_Y p - \mathrm{ACC_{Y.X}(p)}.$$

Therefore condition (5.10) can be written as

$$\int_{X(p)}^{\infty} (g(x) - \mu_Y)dF(x) \leq 0 \quad \text{for all p.} \tag{5.12}$$

We now turn to the proof of property (f).

1. *Sufficiency*. We need to prove that if $ACC_{Y.X}(p) - \mu_Y p > 0$ for all p then $cov(Y, T(X)) < 0$ for all T(X) for which $T'(X) > 0$. Consider first

$$cov(Y,X) = \int_{-\infty}^{\infty}\int_{-\infty}^{\infty} x(y - \mu_Y)f(x,y)dydx = \int_{-\infty}^{\infty} x\left[\int_{-\infty}^{\infty} (y - \mu_Y)f(y|x)dy\right]dF(x)$$
$$= \int_{-\infty}^{\infty} x[g(x) - \mu_Y]dF(x).$$

For any point X(p),

$$cov(Y,X) = \int_{-\infty}^{X(p)} x[g(x) - \mu_Y]dF(x) + \int_{X(p)}^{\infty} x[g(x) - \mu_Y]dF(x). \tag{5.13}$$

Because $0 < \int_{-\infty}^{X(p)} [g(x) - \mu_Y]dF(x)$ for all p we have that

$$cov(Y,X) \leq X(p)\int_{-\infty}^{X(p)} [g(x) - \mu_Y]dF(x) + \int_{X(p)}^{\infty} x[g(x) - \mu_Y]dF(x)$$
$$\leq X(p)\int_{-\infty}^{X(p)} [g(x) - \mu_Y]dF(x) + X(p)\int_{X(p)}^{\infty} [g(x) - \mu_Y]dF(x)$$
$$= X(p)[ACC_{Y.X}(p) - \mu_Y p] + X(p)[\mu_Y p - ACC_{Y.X}(p)] = 0.$$

Hence we have proved that
$ACC_{Y.X}(p) - \mu_Y p > 0$ for all p implies that $cov(Y, X) \leq 0$.
The changes from X to T(X) with $T'(X) > 0$ and from X(p) to T(X(p)) do not alter the proof.

2. *Necessity*. We need to prove that if $cov(Y, T(X)) < 0$ for all T(X) for which $T'(X) > 0$ then ACC is above the LOI for all p. We do that by proving that if the ACC intersects the LOI then there exist two nondecreasing continuous

transformations $T_1(X)$ and $T_2(X)$ for which $T_1'(X)$ and $T_2'(X)$ are nonnegative, such that $\mathrm{cov}(Y, T_1(X)) > 0$ and $\mathrm{cov}(Y, T_2(X)) < 0$. Assume for simplicity that ACC and LOI intersect exactly once. That is,

$$\begin{aligned} ACC_{Y.X}(p) - \mu_Y\, p &> 0 \text{ for } p < p^* \\ ACC_{Y.X}(p) - \mu_Y\, p &= 0 \text{ for } p = p^* \\ ACC_{Y.X}(p) - \mu_Y\, p &< 0 \text{ for } p > p^* \end{aligned}$$

Let $X^* = X(p^*)$ and let a and b be constants with $a \geq 0$. Define

$$T_1(x) = \begin{cases} X^* & x \leq X^* \\ ax + (1-a)X^* & x \geq X^* \end{cases}$$

$$T_2(x) = \begin{cases} ax + b & x \leq X^* \\ aX^* + b & x \geq X^* \end{cases}$$

then

$$\begin{aligned} \mathrm{cov}(Y, T_1(X)) &= \int_{-\infty}^{\infty} T_1(x)[g(x) - \mu_Y]dF(x) \\ &= \int_{-\infty}^{X^*} T_1(x)[g(x) - \mu_Y]dF(x) + \int_{X^*}^{\infty} T_1(x)[g(x) - \mu_Y]dF(x) \\ &= X^*[ACC_{Y.X}(p^*) - \mu_Y p^*] + \int_{X^*}^{\infty} T_1(x)[g(x) - \mu_Y]dF(x) \\ &= X^*[ACC_{Y.X}(p^*) - \mu_Y p^*] + \int_{X^*}^{\infty} [a\,x + \bar{a}\,X^*][g(x) - \mu_Y]dF(x), \end{aligned} \tag{5.14}$$

where $\bar{a} = 1 - a$. The first term above is equal to zero because p^* is the point of intersection. Therefore we obtain that

$$\mathrm{cov}(Y, T_1(X)) = a\int_{X^*}^{\infty} x[g(x) - \mu_Y]dF(x) + \bar{a}X^*[\mu_Y p^* - ACC_{Y.X}(p^*)]. \tag{5.15}$$

The second term is equal to zero, and we get

$$\mathrm{cov}(Y, T_1(X)) = a\int_{X^*}^{\infty} x[g(x) - \mu_Y]dF(x) \geq aX^*\int_{X^*}^{\infty} [g(x) - \mu_Y]dF(x) = 0. \tag{5.16}$$

Similarly,

$$\begin{aligned}
\mathrm{cov}(Y,T_2(X)) &= \int_{-\infty}^{X^*} T_2(x)[g(x) - \mu_Y]dF(x) + \int_{X^*}^{\infty} T_2(x)[g(x) - \mu_Y]dF(x) \\
&= a\int_{-\infty}^{X^*} x[g(x) - \mu_Y]dF(x) + b\int_{-\infty}^{X^*} [g(x) - \mu_Y]dF(x) \\
&\quad + aX^*[\mu_Y p^* - ACC_{Y.X}(p^*)] \\
&= a\int_{-\infty}^{X^*} x[g(x) - \mu_Y]dF(x) \le aX^*[ACC_{Y.X}(p^*) - \mu_Y p^*] = 0
\end{aligned}$$

using the fact that p* is the intersection point.

This property is a modification of Grether (1974). It implies that whenever the ACC intersects the LOI one can divide the data into two sections, conditional on the values of X. In one section $\mathrm{cov}(Y, X) < 0$ and in the other section $\mathrm{cov}(Y, X) > 0$. By applying a monotonic transformation to X the investigator can change the magnitudes of the covariances in these sections, thereby affecting the sign of the overall covariance (and even changing its sign!). Yitzhaki (1990) derives the conditions under which it is possible to change the sign of an OLS regression coefficient by applying a monotonic transformation to one of the variables. This issue will be discussed in Chap. 19. Note, however, that if the $ACC_{Y.X}$ and the $LOI_{Y.X}$ intersect, it does not necessarily imply that $ACC_{X.Y}$ and $LOI_{X.Y}$ intersect. The ACC in Fig. 5.2 intersects the LOI at C.

(g) If Y and X follow a bivariate normal distribution with $\rho \neq 0$ then $ACC_{Y.X}$ and $LOI_{Y.X}$ do not intersect. Thus, a monotonic transformation cannot change the sign of the covariance. This issue will be further developed in Chap. 7 when dealing with the simple regression coefficient.

Additional properties of ACC are presented in Yitzhaki and Olkin (1988). Additional curves that are based on the concentration curve and will be useful for the analyses of the components of the regression coefficient will be presented in Chap. 7.

5.4 The Absolute Lorenz Curve and Second-Degree Stochastic Dominance

Economic theory requires the knowledge or, alternatively, requires imposing assumptions concerning the utility function of the decision-maker. Because the utility function is not directly observed, this requirement imposes a serious obstacle for meaningful economic analysis. Stochastic Dominance (SD) rules intend to provide results that can be applied to *classes* of utility functions, and therefore reduce the need to know the exact utility function. This is the basic advantage of using stochastic dominance. On the other hand, rules and conclusions that are relevant for a large set of possible utility functions may suffer from being too restrictive empirically so that they are not so relevant in practice. The aim of this section is to present the contributions of the ACC and the Gini in forming necessary and sufficient conditions for stochastic dominance and in performing statistical analyses compatible with stochastic dominance. The literature on stochastic dominance (also referred to as Majorization) is huge and we do not survey it in this section. The interested reader is referred to Levy (2006) for a survey of stochastic dominance in finance and to Marshall and Olkin (1979) for a survey in statistics and related fields. To be able to connect between stochastic dominance and concentration curves we start with the presentation of the basic definitions of stochastic dominance and then we will apply the SD conditions using concentration curves.

Consider a decision-maker who decides according to his or her expected utility. The utility functions are classified according to the assumptions imposed on their derivatives.

Definition 5.3 The utility function u belongs to U_1 where $U_1 = \{u|u' \geq 0$ and u is a nondecreasing concave utility function$\}$.

Definition 5.4 The utility function u belongs to U_2 where $U_2 = \{u|u' \geq 0$ and $u'' \leq 0$ and u is a nondecreasing concave utility function$\}$.

Definition 5.5 Given two distributions $F_1 = F(Y_1)$ and $F_2 = F(Y_2)$ we say that F_1 First-degree dominates F_2 (FSD) if $E\{U(Y_1)\} \geq E\{U(Y_2)\}$ for all $u_\varepsilon\ U_1$.

Definition 5.6 Given two distributions F_1 and F_2 we say that F_1 Second-degree dominates F_2 (SSD) if $E\{U(Y_1)\} \geq E\{U(Y_2)\}$ for all $u_\varepsilon\ U_2$.

One can continue adding assumptions on the class of utility functions but we stop here because the Gini is relevant to the SSD criterion only.

We now turn to the propositions that enable one to identify FSD and SSD. We will concentrate on those that relate SSD rules to the ALC and Gini.

Proposition 5.2 *Distribution $F(Y_1)$ FSD dominates $F(Y_2)$ if $F(Y_1) \leq F(Y_2)$ everywhere with at least one strong inequality. If on the other hand $F(Y_1)$ and $F(Y_2)$ intersect, then one can find $u^A(Y)\varepsilon U_1$ and $u^B(Y)\varepsilon U_1$ so that both*

$$E\{U^A(Y_1)\} \geq E\{U^A(Y_2)\} \text{ and } E\{U^B(Y_1)\} \leq E\{U^B(Y_2)\} \text{ hold.}$$

Proof Levy (2006). First-degree stochastic dominance is unrelated to the Gini. However, a basic assumption in economics is that the marginal utility of income declines with income, hence it is natural to impose an assumption of concavity on the utility function. This leads to the connection between stochastic dominance and the ALC. Historically, the Lorenz curve was used as a descriptive measure till Atkinson (1970) turned it into an analytical tool that is useful in detecting SSD. Shorrocks (1983) extended Atkinson's result to cover all distributions, and not necessarily equal-mean distributions. Proposition 5.3 presents the connection between SSD and the ALC.

Proposition 5.3 *The distribution of Y_1 SSD dominates the distribution of Y_2 if ALC(Y_1) $\geq$ ALC(Y_2) everywhere with at least one strong inequality. If on the other hand ALC(Y_1) and ALC(Y_2) intersect, then one can find $u^A(Y)\varepsilon U_2$ and $u^B(Y)\varepsilon U_2$ so that both $E\{U^A(Y_1)\} \geq E\{U^A(Y_2)\}$ and $E\{U^B(Y_1)\} \leq E\{U^B(Y_2)\}$ hold.*

Proof See Shorrocks (1983).

Once the relationship between the ALC and SSD has been established the GMD can be used in order to form necessary conditions for SSD (Yitzhaki 1982a) as given in the next proposition.

Proposition 5.4 *The following are necessary conditions for $F(Y_1)$ to SSD $F(Y_2)$:*

$$\mu_1 \geq \mu_2, \tag{5.17}$$

$$\mu_1 - 2\,\mathrm{cov}(Y_1, F(Y_1)) \geq \mu_2 - 2\,\mathrm{cov}(Y_2, F(Y_2)). \tag{5.18}$$

Note that (5.18) can be expressed as

$$\mu_1 - 0.5\ \Delta_{Y_1} \geq \mu_2 - 0.5\ \Delta_{Y_2}.$$

Proof The proof of (5.17) is trivial because an expected utility maximizer with a linear utility function would prefer μ_1 over μ_2.

The proof of (5.18) is simple when using geometric considerations. If F(Y_1) SSD F(Y_2), then according to proposition 5.3 ALC(Y_1) $\geq$ ALC(Y_2). It follows that the area below ALC(Y_1) is larger than the area below ALC(Y_2). The area below the LOI is $\mu/2$, while the area between the LOI and the ALC is cov(Y, F(Y)). Hence, the area below the ALC is $\mu/2$ − cov(Y, F(Y)).

The next two propositions are not used in the analysis performed in this book. They are given here (without proofs) for the completeness. Details can be found in Yitzhaki (1982a, 1983, 1999).

Proposition 5.5 *Provided that the cumulative distributions intersect exactly once, (5.17) and (5.18) are necessary and sufficient conditions for $F(Y_1)$ to SSD $F(Y_2)$.*

It is interesting to note that if the utility function is increasing and convex (that is, if $u' \geq 0$, and $u'' \geq 0$) then one gets the following result:

Proposition 5.6 *Conditions (5.17) and (5.19) (below) are necessary conditions for the distribution of Y_1 to SSD the distribution of Y_2:*

$$\mu_1 + 2\,\mathrm{cov}(Y_1, F(Y_1)) \geq \mu_2 + 2\,\mathrm{cov}(Y_2, F(Y_2)). \tag{5.19}$$

Proposition 5.4 is the key factor for the advantage of using the GMD over the variance from the point of view of the economist. The use of the variance to reflect risk or inequality is valid only if one restricts the distributions to be normal or restricts the utility function to be quadratic. Proposition 5.4 indicates that although the GMD is not compatible with any utility function (Newberry, 1970), it still can be used to form necessary conditions for SSD. Therefore a researcher who uses the GMD can perform statistical analysis without being ridiculed for violating economic theory. The following example illustrates this point: assume that one has to choose between two possible lotteries. Lottery A gives one dollar with probability p, and 2 dollars with probability (1 − p). Lottery B gives ten dollars with probability p, and a million dollars with probability (1 − p). Needless to say that any rational person would prefer lottery B over lottery A. Note, however, that lottery A presents a lower mean than lottery B but it also presents a lower variability, which implies that anyone who relies on the mean and the variance (or the Gini) is not capable of deciding which lottery is preferred. Clearly, no one would be deceived by this example. However, when choosing a portfolio with more than three assets, our guess is that most people would not be able to avoid such a trap. Proposition 5.4 enables us to discard lottery A as a candidate for being preferred by someone.

5.5 The ACC and Marginal Conditional Stochastic Dominance

The aim of this section is to extend the ideas that connect the SSD to the ALC into the area of multiple variables. When dealing with multiple variables it is worth to distinguish between two cases, according to the problem we are dealing with:

(a) Single output/multi-inputs
(b) Multi-outputs/multi-inputs

The single output/multi-inputs case covers almost all problems in economic theory. It holds whenever one can achieve the same target by alternative sets of inputs, or whenever trade is possible between targets. The multi-outputs/multi-inputs is seldom analyzed in economic theory.

To be concrete let us present several examples. A consumer who maximizes the utility function is a classical case of a single output/multi-inputs. Also, whenever one acts under a budget constraint, we are dealing with a single output/multi-inputs. On the other hand, analyzing the impact of health and income on the consumer, assuming that no amount of income can substitute for health, is a multi-outputs/multi-inputs case. The distinction between the two problems is sometimes not easy. For example, assume that an individual is participating in gambles in the commodity space. Assume also that at one point in time the outcomes of the gambles are revealed. Then, if trade is allowed following the announcement of the final results then we are dealing with a single output/multi-inputs case. If trade is not allowed following the announcement of the state of nature then we are dealing with multi-outputs/multi-inputs case.

In this book we are dealing only with single output/multi-inputs cases. We refer the reader interested in multiple outputs to Taguchi (1981, 1987). For our purpose we will assume that Y, the output (target), is composed of a linear combination of variables, X.

That is

$$Y = \beta_1 X_1 + \cdots + \beta_n X_n, \tag{5.20}$$

which can be viewed as a budget constraint where the βs represent the prices, or as the return on a portfolio with the βs representing the shares of the appropriate components of the portfolio.[3] Note that

$$\mathrm{ALC(Y)} = \sum_{i=1}^{n} \beta_i \mathrm{ACC}_{X_i.Y}. \tag{5.21}$$

Therefore, provided that X_j is not affected by a change in β_i (i, j = 1,. . .,n; i ≠ j), we get

$$\frac{\partial \mathrm{ALC(Y)}}{\partial \beta_i} = \mathrm{ACC}_{X_i.Y}. \tag{5.22}$$

Next we define the concept MCSD (Mayshar & Yitzhaki, 1995, 1996; Shalit & Yitzhaki, 1994).

MCSD is intended to extend the concept of handling SSD criterion in an environment of multi-inputs. To be empirically relevant we have to impose two major constraints on the concept. The rationale behind those constraints will be postponed till we present the concepts.

The first constraint is that we restrict the concept to be applicable only for marginal changes. This means that we are not dealing with SSD in general, but only of a small change (marginal change). The second constraint is a consequence of the first one. Because we are dealing with a small change, we need a starting point from which the change is measured, hence the use of the term conditional, because the analysis is relevant to a small change from the original position.

Let $d\beta_k$ be the marginal change in β_k. Assume that there is a constraint on (5.20) that the change in β has to comply with, as is in the case of having a budget constraint:

$$\mathrm{C}(\beta_1, \ldots, \beta_n) = 0,$$

so that

$$c_k d\beta_k + c_j d\beta_j = 0, \tag{5.23}$$

where c_k is the derivative of (5.20) with respect to β_k. This implies that the change in β_k is restricted by the constraint in (5.23). That is, we consider substituting an infinitesimal portion of X_j by X_k:

$$\mathrm{dE\{U(Y)\}} = \mathrm{E\{U'(Y)dY\}} = \mathrm{E}\,\{\mathrm{U'(Y)}(X_k d\beta_k + X_j\, d\beta_j)\}, \tag{5.24}$$

[3] One of the Xs can be a constant.

where $E\{\cdot\}$ is the expectation over all X's. Inserting (5.23) and assuming that $d\beta_k$ is positive yield

$$\frac{dE\{\ U(Y)\}}{d\beta_k} = E\{\ U'(Y\ )\ (X_k - X_j)\}. \tag{5.25}$$

Definition of MCSD: Given the set $\{\beta\}$ we say that X_k MCS dominates X_j if $\frac{dEU(Y)}{d\beta_k} \geq 0$ for all increasing concave utility functions defined on Y.

Proposition 5.7 *Given the set* $\{\beta\}$, X_k *MCS dominates* X_j *if and only if*

$$ACC_{X_k.Y}(p) \geq ACC_{X_j.Y}(p) \quad \text{for all p, } 0 \leq p \leq 1,$$

with at least one strong inequality. If on the other hand the ACCs intersect, then one can find two legitimate utility functions that will reverse the rankings of the X′s.

Proof

Sufficiency. We have to prove that Y + dY SSD dominates Y. According to proposition 5.3 we have to prove that ALC(Y + dY) is not lower than ALC(Y) everywhere. Using (5.21) and (5.22) that present the derivatives of the ALC with respect to changes in the βs and imposing the constraint complete the proof.

Necessity. Assume that the ACCs intersect. Then the ALCs of Y + dY and Y intersect. We again refer to (5.23) to complete the proof.

Proposition 5.7 is analogous to proposition (5.3), but dominance is conditional on a given Y; hence it is marginal. The explanation to this restriction is that in almost all economic models it is assumed that the optimal point is an interior one. If we do not restrict MCSD to the margin, then variable k would replace variable j entirely, leading to a corner solution. This point will be stressed and elaborated upon when we deal with applications of MCSD in the areas of income distributions and finance (Chaps. 14 and 17).

The MCSD concept can be applied with more than two variables. In this case one substitutes the two inputs by two linear combinations of variables: the dominating and the dominated. The optimal combination in each group is found by numerical optimization. Illustrations of the applications appear in Yitzhaki and Mayshar (2002) and Shalit and Yitzhaki (2003). Illustrations are discussed in Chap. 14 in the area of tax reforms, and in Chap. 17 in the area of finance.

5.6 The ACC and the Monotonicity of the Correlations and the Regression Slopes

An additional use of the concentration curve and curves that are derived from it is to study the effects of certain actions taken by the investigator on the signs of the correlation or the regression coefficients. The actions that can be analyzed by the concentration curves are the following:

1. Throwing extreme observations.
2. Throwing irrelevant observations and using a subgroup of the population—e.g., bounds on observations that participate in the regression.
3. Substituting a continuous variable by a discrete one by dividing the range into nonoverlapping intervals (called bins). The data entries are taken to be either the mid-points or the averages of the intervals.
4. Applying a monotonic nondecreasing transformation to one or more variables.

Note that (3) can be viewed as a special case of (4).

While throwing observations sounds suspicious, using a transformation seems natural and is being used in practice quite often.

Our main goal is to analyze the effect of a transformation on the sign of the estimator of the regression coefficient. The reason for stressing sign change is that it may reverse the conclusion reached. Instead of positive (negative) effect it may turn the effect into a negative (positive) one.

The suggested tool is graphical. To be able to geometrically see the monotonicity of the slope of a regression curve in a Gini or an OLS regression setting we suggest to plot a curve that is based on the vertical differences between the LOI and the ACC. We refer to this curve as the LMA curve and define it below.

Let $g(x) = E\{Y|X = x\}$ be the conditional expectation of Y given X. We will refer to it as the regression curve. We remind the reader the two basic definitions:

Definition of ACC: The ACC of *Y* with respect to *X* denoted by $ACC_{Y.X}(p)$ is

$$ACC_{Y.X}(p) = \int_{-\infty}^{X(p)} g(t)dF_X(t), \quad \text{where } X(p) \text{ is implicitly defined by } p = \int_{-\infty}^{X(p)} dF_X(t) .$$

A special case of the ACC curve is the ALC—$ACC_{Y.Y}(p)$.

For simplicity of exposition, we write ACC instead of $ACC_{Y.X}(p)$.

Definition of LOI: Connect the points (0, 0) and $(1, \mu_Y)$ by a straight line; Yitzhaki and Olkin (1991) call this line the LOI. (If Y and X are independent, then the ACC curve coincides with the LOI.) Because we are interested in "deviations from independence" we will be interested in a curve which is the LOI minus the ACC.

Note that the LOI is also an ACC curve.

Definition of LMA: $LMA_{Y.X}(p) = \mu_Y p - ACC_{Y.X}(p)$ is defined as the *LOI minus the ACC of Y with respect to X*.

The properties of ACC and LMA are as follows (proofs are omitted because they can be easily derived by the properties of ACC curves which are given in Sect. 5.3):

(a) The ACC passes through the points (0, 0) and $(1, \mu_Y)$.
Property (a) enables us to define a variation of the ACC that will make the analysis of the regression curve easier. The LMA starts at (0, 0) and ends at (1, 0).

(b) The derivative of the LMA with respect to F (at X(p)) is $\mu_Y - E_Y(Y|X = X(p))$.

This follows directly from the definition of the LOI and ACC. As a consequence the $LMA_{Y.X}(p)$ is increasing (decreasing, constant) if and only if $\mu_Y - g(X(p)) > (<, =) 0$.

(c) The LMA is concave at F (convex, straight line) if and only if $\partial g(X(p))/\partial X(p) > (<, =) 0$.

(d) If X and Y are independent then $ACC_{Y.X}(p)$ is a straight line which coincides with the LOI, and the LMA curve coincides with the horizontal axis.

Properties (c) and (d) enable the user to identify sections with constant, increasing, and decreasing slopes of the regression curve: linearity of LMA implies a flat regression curve, concavity of LMA means an increasing regression curve, while convexity means a decreasing curve.

(e) The area between the LMA and the horizontal axis is equal to $\text{cov}(Y, F_X(X))$ (Yitzhaki, 1990). Note that if the curve intersects the horizontal axis then the sign of $\text{cov}(Y, F_X(X))$ depends on the magnitudes of the areas above and below the horizontal axis.

(f) The LMA is above the horizontal axis for all F if and only if $\text{cov}(Y, T(X)) > 0$ for any continuous differentiable monotonically increasing functions T(X).

The advantage of using the LMA (instead of the ACC) is that it is easy to detect what will happen to cov(Y, F(X)) (and hence to the sign of the regression coefficient) if sections of observations of X are omitted from the regression, as will be illustrated later.

For the purpose of analyzing the effect on the OLS regression coefficient one needs a modified LMA curve for which the area beneath it will be equal to cov(Y, X). It turns out that a simple transformation can make the curve applicable to OLS: if one substitutes the horizontal axis to be X instead of F_X, then the area between the new curve and the horizontal axis will be equal to cov(Y, X) (see Sect. 5.2 or Yitzhaki (1998)). However, the nature of the curve changes and further research is needed to study its properties. For our purposes it is sufficient that property (f) holds in the transformed curve so that if the sign of the covariance in Gini regression cannot be changed by truncating the distribution of the explanatory variable at the point of the intersection with the horizontal axis then it is impossible to change the sign of the regression coefficient in OLS regression by a monotonic transformation of the explanatory variable. That is, one can change the sign of a regression coefficient in a Gini regression by truncating the distribution at the point of intersection with the horizontal axis if and only if there is a monotonic transformation that can change the sign of the regression coefficient in an OLS regression.

In the rest of this section we limit the discussion to the Gini regression and correlation coefficients. In Chap. 19 we will list the properties that also apply to the OLS regression.

The (semi-parametric) Gini regression coefficient is a ratio of two covariances:

$$\beta^N_{Y.X} = \frac{\text{cov}(Y, F(X))}{\text{cov}(X, F(X))}.$$

The denominator is always positive; hence the sign of the regression coefficient is determined by the numerator. A monotonic transformation of the explanatory variable, X, does not affect F(X). Therefore, unlike the OLS, a monotonic transformation of X cannot change the sign of the Gini regression coefficient. However, it may change its magnitude. By property (e) of the LMA curve one can see whether there are sections with different signs. Whenever there are sections with different signs, one can change the sign of the Gini regression coefficient by truncating the distribution of X.

To ease the analysis of the effect on a regression coefficient, we normalize the LMA curve by dividing it by cov(X, F(X)). We call the curve *NLMA* (Normalized Line Minus ACC). The important (for us) property of this curve is that the area between the curve and the horizontal axis is equal to the Gini regression coefficient. Because the analysis is relatively simple, we will do it by an illustration.

5.7 An Illustration: Labor Force Participation by Gender and Age[4]

Figure 5.3 presents the Normalized Line (of independence) Minus the ACC which is the LMA divided by cov(X, F(X)). As discussed above, the total area enclosed by the curve and the horizontal axis is equal to the Gini regression coefficient. The horizontal axis depicts the cumulative distribution according to age, while the vertical axis depicts the difference between the cumulative value of participation in the labor force if participation is independent of age (i.e., LOI) and the actual cumulative value of participation, divided by one-fourth of the GMD of age. The data are taken from Income Survey, 2005, conducted by the Israeli Central Bureau of Statistics.

Figure 5.3 is based on 12,685 observations for women and 11,213 for men. The curve enables us to detect regions with positive or negative slopes according to whether the curve is concave or convex, and according to whether the regression coefficient of each section is contributing positively or negatively to the overall regression coefficient. If the curve is above (below) the horizontal axis then this range has a positive (negative) contribution to the regression coefficient. In addition a concave (convex) section implies that if we take this section alone we will find a positive (negative) regression coefficient.

Let us concentrate first on the NLMA curve for men.

The curve is below the horizontal line; hence the regression coefficient is negative. Moreover, the curve does not intersect the horizontal axis; therefore there is no monotonic transformation of age that can change the sign of the OLS regression coefficient. However, throwing 70% of the observations with high age

[4] This section is based on Yitzhaki and Golan (2010).

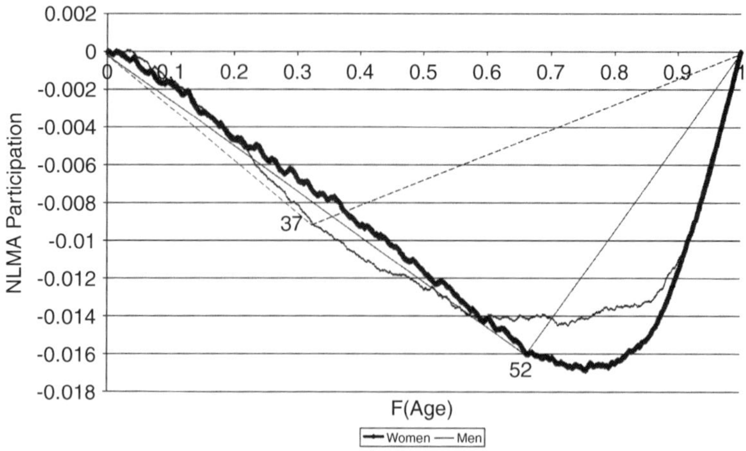

Fig. 5.3 NLMA curve for participation in the labor market vs. age according to gender. *Source*: Yitzhaki and Golan (2010)

from the regression, that is, restricting the group investigated to those below 37, will cause the Gini and OLS regression coefficients to be positive and this holds for any monotonic transformation of age. The reason is that the curve is above the straight line that connects the two extreme points. (Omitting 70% of the observations will turn the line connecting the two extreme points to be the horizontal axis for the relevant graph and the NLMA lies above it, as can be seen in Fig. 5.3).

Substituting a continuous variable by a discrete one with the data entries taken to be either mid-points or averages of the relevant intervals will not be effective in changing the sign of the regression coefficient, because it is equivalent to creating a new LMA curve composed of straight lines. The straight lines connecting the points form the new LMA curve that represents the between-groups regression coefficient. As can be seen it yields the same sign of the regression coefficient, which, in this case, will continue to be negative. (Note, however, that the denominator of the between-groups regression coefficient is different than the one used to normalize the original curve; hence, one cannot learn about the magnitude of the between-groups regression coefficient.)

Applying a monotonic nondecreasing transformation to the explanatory variable is not capable of changing the sign of the OLS regression coefficient because the curve does not cross the horizontal axis. However, omitting 50% of the observations with high ages and applying a monotonic transformation to age that will shrink the distance between the remaining high ages while increasing the differences between low ages (such as the log transformation) may change the OLS regression coefficient to a positive one.

Turning to women we can see that there is a tiny section at low ages with a positive regression coefficient. However, one needs an extreme transformation to change the sign of the regression coefficient to a positive one. Restricting the age

Table 5.1 Regression coefficients of participation on age in different sections

	GINI		OLS	
	Females	Males	Females	Males
Section I	0.0008	0.008	0.0007	0.008
Section II	−0.034	−0.015	−0.034	−0.016
All	−0.009	−0.009	−0.010	−0.010

Source: Yitzhaki and Golan (2010)

to be lower than 52 will make the regression coefficient positive and it will remain so for all possible monotonic transformations of age. Note, however, that one can omit observations for ages that are higher than 52 and still get a regression coefficient that is positive, but a monotonic transformation of age can change its sign.

To complete the illustration, Table 5.1 presents the regression coefficients in each section.

As can be seen in Fig. 5.3 and Table 5.1 the overall Gini and OLS regression coefficients are negative, while the curve is concave for males for the lower 30% of the observations (up to age 37) and for females for the 65% of the observations with smallest ages (up to age 52). As a result in these sections the regression coefficients of both OLS and Gini regressions are positive. On the other hand, in the second section the curves are convex; hence the regression coefficients in this section are negative. Note, however, that the curves do not cross the horizontal axis, implying that the contributions of the two sections are adding up to the overall regression coefficients (and not canceling each other). Also note that for the Gini regression we can evaluate, by adding and subtracting positive and negative areas, how many additional observations of the "wrong" sign we can add without changing the sign of the regression coefficient. We note that the regression coefficients obtained by the two methods are similar. However, using the Gini method and the figures which are derived from the Gini enables the user to make the partition into sections as shown above.

5.8 Summary

The aim of this chapter was to present another advantage of the Gini over the variance method: the relationship with concentration curves. One advantage of using Lorenz and concentration curves is that the curves allow the researcher to learn more about the structure of the relationship between the variables by visual tools. Most of the models used in econometrics are based on structured connections among variables, based on underlying assumptions. As an example, notice the assumption of the linearity of models. The use of curves enables one to check graphically whether the relationship is monotonic over the entire range of the explanatory variable. If it is, then the assumption of linearity is not violated in a crucial way. Although the researcher assumes linearity, the estimators offered by the OLS and Gini methods can still be derived, because they can be described as weighted averages of properties of the data (i.e., slopes between adjacent observations). Those averages may mask the fact that not all the data speak in the same way. To inspect or to calm such fears, the concentration curves enable one to see the contribution of each section of the explanatory variable to the parameter that is inspected. This way we can see “beyond the model.” An alternate way to look at the differences between parameters and concentration curves is that while the parameters supply us with necessary conditions, the curves supply us with necessary and sufficient conditions.

Chapter 6
The Extended Gini Family of Measures

Introduction

The GMD has many alternative presentations. Some of these alternative presentations can be extended into families of variability measures and the GMD can be viewed as one member of such a family. The fact that there are several alternative presentations implies that one can present the GMD and the Gini coefficient as belonging to several alternative families. These families differ in the properties they have. We do not intend to survey the properties of all possible families. We choose to concentrate on one family that is useful in several fields of applications. We will refer to it as the extended Gini family. However, the reader should keep in mind that for different fields of applications one may want to have alternative extensions.

The extended Gini family (hereafter EG) is a family of variability and inequality measures that depends on one parameter, the extended Gini parameter. The investigator can choose a member of the family by assigning a value to the parameter. One advantage of having a family lies in the fact that one can perform a sensitivity analysis and evaluate the robustness of the conclusions by changing the EG parameter. That is, by changing the metric of the variability measure (as will be discussed later). The selection of the parameter can be interpreted in several ways, depending on the area of application and on the objective of the research. The EG family is mainly used in the areas of finance and income distributions. In the area of finance the parameter represents the degree of risk aversion, while in the area of income distribution the parameter represents the social attitude of the investigator. In econometrics, the use of different parameters can be viewed simply as a sensitivity test, without any interpretation that relates it to a concept in economic theory. Several surveys can be found in the econometric literature, each covering a specific application. A survey on hedging theory in finance can be found in Lien and Tse (2002), while Wodon and Yitzhaki (2002b) survey some applications in the areas of tax reforms and income distributions. The use in the stock market is discussed in Gregory-Allen and Shalit (1999) and in Shalit and Yitzhaki (2002), while the use in econometrics will be covered in Chap. 19.

S. Yitzhaki and E. Schechtman, *The Gini Methodology: A Primer on a Statistical Methodology*, Springer Series in Statistics 272, DOI 10.1007/978-1-4614-4720-7_6,

The basic definitions of the members of the EG family used in this book are based on the covariance. In order to simplify the presentation we will use cov(X,F(X)) as the Gini, ignoring the constant (4) that is needed to adjust the definition to the GMD. Even with this simplification, the fact that the extended Gini is being used in different areas resulted in two alternative definitions. Let

$$\Delta(\theta, X) = -\ \theta\text{cov}(X, [1 - F(X)]^{\theta-1}),\ \theta > 0,\ \theta \neq 1. \tag{6.1}$$

Then the two definitions refer to $\theta = v$ and $\theta = v + 1$, where v is the extended Gini parameter. More explicitly, the first definition is

$$\Delta(v, X) = -v\text{cov}(X, [1 - F(X)]^{v-1}), \quad v > 0,\ v \neq 1. \tag{6.1a}$$

This definition is mainly used in the areas of income distribution and finance, due to the need to adjust the definition to the theory of stochastic dominance. The second definition is

$$\Delta(v, X) = -(v + 1)\ \text{cov}(X, [1 - F(X)]^{v}), \quad v > (-1), v \neq 0. \tag{6.1b}$$

Definition (6.1b) is mainly used in the area of econometrics, in which case the term (v + 1) cancels because the parameters are expressed as ratios. The motivation for the different definitions is the need for a simple representation, relevant to the specific application. In this chapter we will use definition (6.1a). In chapters that deal with regression we will use definition (6.1b).

Definition (6.1a) can easily be used in order to define the relative extended Gini (also called the extended Gini coefficient). This is simply done by dividing $\Delta(v, X)$ by the mean of X, provided that the mean is positive. The relative extended Gini is given by

$$G(\ v,X) = \frac{-v\text{cov}(X, [1 - F(X)]^{v-1})}{\mu} \quad v > 0. \tag{6.2}$$

Similar to the Gini coefficient, the range of the extended Gini coefficient for nonnegative distributions is between zero (egalitarian) and one (one member receives all). The only difference is in the weighting scheme that is applied to the vertical distance between the egalitarian line and the Lorenz curve. This weighting scheme will be investigated in Chap. 18, while detailed derivation can be found in Yitzhaki (1996).

The parameter v which is determined by the investigator is restricted in most applications, especially in finance and income distribution, to be greater than 1, because the basic assumption is that we are interested in a risk-averse investor. However, in the area of econometrics, especially when one deals with sensitivity analysis, the range $\nu < 1$ is also used. The implications of the choices of ν are as follows. In the range $\nu > 1$:

1. If $\nu \rightarrow 1$ then the variability index represents the attitude of someone who does not care about variability, i.e., the index tends to zero regardless of the variability of the distribution.
2. On the other extreme, $\nu \rightarrow \infty$ represents variability as viewed by a max–min investigator (i.e., someone who cares only about the lowest portion of the distribution).
3. The case where $\nu = 2$ represents the GMD (up to a constant. GMD is defined as 4cov(X,F(X)), while $\Delta(2,X) = 2\text{cov}(X,F(X))$ by (6.1a)). Note that as opposed to the other members of the family $\Delta(2,X)$ is linear in F(X). Given that $\nu > 1$, the members of the family are always nonnegative, and a mean-preserving spread will always increase their values.

In the range $\nu \in [0, 1]$ the values of the members are always negative and if $\nu \rightarrow 0$ then the index represents variability as viewed by a Max–Max investigator.

As was mentioned above, in almost all the applications in finance and income distribution ν is restricted to be greater than 1 because one is interested in an attitude of a risk averse or inequality-averse behavior. Moreover, allowing for $\nu < 1$ complicates the interpretations of alternative definitions of the EG. For this reason we will restrict the parameter to be $\nu > 1$, and only when we use it in econometric applications and with the covariance definition as in (6.1a) we will allow for $\nu < 1$.

The previous chapters have demonstrated several alternative ways of presenting the GMD. Therefore it is only natural that there are several approaches for extending the GMD into a family of indices. Each approach has its own rationale, depending on the field in which it is used.

We suggest several introductions for this chapter. Each introduction leads to a different definition of the EG, allowing the reader to choose the one suitable for her/him.

In what follows we introduce the different definitions and prove that they are equivalent to the formal one of (6.1a), for $\nu > 1$.

The first introduction is based on the dual approach to moments. This approach is useful in the areas of statistics and econometrics. The second introduction is based on the income inequality approach, which is useful in the areas of income distribution and welfare economics, and the third introduction is based on the dual approach to risk. This approach is useful for modeling in the area of decisions under risk.

The structure of the chapter is as follows: the first section is devoted to the three introductions. We then proceed with alternative definitions (Sect. 6.2), while the properties and the metric used in the EG family are described in Sect. 6.3. Alternative presentations of the extended Gini covariances and correlations are detailed in Sect. 6.4. The decomposition of the extended Gini is presented in Sect. 6.5 and the relationship between the extended Gini and stochastic dominance is discussed in Sect. 6.6. Section 6.7 concludes.

6.1 The Three Introductions

6.1.1 *The "Dual Approach to Moments" Introduction*

One of the basic concepts in statistics is the use of the moments of the distribution in order to characterize it. Moments are classified into general moments and central moments. A general moment is the expected value of a power function of the variate. That is, the nth degree general moment is the expected value of the variable after it is raised to the nth power. The nth degree central moment is based on raising the deviations of the variate from its first general moment to the nth power.

Formally, a general moment of degree n is

$$m(X,n) = E(X^n) = \int x^n f(x)dx = \int x^n dF(x), \tag{6.3}$$

which can be presented for nonnegative variables as

$$m(X,n) = n\int_0^\infty x^{n-1}(1 - F(x))dx.$$

The central moment of degree n is

$$\begin{aligned} mc(X,n) = E\{(X - E(X))^n\} &= \int (x - E(X))^n f(x)dx \\ &= \int (x - E(X))^n dF(x). \end{aligned} \tag{6.4}$$

For example, the variance is the central moment of degree 2, or alternatively, the second general moment minus the square of the first general moment.

It can be seen from the second part of (6.3) that the nth general moment is a combination of a power function applied to the variate and a linear function of the cumulative distribution. A natural alternative approach, although more complicated, is to apply a power function to the cumulative distribution and to integrate over X. This will be referred to as the dual approach to moments. Here two possibilities come to mind.

Let X be a bounded random variable, defined in $[a, \infty)$ and define[1]

---

[1] While writing this book we have encountered an approach that enables to remove the restriction $a > -\infty$, but this approach seems to complicate the presentation. The interested mathematically oriented reader is referred to González-Abril, Valesco Morente, Gavilán Ruiz, and Sánchez-Reyes Fernández (2010). In the classification used in this book it can be referred as the Lorenz version of the coefficient of variation.

$$\mathrm{mg}(v,\mathrm{X}) = \int_{a}^{\infty} [1 - \mathrm{F}^{v}(\mathrm{x})]\mathrm{dx} + \mathrm{a},\ v > 0, \tag{6.5}$$

and

$$\mathrm{meg}(v,\mathrm{X}) = \int_{a}^{\infty} [1 - \mathrm{F}(\mathrm{x})]^{v}\mathrm{dx} + \mathrm{a},\ v > 0. \tag{6.6}$$

In what follows, the integration is over $[a, \infty)$ unless specified otherwise.

Proposition 6.1

$$\Delta(v, \mathrm{X}) = \mu - \mathrm{meg}(v , \mathrm{X})$$

Proof

$$\mathrm{meg}(v,\mathrm{X}) = \int_{a}^{\infty} [1 - \mathrm{F}(\mathrm{x})]^{v}\mathrm{dx} + \mathrm{a}.$$

Integration by parts, with $[1 - \mathrm{F}(\mathrm{X})]^{v} = \mathrm{u}$ and $\mathrm{dx} = \mathrm{dv}$ implies

$$\int_{a}^{\infty} [1 - \mathrm{F}(\mathrm{x})]^{v}\mathrm{dx} + \mathrm{a} = \mathrm{x}[1 - \mathrm{F}(\mathrm{x})]^{v}|_{a}^{\infty} + v\int_{a}^{\infty} \mathrm{x}[1 - \mathrm{F}(\mathrm{x})]^{v-1}\mathrm{f}(\mathrm{x})\mathrm{dx} + \mathrm{a}$$

$$= -\mathrm{a} + v\mathrm{E}(\mathrm{X}[1 - \mathrm{F}(\mathrm{X})]^{v-1}) + \mathrm{a}$$

$$= v\left(\mathrm{cov}(\mathrm{X},[1 - \mathrm{F}(\mathrm{X})]^{v-1}) + \frac{\mu}{v}\right)$$

$$= v\mathrm{cov}(\mathrm{X}, [1 - \mathrm{F}(\mathrm{X})]^{v-1}) + \mu = \mu - \Delta(v, \mathrm{X}).$$

Therefore $\Delta(v, \mathrm{X}) = \mu - \mathrm{meg}(v,\mathrm{X})$.

The advantage of the dual approach to moments over the regular mode lies in the fact that the power function is applied to a term that is bounded between zero and one, $0 \le \mathrm{F}() \le 1$ as opposed to the variate itself, which is not necessarily bounded.

The following well-known facts will be used later.

Fact 1 The cumulative distribution function (cdf) of $\max(\mathrm{X}_1,\ldots,\mathrm{X}_v)$ is $\mathrm{F}^{v}(\mathrm{x})$.

Fact 2 The cdf of $\min(\mathrm{X}_1,\ldots,\mathrm{X}_v)$ is $[1 - [1 - \mathrm{F}(\mathrm{x})]^{v}]$.

Fact 3 $\mathrm{E}(\mathrm{X}) = \mathrm{a} + \int_{a}^{\infty} [1 - \mathrm{F}(\mathrm{x})]\mathrm{dx}$.

Note that $\mathrm{E}(\mathrm{X}) = \mathrm{mg}(1,\mathrm{X}) = \mathrm{meg}(1,\mathrm{X})$.

For completeness, we note that generally (for unbounded X),

$$\mathrm{E}(\mathrm{X}) = \int_{0}^{\infty} [1 - \mathrm{F}(\mathrm{x}) - \mathrm{F}(- \mathrm{x})]\mathrm{dx}.$$

Using these facts, we get that

$$\mathrm{mg}(v, \mathrm{X}) = \mathrm{E}[\max(\mathrm{X}_1, \ldots, \mathrm{X}_v)] \tag{6.7}$$

and

$$\mathrm{meg}(v, \mathrm{X}) = \mathrm{E}[\min(\mathrm{X}_1, \ldots, \mathrm{X}_v)]. \tag{6.8}$$

This approach leads to an alternative definition of the EG, given that v is restricted to be an integer.

Proposition 6.2 *Let X be distributed in [a,∞), then the extended Gini of (6.1a) can be expressed as*

$$\begin{aligned}\Delta(v, \mathrm{X}) &= \mu - \mathrm{E}\{\min(\mathrm{X}_1,\ldots,\mathrm{X}_v)\} \\ &= \int_a^\infty \{[1-\mathrm{F(x)}] - [1-\mathrm{F(x)}]^v\}\, \mathrm{dx},\end{aligned} \tag{6.9}$$

where $\mu = E(X)$.

Proof Starting with definition (6.1a),

$$\begin{aligned}\Delta(v, \mathrm{X}) &= -v\mathrm{cov}(\mathrm{X}, [1-\mathrm{F(X)}]^{v-1}) \\ &= -v\mathrm{E}(\mathrm{X}[1-\mathrm{F(X)}]^{v-1}) + v\mathrm{E(X)E}([1-\mathrm{F(X)}]^{v-1}).\end{aligned}$$

Using integration by parts we get

$$\mathrm{E}(\mathrm{X}[1-\mathrm{F(X)}]^{v-1}) = \int_a^\infty \mathrm{x}[1-\mathrm{F(x)}]^{v-1}\mathrm{f(x)dx} = \frac{\mathrm{a}}{v} + \frac{1}{v}\int_a^\infty [1-\mathrm{F(x)}]^v\mathrm{dx}$$

and

$$\mathrm{E}([1-\mathrm{F(X)}]^{v-1}) = \frac{1}{v}.$$

Combining the two parts together we get that

$$\begin{aligned}\Delta(v, \mathrm{X}) &= \mu - \int [1-\mathrm{F(x)}]^v\mathrm{dx} - \mathrm{a} = \mu - \mathrm{E}(\min(\mathrm{X}_1,\ldots,\mathrm{X}_v)) \\ &= \int ([1-\mathrm{F(x)}] - [1-\mathrm{F(x)}]^v)\mathrm{dx}.\end{aligned}$$

A related measure which is a combination of the two power functions (6.5) and (6.6) was suggested by González-Abril et al. (2010):

$$\mathrm{msg}(v,X) = \mathrm{mg}(v,X) - \mathrm{meg}(v,X). \tag{6.10}$$

Unlike mg of (6.5) and meg of (6.6) which are (generally) asymmetric with respect to the underlying distribution of X (i.e., $\mathrm{mg}(v, X) \neq \mathrm{mg}(v, -X)$ and $\mathrm{meg}(v, X) \neq \mathrm{meg}(v, -X)$), msg of (6.10) is symmetric with respect to the underlying distribution as shown in the following proposition.

Proposition 6.3 *$msg(v, X) = msg(v, -X)$*

Proof We use (6.7) and (6.8) to get

$$\mathrm{mg}(v, -X) = E[\max(-X_1,\ldots,-X_v)] = -E[\min(X_1,\ldots,X_v)]$$

and

$$\mathrm{meg}(v, -X) = E[\min(-X_1,\ldots,-X_v)] = -E[\max(X_1,\ldots,X_v)].$$

Combining these two equations and (6.10) we get that

$$\mathrm{msg}(v, -X) = \mathrm{msg}(v,X).$$

Equation (6.10) can be further generalized by using different extended Gini parameters for mg and meg, but this generalization will not be discussed here. However, it is worth noting that mg, meg, and msg with $v = 2$ are closely related to GMD as will be shown in the next three propositions.

Proposition 6.4 *Let X be distributed in $[a, \infty)$, then $msg(2, X) = GMD$.*

Proof

$$\mathrm{msg}(2,X) = \int \{[1 - F^2(x)] - [1 - F(x)]^2\}\,dx = 2\int F(x)[1 - F(x)]dx = \mathrm{GMD}. \tag{6.11}$$

The last equality above results from presentation (2.9) of the GMD. Recall that GMD can be expressed as the expected value of $\{\max(X_1,X_2) - \min(X_1,X_2)\}$ (see (2.6)), which is simply msg(2,X) (by (6.7), (6.8), and (6.10)).

In what follows we list some more relationships between the measures.

Proposition 6.5 *Let X be distributed in $[a, \infty)$, then*

$$\mathrm{meg}(2, X) = \mu - 0.5 * \mathrm{GMD}. \tag{6.12}$$

Proof See proposition 6.1.

Proposition 6.6 *Let X be distributed in $[a, \infty)$, then*

$$\mathrm{mg}(2, X) = \mu + 0.5 * \mathrm{GMD}. \tag{6.13}$$

Proof

$$\begin{aligned} \mathrm{mg}(2, \mathrm{X}) &= \int (1 - \mathrm{F}^2(\mathrm{x}))\mathrm{dx} + \mathrm{a} = \int (1 - \mathrm{F}(\mathrm{x}))(1 + \mathrm{F}(\mathrm{x}))\mathrm{dx} + \mathrm{a} \\ &= \int (1 - \mathrm{F}(\mathrm{x}))\mathrm{dx} + \int \mathrm{F}(\mathrm{x})(1 - \mathrm{F}(\mathrm{x}))\mathrm{dx} + a = \mu + 0.5^{*}\mathrm{GMD}. \end{aligned}$$

Alternatively,

$$\begin{aligned} \mathrm{mg}(2, \mathrm{X}) &= \mathrm{msg}(2,\mathrm{X}) + \mathrm{meg}(2, \mathrm{X}) = \mathrm{GMD} + \mu - 0.5 * \mathrm{GMD} \\ &= \mu + 0.5 * \mathrm{GMD}. \end{aligned}$$

One of the uses of the EG that is related to this introduction is in the area of characterization of distributions. Aaberge (2000) looks at the Lorenz curve as a cumulative distribution. For this purpose it is required that the distribution is defined for nonnegative random variables only. Kleiber and Kotz (2002) use presentation (6.6), but they also restrict the distribution to be for nonnegative random variables. They show that any distribution F, having a finite first moment, can be characterized by its sequence of absolute extended Gini indices and (up to a constant) by its sequence of relative extended Gini indices (extended Gini coefficients). Using proposition 6.1 this means that the distribution is characterized (for nonnegative variables) by the sequence of meg's, i.e., by

$$\mathrm{meg}(\nu,\mathrm{X}) = \int_0^\infty [1 - \mathrm{F}(\mathrm{x})]^\nu \mathrm{dx},$$

because $\Delta(\nu, \mathrm{X}) = \mu - \mathrm{meg}(\nu, \mathrm{X})$.

Our interest is in applications of the extended Gini in economics. Therefore we will not deal with characterizations of distributions by the different variants of the EG. In order not to spread the discussion to too many branches we will concentrate on applications of (6.6) which are useful for analyzing expectations of concave functions. These uses are predominant in welfare economics and finance. An additional important advantage of relying on (6.6), which is asymmetric, is that in some of the applications in econometrics we will be interested in testing symmetry between distributions (referred to as exchangeability). For this purpose, using a symmetric measure is not useful because it means that symmetry is imposed on the measure.

6.1.2 The "Income Inequality Approach" Introduction

In a path-breaking paper Atkinson (1970) proved several results concerning the ranking of income distributions according to expected values of all increasing concave social welfare functions. (The term increasing concave social welfare function implies that the social evaluation of the marginal utility of income is positive and declining, so that the society is egalitarian.) One of the important results is that for distributions with equal means all social welfare functions show the same order of average social welfares (that is, the same ordering of inequality) if and only if the appropriate Lorenz curves do not intersect. If, on the other hand, the Lorenz curves intersect then it is always possible to find two alternative social welfare functions which rank average social welfares differently (to be discussed in Chap. 14). This finding by Atkinson has opened the way for using the Lorenz curve as a basic tool in applications of the concept of second-degree stochastic dominance in welfare economics. This tool allows the analyses of the effect of tax reforms and decision under risk to be applied to a wide group of welfare functions, freeing the analysis from the need to specify the welfare function exactly. In addition, Atkinson suggested a new index of inequality. Atkinson's index depends on one parameter, ε, referred to as the degree of inequality aversion. The parameter indicates the social evaluation of the marginal utility of income, which means the social attitude of the researcher. Atkinson's result should be viewed as revolutionary. It is the first time that someone actually proves that it is impossible to measure inequality without specifying explicitly or implicitly, through an inequality measure, a social welfare function. As a result, inequality measurement ceases to be viewed as "scientific" or objective and it returns to be in the domain of political economy. In some sense the implication of Atkinson's findings is that one has to state her social preference first in order to be able, in most relevant cases, to evaluate which society is better off.

Some of the applications of the EG can be interpreted as following Atkinson's spirit. The social attitude of the investigator is represented by the choice of the EG parameter.

The facts that the area between the 45° line and the Lorenz curve is one half of the Gini coefficient and that all EG coefficients can also be presented as weighted summations of the areas enclosed between the 45° line and the Lorenz curve (Yitzhaki, 1983, and below) naturally imply that one can use some extension of the Gini coefficient to imitate Atkinson's index. Atkinson's index initiated research in the area that led to several definitions of extensions of the Gini coefficient which depend on one parameter (see Chakravarty, 1988, 1990: Donaldson & Weymark, 1980, 1983; Kakwani, 1980; Yitzhaki, 1983). Because there are different approaches of spelling the Gini, it took some time to investigate the differences and similarities among the different approaches. Shorrocks (1983) extended Atkinson's result to distributions with unequal means. The relationship between the extended Gini and social welfare function will be presented in Chap. 13. Because the social welfare function is assumed to be a concave function, the interest of economists in the field was only in equations of the type suggested in (6.6).

6.1.3 The "Dual Approach to Risk" Introduction

Expected utility theory is the main paradigm that is used in the analyses of decisions under risk. In some sense it is an extension of the consumer theory which assumes that a rational consumer maximizes her utility function subject to a budget constraint. Faced by a risky environment, it is assumed that the agent maximizes her expected utility. Under certainty, the utility function needs to be an ordinal representation of the preferences, which means that any monotonic increasing transformation of the utility function can also serve as a utility function, representing the same preferences.

This extension does not come without a price in terms of the requirement from the theoretical model and the ability to deduce firm conclusions from the model. In consumer theory, one can divide the effect of a change in a price on the demand into two effects: the substitution effect and the income effect. The substitution effect has a negative sign, while the income effect can be either positive or negative, depending on the assumed utility function. This is a weakness of the theory because if as a result of the modeling effort everything is possible, then the theory is not very helpful in shaping our opinion.

Expected utility is a linear combination of utilities at different levels of income with probabilities serving as the weighting coefficients. This means that the use of expected utility instead of just utility requires assumptions on the behavior of the marginal utility of income. In other words, the assumptions concerning the marginal utility of income determine the behavior under risk.

The determination of behavior under risk by the marginal utility of income does not allow for the possibility that individuals have different attitudes toward risk but still have the same attitude toward income. For example, it may be that some individuals suffer more from being under uncertainty than others, although the utility from income is the same. To bring a concrete example imagine the state of mind of a criminal who is tired of running away from the police. It does not necessarily mean that his attitude toward income or freedom has changed. It may be that his abilities to be in a state of running away and to be in a state of alert have changed. Another example is the occupational choice between being an employee or a self-employed. It need not be that a self-employed person has a higher preference toward income. It may be a result of being less stressed from exposure to risk. Note that the verbal term is “risk bearing,” which indicates the suffering from having to deal with the uncertainty. This suffering is not determined only by the attitude toward income, although the motivation to bear the risk may be influenced by the will to get a reward in terms of income. For this purpose, one may want to disassociate risk bearing from the marginal utility of income because they may represent different attributes.

To sum up—there are two caveats in expected utility theory: the need to specify the marginal utility of income and the need to disassociate the marginal utility of income from being the only factor that determines behavior under risk.

The literature offers several ways for moving forward:

(a) Assuming specific utility functions. Usually this is done implicitly by using a measure of variability as a measure of risk. For example the use of the mean and variance to describe behavior under risk.
(b) Deriving results that hold for a subset of all utility functions. This literature appears under the titles of stochastic dominance or welfare dominance.

(c) Defining "dual utility functions" that enable the user to disengage decision under risk from the marginal utility of income (Yaari, 1987, 1988).

The advantage of the extended Gini over other measures of variability is twofold. First, it is a decomposable measure of variability that resembles the variance. Therefore it can be used for statistical analyses which resemble the analysis that relies on the variance. Second, it can be used in order to provide specific utility functions that fall under (a), (b), or (c). Because in the above-mentioned areas the typical utility function is concave, the relevant presentation of the EG for this approach is (6.6). For example, both under Yaari's dual approach and under the expected utility theory, a key parameter is the certainty equivalent. This parameter describes the value under certainty which is equivalent (from the point of view of the expected utility maximizer) to the value attributed to the uncertain distribution of outcomes. In this sense, the use of the EG enables one to set the risk aversion parameter of the investor as the parameter that determines the way variability (that is, riskiness) is defined, and then to proceed with the statistical analysis.

Using a specific v in (6.6) enables the user to define the certainty equivalent of the distribution. Hence, any decision that is based on (6.6) will not contradict Yaari's dual theory. On the other hand, one can perform statistical analyses, construct portfolios, and analyze policies in a manner that is similar to the variance-based statistical theory. In other words, Atkinson (1970) has demonstrated that in certain cases in order to evaluate variability or inequality one needs to state her preference and only then one can carry out the statistical analysis. We show in the empirical applications that the EG approach is an approach that enables the user to test whether the data obey those restrictions. However, as will be shown later, if, for example, the data come from the multivariate normal distribution, or if the regression curve obeys the linearity rules as assumed, then the choice of the EG parameter does not matter and the EG approach is redundant. The procedure will be as follows: we use the EG parameter as a parameter defining the views of the investigator. We then perform the statistical analyses for several choices of the EG parameter, that is, for several such views, and if the conclusions reached do not change dramatically then we can conclude that one does not have to specify her views in advance. If, on the other hand, the conclusions change then the researchers must agree on their views before reaching any conclusion.

6.2 The Alternative Definitions

Having described three different introductions, we now turn to the main objective of this chapter, namely to replicate the alternative presentations of the GMD, Gini correlation, etc. for the EG in order to use them in the chapters that deal with the applications. We concentrate only on a few alternative definitions which are relevant for the applications. Before doing that we remind the reader that we deal with one specific type of extended Gini, the one that is useful for dealing with concave functions—that is, the one that is derived from (6.6):

$$\mathrm{meg}(v, X) = \int_a^\infty [1 - F(x)]^v dx + a = \mu + v \operatorname{cov}(x,[1 - F(x)]^{v-1}),\ v > 0.$$

That is, $\Delta(v, X) = \mu - \mathrm{meg}(v, X)$ (see proposition 6.1).

We start with the basic definition of the extended Gini (see (6.1a) above).

Presentation 6.2.1 The basic definition of extended Gini with extended Gini parameter v is

$$\Delta(v, X) = -v \operatorname{cov}(X,[1 - F(X)]^{v-1})\ v > 0;\ v \neq 1.$$

This is an extension of (2.15) (up to a constant) which stated that

$$\Delta = 4\ E\ \{X\ (F(X) - E[F(X)])\} = 4\operatorname{cov}[X, F(X)].$$

There is a major difference between the members of the family with $v < 1$ and those with $v > 1$. While the former represent an attitude as represented by convex functions, the latter represent concave functions.

Presentation 6.2.2 The following presentation holds only when v is an integer. Let $X_1,\ldots,X_v$ be v i.i.d. random variables. Then the extended Gini can be presented as

$$\Delta(v, X) = \mu - E[\mathrm{Min}(X_1,\ldots,X_v)]. \tag{6.14}$$

This is an extension of (2.5) which stated that

$$\Delta(2, X) = 0.5{*}\mathrm{GMD} = \mu - E\ [\mathrm{Min}\ (X_1, X_2)],\ \text{to all integers } v = 3, \ldots, n.$$

The equivalence between presentations (6.2.1) and (6.2.2) was proved in the previous section (presentation 6.2).

Presentation 6.2.3 The extended Gini is a weighted average of the distances between the line of equality (LOE) and the absolute Lorenz curve (ALC) (see Chap. 2 for the definitions of LOE and ALC).

Specifically,

$$\Delta(v, X) = v(v - 1) \int_0^1 (1 - p)^{v-2} (\mu\, p - \mathrm{ALC}(p))dp, \tag{6.15}$$

where $\mathrm{ALC}(p) = \int_{-\infty}^{x(p)} x dF(x)$ and $p = \int_{-\infty}^{x(p)} dF(x) = F(x(p))$.

Claim Presentation (6.2.3) is equivalent to presentation (6.2.1).

Proof Using integration by parts, with

$$u = \mu p - ALC(p);\ du = (\mu - x(p))dp; \quad dv = (\nu - 1)(1 - p)^{\nu-2}dp;$$
$$v = -(1-p)^{\nu-1}$$

we get that the right-hand side of (6.15) is

$$-\nu(1-p)^{\nu-1}(\mu p - ALC(p))\,|_0^1 + \nu\int_0^1 (1-p)^{\nu-1}(\mu - x(p))dp.$$

The first term is equal to zero. The second term can be handled by a change of variable technique with p = F(x(p)). Using it we get

$$-\nu\int\{(x - \mu)(1 - F(x))^{\nu-1}\}\, f(x)dx = -\nu \mathrm{cov}(X, (1 - F(X))^{\nu-1}).$$

In the last step we used the fact that cov(X,Y) = E[(X − μ_X)Y].

6.3 The Properties of the Extended Gini Family

The properties of the members of the EG family are similar to the properties of the GMD except for two major issues. To save space and proofs, we will concentrate on the differences between the EG and the GMD.

The first difference is concerned with an asymmetric property with regard to the underlying distribution as stated in the following claim.

Claim

(a) $\Delta(2, X) = \Delta(2, -X)$ for all F.
(b) In general

$$\Delta(\nu, X) \neq \Delta(\nu, -X) \quad \text{for} \quad \nu \neq 2. \tag{6.16}$$

(c) Equality between the two sides of (6.16) holds for symmetric distributions.

Proof

(a) $F_{-X}(-x) = P(-X \leq -x) = P(X \geq x) = 1 - F_X(x)$. Therefore,

$$\Delta(2, -X) = -2\mathrm{cov}(-X,[1 - F_{-X}(-X)]) = 2\mathrm{cov}(X,F(X))$$
$$= -2\mathrm{cov}(X,[1 - F(X)]) = \Delta(2,X).$$

(b) We show this by an example. Let $F(X) = X^2$, for X uniformly distributed in [0,1] and let $\nu = 3$. Then

$$\Delta(3,X) = -3\text{cov}(X,[1-F(X)]^2) = -3E(X[1-F(X)]^2) + 3 \times \frac{2}{3} \times \frac{1}{3}$$

Now, $E(X[1-F(X)]^2) = \int_0^1 x(1-x^2)^2 2x dx = \frac{16}{105}$

So $\Delta(3,X) = -3 \times \frac{16}{105} + \frac{2}{3} = \frac{22}{105}$.

On the other hand,

$$\Delta(3,-X) = -3\text{cov}(-X,[1-F_{-X}(-X)]^2) = 3\text{cov}(X,F^2{}_X(X))$$
$$= 3\text{cov}(X,F^2(X)) - 3E(X)E(F^2(X)) = 3\text{cov}(X,F^2(X)) - \frac{2}{3}.$$

$E(XF^2(X)) = \frac{2}{5}$, so we get that

$$\Delta(3,-X) = 3 \times \frac{2}{5} - \frac{2}{3} = \frac{8}{15} \neq \frac{22}{105}.$$

(c) $\Delta(\nu, X) = -\nu E[(X - \mu)(1-F(X))^{\nu-1}]$

and

$$\Delta(\nu,-X) = \nu E[(X - \mu)(1 - F_{-X}(-X))^{\nu-1}].$$

We need to show that when F is symmetric, then $\Delta(\nu, X) = \Delta(\nu, -X)$. Let X be symmetrically distributed in $[\mu - a, \mu + a]$. Using the fact that $F_{-X}(-X) = 1 - F_X(X)$ we need to show that $E[(X - \mu)(F_X(X))^{\nu-1}] + E[(X - \mu)(1 - F_X(X))^{\nu-1}] = 0$.

$$E[(X - \mu)(F_X^{\nu-1}(X))] + E[(X - \mu)(1 - F_X(X))^{\nu-1}]$$
$$= \int_{\mu-a}^{\mu+a} [(x - \mu)[F_X^{\nu-1}(x) + (1 - F_X(x))^{\nu-1}]f_X(x)dx.$$

Using the change-of-variable technique with $y = (x - \mu)$ we get for $Y = (X - \mu)$

$$= \int_{-a}^{a} y[F_Y^{\nu-1}(y) + (1 - F_Y(y))^{\nu-1}]f_Y(y)dy.$$

We now use the change-of-variable technique on the first term, with $y = (-t)$ to get, for $Y = (-T)$

$$\int_{-a}^{a} yF_Y^{\nu-1}(y)f_Y(y)dy = -\int_{a}^{-a} (-t)F_{-T}^{\nu-1}(-t)f_{-T}(-t)dt = -\int_{-a}^{a} t(1-F_T(t))^{\nu-1}f_T(t)dt.$$

We now use the symmetry of Y about zero: $Y \stackrel{d}{=} (-Y) = T$(where $\stackrel{d}{=}$ means "equal in distribution"), which completes the proof.

The other property in which the members of the EG family are different from the GMD is the possibility to be decomposed according to population subgroups. While the GMD of a union of several subpopulations can be decomposed into the individual contributions of the subpopulations (plus some additional terms), such a decomposability of the EG is not available to the best of our knowledge. An intuitive explanation of the difficulty is because one applies a power function to the cumulative distribution function. Therefore the additive property of the union of several subpopulations' cumulative distributions as a function of the individuals' distributions is lost. Future research may shed some light with respect to the question of whether an additional parameter of interest may be hiding behind the decomposition of the EG with respect to subpopulations.

The metric of the extended Gini. We have shown in Chap. 2 that the metric used to derive the variance is the Euclidean metric, while the metric of the GMD is the "city block." The metric that leads to the EG can be referred to as "hilly city block" or alternatively, the "condensed city block." Similar to the GMD case, one is allowed to move east–west or south–north. But while under the GMD one can substitute a centimeter of south–north by a centimeter of east–west along the equal-GMD curve, under the EG, the slope of the equal-EG curve is still a constant, but its magnitude depends on ν. The bigger the value of ν, the bigger the slope of the line representing the equal-EG curve. We now give a geometrical interpretation (Fig. 6.1) similar to the presentation given in Chap. 2 (Fig. 2.2).

The minimum number of observations needed to plot equal-EG curves is three. In order to simplify the presentation it is assumed that all observations are positive. We denote them by $0 < x_1 < x_2 < x_3$. To be able to present the equal-EG curve, we rely on the fact that the EGs are not sensitive to the addition of a constant. Therefore we can define $\delta_1 = x_2 - x_1$ and $\delta_2 = x_3 - x_2$.

It is easy to see that an equal-EG curve will be a linear function of the form:

$$C = c_1(\nu)\delta_1 + c_2(\nu)\delta_2, \tag{6.17}$$

where C determines the level of the curve (i.e., the value of the EG)), while $c_1(\nu)$ and $c_2(\nu)$ are constants determined by the selection of ν. Hence, the equal-EG curves are all linear with the slope determined by ν. Having δ_1 on the horizontal axis and δ_2 on the vertical axis, the larger the value of ν the larger the absolute value of the (negative) slope. For this reason we refer to the metric as "hilly city block" metric.

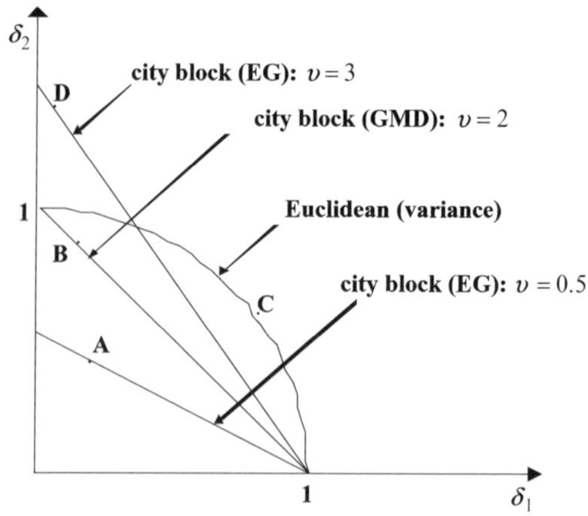

Fig. 6.1 Equal GMD, EG, and variance curves. Source: Yitzhaki and Schechtman (2005), Fig. 1, p. 408. Reprinted with permission by Metron International Journal of Statistics

6.4 Alternative Presentations of the Extended Gini Covariances and Correlations

Let X and Y be two random variables having continuous marginal distribution functions F and G, respectively, and a joint distribution function H(X,Y). Using the covariance presentation of the EG (6.1a), the definitions of the EG equivalents of covariances and correlations follow immediately

$$\Delta C(\nu, X, Y) = -\nu \operatorname{cov}(X,[1 - G(Y)]^{\nu-1}), \tag{6.18}$$

and

$$\Delta C(\nu, Y, X) = -\nu \operatorname{cov}(Y,[1 - F(X)]^{\nu-1}).$$

Two additional presentations of the EG covariance will be used in this chapter. We present them below.

Presentation 6.4.1

$$\Delta C(\nu, X, Y) = \nu(\nu - 1)\iint (H(x,y) - F(x)G(y))(1 - G(y))^{\nu-2} dG(y)dx. \tag{6.19}$$

This presentation is equivalent to the basic definition (6.18). The proof can be found in Appendix 6.1.

The second presentation that will be used, which is similar to the definition of the extended Gini by the ALC, is based on the absolute concentration curve.

Presentation 6.4.2 Let $g(x) = \mu_{Y.X} = E\{Y|X{=}x\}$. Then

$$\mathrm{cov}(Y, -[1-F(X)]^{v-1}) = (v-1)\int_0^1 (1-p)^{v-2}(\mu_Y p - ACC_{Y.X}(p))dp, \quad (6.20)$$

where

$$ACC_{Y.X}(p) = \int_{-\infty}^{x(p)} g(x)dF(x) \quad \text{and } p = \int_{-\infty}^{x(p)} dF(x). \quad (6.21)$$

The proof is similar to the proof of the equivalence between presentations 6.2.1 and 6.2.3 above with the main modification: ALC should be replaced by ACC.

Having defined the EG covariances we now move to define the extended Gini correlations. The extended Gini correlations are defined as

$$\Gamma(v, X,Y) = \frac{\mathrm{cov}(X, [1-G(Y)]^{v-1})}{\mathrm{cov}(X,[1-F(X)]^{v-1})};$$
$$\Gamma(v, Y, X) = \frac{\mathrm{cov}(Y, [1-F(X)]^{v-1})}{\mathrm{cov}(Y,[1-G(Y)]^{v-1})}. \quad (6.22)$$

An alternative definition using (6.19) is

$$\Gamma(v, X,Y) = \frac{(v-1)\int\int (H(x,y) - F(x)G(y))(1-G(y))^{v-2}dG(y)dx}{\mathrm{cov}(X, -[1-F(X)]^{v-1})}. \quad (6.23)$$

Next we list the main properties of the family of correlation coefficients. Proofs which are similar to the ones for the special case v = 2 and are discussed in Chap. 3 will not be repeated here.

The main properties of $\Gamma(v, X, Y)$ are

1. Let F and G be the cumulative distribution functions of X and Y, respectively. Then, for every joint distribution function H(X,Y) and for every v, $\Gamma(v, X,Y) \le 1$ for all (X,Y). (Note: it is not bounded in [−1,1]).
2. If Y is a monotone increasing function of X, then $\Gamma(v, X,Y) = 1$ for all v.
3. If X and Y are independent, then $\Gamma(v, X,Y) = \Gamma(v, Y, X) = 0$ for all v.
4. $\Gamma(v, X, Y)$ is invariant under all strictly monotonic increasing transformations of Y.
5. Let (X,Y) have a bivariate normal distribution with correlation coefficient ρ, then $\Gamma(v, X,Y) = \Gamma(v, Y,X) = \rho$ for all v.
6. Invariance under exchangeability. Let (X,Y) be exchangeable up to a linear transformation, then $\Gamma(v, X,Y) = \Gamma(v, Y,X)$ for all v.

By exchangeability it is meant that there exist a, b, c, and d (a > 0; c > 0) such that (X,Y) and (aY + b, cX + d) are identically distributed. Intuitively, "exchangeability" means that there exists a linear transformation that makes the shapes of the joint distributions identical. In particular, it is meant that the shapes of the marginal and conditional distributions are identical (see Chap. 3 for more details).

The proofs of properties 2, 3, 4, and 5 are very similar to the proofs of the equivalent properties of the Gini correlations given in Chap. 3 and will not be repeated here. The proofs of 1 and 6 are given below.

Proof of property 1 The EG correlation coefficient is defined as

$$\Gamma(v, X, Y) = \frac{\text{cov}(X, -[1 - G(Y)]^{v-1})}{\text{cov}(X, -[1 - F(X)]^{v-1})}.$$

The proof is based on the following claim.

Claim Given the marginal distribution functions of X and Y, and assuming that the densities exist and are positive everywhere, cov(X,Y) is maximal when $E\{Y|X=x\}$ is an increasing function of X.

Proof of the claim (See Sect. 3.3)

We now turn to the proof of property 1.

We need to show that

$$\text{cov}(X, -[1 - G_Y(Y)]^{v-1}) \leq \text{cov}(X, -[1 - F_X(X)]^{v-1}).$$

Note that $G_Y(Y)$ has a uniform distribution over [0,1] (that is, it is a U(0,1) random variable), hence $U = 1 - G_Y(Y)$ is also U(0,1), and

$$E(U^{v-1}) = \int_0^1 u^{v-1} du = \frac{1}{v}.$$

Therefore,

$$E[1 - F_X(X)]^{v-1} = E[1 - G_Y(Y)]^{v-1} = \frac{1}{v}.$$

Note that X and $-[1 - F_X(X)]^{v-1}$ are nondecreasing functions of X. By the claim, $\text{cov}(X, -[1 - G_Y(Y)]^{v-1})$ achieves its maximal value when $-[1 - G_Y(Y)]^{v-1}$ is an increasing function of X. Now, $-[1 - G_Y(Y)]^{v-1}$ is nondecreasing if and only if $G_Y(Y)$ is a nondecreasing function of X, which implies $F_X(X) = G_Y(Y)$.

This means that the maximum is achieved at $\text{cov}(X, -[1 - F_X(X)]^{v-1})$, which completes the proof.

We note that while in the case $\nu = 2$ (that is, the Gini correlation coefficient) the lower bound is (-1) and the proof is similar to the one for the upper bound, in the case of the extended Gini correlation this is not the case, as is shown below by an example.

Proof of property 6 Under exchangeability, H(X,Y) = H(Y,X). The proof is similar to the proof for the GMD (see Chap. 3), except that each F is replaced by $-(1 - F(X))^{\nu-1}$ and $E(F(X)) = 0.5$ is replaced by $E(-[1 - F(X)]^{\nu-1}) = -\frac{1}{\nu}$.

Additional properties

1. An alternative sufficient condition for the equality of the extended Gini correlation coefficients (namely $\Gamma(\nu, X,Y) = \Gamma(\nu, Y,X)$) is that $ACC_{X,Y}(p) = ACC_{Y,X}(p)$ for all p, where $ACC_{X,Y}(p)$ is the absolute concentration curve as defined in Chap. 5 and in (6.21). Using (6.21), the proof is immediate. We call distributions for which $ACC_{X,Y}(p) = ACC_{Y,X}(p)$ for all p "well-behaved" distributions.
2. It is interesting to note that for the special case $\nu = 2$, if $E\{Y|X=x\}$ is a monotonic decreasing function of X, then $\Gamma(\nu, X,Y) = -1$ (the lower bound for the special case), but this does not hold for the general case, as the following example shows

   Let $Y = -X$. For this case, $\Gamma(\nu, X,Y) = -1$ implies that

   $$\text{cov}(X,[F(X)]^{\nu-1}) = -\text{cov}(X, -[1 - F(X)]^{\nu-1}).$$

   For $\nu = 2$ the condition holds.
   For $\nu = 3$ the condition translates into whether or not $\text{cov}(X,F(X)) = \text{cov}(X,F^2(X))$, which is generally not true. For example, choose $F(x) = x^2$, for $0 \le x \le 1$.
   Then $\text{cov}(X, F(X)) = 1/15$, but $\text{cov}(X, F^2(X)) = 4/63$. The lower bound for the general case is discussed below.
   The family of correlation measures $\Gamma(\nu, X, Y)$ differs from the classical correlation (Pearson) ρ in three major properties.

   (a) $\Gamma(\nu, X, Y) = 1$ whenever Y is an increasing function of X, not necessarily linear. (This property holds for the Spearman coefficient as well).
   (b) Let F and G be cumulative distribution functions of X and Y, respectively. Then, there exists a joint distribution function H(X,Y) such that for every ν, $\Gamma(\nu, X, Y) = 1$. The fact that the upper bound of 1 can always be achieved is helpful as a benchmark. See the discussion on the proper bounds of Pearson, Spearman, and Gini correlation coefficients by, for example, Schechtman and Yitzhaki (1999). The proof can be found in Schechtman and Yitzhaki (2003) and in Appendix 6.3.

   (c) The lower bound of $\Gamma(\nu, X,Y)$ is given by

   $$\Gamma(\nu, X, Y) \ge \frac{\int F(x)(F^{\nu-1}(x) - 1)dx}{\int ((1 - F(x))(1 - (1 - F(x))^{\nu-1})dx}, \tag{6.24}$$

and is achieved when $Y = -X$. Two special cases are worth mentioning: the case where X comes from a symmetric distribution and the case with $v = 2$. In these two cases, the lower bound is -1, same as for the classical correlation coefficient. However, in general the lower bound depends on the cumulative distribution—the more concave it is, the lower it can get. The proof can be found in Schechtman and Yitzhaki (2003) and in Appendix 6.3. An intuitive explanation of property (c) will be given following the decomposition of the EG of a sum of random variables (end of Sect. 6.5).

The lower bound can be expressed, for the case where v is an integer, as

$$\frac{-\Delta(v, -X)}{\Delta(v, X)} = \frac{\mu - E[\max(X_1,\ldots,X_v)]}{\mu - E[\min(X_1,\ldots,X_v)]}.$$

We illustrate this lower bound for X exponentially distributed, with a scale parameter of unity. Kleiber and Kotz (2002) show that $\Delta(v, X) = 1 - \frac{1}{v}$. It can be shown that

$$E[\max(X_1,\ldots,X_v)] = \sum_{k=1}^{v} (-1)^{k+1} \binom{v}{k} \frac{1}{k}.$$

Therefore, the lower bound for $v = 3$ is $\frac{1-\frac{11}{6}}{1-\frac{1}{3}} = -\frac{5}{4}$. The lower bound for $v = 4$ is $-13/9$, smaller than the bound for $v = 3$ because the cumulative distribution is more concave.

6.5 The Decomposition of the Extended Gini

Let (Y_1,Y_2) have a continuous bivariate distribution. In what follows we show that if Y_0 is a linear combination of Y_1 and Y_2, then the extended Gini coefficient of Y_0 can be decomposed in a way which is similar to the decomposition of the coefficient of variation, plus an additional term which reflects the asymmetry of the EG correlation coefficient. Note the change in notation. We use (Y_1,Y_2) rather than (X,Y) because the decomposition can easily be extended so that Y_0 is a linear combination of Y_1, $Y_2,\ldots,Y_k$.

Let $Y_0 = \alpha Y_1 + \beta Y_2$, where α and β are given constants. Then

(a) $$G_0^2 - [\alpha D_{10} G_1 + \beta D_{20} G_2] G_0 = \alpha^2 G_1^2 + \beta^2 G_2^2 + \alpha \beta G_1 G_2 (\Gamma(v,Y_1,Y_2) + \Gamma(v,Y_2,Y_1)), \quad (6.25)$$

where G_i i = 0,1,2 are the extended Gini coefficients and

$$\Gamma(v,Y_i,Y_j) = \frac{\mathrm{cov}(Y_i, [1 - F_j(Y)]^{v-1})}{\mathrm{cov}(Y_i, [1 - F_i(Y)]^{v-1})}$$

for i,j = 0,1,2 are the extended Gini correlations, $D_{0i} = \Gamma(v, Y_0, Y_i) - \Gamma(v, Y_i, Y_0)$ for i = 1,2 are the differences between the extended Gini correlations.

(b) Provided that $D_{0i} = 0$ for i = 1,2 and that the two EG correlations between Y_1 and Y_2 are equal the following decomposition holds:

$$G_0^2 = \alpha^2 G_1^2 + \beta^2 G_2^2 + 2\,\alpha\,\beta G_1 G_2 \Gamma, \tag{6.26}$$

where $\Gamma = \Gamma(v, Y_1, Y_2) = \Gamma(v, Y_2, Y_1)$ are the extended Gini correlation coefficients between Y_1 and Y_2 (and between Y_2 and Y_1).

The structure of (6.26) is identical to the decomposition of the variance, with G_i^2 substituting for the variance and Γ substituting for Pearson's correlation. Note that by a proper choice of α and β, (6.25) and (6.26) can be applied both to absolute measures like the EG and to relative measures such as the extended Gini coefficient.

The proof of the decomposition is given by Schechtman and Yitzhaki (2003) and in Appendix 6.4.

Clearly, property (b) of the claim is a special case of property (a). However, because of its similarity to the variance decomposition, the practical importance of case (b) is much greater than that of the general case because it implies that ANY variance-based model can be replicated, using the extended Gini as a substitute for the variance as a measure of dispersion. It is worthwhile to mention that (6.26) is easier to work with than (6.25). The question is, however, how restrictive the assumptions $D_{ij} = 0$ really are. Schechtman and Yitzhaki (1987) showed that a sufficient condition for $D_{ij} = 0$ is that the variables are exchangeable up to a linear transformation. However, this is only one possible sufficient condition. Clearly, this sufficient condition does not exhaust all possibilities. The equality of the two ACCs between Y_i and Y_j is another sufficient condition. Further research is needed to find the necessary and sufficient conditions for $D_{ij} = 0$ if possible. Note also that under bivariate normality $D_{ij} = 0$.

While equation (6.26) enables one to imitate variance-based models, (6.25) opens new possibilities. It is worth stressing that each violation of the condition $D_{ij} = 0$ is reflected in a specific term in the decomposition (6.25). Therefore one can identify the random variables whose distributions are not "well behaved" and attach a quantitative value to the violation.

We conclude this section by using the decomposition to give an intuitive explanation to why the lower bound of the EG correlation may be smaller than (−1). Let us start decomposing the identity Y = X+(−X) = 0. Clearly $\Delta(v, Y) = 0$ by construction.

Also $D_{XY} = D_{-XY} = 0$. Hence

$$0 = \Delta^2(v, X) + \Delta^2(v, -X) + \Delta(v, X)\,\Delta(v, -X)\,\Gamma(v, X, -X) + \Delta(v, X)\,\Delta(v, -X)\,\Gamma(v, -X, X).$$

Denote $C = \Gamma(v, X, -X) + \Gamma(v, -X, X)$ then we can rewrite the equation as

$$0 = \Delta^2(v,X) + \Delta^2(v,-X) + C\,\Delta(v,X)\Delta(v,-X).$$

We now show that $C < -2$, implying that at least one of $\Gamma(v,X,-X)$ and $\Gamma(v,-X,X)$ is smaller than (-1). Assuming $\Delta(v,X) \neq \Delta(v,-X)$, which may occur whenever a distribution is not symmetric around its mean then $\Delta^2(v,X) + \Delta^2(v,-X) > 2\,\Delta(v,X)\Delta(v,-X)$ and we get that $0 > 2\Delta(v,X)\Delta(v,-X) + C\,\Delta(v,X)\Delta(v,-X)$. Therefore $0 > 2 + C$ and $C < -2$.

6.6 Stochastic Dominance and the Extended Gini

In this section we show that a necessary condition for second-degree stochastic dominance (SSD) is that the mean minus the EG of the dominating distribution is not lower than the mean minus the EG of the dominated distribution for all v. However, even if this condition holds for all v it does not form a sufficient condition for dominance. In addition, we note that a comparison of the mean minus the EG as defined in (6.6) is equivalent to a special case of Yaari's (1987, 1988) decision functions. Therefore if distribution A dominates distribution B by at least one EG parameter according to (6.6) then distribution A cannot be dominated by distribution B according to Yaari's criteria. Therefore, each EG can be used to construct a necessary condition for dominance both according to expected utility theory and according to Yaari's dual theory.

We start with a definition of SSD.

Let X, Y be two continuous random variables with cumulative distributions $F_X(\,)$ and $F_Y(\,)$, respectively. It is assumed that the expected values of the distributions are bounded ($\mu_X < \infty$; $\mu_Y < \infty$). Let U be the class of increasing concave functions and let V be the class of increasing convex functions. We say that X dominates (by SSD) Y for concave (convex) functions if

$$E\{u(X)\} \geq E\{u(Y)\} \quad \text{for all } u \in U. \tag{6.27}$$

$$E\{v(X)\} \geq E\{v(Y)\} \quad \text{for all } v \in V. \tag{6.28}$$

The intuition behind the necessary condition is straightforward using simple geometry. It is based on the following three facts.

1. A necessary and sufficient condition for SSD is that the absolute Lorenz curve (ALC) of the dominating distribution lies above (not lower than) the ALC of the dominated distribution.
2. The area below the line of equality (LOE) is equal to the mean of the distribution multiplied by a constant.
3. The extended Gini is a weighted average of the distances between the LOE and the ALC (see Chap. 2 for the definitions of LOE and ALC and presentation 6.2.3 in Sect. 6.2).

Using these facts, the weighted area below the ALC is equal to the mean multiplied by a constant minus the value of the EG. A necessary condition for one curve to be above the other is that the area below the curve is bigger for the dominating curve (i.e., for the dominating distribution). In order not to bombard the reader with redundant proofs, we will state the propositions in this section without proofs.

Proposition 6.7

$$\text{Inequality (6.27) holds iff} \quad \int_{-\infty}^{z} [F_Y(t) - F_X(t)]dt \geq 0 \quad \text{for all z.}$$

$$\text{Inequality (6.28) holds iff} \quad \int_{z}^{\infty} [F_Y(t) - F_X(t)]dt \geq 0 \quad \text{for all z.}$$

A proof of the first part of the proposition (for concave functions) can be found in Hanoch and Levy (1969) and in Rothschild and Stiglitz (1970) and the proof of the second part (for convex functions) can be found in Spencer and Fisher (1992). Yitzhaki (1983, 1999) gives more convenient proofs using the Lorenz curve and Gini.

Conditions (6.27) and (6.28) are related to each other by

$$\int_{-\infty}^{Z} [F_Y(t) - F_X(t)]dt + \int_{Z}^{\infty} [F_Y(t) - F_X(t)]dt = \mu_X - \mu_Y. \tag{6.29}$$

It is easy to see that if $\mu_X = \mu_Y$ then

$$\int_{-\infty}^{Z} [F_Y(t) - F_X(t)]dt \geq 0 \quad \text{for all z iff} \quad \int_{z}^{\infty} [F_Y(t) - F_X(t)]dt \leq 0 \quad \text{for all z,}$$

and we may conclude that for distributions with equal means the rules are symmetric in the sense that "X dominates Y for concave (convex) functions" is equivalent to "Y dominates X for convex (concave) functions." However, if $\mu_X \neq \mu_Y$, this symmetry need not hold. That is, X may dominate Y for concave functions, without Y dominating X for convex functions. To formally prove the asymmetry note that $\mu_X \geq \mu_Y$ is a necessary condition for dominance in both cases. Hence, if $\mu_X > \mu_Y$ then X may dominate Y for concave or convex functions, while Y cannot dominate X for both types of functions.

Further insight into the origin of the asymmetry can be gained by presenting the necessary and sufficient conditions in terms of ALCs. The ALC is defined by

$$\text{ALC(p)} = \int_{-\infty}^{Z} t\, f(t)dt, \tag{6.30}$$

where $z = F^{-1}(p)$ and f( ) is the density function.

Proposition 6.8 states the necessary and sufficient conditions for second-order dominance of concave functions in terms of ALCs.

Proposition 6.8

$$\mathrm{ALC_X(p)} \geq \mathrm{ALC_Y(p)} \quad \text{for all } 0 \leq p \leq 1 \text{ iff } \int_{-\infty}^{z} [F_Y(t) - F_X(t)]dt \geq 0 \quad \text{for all z.}$$

For the proof see Lambert (2001), Shorrocks (1983), and Yitzhaki and Olkin (1991).

To construct necessary and sufficient conditions for dominance of convex functions in terms of ALCs it is convenient to define the ALC in terms of $h = 1 - p$. Under this formulation the observations are sorted in descending (instead of ascending) order. Then, the absolute (descending-order) Lorenz curve portrays the cumulative value of the largest h percent of the population (the appropriate ALC is above the diagonal). In this case, the absolute (descending-order) Lorenz curve, DALC, is a mirror image of the conventional ALC

$$\mathrm{DALC(h)} = \int_{z}^{\infty} tf(t)dt \quad \text{where } z = z(h) = H^{-1}(h) = F^{-1}(1-h). \tag{6.31}$$

Proposition 6.9 states the necessary and sufficient conditions for second-order dominance of convex functions.

Proposition 6.9

$$\mathrm{DALC_X(h)} \geq \mathrm{DALC_Y(h)} \quad \text{for all } h, 0 \leq h \leq 1 \text{ iff } \int_{z}^{\infty} [F_Y(t) - F_X(t)]\, dt \geq 0 \quad \text{for all z.} \tag{6.32}$$

Proof See Appendix 6.2. (The proof is based on Yitzhaki, 1999).

The necessary and sufficient conditions for second-order dominance for convex functions are stated in terms of descending-order ALCs, while those for concave functions are stated in terms of ascending-order ALCs. Both state that dominance means that one curve cannot intersect the other. Hence, the only difference between the conditions for concave and convex functions is the definition of the ALC. This difference also shows up in forming necessary conditions. A necessary condition for dominance is that the area below the dominating curve cannot be smaller than the area below the dominated curve. As shown by Yitzhaki (1982a, 1982b, 1982c) this means that $\mu_X \geq \mu_Y$ and $\mu_X(1 - G_X) \geq \mu_Y(1 - G_Y)$ (G is the Gini coefficient or EG coefficient) are necessary conditions for dominance of X over Y for concave functions.[2]

By exactly the same method one can show that $\mu_X \geq \mu_Y$ and $\mu_X(1 + G_X) \geq \mu_Y(1 + G_Y)$ are necessary conditions for dominance of X over Y for convex

[2] To see the intuition note that $\mu_X \geq \mu_Y$ is the value of the ALC at the upper right-hand corner, while $\mu_X(1 - G_X) \geq \mu_Y(1 - G_Y)$ implies that the area below the ALC for X is bigger than that for Y.

functions. Note that these conditions do not form necessary conditions for first-degree stochastic dominance (FSD). FSD allows dominance by partially convex and partially concave functions.

Our final point is that fulfilling all necessary conditions for SSD by using all members of the EG family does not provide a sufficient condition for dominance. This is proved by supplying a counter example (Yitzhaki, 1983) as follows.

Let $F_1(Y)$ be the following distribution:

$$F_1(y) = \begin{pmatrix} 0 & \text{for} & y<0.49 \\ 0.99 & \text{for} & 0.49 \le y<0.49+c \\ 1 & \text{for} & y \ge 0.49+c \end{pmatrix},$$

where $c > 1$ is a given constant. Let $F_2(y)$ be the uniform distribution on [0,1]. Then for $F_1(y)$, $a = 0.49$, where a is the lower bound of Y, and

$$meg(v,X) = \int_a^{\infty} [1 - F_1(y)]^v dy + a = \int_{0.49}^{0.49+c} (1 - 0.99)^v dy + 0.49 = (0.01)^v c + 0.49$$

while for $F_2(y)$, $a = 0$ and

$$\text{meg}(v,Y) = \int_0^{\infty} [1 - F_2(y)]^v dy = \frac{1}{1+v}.$$

meg for F_1 is greater than meg for F_2 for all $v \ge 1$, although F_1 does not SSD dominate F_2. (Note that $\int_0^{0.99} [F_1(y) - F_2(y)]dy = 0.005$).

Going back to Yaari's theory, an inspection of (6.6) reveals that the meg is a special case of Yaari's decision functions in the areas of risk and income distribution. Hence, dominance by one EG implies dominance by one possible Yaari's decision function.

6.7 Summary

The GMD can be extended in several alternative ways into families of variability and inequality measures that share most of its properties. We have chosen only one such family—a covariance-based extension that is intended to represent expected values according to concave functions. This extension shares many properties with the GMD, except for two: it is asymmetric with respect to the distribution and it does not decompose with respect to population subgroups.

The motivation for such an extension varies according to the area of application: in finance and welfare economics it is intended to adjust the modeling and estimation procedure to the theoretical requirements in each area, while in the area of econometrics the primary motive is sensitivity analysis and investigating the linearity of the model. Each extension to a family implies that some of the properties of the GMD are lost while others are added. For example, one could extend the Gini in a way in which the symmetric property is not lost. But then the adjustment to finance and social welfare analysis is lost. Our extension kept the adjustments to welfare economics and finance and also kept the decomposition properties of a linear combination of random variables. However, the property of neatly decomposing the GMD of a population that is composed of several subpopulations is lost when one moves to the extended Gini. Further research is needed to determine whether this property is lost forever or maybe a new meaningful decomposition will be developed, which will enable the family to reveal another property of the data in the same way that ANOGI reveals more than ANOVA.

Appendix 6.1

Presentation 6.4.1

$$\begin{aligned}\Delta C(\nu, X,Y) &= -\nu\, \mathrm{cov}(X,[1-G(Y)]^{\nu-1})\\ &= \nu(\nu-1)\int\!\!\int\{H(x,y)-F(x)G(y))\}\,(1-G(y))^{\nu-2}dG(y)dx.\end{aligned}$$

Proof of presentation 6.4.1 Let (X_1,Y_1) and (X_2,Y_2) be i.i.d. random variables. As shown by Kruskal (1958),

$$\begin{aligned}2\mathrm{cov}(X,Y) &= E[(X_1-X_2)(Y_1-Y_2)]\\ &= E\{\int\!\!\int[I(u,X_1)-I(u,X_2)][I(t,Y_1)-I(t,Y_2)]dudt\},\end{aligned}$$

where

$$I(u,X) = \left\{\begin{matrix}1 & \mathit{if}\ \ u\le X\\ 0 & \text{otherwise}\end{matrix}\right\}.$$

There are two types of components in the integral:

(a) $I(u,X_1)I(t,Y_1)$, where X_1 and Y_1 are dependent, and
(b) $I(u,X_1)I(t,Y_2)$, where X_1 and Y_2 are independent.

Replacing Y by $(1-G(Y))^{(\nu-1)}$ we get

For (a)

$$\begin{aligned}
&E[I(u,X_1)I(t,(1-G(Y_1))^{(\nu-1)}] = P[u \le X_1, t \le (1-G(Y_1))^{(\nu-1)}] \\
&\quad = p(u \le X_1, G(Y_1) \le 1 - t^{1/(\nu-1)}) = P(u \le X_1, Y_1 \le G^{-1}(1 - t^{1/(\nu-1)})) \\
&\quad = P(Y_1 \le G^{-1}(1 - t^{1/(\nu-1)})) - P(Y_1 \le G^{-1}(1 - t^{1/(\nu-1)}), X_1 \le u) \\
&\quad = 1 - t^{1/(\nu-1)} - H(u, G^{-1}(1 - t^{1/(\nu-1)})).
\end{aligned}$$

For (b)

$$\begin{aligned}
&E[I(u,X_1)I(t,(1-G(Y_2))^{\nu-1})] = P(u \le X_1)P(t \le (1-G(Y_2))^{\nu-1}) \\
&\quad = P(u \le X_1)P(G(Y_2) \le 1 - t^{1/(\nu-1)}) = (1-F(u))(1 - t^{1/(\nu-1)}) \\
&\quad = 1 - F(u) - t^{1/(\nu-1)} + F(u)t^{1/(\nu-1)}.
\end{aligned}$$

Combining the pieces and substituting into the integrals, we get

$$\begin{aligned}
&2\mathrm{cov}(X,(1-G(Y))^{\nu-1}) = 2\int_0^\infty\int_0^1 \{[1 - t^{1/(\nu-1)} - H(u,G^{-1}(1 - t^{1/(\nu-1)}))] \\
&- [1 - F(u) - t^{1/(\nu-1)} + F(u)t^{1/(\nu-1)}]\}\, dudt = 2\int_0^\infty\int_0^1 [F(u) - H(u,G^{-1}(1 - t^{1/(\nu-1)})) \\
&- F(u)t^{1/(\nu-1)}]dudt.
\end{aligned}$$

Substituting $t^{1/(\nu-1)} = 1 - G(y)$ and $dt = -(\nu-1)(1-G(y))^{\nu-2}dG(y)$, we get

$$\begin{aligned}
&-2\int_0^\infty\int_\infty^0 [F(u) - H(u,y) - F(u)(1-G(y))](1-G(y))^{\nu-2}(\nu-1)dudG(y) \\
&\qquad = -2(\nu-1)\int_0^\infty\int_0^\infty (H(u,y) - F(u)G(y))(1-G(y))^{\nu-2}dudG(y)
\end{aligned}$$

and thus,

$$\begin{aligned}
&-\mathrm{cov}(X,(1-G(Y))^{\nu-1}) \\
&= (\nu-1)\int\int [H(u,y) - F(u)G(y)](1-G(y))^{\nu-2}dudG(y).
\end{aligned}$$

Appendix 6.2

Proposition 6.9

$$\mathrm{DALC_X(h)} \geq \mathrm{DALC_Y(h)} \quad \text{for all h}, 0 \leq h \leq 1 \text{ iff } \int_z^{\infty} [F_Y(t) - F_X(t)]\, dt \geq 0$$

for all z.

Proof

(a) Sufficiency: We have to prove that if the integral is nonnegative for all z, then

$$\mathrm{DALC_X(h)} \geq \mathrm{DALC_Y(h)} \quad \text{for all h}, 0 \leq h \leq 1.$$

Using integration by parts one can write

$$\int_z^{\infty} [1 - F(t)]dt = -H(z)z + \int_z^{\infty} tf(t)dt = \mathrm{DALC(h)} - zH(z),$$

where $H(z) = 1 - F(z) = h$. Equivalently, $z = z(h) = F^{-1}(1 - h)$. Therefore

$$\int_z^{\infty} [F_Y(t) - F_X(t)]dt = \mathrm{DALC_X}(h_1) - zh_1 - \mathrm{DALC_Y}(h_2) + zh_2, \tag{A6.1}$$

where $h_1 = H_X(z)$ and $h_2 = H_Y(z)$ and $\mathrm{DALC_X}(h_1)$ denotes the absolute (decreasing) Lorenz curve of X, evaluated at $h_1 = H_X(z)$.
Two additional properties of the absolute Lorenz curves (increasing and decreasing) are required in order to complete the proof

$$\partial \mathrm{ALC(p)}/\partial p|_{P=F(z)} = z\ ;\ \partial \mathrm{DALC(h)}/\partial h|_{h=H(z)} = z.$$

$$\partial^2 \mathrm{ALC(p)}/\partial p^2|_{p=F(z)} \geq 0\ ;\ \partial^2 \mathrm{DALC(h)}/\partial h^2|_{h=H(z)} \leq 0.$$

Using these properties (A6.1) can be written as

$$\int_z^{\infty} [F_Y(t) - F_X(t)]\, dt = \mathrm{DALC_X}(h_1) - h_1\partial(\mathrm{DALC_X})/\partial h|_{h=h_1}$$
$$+ h_2\partial(DALC_Y)/\partial h|_{h=h2} - \mathrm{DALC_Y}(h_2), \tag{A6.2}$$

where $h_1 = H_X(z)$ and $h_2 = H_Y(z)$. That is, the integral on the left hand side of (A6.2) is equal to the difference between the two absolute (decreasing) Lorenz

Fig. 6.2 Generalized (absolute) Lorenz curve. Source: Yitzhaki (1999), Fig. 19.1, p. 363. Reprinted with permission by Physica Verlag Heidelberg

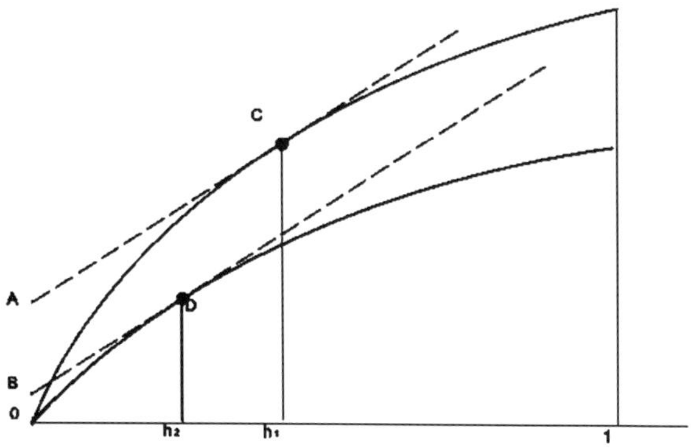

curves evaluated at equal-slope points *plus* a term which is based on the derivatives of the curves. Figure 6.2 portrays the term of equation (A6.2). Let OC be the curve $DALC_X$ and let OD be the curve $DALC_Y$. The curves are concave because the figure plots the absolute (descending) Lorenz curves, that is, h_1 represents a percentage of the highest observations and h_1C is the cumulative value of the variate of those observations. The points C and D have the same slopes. Hence the first term represents OA, while the second term is OB.

Because at the points D and C the derivatives of the curves are equal (and equal to z) we may write (A6.2) as

$$\int_z^{\infty} [F_Y(t) - F_X(t)]dt = DALC_X(h_1) + \partial(DALC_Y)/\partial h|_{h=h2}[h_2 - h_1] - DALC_Y(h_2). \tag{A6.3}$$

Note that $DALC_Y(h_2) + \partial(DALC_Y)/\partial h\,|_{h=h2}[h_1-h_2]$ is the first-order approximation of $DALC_Y(h_1)$, approximated from h_2. Because $DALC_Y$ is concave, $DALC_Y(h_1)$ is smaller than its first-order approximation. Therefore,

$$0 \leq \int_z^{\infty} [F_Y(t) - F_X(t)]dt \leq DALC_X(h_1) - DALC_Y(h_1). \tag{A6.4}$$

The fact that the relationship holds for all z implies that it also holds for all h.

(b) Necessity:

We have to prove that if $DALC_X(h) \geq DALC_Y(h)$ for all h then

$$\int_z^{\infty} [F_Y(t) - F_X(t)]dt \geq 0 \quad \text{for all z.}$$

Using (A6.3) it is sufficient to prove that if $DALC_X(h) \geq DALC_Y(h)$ for all h then

$$\mathrm{DALC_X(h_1)} + \partial(\mathrm{DALC_X})/\partial \mathrm{h}\ [\mathrm{h_2} - \mathrm{h_1}] - \mathrm{DALC_Y(h_2)} \geq 0$$

for all z where $h_1 = H_X(z)$ and $h_2 = H_Y(z)$. Since DALC() is concave

$$\mathrm{DALC_X(h_1)} + \{\partial(\mathrm{DALC_X})/\partial \mathrm{h}\}[\mathrm{h_2} - \mathrm{h_1}] \geq \mathrm{DALC_X(h_2)},$$

hence

$$\begin{aligned}&\mathrm{DALC_X(h_1)} + \partial(\mathrm{DALC_X})/\partial \mathrm{h}|_{\mathrm{h2}}[\mathrm{h_2} - \mathrm{h_1}] - \mathrm{DALC_Y(h_2)}\\ &\quad\geq \mathrm{DALC_X(h_2)} - \mathrm{DALC_Y(h_2)} \geq 0.\end{aligned}$$

Appendix 6.3

Property (b) Let F and G be the cumulative distribution functions of X and Y, respectively.
Then there exists a joint distribution function H(X,Y) such that for every ν,

$$\Gamma(\nu, \mathrm{X}, \mathrm{Y}) = 1.$$

Proof of property (b) (taken from Schechtman and Yitzhaki (2003)) Fréchet (1951) has shown that there exist bivariate distributions $H_0(x,y)$ and $H_1(x,y)$ with marginals (F,G) such that for ANY bivariate distribution H(x,y) with the same marginals,

$$\mathrm{H_0(x, y)} \leq \mathrm{H(x, y)} \leq \mathrm{H_1(x, y)},$$

where $H_0(x,y) = \max\{F(x)+G(y)-1,0\}$ and $H_1(x,y) = \min\{F(x),G(y)\}$ are the Frechet minimal and Frechet maximal distributions, respectively (De Veaux, 1976).

Using Frechet's results we obtain the upper bound as follows (all integrals are from $-\infty$ to ∞ unless stated otherwise):

$$\begin{aligned}&-\mathrm{cov}(\mathrm{X},(1-\mathrm{G(Y)})^{\nu-1}) = (\nu-1)\int\int(\mathrm{H(x,y)} - \mathrm{F(x)G(y)})(1-\mathrm{G(y)})^{\nu-2}\mathrm{dG(y)dx}\\ &= (\nu-1)\int\int \mathrm{H(x,y)}(1-\mathrm{G(y)})^{\nu-2}\mathrm{dG(y)dx} - (\nu-1)\int\int \mathrm{F(x)G(y)}(1-\mathrm{G(y)})^{\nu-2}\mathrm{dG(y)dx}.\end{aligned}$$

Using Frechet minimal distribution and using integration by parts, we get that the first integral on the right-hand side can be bounded as follows

$$\int\int H(x,y)(1-G(y))^{\nu-2}dG(y)dx \le \int\int \min(F(x), G(y))(1-G(y))^{\nu-2}dxdG(y)$$

$$= \int\int_{-\infty}^{G^{-1}(F(x))} G(y)(1-G(y))^{\nu-2}dxdG(y) + \int\int_{G^{-1}(F(x))}^{\infty} F(x)(1-G(y))^{\nu-2}dxdG(y)$$

$$= \frac{1}{(\nu-1)}\int F(x)(1-F(x))^{\nu-1}d(x) + \frac{1}{(\nu-1)}\int -G(y)(1-G(y))^{\nu-1}\Big|_{-\infty}^{G^{-1}(F(x))}$$

$$+ \frac{1}{(\nu-1)}\int\int_{-\infty}^{G^{-1}(F(x))} (1-G(y))^{\nu-1}dG(y)$$

$$= \frac{1}{(\nu-1)}\int\int_{-\infty}^{G^{-1}(F(x))} (1-G(y))^{\nu-1}d(G(y))$$

$$= \frac{1}{\nu(\nu-1)}\int (1-(1-F(x))^{\nu})dx.$$

The second integral, again by integration by parts, can be expressed as

$$\int\int F(x)G(y)(1-G(y))^{\nu-2}dG(y)dx = \int F(x)\{\int \frac{1}{(\nu-1)}(1-G(y))^{\nu-1}dG(y)\}\ dx$$
$$= \frac{1}{\nu(\nu-1)}\int F(x)dx.$$

Combining the two parts and multiplying by $(\nu-1)$ we get

$$-\operatorname{cov}(X,(1-G(Y))^{\nu-1}) \le \frac{1}{\nu}\int \{(1-F(x)) - (1-F(x))^{\nu}\}\ dx.$$

Similar arguments show that

$$-\operatorname{cov}(X,(1-F(X))^{\nu-1}) = \frac{1}{\nu}\int \{(1-F(x)) - (1-F(x))^{\nu}\}\ dx,$$

Because in this case $H(x,y) = \min(F(x),G(y))$. Therefore by choosing $H(x,y)$ to be Frechet minimal distribution, the upper bound of 1 is achieved.

Property (c) The lower bound of $\Gamma(\nu, X,Y)$ is given by

$$\Gamma(\nu, X,Y) \ge \frac{\int F(x)(F^{\nu-1}(x)-1)dx}{\int((1-F(x))(1-(1-F(x))^{\nu-1})dx},$$

and is achieved when $Y = (-X)$.

Proof of property (c) (taken from Schechtman and Yitzhaki (2003))
Using Fréchet (1951) results and integrating by parts we get

$$\int\int H(x,y)(1-G(y))^{\nu-2}dG(y)dx \geq \int\int \max(F(x)+G(y)-1,0)(1-G(y))^{\nu-2}dG(y)dx$$
$$= \int\int_{G^{-1}(1-F(x))}^{\infty} F(x)(1-G(y))^{\nu-2}dG(y)dx + \int\int_{G^{-1}(1-F(x))}^{\infty} G(y)(1-G(y))^{\nu-2}dG(y)dx$$
$$-\int\int_{G^{-1}(1-F(x))}^{\infty} (1-G(y))^{\nu-2}dG(y)dx$$
$$= \frac{\int F(x)F^{\nu-1}(x)dx}{(\nu-1)} + \frac{\int (1-F(x))F^{\nu-1}(x)dx}{(\nu-1)}$$
$$+ \frac{\int -(1-G(y))^{\nu}|_{G^{-1}(1-F(x))}^{\infty}dx}{\nu(\nu-1)} - \frac{\int F^{\nu-1}(x)dx}{(\nu-1)} = \frac{F^{\nu}(x)}{\nu(\nu-1)}.$$

Combining the pieces together, we get

$$-\operatorname{cov}(X,(1-G(Y))^{\nu-1}) = (\nu-1)\int\int (H(x,y)-F(x)G(y))(1-G(y))^{\nu-2}dG(y)dx$$
$$\geq \frac{\int F^{\nu}(x)dx}{\nu} - \frac{\int F(x)dx}{\nu} = \frac{1}{\nu}\int (F^{\nu}(x)-F(x))dx.$$

Therefore we get that

$$\Gamma(\nu, X, Y) \geq \frac{\int F(x)(F^{\nu-1}(x)-1)dx}{\int ((1-F(x))(1-(1-F(x))^{\nu-1})dx}.$$

Appendix 6.4

The decomposition

Let $Y_0 = \alpha Y_1 + \beta Y_2$, where α and β are given constants. Then

(a)
$$G_0^2 - [\alpha D_{10}G_1 + \beta D_{20}G_2]G_0 = \alpha^2 G_1^2 + \beta^2 G_2^2 + \alpha\beta G_1G_2(\Gamma(\nu,Y_1,Y_2) + \Gamma(\nu,Y_2,Y_1)),$$

where G_i $i = 0,1,2$ are the extended Gini coefficients and

$$\Gamma(\nu,Y_i,Y_j) = \frac{\operatorname{cov}(Y_i,[1-F_j(Y)]^{\nu-1})}{\operatorname{cov}(Y_i,[1-F_i(Y)]^{\nu-1})}$$

for $i, j = 0,1,2$ are the extended Gini correlations, $D_{i0} = \Gamma(\nu,Y_0,Y_i) - \Gamma(\nu,Y_i,Y_0)$ for $i = 1,2$ are the differences between the extended Gini correlations.

(b) Provided that $D_{i0} = 0$ for $i = 1,2$ and that the two EG correlations between Y_1 and Y_2 are equal the following decomposition holds:

$$G_0^2 = \alpha^2 G_1^2 + \beta^2 G_2^2 + 2\alpha\beta G_1 G_2 \Gamma,$$

where $\Gamma = \Gamma(v, Y_1, Y_2) = \Gamma(v, Y_2, Y_1)$ are the extended Gini correlation coefficients between Y_1 and Y_2 (and between Y_2 and Y_1).

Proof of the decomposition (taken from Schechtman and Yitzhaki (2003)).

Proof of (a) Using the properties of the covariance we can write

$$\begin{aligned} G_0 &= -v\,\mathrm{cov}(\alpha Y_1 + \beta Y_2, [1 - F_0(Y)]^{v-1}) \\ &= -v(\alpha\,\mathrm{cov}(Y_1, [1 - F_0(Y)]^{v-1}) + \beta\,\mathrm{cov}(Y_2, [1 - F_0(Y)]^{v-1}) \\ &= \alpha\,\Gamma(v, Y_1, Y_0) G_1 + \beta\,\Gamma(v, Y_2, Y_0) G_2. \end{aligned}$$

Define now the identity $\Gamma(v, Y_i, Y_0) = D_{i0} + \Gamma(v, Y_0, Y_i)$ for $i = 1,2$, where D_{i0} is the difference between the two Gini correlations. Using the identity we get

$$G_0 = \alpha(\Gamma(v, Y_0, Y_1) + D_{10}) G_1 + \beta(\Gamma(v, Y_0, Y_2) + D_{20}) G_2.$$

Rearranging the terms, we see that

$$G_0 - \alpha D_{10} G_1 - \beta D_{20} G_2 = \alpha\,\Gamma(v, Y_0, Y_1) G_1 + \beta\,\Gamma(v, Y_0, Y_2) G_2.$$

Using the properties of the covariance we now get that

$$\begin{aligned} \Gamma(v, Y_0, Y_1) &= \frac{\mathrm{cov}(Y_0, [1 - F_1(Y)]^{v-1})}{\mathrm{cov}(Y_0, [1 - F_0(Y)]^{v-1})} \\ &= \frac{\alpha\,\mathrm{cov}(Y_1, [1 - F_1(Y)]^{v-1}) + \beta\,\mathrm{cov}(Y_2, [1 - F_1(Y)]^{v-1})}{\mathrm{cov}(Y_0, [1 - F_0(Y)]^{v-1})} \\ &= \frac{\alpha G_1 + \beta G_2 \Gamma(v, Y_2, Y_1)}{G_0}. \end{aligned}$$

Writing $\Gamma(v, Y_0, Y_2)$ in a similar manner we get

$$\begin{aligned} G_0^2 - [\alpha D_{10} G_1 + \beta D_{20} G_2] G_0 &= \alpha G_1(\alpha G_1 + \beta G_2 \Gamma(v, Y_2, Y_1)) \\ &\quad + \beta G_2(\alpha \Gamma(v, Y_1, Y_2) G_1 + \beta G_2) \\ &= \alpha^2 G_1^2 + \beta^2 G_2^2 + \alpha\beta G_1 G_2(\Gamma(v, Y_1, Y_2) + \Gamma(v, Y_2, Y_1)). \end{aligned}$$

Note that one can substitute D_{i0} by the difference in correlations and get

$$\begin{aligned} &G_0^2 - [\alpha(\Gamma(v,Y_1,Y_0) - \Gamma(v,Y_0,Y_1))G_1 + \beta(\Gamma(v, Y_2,Y_0) - \Gamma(v,Y_0,Y_2))G_2]G_0 \\ &\quad = \alpha^2 G_1^2 + \beta^2 G_2^2 + \alpha\beta G_1 G_2(\Gamma(v,Y_1,Y_2) + \Gamma(v,Y_2,Y_1)). \end{aligned}$$

Proof of (b) Assuming equality of the two Gini correlation coefficients between Y_0 and Y_1 sets $D_{10} = 0$. A similar assumption with respect to Y_0 and Y_2 sets $D_{20} = 0$. The assumption of $\Gamma = \Gamma(v, Y_1, Y_2) = \Gamma(v, Y_2, Y_1)$ completes the proof of part (b).

Chapter 7
Gini Simple Regressions

Introduction

The basic building block in regression is the covariance between the dependent variable and the explanatory variable(s). There are two regression methods that can be interpreted as based on Gini's Mean Difference (GMD). The first method is based on the fact that one can present the Gini-covariance between the dependent variable and the explanatory variable as a weighted sum of slopes of the regression curve (a semi-parametric approach). The second method is based on the minimization of the GMD of the residuals. The semi-parametric approach is similar in its structure to the Ordinary Least Squares (OLS) method. That is, the regression coefficient in the OLS has an equivalent term in the Gini semi-parametric regression. The equivalent term is constructed by substituting the covariance and the variance in the OLS regression by the Gini-covariance (hereafter co-Gini) and the Gini, respectively. However, unlike the OLS, the Gini regression coefficient and its estimator are not derived by solving a minimization problem. Therefore they do not have optimality properties and cannot be described as "the best," at least not with respect to a simple target function. On the other hand, the second method, the minimization of the GMD of the residuals implies optimality but it has its drawbacks. Like Mean Absolute Deviation (MAD) and quantile regressions, the regression coefficient does not have an explicit presentation and can be calculated only numerically. The combination of the two methods of Gini regression enables the user to investigate the appropriateness of the assumptions that lie behind the OLS and Gini regressions (e.g., the linearity of the relationship) and therefore can improve the quality of the conclusions that are derived from them. Moreover, when dealing with a multiple regression one can combine the semi-parametric regression method with the OLS regression method. That is, several explanatory variables can be treated as in the OLS, while others are treated using the Gini method. This flexibility enables one to evaluate the effect of the choice of a regression method on the estimated coefficients in a gradual way by substituting the methodology of the

S. Yitzhaki and E. Schechtman, *The Gini Methodology: A Primer on a Statistical Methodology*, Springer Series in Statistics 272, DOI 10.1007/978-1-4614-4720-7_7,

estimation for each explanatory variable in a stepwise way rather than in an "all or nothing" way. This issue will be discussed in Chap. 8.

This chapter concentrates on the alternative presentations of the regression coefficient in a simple regression framework and on the interaction between them. The structure of the chapter is the following: Sect. 7.1 introduces alternative presentations using the semi-parametric approach. The minimization approach is described in Sect. 7.2. The two approaches are combined in Sect. 7.3. In Sect. 7.4 we introduce goodness of fit measures, while a test for normality is discussed in Sect. 7.5. In Sect. 7.6 we discuss the instrumental variable method, both for OLS and for Gini regressions. The extended Gini regression is detailed in Sects. 7.7 and 7.8 concludes.

7.1 Alternative Presentations: The Semi-Parametric Approach

The semi-parametric approach is based on the fact that the regression coefficient can be presented as a weighted sum of slopes of the regression curve, as will be shown below. The Gini-based parameter and estimator resemble the OLS in the sense that all the expressions used have parallels in OLS regression. We note that both the OLS and Gini semi-parametric regression methods do not require the specification of a functional form of the model. They can be used whenever the investigator is interested in estimating average slopes or arc-elasticities without requiring a formal model, resembling the method suggested by Härdle and Stoker (1989) and Rilstone (1991). Gini parameters and estimators that are derived according to this approach will be denoted by the subscript N. We refer to this method as semi-parametric because it does not rely on the linearity assumption nor on any distributional assumptions. It is worth noting that the OLS regression coefficient shares these properties. That is, it can be presented as a weighted sum of the slopes of the regression curve. The only difference between the OLS and the Gini regression coefficients lies in the weights attached to the slopes.

Formally, let (Y, X) be a bivariate random variable that follows a continuous distribution with finite first and second moments. At this stage we do not impose additional assumptions. In particular, we do not assume that X is fixed nor that there is a linear relationship between the two variables.

The objective: the investigator is interested in constructing a linear predictor of Y that is based on X. The theoretical linear predictor is denoted by

$$\hat{\mathrm{Y}} = \alpha + \beta \mathrm{X} \tag{7.1}$$

where at this stage α and β are arbitrary constants imposed by the investigator. We define the residual as

$$\varepsilon = \mathrm{Y} - \hat{\mathrm{Y}} \equiv \mathrm{Y} - \alpha - \beta \mathrm{X}. \tag{7.2}$$

Note that (7.2) is an identity and that the residuals are not assumed to have properties of their own at this stage. All their properties are derived from the properties of (Y, X). Using the properties of the covariance one gets

$$\mathrm{cov}(\mathrm{Y},\mathrm{X}) \equiv \mathrm{cov}(\alpha + \beta X + \varepsilon, \mathrm{X}) \equiv \ \beta \ \mathrm{cov}(\mathrm{X},\mathrm{X}) + \mathrm{cov} \ (\varepsilon, \mathrm{X}). \qquad (7.3)$$

We now add an assumption. Imposing the orthogonality restriction $\mathrm{cov}(\varepsilon, \mathrm{X}) = 0$ changes (7.3) from an identity to an equation that can be solved to give the following formula

$$\beta = \frac{\mathrm{cov(Y,X)}}{\mathrm{cov(X,X)}}. \qquad (7.4)$$

β of (7.4) is equivalent in its structure to β_{OLS} which is obtained from the normal equations, one of which is $\mathrm{cov}(\varepsilon, \mathrm{X}) = 0$.

In OLS the constraint that $\mathrm{cov}(\varepsilon, \mathrm{X}) = 0$ is derived from the minimization of the variance of the error term. We now turn to the Gini world. Starting with (7.4) and replacing each term by the equivalent term from the Gini method, cov(Y, F(X)) and cov(X,F(X)) replace cov(Y,X) and cov(X, X) yielding the equivalent parameter for the Gini semi-parametric regression:

$$\beta_{\mathrm{N}} = \frac{\mathrm{cov(Y,F(X))}}{\mathrm{cov(X,F(X))}}. \qquad (7.5)$$

Note that by using the properties of the covariance we get

$$\mathrm{cov}(\varepsilon_{\mathrm{N}}, \mathrm{F(X)}) = 0, \qquad (7.6)$$

where ε_{N} is the residual of the semi-parametric Gini regression. Equation (7.6) is the equivalent of the normal equation in OLS regression. (An alternative way to derive (7.5) is to start from the equivalent of (7.3) and impose (7.6) which is an orthogonality condition in the Gini world).

Once the slope of the linear predictor (7.1) is determined, one can use an additional constraint to determine α. If one wishes that the regression line will pass through the means of the variables then α will be determined as a solution to (7.7):

$$\mu_{\mathrm{Y}} = \alpha + \beta_{\mathrm{N}} \ \mu_{\mathrm{X}}. \qquad (7.7)$$

However, one can use other criteria to determine α such as minimizing the sum of the absolute deviations of the residuals from a constant in which case α causes the regression line to pass through the median, or any quantile of the residual distribution, as in quantile regression (with some modifications). The important point here is that one can separate between the criterion that is used to determine the slope and the criterion used to determine the constant term, which is determined as a location parameter.

7.1.1 *The Ordinary Least Squares Regression Coefficient*[1]

The purpose of this section is to interpret the OLS regression coefficient as a weighted average of the slopes of the regression curve. This interpretation is useful because it stresses the fact that the regression coefficient in the linear predictor (7.1) does not depend on the underlying distribution of the dependent variable and that the linearity assumption in the regression model plays no role in the derivation of the coefficient.

Let (Y, X) be a bivariate random variable with a density function f(y, x). Let f_X, F_X, μ_X, and $\sigma_X{}^2$ denote the density, the cumulative distribution, the expected value, and the variance of X, respectively. Assume that the first and the second moments exist. Let $g(x) = E\{Y|X = x\}$ be the regression curve, where $g'(x)$ is the slope of the regression curve defined as

$$g'(x) = \frac{\partial E\{Y|X = x\}}{\partial x}. \tag{7.8}$$

Proposition 7.1 *Let $\hat{Y} = \alpha + \beta X$ denote a linear predictor of Y given X. Then β_{OLS} can be expressed as a weighted average of the slopes of the regression curve:*

$$\beta_{OLS} = \int w(x) g'(x) dx, \tag{7.9}$$

where $w(x) > 0$ denotes the weight at x and $\int w(x)dx = 1$. The weights are

$$w(x) = \frac{1}{\sigma_X^2}[\mu_X F_X(x) - ALCV(x)] = \frac{F_X(x)}{\sigma_X^2}(\mu_X - E\{X|X \le x\}), \tag{7.10}$$

where

$$ALCV(x) = \int_{-\infty}^{x} t f_X(t)\, dt = F_X(x) E\{X|X \le x\}. \tag{7.11}$$

ALCV(x) is the absolute Lorenz curve as a function of X, while $[\mu_X Fx(X)]$ is the transformed line of independence (LOI) (see Chap. 5). Hence the term in the squared brackets in (7.10) is actually the vertical difference and the (transformed) absolute Lorenz curve at x.

[1] This section is based on Yitzhaki (1996).

Proof (Yitzhaki, 1996). Note that $\beta_{OLS} = \mathrm{cov(Y, X)/cov(X, X)}$. The numerator can be expressed as

$$\begin{aligned}\mathrm{cov(Y,X)} &= \mathrm{E_X E_Y[(Y - \mu_Y)(X - \mu_X)] = E_X E_Y[(X - \mu_X)Y]} \\ &= \mathrm{E_X[(X - \mu_X)E(Y|X = x)]} \\ &= \int \mathrm{(x - \mu_X)g(x)f_X(x)dx},\end{aligned} \tag{7.12}$$

where $\mathrm{g(x) = E\{Y|X = x\}}$ is the conditional expectation.

Using integration by parts, with $\mathrm{v'(x) = (x - \mu_X)f_X(x)dx, v(x) = \int_{-\infty}^{x} (t - \mu_X) f_X(t)dt, u(x) = g(x),}$ and $\mathrm{u'(x)dx = g'(x)dx}$ we get

$$\mathrm{cov(Y,X)} = \left\{ \int_{-\infty}^{x} \mathrm{(t - \mu_X)f_X(t)dt} \right\} \mathrm{g(x)}|_{-\infty}^{\infty} + \int_{-\infty}^{\infty} \left[\int_{-\infty}^{x} \mathrm{(\mu_X - t)f_X(t)dt} \right] \mathrm{g'(x)dx}. \tag{7.13}$$

Given that the second moments exist, the first term equals zero. Hence, (7.13) can be written as

$$\mathrm{cov(Y,X)} = \int_{-\infty}^{\infty} [\,\mu_X \mathrm{F(x) - ALCV(x)]g'(x)dx}. \tag{7.14}$$

Therefore

$$\beta_{OLS} = \frac{\mathrm{cov(Y,X)}}{\mathrm{cov(X,X)}} = \frac{1}{\sigma_X^2} \int_{-\infty}^{\infty} [\mu_X \mathrm{F_X(x) - ALCV(x)]\ g'(x)dx} = \int \mathrm{w(x)g'(x)dx}.$$

To show that the sum of the weights equals one we apply the same procedure to the denominator of the OLS regression coefficient with $\mathrm{g'(x)} \equiv 1$. That is,

$$\sigma^2{}_X = \mathrm{cov(X,X)} = \int_{-\infty}^{\infty} [\,\mu_X \mathrm{F_X(x) - ALCV(x)]dx} \tag{7.15}$$

and we get $\int \mathrm{w(x)dx} = \frac{1}{\sigma_X^2} \int [\,\mu_X \mathrm{F_X(x) - ALCV(x)]dx} = 1$.

In order to illustrate the effect of the distribution of the explanatory variable on the weighting scheme we consider three specific examples: the uniform, the normal, and the lognormal distributions. The first two examples illustrate interesting cases; the third can represent the distribution of income.

(a) The uniform distribution. Let X have a uniform distribution on [a, b]. Applying (7.10), the weight attached to the slope at x is

$$\mathrm{w(x)} = \frac{\mathrm{6(b - x)(x - a)}}{\mathrm{(b - a)^3}}. \tag{7.16}$$

This weighting scheme, which represents a case where the observations of X are equally spaced, is symmetric around the median, and the closer the observation is to the median the larger the weight it gets. An interesting feature of this weighting scheme is that its shape remains unchanged regardless of whether it is viewed as a function of X or F_X, implying that the weighting schemes of the OLS and Gini regressions are identical (to be seen later). That is, provided that the distribution of the explanatory variable is uniform (or equidistant) then the OLS and Gini regressions yield the same regression coefficient, regardless of the shape of the regression curve. We note in passing that the EG regression can be used to examine the curvature of the regression curve (see Sect. 7.7).

(b) The normal distribution. Let X have a standard normal distribution. By (7.10)

$$w(x) = -\frac{1}{\sqrt{2\pi}} \int_{-\infty}^{x} t e^{-t^2/2} dt = \frac{1}{\sqrt{2\pi}} e^{-x^2/2}. \tag{7.17}$$

The weight for this case is equal to the value of the density of the standard normal distribution at x. Hence, each percentile of the population receives an equal weight. This result offers an intuitive explanation to the weighting scheme of the OLS. It is adjusted to the case where the explanatory variable has a normal distribution, which explains its efficiency in this case. Shalit (2010) uses this property in order to suggest a test for normality.

(c) The lognormal distribution. Let X have a lognormal distribution with parameters μ and σ^2 (that is, ln(X) is normally distributed with mean μ and variance σ^2). The mean of X is $\mu_X = e^{\mu+\frac{1}{2}\sigma^2}$ and the variance of X is $\sigma_X^2 = (e^{\sigma^2} - 1)e^{2\mu+\sigma^2}$. It is convenient to rewrite the right-hand side of (7.10) as

$$w(x) = \frac{\mu_x}{\sigma_x^2}[F_x(x) - F_1(x)] = \frac{F_x(x)}{\sigma_x^2}[\mu_x - E(X|X \le x)] \tag{7.18}$$

where $F_1(x) = (1/\mu_X)\int_0^x t dF_X(t)$ is the first moment cumulative distribution of X.[2] The first moment distribution is also lognormal, with parameters μ + σ^2 and σ^2 (Aitchison & Brown, 1963).

Hence, the weight at x is the difference between two cumulative lognormal distributions. Using the usual transformation we can write the weight as

$$w(x) = \frac{\mu_X}{\sigma_X^2}\left[\Phi\left(\frac{\ln(x) - \mu}{\sigma}\right) - \Phi\left(\frac{\ln(x) - \mu - \sigma^2}{\sigma}\right)\right], \tag{7.19}$$

where Φ( ) is the standard normal cumulative distribution. This term can be evaluated numerically.

[2] Note that F_1 is the Lorenz in terms of X. It is defined as LCV (x) in Sect. (5.3). It is called first moment distribution by Hart (1975).

Table 7.1 The OLS weighting schemes for a lognormal distribution

Parameters								
μ	0.	0.	0.	0.	7.0	0.	0.	0.
σ	0.45	0.6	0.7	0.75	0.75	0.8	1.0	1.2
Gini coef.	0.25	0.33	0.38	0.4	0.4	0.43	0.52	0.60
Column no.	(1)	(2)	(3)	(4)	(5)	(6)	(7)	(8)
Decile								
1	2.5	2.1	1.0	0.8	0.8	0.7	0.3	0.1
2	3.9	3.4	1.9	1.6	1.6	1.4	0.7	0.3
3	5.0	4.5	2.8	2.4	2.5	2.1	1.1	0.5
4	6.0	5.5	3.7	3.3	3.3	2.9	1.7	0.9
5	7.1	6.6	4.8	4.3	4.4	3.9	2.4	1.3
6	8.8	8.3	6.5	6.0	6.0	5.5	3.6	2.1
7	10.2	9.9	8.3	7.8	7.9	7.3	5.2	3.3
8	12.6	12.5	11.4	11.0	11.1	10.5	8.2	5.7
9	16.1	16.4	16.9	16.7	17.0	16.5	14.4	11.2
10	27.9	30.4	42.8	46.0	45.3	49.3	62.6	74.7

Source: Yitzhaki (1996), Table 2, p. 481

Table 7.1 presents the weighting scheme w(x) when X has a lognormal distribution for different values of the parameters μ and σ. It turns out that the weighting scheme is not sensitive to μ (note the small differences between columns 4 and 5), but it is sensitive to changes in σ.

For ease of reference Table 7.1 also presents the Gini coefficient corresponding to each value of σ. As seen in column 4, when the Gini coefficient is 0.4 (a typical value for before-tax income), the expected weight of the top decile is around 45%; and for the highest quintile it is over 60%. If wealth were used as an explanatory variable, a Gini coefficient of 0.55 could be considered typical. In this case, the weight of the top decile may well exceed 60%. Calculations for other distributions (the Pareto and the exponential) show that the share of the top decile is never less than 28%.

As shown above, the OLS regression coefficient can be presented as a weighted average of slopes. There are two possible presentations of the weighting scheme. The first presentation is as a function of the explanatory variable, namely w(x) (as seen above), and the second one presents the weight as a function of F_X, the cumulative distribution of the explanatory variable. The second presentation is useful whenever one is interested in the shares of the weights assigned to portions of the population. The transformation from one presentation to the other can be done by letting $F_X(x) = p$, $0 \leq p \leq 1$, and defining a weighting scheme $v(p) = w[F_X^{-1}(p)]$, where F_X^{-1} is the inverse of the cumulative distribution. ($F_X(x)$ is a monotonic increasing differentiable function).

Proposition 7.1 can be replicated for a discrete distribution. We present it because it adds some intuition to the interpretation. The proof is similar in nature to the proof of proposition 7.1 and is given in Appendix 7.1.

Proposition 7.1′ *The OLS regression coefficient for a discrete distribution of X is a weighted sum of slopes defined by adjacent observations. That is,*

$$\beta_{\mathrm{OLS}} = \sum\nolimits_{\mathrm{i}=1}^{\mathrm{n}-1} \mathrm{w_i b_i} \tag{7.20a}$$

where $w_i > 0$, $\Sigma w_i = 1$, $b_i = \Delta y_i/\Delta x_i$, $\Delta x_i = x_{i+1} - x_i$ *and where the observations are arranged in increasing order of X. The weights are given by*

$$\mathrm{w_i} = \frac{\left(\sum\limits_{j=i}^{n-1} i(n-j)\Delta x_j + \sum\limits_{j=1}^{i-1} j(n-i)\Delta x_j\right)\Delta x_i}{\sum\limits_{k=1}^{n-1}\left(\sum\limits_{j=k}^{n-1} k(n-j)\Delta x_j + \sum\limits_{j=1}^{k-1} j(n-k)\Delta x_j\right)\Delta x_k}, \tag{7.21a}$$

(where $\sum\limits_{j=1}^{0} j(n-1)\Delta x_j$ *is defined to be zero).*

The estimator follows the same structure.

It proves convenient to represent the weights by the ALCV. In this case the ith weight can be expressed as

$$\mathrm{w_i} = \frac{\mathrm{i}}{\mathrm{n}} \frac{(\bar{\mathrm{x}}_\mathrm{n} - \bar{\mathrm{x}}_\mathrm{i})\Delta \mathrm{x_i}}{\hat{\sigma}_\mathrm{x}^2}, \tag{7.21b}$$

where $\bar{x}_i$ is the mean of the i smallest observations of X, while $\hat{\sigma}_x^2$ is the estimate of the variance of X. Note that the weight is the vertical distance between LOI and the absolute Lorenz curve *multiplied by* the appropriate Δx and *divided by* the whole area between the shifted LOI and the shifted absolute Lorenz curve (which is represented by the variance in the denominator).

The two types of components in (7.20a) are the slopes b_i and the weights w_i.

The slopes are determined by the data and the weighting scheme w_i depends on the distribution of the explanatory variable X alone, similar to the continuous case. Therefore one can interpret the differences between the OLS and the Gini estimators as originating from the weighting schemes employed. Under the present scheme, the contribution of each observation to the estimate is decomposed into (1) the effect of the weighting scheme at the observation and (2) the slope defined by the observation and the one that precedes it. Adding up the weights of several observations yields the contribution of a region of the explanatory variable to the estimate.

Equation (7.21a) reveals that the weight w_i depends both on the rank of the observation (through i) and on its distance from the adjacent observation as defined by the difference in the explanatory variable (through Δx_i). Similar to the continuous case, the weight increases as the observation gets closer to the median and as the distance between observations increases. To control for the latter effect, consider the case where the observations of the explanatory variable are equally spaced, that is, $\Delta x_i = c$ for all i. In this case the weights in (7.21a) can be simplified and written as

$$\mathrm{w(i) = K\ i(n-i), \quad where\ K = 6/[n(n-1)(n+1)].}$$

Table 7.2 OLS weighting schemes of adjacent slopes according to income deciles[a]

		Empirical weights	
Income decile	Equidistant weights[b] (1)	Whole sample (2)	Truncated sample[c] (3)
1 (lowest)	0.028	0.005	0.014
2	0.76	0.013	0.036
3	0.112	0.024	0.058
4	0.136	0.032	0.091
5	0.148	0.037	0.093
6	0.148	0.050	0.113
7	0.136	0.072	0.150
8	0.112	0.080	0.169
9	0.076	0.099	0.168
10	0.028	0.588	0.109
All	1.0	1.0	1.0

Source: Yitzhaki (1996), Table 1, p. 480

Source: Tabulation from the Israeli Survey of Family Expenditures, 1979/80, households with two members only (531 observations)

[a] Income is defined as before-tax income, including imputed rent on own housing

[b] The weighting scheme for a sample with observations with equal distance

[c] Only 500 observations: the 31 highest observations were deleted from the sample

It is easy to see that the weighting scheme is symmetric around the median, and the closer the observation gets to the median, the larger its weight becomes (i.e., the weighting scheme describes a parabola with a maximum at $i = n/2$). We illustrate the weighting scheme by an empirical example. Table 7.2 presents results for the Israeli Survey of Family Expenditures, 1979/80, taken from Yitzhaki (1996).

The regression model uses expenditure as the dependent variable and the explanatory variable is the income. The original sample used is the *Israeli Survey of Family Expenditures 1979/80*. To insure conservative estimates, one has to decrease the variance of income. Therefore only observations of households with two members are used, a group that includes the largest number of observations (531) among households of different sizes. (Other groups with different family sizes have similar patterns of weighting schemes.) Column (1) presents the weighting scheme according to deciles assuming that the distribution of the explanatory variable is uniform. As can be seen, the top decile receives less than 3% of the weight. However, if Δx_i varies along the distribution, then the larger Δx_i gets, the larger the weight attached to the respective slope. To see the effect of the variability of Δx_i on the OLS weighting scheme, column (2) in Table 7.2 presents the empirical weighting scheme.

As can be seen, the OLS estimator assigns around 60% of the weight to the top decile; over three-quarters of the total weight accrues to the top three deciles. To see the implication of this empirical weighting scheme on the estimator consider the following hypothetical case: suppose that one estimates a linear Engel curve which presents the consumption of a commodity as a function of income. The model is misspecified because in the population, instead of the relationship being linear over the entire range it is composed of two linear sections. The marginal propensity to spend (i.e., the slope of the regression curve) is a constant +0.1 for the poorest 90% of the population and −0.1 for the top decile. In this case, the OLS estimate of the marginal propensity to spend will be (−0.0176). That is, the commodity will be considered as an inferior commodity, although for 90% of the population the marginal propensity to spend is positive. (Had the observations been equally spaced, the OLS estimate would have been 0.094.)

The same weighting scheme percolates down to other statistics calculated from the regression, to misspecification tests and to other tests based on the distribution of the error term. This is so because any misspecification test that will be based on the OLS will rely on the same weighting scheme, so that there is no guarantee that low-income groups are not ignored. Testing for undue influence may therefore be helpful. Note that the influence of an observation takes two forms: its weight and the deviation of its slope from the average slope. If only observations with low weights deviate from the average slope, the influence of each observation may be small, either because of its low weight or because of the small deviation of its slope from the average slope.

To sum up: it was shown that for the OLS method the weighting scheme is affected by (1) the cumulative distribution function of the explanatory variable (represented by the rank of each observation in the sample) and (2) the distance (in terms of the explanatory variable) between each observation and the adjacent one. This difference is raised to the power two (see (7.21a)), which exacerbates its effect. (The use of a quadratic difference can be traced to the variance.)

In the next section we show that the Gini regression coefficient can also be presented as a weighted average of (the same) slopes. The only difference between the two coefficients lies in the weights attached to the slopes.

7.1.2 The Gini Semi-Parametric Regression Coefficient

In this section we replicate the methodology of Sect. 7.1.1 and show that the Gini regression coefficient can also be presented as a weighted sum of slopes (same slopes as in the OLS) of the regression curve. The only difference between the presentations lies in the weighting scheme. In what follows we use the properties of the GMD and the Lorenz curve.

The GMD semi-parametric regression coefficient was defined in (7.5) as

$$\beta_{\mathrm{N}} = \frac{\mathrm{cov}(\mathrm{Y},\mathrm{F_X}(\mathrm{X}))}{\mathrm{cov}(\mathrm{X},\mathrm{F_X}(\mathrm{X}))}. \tag{7.22}$$

Proposition 7.2 *The Gini semi-parametric regression coefficient is a weighted sum of the slopes of the regression curve. That is,*

$$\beta_N = \int v(x)\ g'(x)dx, \tag{7.23}$$

with $v(x) > 0$ *and* $\int v(x)dx = 1$, *where*

$$v(x) = \frac{[1 - F_X(x)]\,F_X(x)}{\int_{-\infty}^{\infty} [1 - F_X(t)]\,F_X(t)dt}. \tag{7.24}$$

Proof We start with the numerator of (7.22a).

$$\begin{aligned}\mathrm{cov}\{Y, F_X(X)\} &= E_X E_Y\{(Y - \mu_Y)(F_X(X) - 1/2)\}\\ &= E_X\{(F_X(X) - 1/2)g(X)\} = \int (F_X(x) - 1/2)g(x)f_X(x)dx.\end{aligned}$$

Using integration by parts with $u = g(x)$ and

$$v' = [F_X(x) - 1/2]f_X(x)dx;\ v = -\tfrac{1}{2}\left(1 - F_X(x) - [1 - F_X(x)]^2\right)$$

yields

$$\begin{aligned}2\mathrm{cov}(Y, F_X(X)) = &-\left([1 - F_X(x)] - [1 - F_X(x)]^2\right)g(x)\big|_{-\infty}^{\infty}\\ &+ \int [1 - F_X(x)]F_X(x)\ g'(x)dx.\end{aligned}$$

For a bounded g(x), the first term is equal to zero. Hence

$$2\mathrm{cov}(Y, F_X(X)) = \int (1 - F_X(x))F_X(x)g'(x)dx.$$

Applying the same procedure to the denominator with $g'(x) \equiv 1$ completes the proof of proposition 7.2.

Proposition 7.2 can be replicated for a discrete distribution. We present it because it adds some intuition to the interpretation. The proof is similar in nature to the proof of 7.2 and will be omitted.

Proposition 7.2′ *The Gini regression coefficient for a discrete distribution of X is a weighted sum of slopes defined by adjacent observations. That is,*

$$\beta_N = \sum_{i=1}^{n-1} v_i b_i \tag{7.25}$$

where $v_i > 0$, $\Sigma v_i = 1$, $b_i = \Delta y_i/\Delta x_i$, $\Delta x_i = x_{i+1} - x_i$ *and where the observations are arranged in an increasing order according to X. The weights are given by*

$$v_i = \frac{(n - i)i\ \Delta x_i}{\sum_{k=1}^{n-1} [(n - k)k]\ \Delta x_k}. \tag{7.26}$$

The estimator follows the same structure.

Let us now discuss the intuition that lies behind the weighting schemes. The weighting schemes depend on two factors. The first factor is the rank of the observation—the maximum weight is attached to the median observation of the explanatory variable, and then the weight declines symmetrically as the observation gets farther away from the median. This property is shared both by the OLS weighting scheme (7.15) and by the GMD weighting scheme (7.24). The other factor affecting the weight is the distance between adjacent observations which is embodied in Δx. The difference between the OLS and Gini methods is in the weight attached to Δx. While the weight in Gini regression is based on Δx itself, the weight in OLS is based on $(\Delta x)^2$. This difference explains the fact that OLS is more sensitive to outliers than the GMD regression coefficient.

Two special cases are of interest. The first case is when X comes from a normal distribution. As shown earlier, in the OLS method the weights given to all the percentiles are the same (see (7.17)). That is, the decline in the weight as a result of the fact that the observation gets farther from the median is compensated by the increase in Δx between adjacent observations, which gets larger as the observations are getting farther from the median. As a result, the weight is left unchanged and all observations get the same weights. The second case is when X comes from a uniform distribution. In this case the shape of the weighting scheme remains unchanged regardless of whether it is viewed as a function of X or F_X, implying that the weighting schemes of the OLS and Gini regressions are identical (see (7.16) above). Note that in this case the two methods yield the same regression coefficients regardless of the shape of the regression curve. Propositions 7.1 and 7.2 demonstrate that the linearity assumption of the regression curve plays no role in both methods, because both in OLS and in Gini regressions the regression coefficient is actually a weighted average of slopes defined between adjacent observations. When X has a uniform distribution the two methods yield the same coefficient no matter what the shape of the regression curve is.

The presentation of the regression coefficient as a weighted average of slopes enables the user to perform a sensitivity analysis and learn about the influence of specific observations on the estimators. In this sense it is similar to the idea of the influence curve (Belsley, Kuh, & Welsch, 1980; Hampel, 1974), according to which one derives the effect of dropping an observation from the sample on the estimates. The presentation of the estimator as a weighted average of slopes divides the contribution of each observation to the estimate into two components: (1) the effect of the weighting scheme which depends only on the distribution of the explanatory variable and the regression method and (2) the slope defined by the observation and the one adjacent to it. An observation may be influential because the weight attached to it is large (Δx is relatively large), or due to having an extreme slope, or both. Adding up the weights of several adjacent observations yields the contribution of a region of the explanatory variable to the estimate. This issue will be discussed in the next section, while applications will be presented in Chap. 20 when we analyze the effects of regions of the explanatory variable on the regression coefficient.

Olkin and Yitzhaki (1992) show that the GMD estimator is closely related to Sievers's (1978, 1983) robust estimator of the regression coefficient and to Scholz's (1978) weighted median regression estimator. In addition, Scholz, Sievers, and Olkin and Yitzhaki showed that the sampling distributions of those estimators of the regression coefficients converge to the normal distribution, and suggested estimators for the variances of the estimators. The properties of the GMD-based estimators are similar to those of the weighted average derivative estimator (Härdle & Stoker, 1989; Powell, Stock, & Stoker, 1989). Because both are weighted averages of slopes, they are not derived by optimization and do not require the assumption that the regression curve is linear. Both estimators are based on U-statistics, and hence, for large samples the distributions of the estimators converge to the normal distribution (for details, see Chap. 9 on estimation and Chap. 10 on the asymptotic distribution). The main difference between all these approaches lies in the motivation for the weighting scheme. In Sievers (1978) and Olkin and Yitzhaki (1992) the motivation is robust estimation. An alternative view that will be elaborated on in Chaps. 15 and 16 is the need to adjust the weighting scheme to the social attitude toward inequality or to the risk aversion of the decision maker (Chap. 17).

Finally, the parameters associated with the GMD (the equivalents of the variance and covariance) can also be presented as areas enclosed between the line of independence (LOI) and the absolute concentration curve (ACC) (Yitzhaki, 1998, 2003; Yitzhaki & Olkin, 1991; Yitzhaki & Schechtman, 2004). The properties of the ACC curves enable one to check the monotonicity of the regression curve and to visually observe whether omitting observations that are located in a given section of the range of the explanatory variable could change the sign of the regression coefficient. To avoid repetition this presentation is only shown for the instrumental variable estimator (see Sect. 7.6.2). The adaptation to the simple regression coefficient is immediate. See Chap. 20.

7.1.3 A Presentation Based on the Decomposition to Subpopulations

The use of the properties of the covariance enables us to decompose the Gini and the OLS regression coefficients into the contributions of any grouping of observations (not necessarily adjacent observations).

We start with the decomposition of the Gini regression coefficient. Let $(Y, X) = (Y_0, X_0)$ be a bivariate random variable representing the overall population (distribution) and let (Y_m, X_m) (m=1,. . .,M) be M bivariate random variables representing M disjoint subpopulations (distributions). Let β_m be the regression coefficient of subgroup m, and let $(\bar{Y}., \bar{X}.)$ be the vector of groups' averages, that is, a vector of length M with the elements $(\bar{Y}_1,\bar{X}_1), (\bar{Y}_2,\bar{X}_2),..., (\bar{Y}_M, \bar{X}_M)$. We define

$\overline{F}_{\cdot 0}(X)$ to be the vector of the averages of the ranks of the members of the groups in the overall population (according to X). That is—the observations are ranked within the entire population (with respect to X) and then the average rank per group is taken. For example, the ith element of the vector is $\overline{F}_{\cdot i0}$ which is the average of the ranks of the members of group i when they are ranked within the entire distribution of X (i.e., within the distribution of X_0).

Claim The Gini semi-parametric regression coefficient of the overall population, β_N, can be decomposed as follows

$$\beta_N = \sum_{m=1}^{M} w_m\, \beta_m + w_B \frac{\mathrm{cov}_B(\overline{Y}_{.},\overline{F}_{\cdot o}(X))}{\mathrm{cov}_B(\overline{X}_{.},\overline{F}_{\cdot o}(X))}, \tag{7.27}$$

where $w_m = \frac{p_m O_m \Delta_m}{\Delta_0}$ is the contribution of the mth group to the overall GMD of X, namely Δ_0 (O_m,P_m and Δ_m are defined below), and where

$$w_B = \frac{\mathrm{cov}_B(\overline{X}_{.},\overline{F}_o(X))}{\mathrm{cov}(X,F(X))}, \tag{7.28}$$

where $X = X_o$ and $F = F_o$. The first term of the right-hand side of (7.27) can be interpreted as the intra (within)-group component, while the second term is the inter (between)-groups component. β_m is a kind of a Gini regression coefficient to be defined later.

Proof The decomposition of the Gini covariance is based on $\mathrm{cov}(Y,F_o(X)) = \sum_{m=1}^{M} p_m \mathrm{cov}_m(Y,F_o(X)) + \mathrm{cov}_B(\overline{Y}_{.},\overline{F}_{\cdot o}(X))$,
where $F_o(X)$ is the overall cumulative distribution of the explanatory variable and p_m is the share of subgroup m in the population (see Chap. 4 for details). The next step is dividing and multiplying by the same factors. That is,

$$\mathrm{cov}(Y,F_o(X)) = \sum_{m=1}^{M} p_m \frac{\mathrm{cov}_m(Y,F_o(X))}{\mathrm{cov}_m(X,F_o(X))} \frac{\mathrm{cov}_m(X,F_o(X))}{\mathrm{cov}_m(X,F_m(X))} \mathrm{cov}_m(X,F_m(X))$$

$$+ \mathrm{cov}_B(\overline{Y}_{.},\overline{F}_{\cdot o}(X)) =$$

$$= \sum_{m=1}^{M} p_m \frac{\mathrm{cov}_m(Y,F_o(X))}{\mathrm{cov}_m(X,F_o(X))} O_m \Delta_m + \mathrm{cov}_B(\overline{X}_{.},\overline{F}_{\cdot o}(X)) \frac{\mathrm{cov}_B(\overline{Y}_{.},\overline{F}_{\cdot o}(X))}{\mathrm{cov}_B(\overline{X}_{.},\overline{F}_{\cdot o}(X))},$$

where $O_m = \frac{\mathrm{cov}_m(X,F_o(X))}{\mathrm{cov}_m(X,F_m(X))}$ is the overlapping between group m and the overall population, $\Delta_m = \mathrm{cov}_m(X,F_m(X))$ is (a quarter of) the Gini mean difference of X_m, and $\beta_m = \frac{\mathrm{cov}_m(Y,F_0(X))}{\mathrm{cov}_m(X,F_0(X))}$ is a kind of Gini regression coefficient denoted by β_m. The reason we use the term "kind of" is because the covariance is taken over subgroup m, while $F_o(X)$ refers to the cumulative distribution of the entire population.

Using the above notation we get

$$\beta_N = \sum_{m=1}^{M} w_m \, \beta_m + w_B \frac{cov_B(\bar{Y}_{.}, \bar{F}_{.o}(X))}{cov_B(\bar{X}_{.}, \bar{F}_{.o}(X))},$$

where $w_m = \frac{p_m O_m \Delta_m}{\Delta_0}$ is the contribution of the group to the overall variability, $\Delta_0 = cov(X, F(X))$ is (one-fourth of) GMD of X_0 and $w_B = \frac{cov_B(\bar{X}_{.}, \bar{F}_{.o}(X))}{cov(X, F(X))}$.

The proof for the OLS regression coefficient follows similar lines. The final equation is given in (7.29).

$$\beta^{OLS} = \sum_{m=1}^{M} v_m \, \beta_m^{OLS} + v_B \, \beta_B^{OLS}, \tag{7.29}$$

where $v_m = \frac{p_m \sigma_m^2}{\sigma_o^2}$, $v_B = \frac{\sigma_B^2}{\sigma_o^2}$,
and σ_B^2 is the variance of the vector of the sample means.

Equation (7.29) and its equivalent for the Gini (7.27) offer a connection between the Wald estimator which is actually the between-groups regression coefficient (Pakes, 1982; Wald, 1940) and the overall regression coefficient. In addition, they offer the possibility to evaluate the effect of binning (Wainer, Gessaroli, & Verdi, 2006) on the regression coefficient. Binning is the process of substituting a group of observations by its average. As can be seen from (7.26) its implication is the omission of the intra-group component from the regression coefficient.

It is easy to see that if the grouping is according to nonintersecting sections of X, then the decomposition of the Gini regression coefficient is identical in its structure to the decomposition of the OLS regression coefficient. On the other hand, whenever there is overlapping between the groups according to X, then the share of between-groups co-Gini can be totally different from the share of the between-groups (ordinary) covariance. As shown in Chaps. 4 and 22, the higher the overlapping is, the lower is the share of the between-groups component in the Gini of the overall distribution, while the share of the between-groups in the variance decomposition remains unaffected. (Frick, Goebel, Schechtman, Wagner, & Yitzhaki, 2006 showed it for the decomposition of the Gini, while here we decompose the co-Gini. The extension is straight forward).

There are various reasons for grouping of observations. One possibility is that the grouping is intended to reduce the variability in order to achieve a higher portion of explained variability. Another possibility is that the grouping results in omitting some negating effects that happen to occur in the subgroup. The decomposition proposed here enables the reader to see which effect has led to the result: is it the innocent need to make the results more robust or is it caused by an overzealous investigator who wants to prove his point.

7.1.4 A Presentation Based on Concentration Curves

A special case of decomposition which is applicable only for nonoverlapping groups can be presented graphically. It is based on an alternative, a bit complicated but useful presentation of the semi-parametric Gini regression coefficient as related to areas enclosed by concentration curves. This approach was initiated by Taguchi (1981, 1987). It is based on the tools developed by Blitz and Brittain (1964) for presenting the Gini correlation as based on areas enclosed by absolute concentration curves (ACC). Actually, Taguchi was the first one to apply Gini regression in a multiple regression framework. However his presentation is based on relative concentration curves and the details are in a book written in Japanese. As far as we know this presentation did not get the publicity which we believe it deserves.

In order to make the presentation user-friendly we derive two additional curves that are based on the ACCs. The objectives of this section are twofold. First we present the semi-parametric Gini regression coefficient as the area between two curves. Second, we use the graphical presentation to suggest a way that enables the investigator to check whether the relationship between the dependent variable and the explanatory variable is monotonic over the entire range of the explanatory variable or not. Specifically, assume that we truncate the distribution of the explanatory variable from above or below. Would the sign of the Gini regression coefficient (and the OLS) stay the same? The use of ACC enables us to see whether the sign of the truncated distribution changes, whether a monotonic transformation of the explanatory variable can change the sign of the OLS regression coefficient, and whether a monotonic transformation of the dependent variable can change the sign of the OLS regression coefficient. Being able to change the sign of the coefficient (and the conclusions that follow) by a legitimate transformation makes the conclusions drawn from the data questionable. An advantage of the Gini semi-parametric regression coefficient is that a monotonic transformation of the explanatory variable cannot change its sign. It can only change its magnitude.

Concentration curves play an important role in performing sensitivity analysis in the regression. Therefore we repeat their properties here (without proofs). Readers who are interested in the proofs can find them in Chap. 5.

We start by introducing two additional curves: the LMA and the NLMA curves.

The LMA curve represents the area between the line of independence (or: equality, denoted by LOI) and the ACC and is formally defined below. The NLMA curve is simply a normalized version of LMA, after dividing it by GMD. These two curves are needed when one is interested in the co-Ginis and in Gini correlations or in cases when one is interested in a regression coefficient. In these cases it is convenient to make some adjustments to the ACC to make the properties of the covariance easier to follow and visualize.

Formally, let $g(x) = E\{Y|X = x\}$ be the conditional expectation of Y given X and recall that $p = F(X(p))$. We will refer to $g(x)$ as the regression curve.

Definition of LMA LMA is defined as the LOI minus the ACC of Y with respect to *X*. Formally, $LMA_{YoX}(p) = \mu_Y p - ACC_{YoX}(p)$. The properties of LMA are the following:

(a) The LMA starts at (0,0) and ends at (1,0).
(b) The derivative of the LMA with respect to p (at x(p)) is $\mu_Y - E(Y|X = x(p))$. This follows directly from the definition of the LOI and the ACC. As a consequence, the $LMA_{YoX}(p)$ is increasing (decreasing, constant) if and only if $\mu_Y - g(x(p))) > (<, =)\ 0$.
(c) The LMA is concave at p (convex, straight line) if and only if $\partial g(x(p))/\partial x(p) > (<,=)\ 0$.
(d) If X and Y are independent then $ACC_{YoX}(p)$ is a straight line which coincides with the LOI implying that the LMA curve coincides with the horizontal axis. Properties (c) and (d) enable the user to identify sections with constant, increasing and decreasing slopes of the regression curve. Linearity of LMA implies that its derivative (wrt p) which is equal to $\mu_Y - E(Y|X = x(p))$ (by (b)) is constant (say equals to c). That means that the regression curve is constant (flat) at $\mu_Y - c$, and the slope of the regression curve is zero. An increasing (decreasing) LMA implies that g(x) is below (above) its mean value, concavity (convexity) of LMA means an increasing (decreasing) slope.
(e) The area between the LMA and the horizontal axis is equal to $cov(Y, F_X(X))$ (Yitzhaki, 1990). Note that if the curve intersects the horizontal axis then the sign of $cov(Y, F_X(X))$ depends on the magnitudes of the areas above and below the horizontal axis (and these areas add up to $cov(Y, F_X(X))$).
(f) The LMA is above the horizontal axis for all p if and only if $cov(Y, T(X)) > 0$ for all continuous differentiable monotonically increasing functions T(X).
(g) The LMA enables us to see whether deleting observations from the population will affect the sign of cov(Y,F(X)). To see that, note that for a given value x_u, deleting all the observations with $X > x_u$ will result in truncating the curve at $F(X_u)$ and connecting the point (0,0) with the point at the end of the new curve. The same will happen with truncation from below and truncations from both above and below. We can evaluate the sign of cov(Y,F(X)) of the truncated distribution by simply looking at the curvature of LMA in the sections that were not deleted.

The advantage of using the LMA (instead of the ACC) is that it is easy to detect what will happen to cov(Y,F(X)) (and hence to the sign of the regression curve) if sections of observations of X are omitted from the regression, as will be illustrated in Chap. 19).

For the purpose of analyzing the effect of a monotonic transformation (or truncation) on the sign of the OLS regression coefficient, one needs a modified LMA curve for which the area beneath it will be equal to cov(Y,X) (rather than to cov(Y,F(X))). A simple transformation can make the curve applicable to OLS: if one substitutes the horizontal axis to be X instead of F_X in the ACC curve, then the area between the new curve and the horizontal axis will be equal to cov(Y,X)

(Yitzhaki, 1998). However, the properties of the curve change and further research is needed to study them.

A necessary and sufficient condition for the ability to change the sign of the OLS regression coefficient by applying a monotonic transformation to the explanatory variable is that the LOI and the ACC curves intersect. The explanation is that an intersection means that the overall covariance is composed of negative and positive intra-group covariances, while the between-groups component is equal to zero. An intersection will continue to hold even if we change the LOI and ACC to those that are relevant for cov(Y,X). By applying a monotonic increasing transformation T(X) we can shrink or expand the negative and positive covariances as we please. For example, if the overall covariance is negative and one wants to change it to be positive, then by choosing T(X) with a derivative greater than one in the sections of X with a positive covariance, and smaller than one in the sections of X with a negative one we will be able to change the sign of cov(Y,T(X)) from a negative to a positive sign. Note that if the LOI does not intersect the ACC, then it is impossible to decompose the covariance into negative and positive components with a zero between-groups component. Therefore, the sign of cov(Y,X) cannot be changed by a monotonic transformation applied to X. In the case of the Gini regression coefficient, it is simpler to see why a monotonic transformation on X will not change the sign: the relevant covariance is cov(Y, F(X)) which is not sensitive to a monotonic increasing transformation of X. Hence a monotonic transformation of X cannot change the sign of the Gini regression coefficient. However, it can change its magnitude. Therefore, we may conclude that the Gini regression coefficient is less sensitive than the OLS with respect to monotonic transformations applied to the explanatory variable.

To enable graphical illustration of the components of the Gini regression coefficient, the LMA curve is normalized by dividing it by cov(X, F(X)). The normalized curve is denoted by NLMA. As a result, the total area that is enclosed between the NLMA curve and the horizontal axis is equal to the value of the Gini regression coefficient. The areas above the horizontal axis represent positive contributions, while negative contributions are represented by the areas below it.

To illustrate the usefulness of the NLMA curve consider the curve in Fig. 7.1, then

(a) The vertical axis is the value of LOI minus the ACC (divided by cov(X,F(X)). The total area between the curve and the horizontal axis is β_N = cov(Y,F(X))/cov(X,F(X)) which is the Gini regression coefficient. As a result we can immediately see that the area OAB is positive, contributing towards a positive value to the Gini regression coefficient, while the area DCB is negative, hence it is reducing the value of the coefficient. The total area which is the value of the Gini regression coefficient is positive, because the positive contribution is larger than the negative one.
(b) The same sections (in terms of the transformed values on the horizontal axis) are contributing toward the OLS regression coefficient. However, the

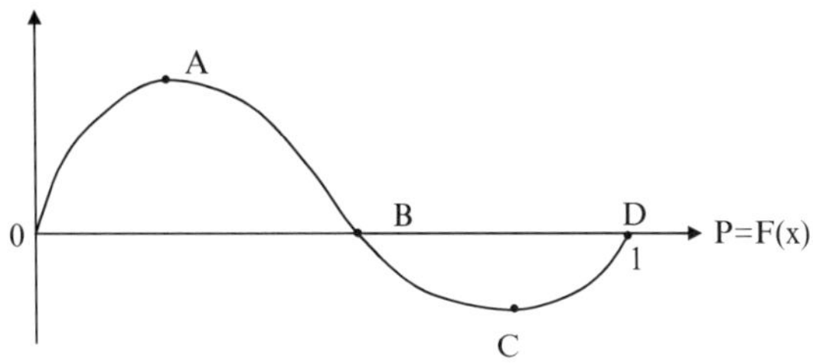

Fig. 7.1 The NLMA curve

magnitudes (but not the signs) may be different. Therefore, the overall sums of the areas may differ resulting in different signs of OLS and Gini regressions.

(c) The slope of the curve is equal to $[\mu_Y - E(Y|X = x(p))]/\text{cov}(X,F(X))$ and the concavity (convexity) of the curve is determined by the sign of the local regression coefficient, that is, $\text{Sign}\left\{\frac{\partial E[Y|X=x(p)]}{\partial x(p)}\right\} = -\text{Sign}\left\{\frac{\partial^2 \text{NLMA}}{\partial^2 p}\right\}$. For example, the curve is concave between zero and B implying that the local regression coefficient is positive, and the opposite occurs between B and D.

(d) By truncating the range of the explanatory variable from below or above (i.e., along the horizontal axis), connecting the extreme points of the truncated curve by a line we can evaluate the area between the curve and the new line and learn about the sign of the regression coefficient in the truncated distribution. For example, truncating the distribution of X from below at A causes cov(Y,F(X)) of the truncated distribution to be negative. (The positive contribution to the area will be smaller than the negative one.)

(e) Consider the point B where the curve intersects with the horizontal axis, as illustrated in Fig. 7.1. Then we can read the following information from the curve:(e.1) The overall Gini regression coefficient is positive (because the area above the horizontal axis is larger than the one below it).(e.2) One can divide the data into two groups according to the value of X at the point B. Both groups have equal mean values of the Y variable (because the deviation from the mean value of Y is zero for both groups).The OLS and Gini regression coefficients for the group of Xs below B (above B) is positive (negative) and it will have the same sign for all monotonic transformations applied to X.

(f) Draw two lines tangent to the curve: one that starts at 0 (0F) and the other starts at 1 (ED) and both are tangents to the curve as illustrated by the lines 0F and ED in Fig. 7.2. Then one can truncate the distribution from above to get a new distribution limited to 0H or truncate it from below to get a truncated distribution on GD. The LMA of the former (truncated) distribution is above the (new) horizontal line, while the latter is below it. Then, by property (f) of the LMA curve all monotonic transformations of X will result in positive Gini and OLS regression coefficients on the section 0H and negative Gini and OLS regression coefficients along GD.

Fig. 7.2 The NLMA curve

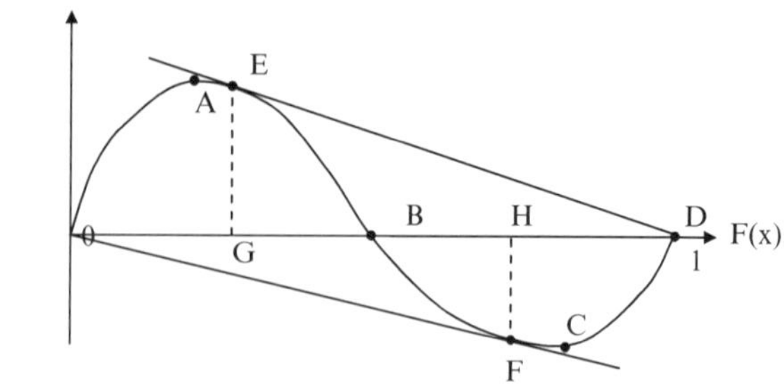

Fig. 7.3 The NLMA curve

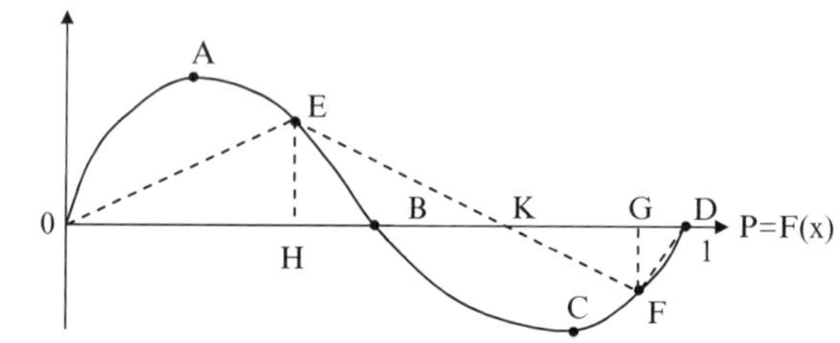

Figure 7.3 demonstrates the connection between the decomposition according to the LMA curve (Sect. 7.1.4) and the decomposition of the covariance (Sect. 7.1.3). Assume now that we arbitrarily divide the observations of X into three nonoverlapping groups: 0H, HG, and GD as shown in Fig. 7.3. Then the Gini and OLS regression coefficients can be decomposed into three intra-group components and one between-groups component. Out of the three intra-group components the one on section 0H is positive and will remain positive for all monotonic transformations of X (the curve lies above the line connecting the extreme points), while those on HG and GD are negative (the curves are below the lines connecting the extreme points). The between-groups component is based on the line segments connecting 0E, EF, and FD. It can be either negative or positive. Clearly, monotonic transformations can change the sign of the regression coefficient both in OLS and in Gini regressions.

To sum up: the semi-parametric Gini regression coefficient can be presented as the area between two curves. The graphical presentations enable the investigator to check whether the relationship between the dependent variable and the explanatory variable is monotonic over the entire range of the explanatory variable or not. It can help the user answer questions such as if I truncate the distribution of the explanatory variable from above or below, would the sign of the Gini regression coefficient (and the OLS) stay the same? The use of ACC enables us to see whether the sign of the truncated distribution changes, and whether a monotonic transformation of the explanatory variable can change the sign of the OLS regression coefficient.[3]

[3] To see if a monotonic transformation of the dependent variable can change the sign of the OLS regression coefficient, one can simply plot the ACC of X with respect to Y.

One of the commonly used assumptions in regression is that the model is linear over the entire range of the explanatory variable. This assumption cannot be tested as part of the regression method used (unless there are repeated observations at least at one level of the explanatory variable or by using two different methodologies) because by construction the error term is orthogonal to the explanatory variable. The use of the concentration curve enables the user to visually see whether the assumption of the linearity of the regression holds for the entire range or maybe large sections of the data violate the linearity assumption or even the monotonicity. Properties (c) and (d) of the LMA enable the user to identify sections with constant, increasing and decreasing slopes of the regression curve: linearity of LMA implies that the slope of the regression curve is zero, concavity of LMA means a positive slope, while convexity means a negative slope. In this sense the uses of the GMD and its Lorenz presentation enable the user to examine the linearity assumption on one hand, or alternatively, to give up on the linearity assumption and to derive the weighting scheme of the slopes from economic theory. In the latter case the interpretation of the Gini regression coefficient would be a weighted average of slopes, weighted according to economic theory. This alternative approach is demonstrated in Chaps. 15 and 18 which are dealing with applications in welfare economics and finance. In addition, this alternative interpretation indicates that the usual assumption that there exists a linear relationship should not be taken for granted. In some cases, mainly in welfare economics and finance, it is not needed, as will be shown in Chaps. 15 and 18 where economic theory requires only a weighted average of slopes, weighted by the marginal utility of income. In other cases the linearity assumption is simply not supported by the data. In these cases— the linear regression model can still be interpreted as a linear approximation to the true regression curve (Chap. 21). Note, however, that in those cases the linear regression model is not useful for prediction. We discuss the validity of the linearity assumption further in Sect. 7.3.

7.1.5 Similarities and Differences Between OLS and Gini Semi-Parametric Regression Coefficients

The properties of the OLS and the Gini regression coefficients were discussed in detail in the previous sections. In this section we summarize the main similarities and differences between the two. We focus on four issues: the presentations as weighted averages of slopes, the use of curves to view the contributions of the sections of the explanatory variable to the regression coefficient, the decomposition of a slope to the contributions of subgroups, and the relationship between direct and reverse regressions.

1. Weighted averages of slopes. Similarity: both can be expressed as weighted averages of (the same) slopes. Because the slopes between adjacent observations are determined by the data, the choice of the regression method is actually a

choice of the weighting scheme. The weighting scheme (for both) is affected by properties of the distribution of the explanatory variable: (a) the rank of each observation and (b) the difference between each observation and the adjacent one. Both methods do not actually rely on the linearity assumption of the regression curve and, as such, can be viewed as linear approximations of the regression curve.

Difference: the difference lies in the weighting scheme. While in the Gini method the distance between adjacent observations is taken as is, it is squared under the OLS regime.

2. The use of curves. Similarity: both can be expressed as areas between two curves. Difference: the difference is in the curves used. As we have shown above, the weighting schemes of the Gini regressions are based on the absolute Lorenz curve, while for OLS the equivalent of the Lorenz curve for the variance is used. However, it is easier to draw conclusions from the curves in the case of Gini than in the OLS.
3. Decomposition by subgroups. Similarity: both can be decomposed into the contributions of subgroups. Difference: Gini takes into account the overlapping while OLS does not. It was shown that if the grouping is according to nonintersecting sections of X, then the decomposition of the GMD is identical in structure to the decomposition of the variance. On the other hand, whenever there is overlapping between the groups according to X, then the share of the between-groups Gini can be totally different from the share of the between-groups variance. It can be easily shown (Frick et al., 2006) that the higher the overlapping the lower the share of the between-groups component in the overall Gini is, while the share of the between-groups component in the variance decomposition remains unaffected.
4. The relationship between direct and reverse regression (Goldberger, 1984). Goldberger introduced the reverse regression in the area of discrimination. Instead of asking whether, given identical characteristics, men earn more than women, he asks whether, given the same salaries, women are more qualified than men. Similarity: The similarity occurs when we look at the multiplication between the direct and reverse regression coefficients. Note that

$$\beta_{\mathrm{YX}}^{\mathrm{OLS}}\ \beta_{\mathrm{XY}}^{\mathrm{OLS}} = \rho^2, \tag{7.30}$$

$$\beta_{\mathrm{YX}}^{\mathrm{N}}\ \beta_{\mathrm{XY}}^{\mathrm{N}} = \Gamma_{\mathrm{XY}}\,\Gamma_{\mathrm{YX}}. \tag{7.31}$$

(Proofs are immediate hence they are not presented here.)

However, although formally similar, the Gini correlations can have different signs, and so can the direct and indirect regression coefficients. Difference: the dissimilarities occur when we look at the ratio of the two:

$$\frac{\beta_{\mathrm{YX}}^{\mathrm{OLS}}}{\beta_{\mathrm{XY}}^{\mathrm{OLS}}} = \frac{\sigma_{\mathrm{Y}}^2}{\sigma_{\mathrm{X}}^2}, \tag{7.32}$$

$$\frac{\beta^{N}_{YX}}{\beta^{N}_{XY}} = \frac{\Gamma_{YX}}{\Gamma_{XY}} \frac{\Delta^2_Y}{\Delta^2_X}. \tag{7.33}$$

That is, the ratio in the Gini regression depends on the levels of the Gini correlations as well as on their signs.

The message that should be kept in mind is that although seemingly similar, one has to be careful in translating the intuition from OLS into Gini. To see that, note that under the Gini regime one can get different signs between the direct and reverse regressions, leading to more complicated interpretations of the results. This is another demonstration of the argument that "the Gini reveals more."

7.2 The Minimization Approach

The minimization approach is based on minimization of the GMD of the residuals. In order to be able to optimize one has to specify the model and the target function, which means in this case to assume that the model is linear. The approach is similar to least absolute deviation (LAD) regressions (Bassett & Koenker, 1978) or quantile regression (Koenker & Bassett, 1978). Here, instead of minimizing the sum of absolute deviations of the residuals, or the weighted sum of the absolute deviations from a quantile of the residuals, the GMD of the residuals (which is the mean of the absolute differences between all pairs of residuals) is minimized. Parameters and estimators derived following the minimization approach will be denoted by the subscript M. (This kind of regression has been developed by Jurečková (1969, 1971); Jaeckel (1972); McKean and Hettmansperger (1978); and Hettmansperger (1984)). Hettmansperger refers to the method as R-regression because the sum of the products of the ranks of the residuals and the residuals themselves is minimized. Because our main target is to imitate the OLS with the Gini, the properties of the R-regression approach will not be repeated here.[4] An application in economics can be found by Chaudhury and Ng (1992). For our argument, only the orthogonality condition is needed. Note, however, that R-regression requires the specification of a model.

Consider the following model:

$$Y \equiv \alpha + \beta X + \varepsilon. \tag{7.34}$$

Note that we do not impose any assumptions on ε.

[4] It is worth emphasizing that the connection between R-regression and GMD was not recognized in the literature mentioned above. Many of the properties of those regressions can be traced to the properties of GMD. Bowie and Bradfield (1998) compare the robustness of several alternative estimation methods in the simple regression case and find the minimization of the GMD among the most robust methods.

Proposition 7.3 *The β which minimizes the GMD of the residuals forms the normal equation of the type cov(X, F(ε)) = 0.*

Proof Denote the ordered residuals by $\varepsilon_1 \leq \varepsilon_2 \leq ... \leq \varepsilon_n$. Then

$$G_\varepsilon = \sum_{i,j} |(y_j - y_i) - \beta(x_j - x_i)| = 2\sum_{i<j} [(y_j - y_i) - \beta(x_j - x_i)], \qquad (7.35)$$

from which we get that

$$\frac{\partial G_\varepsilon}{\partial \beta} = -2\sum_{i<j}(x_j - x_i) = 4\sum_{i=1}^{n} x_i\left[i - \frac{n+1}{2}\right] = 4\mathrm{cov}(X, \frac{R_\varepsilon}{n}) \qquad (7.36)$$

where R_ε is the rank of ε (hence $\frac{R_\varepsilon}{n}$ is the empirical cumulative distribution function of ε). Recall that the residuals are ordered. Hence the rank of ε_i is i. At the minimum, the derivative (if it exists) is equal to zero. The proof that

$$-2\sum_{i<j}(x_j - x_i) = 4\sum_{i=1}^{n} x_i\left[i - \frac{n+1}{2}\right]$$

can be found in Appendix 7.2. (See Olkin and Yitzhaki (1992) for details).

In the sample, (7.36), which is the derivative of (7.35) with respect to β, is a step function because (7.35) is piecewise linear. Hence, it may happen that the solution for β is determined up to a range.

In addition, we point out that

1. The Gini estimator derived under the minimization approach is similar to the minimization of the average of all possible quantile-regression target functions (Koenker & Bassett, 1978). Hence, one can view it as an extension of quantile regressions (see (2.3) in the alternative presentations of the GMD). Similar to those regressions, the estimator cannot be explicitly expressed and it is derived numerically.
2. The orthogonality condition for the minimization of the GMD of the residuals is a co-Gini between the residuals and the explanatory variable. As such, we can visualize it using an ACC or an LMA curve. This graphical presentation will be useful in order to examine the linearity of the regression curve. If the regression curve is linear, we should expect the ACC of X with respect to ε to oscillate randomly along the horizontal axis, as we discuss in the next section. In Chap. 20 we apply the concept of the orthogonality condition to check for the linearity of the regression curve. However, we do not actually estimate this regression, because it is a deviation from the main target of the book.

7.3 The Combination of the Two Gini Approaches

The classical regression model is based on two underlying assumptions. The first assumption is that the relationship between Y and X is linear. The second assumption is that the residual is independent of the explanatory variable. The fact that the

GMD has two covariances that are defined between each pair of variables can be useful in order to examine whether the implicit assumptions assumed in the classical regression hold in the data. To see that, note that the two regressions produce two normal equations.

The two normal equations are

$$\mathrm{cov}(\varepsilon_N, F(X)) = 0 \text{ and } \mathrm{cov}(X, F(\varepsilon_M)) = 0, \tag{7.37}$$

where ε_N and ε_M are the residuals from the semi-parametric and the minimization regression methods, respectively. If the specification of the model is correct so that the residual is independent of the explanatory variable, then we should expect that in addition to (7.37) we will get

$$\mathrm{cov}(\varepsilon_M, F(X)) = 0 \text{ and } \mathrm{cov}(X, F(\varepsilon_N)) = 0. \tag{7.38}$$

That is, the normal equation of each regression method should hold when applying it to the alternative set of residuals. Note that while the equations in (7.37) are produced by the regression method and hence cannot be challenged, the equations in (7.38) can be tested to see whether they hold or not.

A possible specification test for the model is whether the two Gini regressions lead to statistically similar residuals, and as a result, to the same regression coefficient (to be detailed below).

We note that we could do the same exercise by comparing the OLS and the Gini regression coefficients. That is, if the specification of the model is correct then we should expect the OLS and the Gini residuals to be identical, except for a random variation. To see that, recall that both the Gini semi-parametric and OLS regression coefficients can be expressed as weighted sums of slopes between adjacent observations. If the relationship between Y and X is linear along the entire range of the explanatory variable then the slopes between adjacent observations are all equal to one constant. Therefore the two weighting schemes will give the same (constant) slope. Hence the residuals in the sample will be the same (up to random variation). We remind the reader that the OLS and Gini semi-parametric regressions will be identical for any regression curve when the explanatory variable is uniformly distributed (as was shown in Sect. 7.1.1).

The advantage of using the two Gini regression methods is that we are using the same methodology and the same definition of distance. This last advantage will be more apparent in the next chapter, when we deal with multiple regressions. In order to check whether the two methods give similar residuals, we do not need to estimate the two regressions. It is sufficient to use just one regression.

The recommended steps to follow are

1. Estimate the semi-parametric regression—by construction one gets $\mathrm{cov}\ (\varepsilon_N,\ F(X)) = 0$.
2. Test whether $\mathrm{cov}\ (X, F(\varepsilon_N)) = 0$. If the covariance is significantly different from zero, then the specification test fails. Note that having $\mathrm{cov}\ (X, F(\varepsilon_N)) = 0$ is a

necessary condition but not a sufficient one. Only if the vectors ε_N and ε_M are identical we can claim that we have a necessary and sufficient condition.

Alternatively, one can reverse the order. That is, one can apply the minimization approach to the GMD regression and derive the residuals. Then one can check whether the normal equation of the semi-parametric Gini regression holds. Needless to say that because we are dealing with aggregated results, it may happen that the results of the two alternative procedures contradict each other. That is, one test results in failing to reject the null hypothesis and the other rejects it. Our preference is to use the former method because the first stage does not require the specification of a model.

The intuition behind the above-mentioned procedure is as follows. If the population exhibits a linear model with a residual that is independent of the explanatory variable, then the normal equations of the OLS and the two Gini regressions all yield approximately the same vector of residuals.

Note that if the vector of the residuals in the population forms a linear relationship with the explanatory variable then the three methods mentioned above will not be able to estimate the true regression coefficient correctly. Formally if the model in the population is

$$Y = \alpha + \beta X + \varepsilon$$

and $\varepsilon = \alpha_1 + \beta_1 X + \varepsilon_1$ with ε_1 and X being independent, then the model becomes

$$Y = \alpha + \beta X + \alpha_1 + \beta_1 X + \varepsilon_1 = (\alpha + \alpha_1) + (\beta + \beta_1) X + \varepsilon_1,$$

and all the methods will estimate $(\beta + \beta_1)$ rather than β. In other words, whenever X is connected to Y through a linear relationship, and ε is connected to X through a linear relationship, one cannot distinguish between the direct connection between X and Y and the indirect connection through ε. In order to correctly estimate β an IV method is needed.

Further insight can be gained by looking at the NLMA curve of the residual (see Sect. 7.1.4 above for the definition of the NLMA curve). Consider the case of the semi-parametric regression. Assume that we have derived the vector ε_N. Then, by construction $\text{cov}(\varepsilon_N, F(X)) = 0$. Assuming that X and the residual are statistically independent implies that the theoretical NLMA curve of X with respect to ε should coincide with the horizontal axis. Therefore we should expect the NLMA curve to oscillate randomly around the horizontal axis. The curve offers a visual inspection and as we show later one can also develop a statistical test.

7.4 Goodness of Fit of the Regression Model

The objective of this section is to discuss measures of goodness of fit for the Gini regression method.

Let the linear predictor of the regression curve be $\hat{Y} = a + bX$ and define the residual by $e = Y - \hat{Y}$.

In OLS the commonly used measure of the goodness of fit is R^2 which measures the proportion of variability in the data that is explained by the model. That is

$$R^2 = \frac{SSR}{SSTO} = 1 - \frac{SSE}{SSTO}, \tag{7.39}$$

where SSR, SSE, and SSTO are the model sum of squares, the error sum of squares, and the total sum of squares, respectively. It is based on the partitioning of the total sum of squares into two components: a sum of squares due to the model fitted and an error sum of squares. We now turn to Gini regression and use the decomposition of a Gini of a linear combination of random variables (see Chap. 4 for details).

Let $Y = \hat{Y} + e$.

We start with a simple decomposition as follows

$$\Delta_Y = \text{cov}\left(\hat{Y} + e, F(Y)\right) = \text{cov}\left(\hat{Y}, F(Y)\right) + cov(e, F(Y)) =$$
$$= \frac{\text{cov}(\hat{Y}, F(Y))}{\Delta_{\hat{Y}}} \Delta_{\hat{Y}} + \frac{\text{cov}(e, F(Y))}{\Delta_e} \Delta_e = \Gamma_{\hat{Y}Y}\, \Delta_{\hat{Y}} + \Gamma_{eY}\, \Delta_e.$$

Using the decomposition of the square of the GMD of a linear combination of random variables (4.6) we can write

$$\Delta_Y^2 = (D_{Y\hat{Y}}\, \Delta_{\hat{Y}} + D_{Ye}\, \Delta_e)\, \Delta_Y + \Delta_{\hat{Y}}^2 + \Delta_e^2 + \Delta_{\hat{Y}}\, \Delta_e(\Gamma_{\hat{Y}e} + \Gamma_{e\hat{Y}}),$$

where Δ_Y denotes the GMD of Y, $\Gamma_{\hat{Y}e}$ is the Gini correlation between $\hat{Y}$ and e, and $D_{Y\hat{Y}} = (\Gamma_{Y\hat{Y}} - \Gamma_{\hat{Y}Y})$, i.e., it is the difference between the two Gini correlations of Y and $\hat{Y}$.

In the Gini regression, one of the Gini correlations between $\hat{Y}$ and e is zero by construction and the other one is zero if the specification of the model is correct. Hence, if the two correlations between e and $\hat{Y}$ are zero then we get

$$\Delta_Y^2 = (D_{Y\hat{Y}}\, \Delta_{\hat{Y}} + D_{Ye}\, \Delta_e)\, \Delta_Y + \Delta_{\hat{Y}}^2 + \Delta_e^2. \tag{7.40}$$

Furthermore, if the two Gini correlations between Y and $\hat{Y}$ are equal ($\Gamma_{Y\hat{Y}} = \Gamma_{\hat{Y}Y}$) and the same holds for the two Gini correlations between Y and e then we get

$$\Delta_Y^2 = \Delta_{\hat{Y}}^2 + \Delta_e^2, \tag{7.41}$$

which is identical in structure to the OLS decomposition of the total sum of squares. In general the two additional terms in the GMD decomposition (7.40) can be negative or positive. If the distribution of $(\hat{Y}, e)$ is bivariate normal, then their sum, Y, is normally distributed, implying that Y, $\hat{Y}$, and e are exchangeable up to a linear transformation. In this case both Ds in (7.40) are equal to zero and we are left with the same elements as in the OLS. However if $\hat{Y}$ and e have different distributions, then the distribution of Y will be different than the distributions of $\hat{Y}$ and e, and the smaller of the two additional terms will indicate the distribution that is closer in shape to the distribution of Y. Further research is needed to get a better explanation of the role of the additional elements.

Starting with the simplest case in which the specification of the model is correct, the two Gini correlations between Y and $\hat{Y}$ are equal ($\Gamma_{Y\hat{Y}} = \Gamma_{\hat{Y}Y}$) and the same holds for the two Gini correlations between Y and e, and we get (7.41) which is similar in structure to the partitioning of the total sum of squares (see (7.39)). Therefore a natural way to define an R^2—equivalent is by

$$GR^2 = \frac{\Delta_{\hat{Y}}^2}{\Delta_Y^2} = 1 - \frac{\Delta_e^2}{\Delta_Y^2}. \tag{7.42}$$

(Olkin & Yitzhaki, 1992). However as can be seen from the more general decomposition of Δ_Y^2, GR^2 is not parallel to R^2.

If the specification of the model is correct, then following the decomposition of (7.40), we get

$$GR^2 = \frac{\Delta_{\hat{Y}}^2}{\Delta_Y^2} = 1 - \frac{\Delta_e^2}{\Delta_Y^2} - \frac{D_{Y\hat{Y}}\,\Delta_{\hat{Y}} + D_{Ye}\,\Delta_e}{\Delta_Y}. \tag{7.43}$$

Note that the third term on the right-hand side can be either positive or negative, implying that the measure of goodness of fit for this case can be larger or smaller than the measure obtained under the assumptions that led to (7.41). Further research is needed to see if (7.43) can add additional insight.

Additional measures of the quality of the fit of the model to the data in the Gini regression are the Gini correlations between the dependent variable and the predictor. Formally

$$\Gamma_{Y\hat{Y}} = \frac{\mathrm{cov}(Y,F(\hat{Y}))}{\mathrm{cov}(Y,F(Y))} \quad \text{and} \quad \Gamma_{\hat{Y}Y} = \frac{\mathrm{cov}(\hat{Y},F(Y))}{\mathrm{cov}(\hat{Y},F(\hat{Y}))}, \tag{7.44}$$

where $\hat{Y}$ is the predicted variable. In other words, we substitute the R^2 of OLS by three measures: the one in (7.43) and the two in (7.44). Note, however, that in the OLS, the parallels to these three measures are numerically equal.

To sum up: in the general case one can use several goodness of fit indicators such as the two correlation coefficients between the fitted and observed values, and/or one minus the (square of the) ratio of the GMD of the residuals divided by the GMD of the dependent variable. In OLS these three measures are identical. Further research is needed to evaluate the practical contributions of the different measures of fit offered by the Gini.

7.5 A Test of Normality

One of the underlying assumptions in the OLS regression is the normality of the residuals. The aim of this section is to describe a test for normality that can be offered by the Gini methodology. This is an omnibus test for normality. It is appropriate for detecting deviations from normality due to either the skewness or the kurtosis. The following test statistic was suggested by D'Agostino (1971) and the literature based on it.

Let $X_1,\ldots,X_n$ represent a random sample of size n and let $X_{(1)},\ldots,X_{(n)}$ represent the ordered observations. The test statistic D for testing that the sample comes from a normal distribution is

$$D = \frac{T}{n^2 S} \tag{7.45}$$

where

$$T = \sum_{i=1}^{n} \left\{ i - \frac{n+1}{2} \right\} X_{(i)}, \quad \text{and } S^2 = \frac{\sum_{i=1}^{n} (X_i - \overline{X})^2}{n}.$$

As will be shown in Chap. 9, T is a function of the estimator of the GMD. More precisely,

$T = n(n-1)\hat{\Delta}_X$, where $\Delta_X = E(|X_1 - X_2|) = 4\mathrm{cov}(X,F(X)) = \mathrm{GMD}$ and $\hat{\Delta}_X = U(\Delta_X)$ is the U-statistic for estimating the GMD. (For details see (9.4)). Hence one can write D'Agostino test statistic as

$$D = \left(\frac{n-1}{n}\right) \frac{\hat{\Delta}_X}{S}.$$

If the sample is drawn from a normal distribution the expected value of D and its asymptotic standard deviation (asd) are, respectively,

$E(D) = \frac{(n-1)}{2\sqrt{2n\,\pi}} \frac{\Gamma(\frac{n}{2} - \frac{1}{2})}{\Gamma(\frac{n}{2})}$, or approximately $(2\sqrt{\pi})^{-1} = 0.282$, and

$$\mathrm{asd}(D) = \frac{0.03}{\sqrt{n}}.$$

An approximate standardized variable, possessing asymptotically mean zero and variance unity, is

$$Z = \frac{D - (2\sqrt{\pi})^{-1}}{\mathrm{asd}(D)}.$$

If the sample is drawn from a non-normal distribution the expected value of Z tends to differ from zero. The direction of the difference depends on the alternative distribution. Simulation suggests that if the alternative distribution has greater than normal kurtosis, then Z tends to be on the average less than zero, while if the kurtosis is less than the normal kurtosis then Z tends to be greater than zero. D'Agostino provides a table of percentile points for small sample sizes (D'Agostino, 1972) as well as for moderate to large sample sizes (D'Agostino, 1971).

The D'Agostino's test is actually based on the ratio of the GMD to the standard deviation. Shalit (2010) suggests an alternative test of normality that is based on the shape of the ALC. Under the normal distribution, each percentile of the distribution contributes an equal share to the variance (see Sect. 7.1.1). Shalit's test (2010) is based on checking whether the contributions of all percentiles of the data are equal by looking at the sum of squares of the deviations of the contributions from equality. Sufficiently large sum of squares will lead to a rejection of the normality hypothesis. It seems that the difference between D'Agostino and Shalit is the following: both are based on the difference between the GMD and the standard deviation; but while D'Agostino calculates summary statistics first and then performs the test using the summary statistics, Shalit aggregates the deviation of each observation from its value under the equality assumption and then performs the test. Intuitively, Shalit's approach is likely to have more power because it is based on the deviation of each individual observation from its expected value under normality.

7.6 The Instrumental Variable Method

The method of instrumental variables (IV) is widely used to estimate parameters when (some of) the regressors are endogenous. Its popularity has increased as a result of its applicability to the evaluation of the impact of social programs (Angrist, 1990, and others). Recent investigations of its properties include, among others, Bound, Jaeger, and Baker (1995), Heckman and Smith (1995), Angrist, Imbens, and Rubin (1996), and Angrist and Evans (1998), who pointed out its main advantages and drawbacks.

At this point the method is fiercely debated—see, for example, Heckman (1992, 1997, 1999, 2000, 2001), Angrist and Imbens (1999), and Deaton (2009). In some sense the debate is philosophical. It is concerned with issues such as whether it is reasonable to view a variable as exogenous or not, whether one is allowed to view statistical correlation as representing causal relationship or not, and other issues such as how to reach evidence-based conclusions.

The use of the Gini method or other methods as alternatives to OLS cannot contribute to the philosophical debates that determine the approach of the researcher. Almost any concern that is raised with respect to the OLS regression can be raised with respect to the Gini (or any other) regression as well. Therefore we will not deal with those issues in this section. However because in general "the Gini reveals more" and this property carries through to regression, the use of the Gini method can shed light on some of the issues that are hotly debated. This is done by allowing the user to test whether some of the implicit assumptions made in order to reach the conclusion hold in the data. In other words, the use of the Gini methodology enables one to see whether the conclusion reached suffers from deficiencies that originate from some of the hidden assumptions of the OLS. To be specific, we rely heavily on the ability of the Gini methodology to find out whether the relationships between the variables are monotonic over the entire range or not, and whether the IV and the residual are correlated or not. Non-monotonic relationship between the IV and the explanatory variable it replaces means that one can get almost any result by using an IV method. Our method of investigation is a "bottom line" approach. We try to reveal how the estimators are constructed and by this way check whether the implicit assumptions hold or not. We show through the "bottom line" approach that while the OLS is based on the decomposition of the variance, the IV method is based on the decomposition properties of the covariance. We start by presenting the properties of the OLS-IV and later present its parallel concept under the Gini and EG methods.

7.6.1 The OLS Instrumental Variable Method

We start this section with a short technical reminder of the instrumental variables (IV) methodology. The usual scenario concerning the use of IV is to assume the following model:

$$Y \equiv \alpha + \beta X + \varepsilon. \tag{7.46}$$

Note that we present (7.46) as an identity because at this stage we have not assumed anything about the properties of ε except that it is a slack variable intended to close the identity. The next step is to assume that X and ε are correlated so that the OLS estimator is biased, and the direction of the bias depends on the sign of the correlation. To see this note that

$$\frac{\text{cov(Y,X)}}{\text{cov(X,X)}} = \beta + \frac{\text{cov}(\varepsilon\text{,X})}{\text{cov(X,X)}}.$$

To correct for the bias, a variable Z that is correlated with X but uncorrelated with ε is chosen to replace X, and a decomposition of cov(Y,Z) is performed instead of the decomposition of cov(Y,X).

That is,

$$\text{cov}(Y,Z) \equiv \beta\text{cov}(X,Z) + \text{cov}(\varepsilon,Z). \tag{7.47}$$

Because cov (ε, Z) is assumed to be equal to zero (a constraint imposed on the data in the estimation process, which changes (7.47) from an identity to an equation intended to derive β), the OLS-IV population's parameter, β_{OIV}, is defined as

$$\beta_{OIV} = \frac{\mathrm{cov}(Y, Z)}{\mathrm{cov}(X, Z)} = \frac{\beta_{Y,Z}}{\beta_{X,Z}} . \tag{7.48}$$

Proposition 7.4 is the equivalent of proposition (7.1) for an IV regime.

Proposition 7.4 *Given the model in (7.46), the OLS-IV parameter defined in (7.48) is a weighted sum of slopes of the regression curve $g'(x)$. That is,*

$$\beta_{OIV} = \int w_Z(x) g'(x) dx, \tag{7.49}$$

where the weights $w_Z(x) = w(x,z)$ represent the contribution of each segment dx to the covariance cov(X, Z). Thus, $\int w_Z(x)dx = 1$, and

$$w_Z(x) = \frac{1}{\mathrm{cov}(Z, X)} [\, \mu_Z F_X(x) - ACCV_{ZX}(x)], \tag{7.50}$$

where ACCV is the shifted ACC (see proposition 7.1 for an equivalent derivation).

Proof The proof is given by Yitzhaki and Schechtman (2004).

Note that while in proposition 7.1, for the OLS regime, the weights are all positive (because they are based on the Lorenz curve and represent contributions to the variance), in the OLS-IV regime (proposition 7.4) the weights can be both positive and negative (because they are based on ACC and represent contributions to the covariance). Therefore from here on, the terms "weighted average" and "weighted sum" are used for both cases.

The presentation for a discrete distribution (and in the sample) is given in proposition 7.4′.

Proposition 7.4′ *The OLS-IV estimator of the slope of the regression coefficient β is a weighted sum of slopes defined by adjacent observations of X. That is,*

$$b_{OIV} = \sum_{i=1}^{n-1} w_i^{IV} b_i \quad \text{where} \sum_{i=1}^{n-1} w_i^{IV} = 1, \tag{7.51}$$

$$w_i^{IV} = \frac{\left(\sum_{j=i}^{n-1} i(n-j)\, \Delta z_j + \sum_{j=1}^{i-1} j(n-i)\, \Delta z_j \right) \Delta x_i}{\sum_{k=1}^{n-1} \left(\sum_{j=k}^{n-1} k(n-j)\, \Delta z_j + \sum_{j=1}^{k-1} j(n-k)\, \Delta z_j \right) \Delta x_k} \tag{7.52}$$

and b_i are the slopes defined between adjacent observations of X.

The proof is identical to the proof of proposition 7.4. Note, however, that the denominator is equal to cov(Z, X) and that the weight is therefore the contribution of each section Δx_i to cov(Z, X). Note also that the weights are not restricted to be positive here.

As in the case of proposition 7.1, the weights can be expressed in terms of the vertical distance between the LOI and the ACC. That is,

$$w_i^{IV} = \frac{i}{n} \frac{(\bar{z}_n - \bar{z}_i)\ \Delta x_i}{\text{cov}(z, x)} \tag{7.53}$$

where $\bar{z}_i$ is the average of the i smallest observations of Z.

As seen above, the OLS-IV estimator of the slope of the regression coefficient β is a weighted sum of slopes defined by adjacent observations of X. Note, however, that for OLS-IV to be a weighted sum of the true impact of X on Y (i.e., of β in (7.46)), it must be assumed that Z and ε are independent. In some sense the difference between OLS and OLS-IV is relatively minor: both are based on the same slopes, the original slopes defined between Y and X. The only difference is that the weights in OLS are derived from the absolute Lorenz curve for the variance of the explanatory variable X, while the weights of the OLS-IV are based on the ACC of Z with respect to X. However this is a nontrivial difference. The nature of the weighting scheme has been drastically changed: the weights in the OLS are all nonnegative, while the weighting scheme for the OLS-IV is based on the ACC, which in theory can be negative. To see the implications of this difference consider a perfect line between Y and X, so that the slope is a constant. If more than 50% of the weights defined by using Z are positive then the OLS-IV will give a positive slope, while if a lion's share of the weights are negative, then the estimated slope will be negative. Moreover, generally if in the range with negative slopes we also have negative weights, then the contribution to the slope is positive although the slopes are negative. Therefore it is crucial to assure that the weights are all positive because otherwise strange things can happen.

To sum up: both OLS and OLS-IV estimators are weighted sums of b_i and both weighting schemes are based on the vertical distances between LOI and ACC. The only difference between the schemes is in the ACC used—the OLS weighting scheme relies on the shifted absolute Lorenz curve of X, which describes the contribution of each segment of X to the variance of X, while the weights in the OLS-IV estimators are based on the shifted ACC of Z with respect to X, which describes the contribution of each segment of X to cov(X, Z). Weights that are based on the variance are nonnegative, while those based on the covariance can be both positive and negative. An estimator with positive weights can have totally different properties from an estimator with a combination of positive and negative weights. In the former case, the estimator is a convex combination of slopes while in the latter case it is not. To see the implications, note that in the former case the estimate is bounded by Min(b_1,..., b_{n-1}) and Max(b_1, ..., b_{n-1}), where b_i are the slopes defined between adjacent observations of X, whereas in the latter case the estimate is not bounded and can fall outside the range of the observed slopes. (See the example in Sect. 7.6.4).

In addition, negative weights can cancel out positive weights, thereby in effect reducing the effective size of the sample. Imagine a case in which 98% of the slopes are equal to b_1 and only 2% are equal to b_2. Assuming also that the sum of the weights of the b_1 slopes equals to zero, we get that the OLS-IV method estimates the slope to be equal to b_2, although 98% of the observations form a perfect line with b_1 slope. Because of the opposite signs of the weights, the estimate is actually determined by the remaining two percent of the sample. The point that even large samples may be insufficient for the OLS-IV method has been raised by Bound et al. (1995).

It is important to note that the weighting scheme is composed of weights with both signs if and only if $LOI_{Z.X}$ and $ACC_{Z.X}$ intersect. Property (f) of the ACC indicates that this condition is identical to the condition of whether there exists a monotonic transformation of X that can change the sign of $b_{X.Z}$, the regression coefficient in the first-stage regression or not. If the ACC and the LOI intersect, one can split the data into two sets, composed of all observations below (or above) the intersection. Then, the values of cov(X,Z) in the two sets will have opposite signs. A monotonic transformation can change the magnitudes of the two covariances and therefore can change the sign of the regression coefficient between X and Z. This implies that an additional property is required from a good instrument: it should have monotonic relationship with the explanatory variable. The test of this property is based on whether the concentration curve and the LOI intersect or not (see Chap. 11 for tests of intersection). A failure of the instrument to have only positive weights implies that one can change the sign of the OLS-IV estimator by applying a monotonic transformation.

One possible reason for an instrument Z to have a low correlation with X is sampling variability; i.e., the random deviation of the estimate from the population parameter. Ignoring sampling variability, there can be two additional reasons for an instrument Z to have a low correlation with X. One possibility is that although the population's concentration curve of Z with respect to X is located on one side of the LOI, it is close to it. This means that although the expected values of all the weights in the weighting scheme of the slopes of the regression of Z with respect to X are positive, they tend to be close to zero. Hence the correlation between the two variables is weak. Another possibility for low correlation between Z and X is when the concentration curve of Z with respect to X (or X with respect to Z) intersects with the LOI, although there can be sections where it is far away from the LOI. This means that the conditional correlation, conditional on the segment of X we are looking at, changes signs. In this case there are segments of the range of X where the correlation is positive and large and, at the same time, there are other segments where the correlation is negative and large. This case will be reflected by having positive and negative weights. The former type of an instrument should be preferred to the latter because by increasing the sample, the weakness of the former disappears, while the latter will continue to be a weak instrument even for large samples.[5] In other words, a

[5] One of the properties of the ACC (property (g) in Sect. 5.3) says that provided that X and Z are drawn from a bivariate normal distribution then the ACC and LOI in the population do not intersect. This is a sufficient condition for the weights to converge to positive values in large samples.

weak correlation between the instrument and the explanatory variable can cause the weighting scheme to be with mixed signs. However, if the appropriate concentration curve in the population does not intersect with LOI, then a sufficiently large sample can mitigate the impact of a weak correlation as is the case discussed by Bound et al. (1995) and Staiger and Stock (1997). We note that the weakness of the instrument that we stress in this chapter is caused by the non-monotonic relationship between the instrument and the explanatory variable. If the concentration curve in the population intersects with LOI, then this weakness holds in the population and therefore will not disappear even if we rely on the entire population. It implies that the investigator can force the data to deliver the sign of the coefficient he wishes to get. The non-monotonic relationship is identified by the properties of the ACC of Z with respect to X (see Chap. 11).[6]

Because both OLS and OLS-IV estimators are weighted sums of the same slopes, the difference between them can be expressed explicitly. This procedure enables us to trace the sources of the difference between the two estimates. Equation (7.54) details the effect of applying an instrumental variable on the weighting scheme

$$b_{OIV} - b_{OLS} = \frac{1}{b_{Z.X}} \sum_{i=1}^{n-1} \left(\frac{(\bar{z}_n - \bar{z}_i)}{(\bar{x}_n - \bar{x}_i)} - b_{Z.X} \right) w_i b_i, \tag{7.54}$$

where $b_{Z.X}$ is the OLS estimator of the slope of the regression of Z on X, $\bar{x}_i$ and $\bar{z}_i$ are the averages of the i smallest observations of X and Z, respectively, and w_i are the OLS weights of (7.21a). The weight attached to b_i depends on w_i and on the difference shown in the brackets. The first term in the brackets is the difference between the LOI and the ACC of Z with respect to X *divided by* the difference between the LOI and ALC of X. If this ratio at i is larger (smaller) than the slope of the regression of Z on X, the weight attached to b_i is proportionally increased (decreased).

More specifically, the impact of using IV on the estimate is channeled in three possible ways. Consider the case where $b_{OIV} < b_{OLS}$ and recall that the weights, w_i, are all positive and add up to one. The difference between the estimates is determined by the terms inside the brackets and the sign and magnitude of $w_i b_i$ that accompany them. The change in the estimate can be caused by (a) a decline in the terms in the brackets accompanied by a large $w_i b_i$ and an increase in the terms in the bracket accompanied by a small $w_i b_i$, (b) a decrease in the terms in the brackets of both small and large $w_i b_i$, or (c) an increase in both. (Cases (b) and (c) are possible because some terms in the brackets are positive while others are negative. The decrease/increase is in terms of absolute value). Note that (a) represents a good instrument while (b) and (c) represent bad ones. Therefore the decomposition of the

[6] Note that some types of non-monotonicity can be tolerated. For example, if the ACC is concave in some sections and convex in others but does not cross the LOI, then the conditional correlation over those segments can be negative or positive, but the weighting scheme does not change its sign.

sources of the change in the estimate enables one to test whether the identifying assumptions, i.e., the assumptions used to construct the model, are supported by the data or not. For example, one possible reason for using an IV is that economic theory leads us to suspect that X is positively correlated with the residual, and therefore the OLS estimate is biased upward. Let us divide the set of slopes $\{b_i\}$ into two groups: the above-average (bad) group and the below-average (good) group. We refer to the sets as good and bad because chances are that high values of slopes will be more contaminated than low values of slopes due to the positive spurious correlations. If so, one can test whether the decline in the estimate due to the use of an IV is caused by a decline in the weights of bad slopes or by an increase in the weights of good slopes. Increasing the weights of good slopes together with a decline of the weights of bad slopes (case (a) above) should indicate consistency with the economic model, while diminishing (case (b)) or increasing (case (c)) the weights of both good and bad slopes should be viewed as data manipulation and/or a search for spurious correlation. Note that the sum of the weights equals one. Therefore if all the weights are positive, an increase in the weights of one group implies a decrease in the weights of the other. Therefore an instrument that produces only positive weights will never be found as spurious, so that the suggested test is actually a test on the properties of the ACC. Finally, note that the OLS weighting scheme continues to serve under the OLS-IV regime. Hence, the OLS-IV estimator is sensitive to extreme observations, just like the OLS.

Additional implications of the differences between the weighting schemes are discussed after we introduce the Gini IV estimator.

7.6.2 The Gini Instrumental Variable Method

This section replicates Sect. 7.6.1 with an IV method that is based on the GMD. As mentioned in the earlier section, substitution of one technique by another does not eliminate all the theoretical arguments against or in favor of an IV method, but the additional tools that the GMD method provides reduce the ability of investigators to manipulate the results (purposely or by pure ignorance).

The aim of this section is to present the parameter of the Gini-IV method as a weighted sum of slopes defined by adjacent observations in a way that is similar to the OLS-IV estimate. Note that the four estimates: OLS, Gini, OLS-IV, and Gini-IV are based on weighted sums of (the same) slopes. The weighting schemes are based on the vertical distances between an LOI and an ACC. The differences lie in the choices of the ACCs used. The ACCs used for OLS are based on the cumulative value of the variate, while the ACCs used in the GMD approach are based on the cumulative value of the cumulative distribution of the variate. As a result, the GMD estimators are more robust with respect to outliers than the OLS estimators because the weighting scheme is a function of Δx (rather than of $(\Delta x)^2$). As will be seen below, the Gini-IV is based on the ACC of the cumulative distribution of the instrumental variable, which means that the Gini-IV is more robust than OLS-IV because Gini-IV, unlike OLS-IV, is not sensitive to monotonic transformations of the instrument. Alternatively, one can use Gini-IV to check how robust OLS-IV estimators are to a small perturbation in the weighting scheme.

The Gini semi-parametric regression coefficient in a simple regression framework is defined as

$$\beta_N = \frac{COV\ (\ Y,F_X\ (X))}{COV\ (X,\ F_X\,(X))}.$$

Note that in the OLS framework, it is actually an IV with the cumulative distribution serving as an instrument. However, a good instrumental variable should be independent of Y|X, which is not the case here. The instrumental variable population's parameter in the Gini regression framework can be defined in a way that resembles the OLS-IV definition as follows

$$\beta_{GIV} = \frac{COV(Y,\ F(Z))}{COV(X,\ F(Z))} = \frac{\beta_{N,Y.Z}}{\beta_{N,X.Z}}. \tag{7.55}$$

Its derivation is identical to the derivation of an IV estimator under the OLS framework. That is, starting from a decomposition of cov(Y, F(Z)) one derives (7.55). Comparison of (7.48) and (7.55) reveals that the difference between OLS-IV and Gini-IV regression coefficients is that the Gini-IV relies solely on the ranks of the instrumental variable, while OLS-IV relies on the variate itself. Like the OLS and OLS-IV estimators, the GMD and Gini-IV regression coefficients are weighted averages (sums) of slopes between adjacent observations. The only difference is in the weighting scheme. Formally, in the GMD framework, the weights in the population are

$$v_Z(X) = v(Z,X) = \frac{1}{cov(X,\ F(Z))}[0.5F_X(x) - ACCV_{F_Z(Z).X}(x)]. \tag{7.56}$$

Note that because we are dealing with a concentration curve and not with a Lorenz curve, it may intersect the LOI and therefore produce negative weights.

In the sample, or in a discrete distribution, the weight of observation x_i is

$$v_i^{IV} = \frac{p\Delta x_i}{n\ cov(x,\ r^z)}\left[\frac{n+1}{2} - \bar{r}_i^z\right], \tag{7.57}$$

where $\bar{r}_i^z$ is the average rank of the i observations of Z that correspond to the i smallest observations of X (i.e., rank concomitants) and p is i/n. The derivations of the above formulas are variations of a proof in Yitzhaki and Schechtman (2004).

The statements above show that Gini-IV is identical in structure to OLS-IV, where the instrumental variable is replaced by its cumulative distribution. Therefore in the Gini-IV framework an investigator cannot affect the estimate by using a monotonic transformation of the instrumental variable. As for other properties, it is easy to see that the weights in Gini-IV and OLS-IV have the same signs, so there is no difference between the two with respect to transformations of X. Therefore the same condition that identifies a spurious instrument under OLS will do so under the GMD framework. However, because Gini-IV weights are based on Δx, while OLS weights are based on $(\Delta x)^2$, we expect the former to be more robust.

When summarizing the intuitive idea that leads to the Gini-IV, it is worth to recall Angrist and Evans' explanation for the IV method: "The IV method attributes any effect of Z_i on Y_i to the effect of Z_i on X_i." (Angrist & Evans, 1998, p. 458). The Gini regression can be interpreted as attributing any effect of the rank of Z_i on Y_i to the effect of the rank of Z_i on X_i. In the Gini-IV, the IV method is used twice so that the final result is to attribute any effect of change in the rank of Z_i on Y_i to the effect of a change in the rank of Z_i on X_i. This property reduces the sensitivity to Z. The double use of the IV method when using Gini-IV explains the title of Yitzhaki and Schechtman's paper (2004).

7.6.3 *The Similarities and Differences Between OLS and Gini Instrumental Variable Methods*

Having described the two IV approaches, we now list the major similarities and differences between the OLS-IV and the Gini-IV estimators.

(a) A monotonic transformation of Z does not affect the Gini-IV estimate, unlike the case of OLS-IV estimates, where a monotonic transformation of Z does have an effect and may even change its sign. In this sense, the Gini-IV method reduces the possibility of data manipulation. However, as a result of this property, one cannot use two IVs that are monotonic transformations of each other (having Spearman's correlation coefficient of one) because the ranks will be identical which will result in multi-colinearity. This deficiency can be mitigated by using the extended Gini with different EG parameters attached to different explanatory variables. This issue will be discussed in Chap. 8 where the Gini multiple regression is presented.
(b) The Gini-IV attaches less weight to extreme observations than does OLS-IV. Therefore it is more robust to outliers than OLS-IV.
(c) Both the OLS-IV and the Gini-IV can be written explicitly and they rely on the same terminology.
(d) The Gini-IV can be used as a sensitivity test for OLS-IV. Presumably a minor change such as slightly altering the metric of variability should not drastically affect the estimates.
(e) It is well known that one can either estimate the IV directly, as is done above, or use a two-stage least squares procedure. In a Gini framework, those two methods are not equivalent and can yield different estimators. This difference will be discussed when dealing with IV in a multiple regression framework (see Chap. 8).

Table 7.3 The data

X	Y	Z	F_X	F_Z
1	0	9	0.25	1
2	−1	0	0.5	0.25
3	0	8	0.75	0.75
4	1	7	1	0.5

Source: Yitzhaki and Schechtman (2004), p. 303
Reprinted with permission by Metron International Journal of Statistics

Table 7.4 The variance–covariance matrix

	X	Y	Z	F_X	F_Z
X	1.25	0.5	0.25	1.25	−0.5
Y		0.5	1.75	0.5	0.25
Z			12.5		

Table 7.5 Weighting schemes and slopes

I	w_{OLS}	v_G	w_{OIV}	v_{GIV}	b_i
1	0.3	0.3	−3	0.75	−1
2	0.4	0.4	3	0	1
3	0.3	0.3	1	0.25	1

Source of Tables 7.3–7.5: Yitzhaki and Schechtman (2004), pp. 303, 304, called Table 1, 2, 3
Reprinted with permission by Metron International Journal of Statistics

Before we proceed with an example let us remind the reader that a good IV is one that is highly correlated with the explanatory variable and uncorrelated with the residual. The artificial example below is based on four artificial observations and it does not represent a good IV.

7.6.4 An Example: The Danger in Using IV

The aim of the following example is to illustrate the conditions under which the IV method fails to produce a reasonable estimate of the slope. The artificial data set is composed of four observations and three variables, X, Y, and Z, which represent the explanatory, dependent, and instrumental variables, respectively. Table 7.3 presents the data together with the cumulative distributions of X and Z.

The variance–covariance matrix is given in Table 7.4.

As shown, the covariance between X and Z is positive and equals 0.25, and the correlation coefficient is equal to 0.06. (This is a poor IV, intended only for illustration.)

Table 7.5 presents the weighting schemes according to the different methods and the slopes between adjacent observations.

As can be seen from the first column of Table 7.3, the distance between adjacent observations of X equals 1, which causes the OLS and Gini estimators of the slope of the regression curve to be equal (because the weights are equal). Thus, $b_{OLS} = b_G = 0.4$. On the other hand, the IV estimates differ $b_{OIV} = 7$, while $b_{GIV} = -0.5$.

As seen in the last column of Table 7.5, the slopes are either 1 or −1, which means that the OLS-IV estimator is far above the maximum slope observed in the data. This result is explained by the large (in absolute value) negative weight attached by OLS-IV to the negative slope between the first and second observations, which translates into a positive contribution to the estimated slope.

Applying monotonic transformations to Z would leave the Gini-IV estimates unaffected, but may affect the OLS-IV estimate through a change in its weighting scheme and may even change its sign. The effect of applying a transformation to Z on the estimate depends on its effect on the absolute values of the negative and positive weights. Whether a transformation reduces or increases the absolute value of the weight depends on whether its derivative in the relevant range is larger or smaller than one. This fact that a monotonic transformation of the instrument can change the sign of the regression coefficient is illustrated in our example: one can view the use of F(Z) in Gini-IV as using a monotonic transformation of Z. This means that Gini-IV is the OLS-IV of the monotonic transformation, so that $b_{OIV} = 7$ while $b_{GIV} = b_{OIV(\text{of } F(z))} = -0.5$.

The advantage of the Gini method over the OLS is in providing the investigator a way to check through the use of the concentration curve whether the conditions for safe conclusions (i.e., nonintersection of the LOI and the ACC) hold. Note, however, that this test is not capable to resolve the debate concerning the issue of whether the IV used is an exogenous variable or an endogenous one.

7.7 The Extended Gini Simple Regression

Because the GMD is a member of the EG family, one can replicate almost all the previous parts of this chapter with the extended Gini. All one has to do is substitute F(X) by $-[1 - F(X)]^{v-1}$ or $F[(X)]^{v-1}$ or a combination of the two, depending on whether one is interested in applications that fit concave, convex, or symmetric weighting schemes. The only part that cannot be replicated is the decomposition by population subgroups because the EG, unlike the GMD, is not additive over overlapping subpopulations.

For example, (7.22) turns out to be

$$\beta_N(\nu) = \frac{\text{cov}\,(\,Y,\,[1 - F_x(X)]^{\nu-1})}{\text{cov}(X,\,[1 - F_x(X)]^{\nu-1})}, \tag{7.22a}$$

while proposition 7.2 will be

Proposition 7.2′ *The extended Gini semi-parametric regression coefficients are weighted sums of the slopes of the regression curve. That is,*

$$\beta_N(v) = \int w(x,v)\, g'(x)dx,$$

with $w(x,v) > 0$ *and* $\int w(x,v)dx = 1$*, where*

$$w(x,v) = \frac{[1 - F_x(x)] - [1 - F_x(x)]^v}{\int_{-\infty}^{\infty} \{[1 - F_x(t)] - [1 - F_x(t)]^v\}dt}.$$

Proof The proof is given by Yitzhaki (1996, proposition 3, p. 483).

Proposition 7.2′ gives an interesting interpretation to the roles of the different components in determining the regression coefficient. The basic components, based on the raw data, are the slopes and the associated distribution of the explanatory variable. The methodology used determines the weights given to different sections of the distribution of the explanatory variable by choosing the extended Gini parameter, ν. In determining the method used the investigator is actually choosing a weighting scheme to apply to the data. The effect of the choice of the weighting scheme depends on the distribution of the explanatory variable and on the slopes along the regression curve. In some sense, one can replicate the entire chapter substituting the GMD by the EG. If the regression model is well-behaved, i.e., if the model in the population is linear, then for large samples, there should not be any major difference in the estimates. On the other hand, if the relationship between the variables is not linear, then by stressing different sections of the explanatory variable the sign and the magnitude of the estimate can change. In Chap. 8 we present the mixed regression, with some explanatory variables treated by GMD and others by EG or OLS.

7.8 Summary

Two regression methods can be described as based on the GMD. One is based on describing the GMD as a covariance between the dependent variable and the rank of the explanatory variable, while the other is based on minimization of the GMD of the residuals. One advantage of the fact that there are two alternative methods is that one gets two sets of "normal equations." This fact is used in order to evaluate the specification (linearity) of the model and check the underlying assumptions.

Because we are interested in mimicking the OLS, we have concentrated on the regression that is based on the covariance presentation of the Gini. There are several similarities and differences between the OLS and the Gini regression. We focus on four issues: the presentations as weighted averages (sums) of slopes, the use of curves, the decomposition of a slope to the contributions of subgroups, and the relationship between direct and reverse regressions.

More specifically, relying on the covariance-based regression, it is shown that the regression coefficients in OLS, OLS-IV, Gini, and Gini-IV regressions can all be expressed as weighted averages (sums in the case of an IV) of slopes of the regression curve. The difference between the methods is in the weighting schemes. The weights in the case of the OLS and Gini regressions are based on the properties of the absolute Lorenz curve of the explanatory variable, while the weights in the case of IV, under both methods, are based on the properties of the ACC of the instrument with respect to the explanatory variable. The absolute Lorenz curve cannot intersect the LOI, which means that under OLS and Gini regressions all weights are positive. On the other hand, the ACC can intersect the LOI, causing the weights of the IV regression, under both methods, to be with both negative and positive signs. This means that a monotonic transformation of the instrument can change the sign of the regression coefficient. In such cases the estimators may be inconsistent.

One recommendation is to plot the concentration curve (or LMA curve) of the instrument with respect to the explanatory variable in order to see whether a sign change in the weighting scheme can occur. This recommendation is a bit complicated to apply in a multiple regression framework (as will be discussed in Chap. 8), and more work in the extension to multiple regression is still needed.

An implication of the analysis presented in this chapter is that one can interpret switching from one method to the other as a decision to change the weighting scheme applied to the slopes of the regression curve. Switching from a weighting scheme with positive weights to a weighting scheme with both negative and positive weights should be reported because it changes the properties of the estimation procedure.

A new direction for further research is the comparison of the efficiencies of the different methods and the relative advantages of each. The convergence theorems of Davydov and Egorov (2000a, 2000b) seem to be a promising direction in the investigation of the properties of the concentration curves, i.e., the weighting schemes of the different types of regressions.

Finally, the entire chapter can be replicated using the EG instead of the GMD.

Appendix 7.1

Proposition 7.1′ *The OLS regression coefficient for a discrete distribution of X is a weighted sum of slopes defined by adjacent observations. That is,*

$$\beta_{OLS} = \sum_{i=1}^{n-1} w_i b_i \tag{7.20a}$$

where $w_i > 0$, $\Sigma w_i = 1$, $b_i = \Delta y_i/\Delta x_i$, $\Delta x_i = x_{i+1} - x_i$ and where the observations are arranged in an increasing order according to X. The weights are given by

$$\mathrm{w_i} = \frac{\left(\sum\limits_{j=i}^{n-1} i(n-j)\Delta x_j + \sum\limits_{j=1}^{i-1} j(n-i)\Delta x_j\right)\Delta x_i}{\sum\limits_{k=1}^{n-1}\left(\sum\limits_{j=k}^{n-1} k(n-j)\Delta x_j + \sum\limits_{j=1}^{k-1} j(n-k)\Delta x_j\right)\Delta x_k}, \tag{7.21a}$$

(where $\sum\limits_{j=1}^{0} j(n-1)\Delta x_j$ *is defined to be zero).*

Proof of proposition 7.1′ The OLS regression coefficient is

$$\mathrm{b_{OLS}} = \mathrm{cov(Y, X)/cov(X, X)}.$$

For our purposes it is convenient to express the numerator and the denominator in an alternative way. The numerator can be rewritten as

$$\mathrm{cov(Y, X)} = \tfrac{1}{2}\mathrm{E_1 E_2((Y_1 - Y_2)(X_1 - X_2))},$$

where $(\mathrm{Y_j, X_j})$ (j = 1, 2) are i.i.d. variables, and E denotes expectation. Ignoring multiplicative constants (which cancel out when both the numerator and the denominator are considered), the application of this formula to a discrete distribution (and in the sample when dealing with estimation) leads to

$$cov(y, x) = \sum_{i=1}^{n}\sum_{j=1}^{n}(x_i - x_j)(y_i - y_j) = 2\sum_{i=1}^{n}\sum_{j=1}^{i-1}(x_i - x_j)(y_i - y_j),$$

and by substituting $y_i - y_j = \sum\limits_{k=s}^{t-1} b_k\,\Delta x_k$ and $x_i - x_j = \sum\limits_{p=s}^{t-1}\Delta x_p$ for i > j, where s = min(i, j) and
t = max(i, j), we get

$$cov(y, x) = \sum_{i=1}^{n}\sum_{j=i}^{n}\sum_{k=s}^{t-1}\sum_{p=s}^{t-1} b_k\,\Delta x_k\,\Delta x_p\,.$$

After some tedious algebra we get

$$cov(y, x) = \sum_{i=1}^{n-1}\left(\sum_{j=i}^{n-1} i(n-j)\Delta x_j + \sum_{j=1}^{i-1} j(n-i)\Delta x_j\right)\Delta x_i\, b_i\,. \tag{A7.1}$$

Applying the same procedure to the denominator we get

$$cov(x, x) = \sum_{i=1}^{n-1}\left(\sum_{j=i}^{n-1} i(n-j)\Delta x_j + \sum_{j=1}^{i-1} j(n-i)\Delta x_j\right)\Delta x_i\,. \tag{A7.2}$$

Dividing (A7.1) by (A7.2) yields (7.20a) and (7.21a).

Appendix 7.2

Claim

$$-\sum_{i<j}(x_j - x_i) = 2\sum_{i=1}^{n} x_i\left[i - \frac{n+1}{2}\right].$$

Proof

$$\sum_{i<j}(\mathrm{x_j} - \mathrm{x_i}) = \sum_{i=1}^{n-1}(\mathrm{n} - \mathrm{i})\mathrm{x_i} - \sum_{j=2}^{n}(\mathrm{j} - 1)\mathrm{x_j} = \sum_{i=1}^{n-1}\mathrm{nx_i} - \sum_{i=1}^{n-1}\mathrm{ix_i} - \sum_{j=2}^{n}\mathrm{jx_j} + \sum_{j=2}^{n}\mathrm{x_j}$$

$$= \mathrm{nx_1} + \mathrm{n}\sum_{i=2}^{n-1}\mathrm{x_i} + \sum_{j=2}^{n-1}\mathrm{x_j} + \mathrm{x_n} - \mathrm{x_1} - \sum_{i=2}^{n-1}\mathrm{ix_i} - \sum_{j=2}^{n-1}\mathrm{jx_j} - \mathrm{nx_n}$$

$$= (\mathrm{n} - 1)\mathrm{x_1} - (\mathrm{n} - 1)\mathrm{x_n} + (\mathrm{n} + 1)\sum_{i=2}^{n-1}\mathrm{x_i} - 2\sum_{i=2}^{n-1}\mathrm{ix_i}$$

$$= (\mathrm{n} - 1)(\mathrm{x_1} - \mathrm{x_n}) + \sum_{i=2}^{n-1}[(\mathrm{n} + 1)\mathrm{x_i} - 2\mathrm{ix_i}] = 2\sum_{i=1}^{n}\mathrm{x_i}\left[\frac{\mathrm{n} + 1}{2} - \mathrm{i}\right].$$

Chapter 8
Multiple Regressions

Introduction

The purpose of the simple regression is to study the relationship between one explanatory variable and one dependent variable. The purpose of a multiple regression (the term was first used by Pearson and Lee (1908)) is to learn about the relationships between several explanatory variables and a dependent variable. The extension of the model from one explanatory variable into several explanatory variables introduces several complications. For example, in a multiple regression setting one has to consider the effects of the relationships among the explanatory variables on the estimates. On the other hand, an advantage is that one can mix the regression methodologies used (i.e., apply different regression methodologies to different explanatory variables). In this chapter we will be mainly interested in methods of multiple regressions that are based on the simple regression coefficients. By "based on" we mean not only that the multiple regression coefficients are derived by the same principle that is used to derive the simple regression coefficients but also that the simple regression coefficients are used as the building blocks of the multiple regression coefficients. As such, one can learn about their properties from the properties of the simple coefficients. In particular, we have shown in Chap. 7 that the Ordinary Least Squares (OLS) and semi-parametric Gini regression estimators can be interpreted as the slopes of the linear approximations to a regression curve, because they are based on weighted averages of slopes defined between adjacent observations. In other words, the linearity assumption on the regression curve is not used in the estimation stage. This property continues to hold in our extension into the multiple regression case. However, we do introduce some kind of a linearity requirement. The linearity requirement differs from the linearity assumption on the model because it is imposed on the set of equations that are used to derive the multiple regression coefficients, as will be seen below.

Within the class of multiple regressions we will refer to a regression method as a "covariance-based" method if the set of normal equations (e.g., the first-order conditions of the optimization) can be written as a set of *linear* equations, where

S. Yitzhaki and E. Schechtman, *The Gini Methodology: A Primer on a Statistical Methodology*, Springer Series in Statistics 272, DOI 10.1007/978-1-4614-4720-7_8,

the unknown parameters (to be estimated) are the regression coefficients (or covariances) between the pairs of random variables. This family includes OLS regressions, Instrumental Variable Least Squares (OLS-IV), Gini regressions, Instrumental Variable Gini regression (Gini-IV), and the Extended Gini (EG) family of regressions (Schechtman, Yitzhaki & Artsev, 2008). Those regression methods share several properties that can be traced to the properties of the covariance and of the linearity of the normal equations, hence the advantage of incorporating them into one family.

One important shared property is a geometric property, already known for OLS, which we will heavily rely on in this chapter. DeLaubenfels (2006) who reviews the victory of the OLS over alternative methods such as the Least Absolute Deviation (LAD) which preceded it points out that "Geometry—specifically an inner product being used to produce angles and orthogonality—is offered as the reason for least squares becoming preferable." (p. 315).

It is important to stress two fundamental issues in the definition of the covariance-based regressions: the linearity of the set of equations and the fact that the simple regression coefficients are serving as constants (to be shown later). Those two issues determine the properties that distinguish these regressions from other regressions. The linearity requirement implies that the multiple regression coefficients have explicit representations, unlike other regression methods such as quantile regression, Mean Absolute Deviation regressions (MAD), and the regression that is based on the minimization of the Gini of the residuals. The fact that the multiple regression coefficients are based on the simple ones implies that the properties that we have listed in the previous chapter on simple regression coefficients carry through to the multiple regression framework, although it is done in a more complicated way than in the simple regression due to the possible interaction of each explanatory variable with other explanatory variables. A third property can be traced to the above two requirements: one can derive the estimates without using an optimization procedure. (In the OLS one can derive the estimators through an optimization, but one can also derive them without relying on optimization.)

The aim of this chapter is to develop the properties of GMD and EG multiple regressions coefficients. However, instead of concentrating on the properties of those regressions alone, we develop the properties of a wider family that includes OLS as a member. In this sense we follow and further develop the argument by DeLaubenfels (2006, p. 315) who argues "...definitions and fundamental results in the general linear model, analysis of variance, conditional probability, independence, sufficiency, and time series can be unified and clarified as deriving from the inner product." The advantage of looking at the entire family is that this approach enables the user to mix different regression methods such as OLS and Gini regressions in one analysis. We note that we do not detail the estimation procedures and the calculations of the variances of the estimators in this chapter. This will be done in Chaps. 9 and 10.

Also, we do not repeat the concepts that are applicable both in the simple and multiple regressions such as the equivalents of R^2.

The structure of the chapter is the following: in Sect. 8.1 we show that the regression coefficients in a covariance-based multiple regression are derived by solving a system of linear equations with the simple regression coefficients serving as constants and describe the restrictions and properties imposed on GMD or EG regressions. Section 8.2 offers an alternative interpretation of Gini regression as a linear approximation of a general regression curve, Sect. 8.3 deals with combining the minimization and the semi-parametric approaches, Sect. 8.4 compares between OLS-IV and Gini-IV regressions, Sect. 8.5 discusses the effects of commonly used practices in the multiple regression framework, and Sect. 8.6 concludes.

8.1 Multiple Regression Coefficients as Composed of Simple Regression Coefficients

The aim of this section is to show that the multiple regression coefficients in covariance-based regressions are derived by solving a set of linear equations that are composed of simple regression coefficients. As in the previous chapter, the presentation is restricted to population parameters. The estimators and the inference about them will be presented in Chaps. 9 and 10, respectively. However, it is worth to keep in mind that except for corrections for degrees of freedom, all estimators are sample's analogues of the population parameters.

Let $(Y, X_1, \ldots, X_K)$ be continuous random variables that follow a multivariate distribution with finite second moments. For every choice of constants α, $\beta_1, \ldots, \beta_K$ define the random variable ε by the following identity:

$$Y \equiv \alpha + \beta_1 X_1 + \cdots + \beta_K X_K + \varepsilon. \tag{8.1}$$

At this stage, $\alpha, \beta_1, \ldots, \beta_K$ are arbitrary constants ($\beta_1, \ldots, \beta_K$ will later stand for the multiple regression coefficients, while α will be a location parameter). The random variable ε is defined as a slack variable, intended to fulfill identity (8.1). The symbol $\equiv$ is used to indicate that at this stage there are no assumptions imposed on ε and all its properties are determined by the properties of the distribution of $(Y, X_1, \ldots, X_K)$. Equation (8.1) is a tautology, which means that no assumptions have been imposed.

Let $T_1, \ldots, T_K$ be K random variables. The covariances between Y and these variables define a set of identities as follows:

$$\begin{aligned}
\mathrm{cov}(Y, T_1) &\equiv \beta_1 \mathrm{cov}(X_1, T_1) + \cdots + \beta_K \mathrm{cov}(X_K, T_1) + \mathrm{cov}(\varepsilon, T_1) \qquad (8.2)\\
\mathrm{cov}(Y, T_k) &\equiv \beta_1 \mathrm{cov}(X_1, T_k) + \cdots + \beta_K \mathrm{cov}(X_K, T_k) + \mathrm{cov}(\varepsilon, T_k)\\
\mathrm{cov}(Y, T_K) &\equiv \beta_1 \mathrm{cov}(X_1, T_K) + \cdots + \beta_K \mathrm{cov}(X_K, T_K) + \mathrm{cov}(\varepsilon, T_K).
\end{aligned}$$

Dividing each line by the appropriate covariance, subject to the assumption that $cov(X_k, T_k) \neq 0$, $(k = 1, \ldots, K)$ we get:

$$\beta_{01} \equiv \beta_1 1 + \cdots + \beta_K \beta_{K1} + \beta_{\varepsilon 1} \tag{8.3}$$

$$\beta_{0k} \equiv \beta_1 \beta_{1k} + \cdots + \beta_k 1 + \cdots + \beta_K \beta_{Kk} + \beta_{\varepsilon k}$$

$$\beta_{0K} \equiv \beta_1 \beta_{1K} + \cdots + \beta_K 1 + \beta_{\varepsilon K}$$

where the index 0 indicates the dependent variable, $\beta_{\varepsilon j} = \dfrac{cov(\varepsilon, T_j)}{cov(X_j, T_j)}$, $\beta_{kj} = \dfrac{cov(X_k, T_j)}{cov(X_j, T_j)}$ are the regression coefficients in the simple regressions of X_k on T_j, $k,j = 1, \ldots, K$, and $\beta_{0j} = \dfrac{cov(Y, T_j)}{cov(X_j, T_j)}$.

Two special cases are the OLS (iff $T_j = X_j$) and the Gini (iff $T_j = F(X_j)$). Provided that the rank of the matrix of the coefficients composed of the β_{kj}'s is K we get the following "solution" of the identities in (8.3):

$$\begin{pmatrix} \beta_1 \\ \\ \beta_K \end{pmatrix} \equiv \begin{pmatrix} 1 & \beta_{21} & \beta_{K1} \\ & & \\ \beta_{1K} & \beta_{2K} & 1 \end{pmatrix}^{-1} \begin{pmatrix} \beta_{01} - \beta_{\varepsilon 1} \\ \\ \beta_{0K} - \beta_{\varepsilon K} \end{pmatrix} \equiv \mathbf{A}^{-1}[\beta_0 - \beta_\varepsilon] \tag{8.4}$$

where $\mathbf{A}^{-1}$ is a $K \times K$ matrix, while the $\boldsymbol{\beta}$'s are $K \times 1$ vectors. The set of identities (8.4) is the basic structure of the identities that hold in an arbitrary linear model.

So far no assumption has actually been imposed, except that $cov(X_k, T_k) \neq 0$, $k = 1, \ldots, K$, and that the rank of the matrix A is equal to K.

We now impose a set of restrictions (assumptions, in politically correct terms). We impose them on the data in the sample (without imposing any restriction in the population). The restrictions hold in the sample by construction, and therefore cannot be verified nor tested without additional information.

The set of restrictions to be imposed, referred to as "orthogonality conditions," is given by

$$\beta_{\varepsilon k} = 0, \quad \text{for} \quad k = 1, \ldots, K. \tag{8.5}$$

Note that for convenience we keep the notation although these are sample values. One possible interpretation of (8.5) can be that it represents first-order conditions for an optimization with respect to a target function. This is the case for a specific choice of the variables T_k. For example, if $T_k = X_k$ then we are in the OLS regression case. Alternatively, one can follow DeLaubenfels' (2006) geometric interpretation that the inner products of the vectors of explanatory variables and the residual are zero. That is, the explanatory vectors are orthogonal to the residual. In both cases it should be remembered that those conditions are *imposed* on the data and there is no a priori reason to believe that they exist in the population.

(Some implications of these conditions in the population can be tested, as was shown in Chap. 7.) The consequence of imposing the orthogonality conditions is that (8.4) now turns from an identity to a solution of a set of linear equations, so that $\beta_k(k = 1, \ldots, K)$ cease to be arbitrary constants but become the solutions of a set of linear equations.

Formally, using the restrictions (8.5), the identities of (8.4) turn into (8.6):

$$\begin{pmatrix} \beta_1 \\ \\ \beta_K \end{pmatrix} = \begin{pmatrix} 1 & \beta_{21} & \beta_{K1} \\ & & \\ \beta_{1K} & & 1 \end{pmatrix}^{-1} \begin{pmatrix} \beta_{01} \\ \\ \beta_{0K} \end{pmatrix} = \mathbf{A}^{-1}\beta_0. \tag{8.6}$$

The structure given in (8.6) is general, and it corresponds to all members of the covariance-based regressions, depending on the choice of T_k, $k = 1, \ldots, K$. Special cases include:

(a) $T_k = X_k$ for all k, $k = 1, \ldots, K$. Then it is easy to see that (8.6) represents the OLS.
(b) $T_k = F(X_k)$ for all k, $k = 1, \ldots, K$. Then (8.6) represents the semi-parametric Gini regression.
(c) $T_k = -[1 - F(X_k)]^{\nu_k}$ for all k, $k = 1, \ldots, K$, and ν_k are given parameters supplied by the researcher. Then (8.6) represents the structure of the extended Gini regression.
(d) $T_k = Z_k$ for some k, $k = 1, \ldots, K$. Then (8.6) represents the structure of an OLS-IV regression.
(e) $T_k = F(Z_k)$ for some k, $k = 1, \ldots, K$. Then (8.6) represents the structure of a Gini-IV regression.

Several additional properties of (8.6) are worth mentioning.

By choosing T_k one is choosing a transformation to be applied to the data, which is actually a choice of the variability measure used (variance in OLS (a), Gini or extended Gini in the regressions defined in (b) and (c), respectively, covariances between the variables and the variable Z (or a function of it) in the cases of instrumental variables, i.e., (d) and (e)). As a result, this choice determines the metric used (Euclidean in the case of OLS, city block in the case of Gini) and the "orthogonality conditions" applied. In the case of OLS the orthogonality conditions are $\text{cov}(X_k, \varepsilon) = 0$ for all k, under the Gini regression they are $\text{cov}(F(X_k), \varepsilon) = 0$ for all k, etc. As we have shown in the simple regression case, the choice of the variability measure is actually a choice of the weighting scheme to be applied to the slopes defined between adjacent observations of the explanatory variables. This is an important point that seems to be forgotten: the basic building blocks in the regression are the slopes defined between adjacent observations. Hence the difference between the alternative methods in the covariance-based family is in the weighting schemes used in constructing the weighted average of the slopes defined between pairs of adjacent observations.

Having realized that, it is clear that each of the K equations in (8.4) can be defined with different T_k so that one can have mixed regression methods: some equations can be defined as based on GMD, others on OLS, etc. The advantage of a mixed method is that it enables the user to check the robustness of each *imposed* linear normal equation with respect to different regression methodologies, so that only the conditional regression curves for which the linear approximations are not seriously affected by the choice of the methodology will be considered as linear regressions.

An additional important advantage of the mixed regression method is that one does not switch from one regression method to the other in an "all or nothing" way. That is, one can change the methodology of regression for each explanatory variable without changing the methodology with respect to other explanatory variables. This enables the user to distinguish between linear and nonlinear equations. On the other hand, the mixed regression is a bit complicated because the order in which one moves from one method to the other may affect the conclusions.

One disadvantage of the mixed approach is that the target function for which the orthogonality conditions play the role of first-order conditions for optimization is generally not a clear or easy to understand target function. For example, one cannot describe the mixed regression estimators as Best Linear Unbiased Estimators (BLUE). To belittle this argument we remind the reader that the difference between the GMD and the variance is mainly caused by the metric chosen to measure distance. Because it is generally not clear which metric is better suited for the social science and the choice of the metric clearly affects the conclusion as to which estimator is a better one, we believe that the jury has not decided yet which one is a better methodology.

It is most likely that there is no one *best* method. Rather, each subject matter has its own best method. Note that the above list of possible choices of T_k does not cover all possibilities. For example, $T_k = F(X_k)^{v_k}$ is an alternative choice of a regression that has not been investigated yet. Our guess is that it will be similar in its properties to the EG regression, except that by increasing v_k one increases the weight attached to slopes that are located at higher values of X (rather than to the lower values of X, which is what EG does). Also, as we have shown in the EG regression, one can apply a symmetric version of T_k, namely, $T_k = \frac{1}{2}\{ F(X_k)^{v_k} +[1 - F(X_k)]^{v_k}\}$. It is easy to see that even this extension does not cover all possibilities because one can use different powers in the constructions of the different T_k's.

Another point to bear in mind is that would we minimize the GMD of the residuals, as is the case in R-regression (Hettmansperger, 1984; Olkin & Yitzhaki, 1992)[1], then the orthogonality conditions would be $cov(X_k, F(\varepsilon)) = 0, k = 1, \ldots, K$. Hence, one may think that this regression also belongs to the family of covariance-based

[1]Not to be confused with rank regressions in Fortin and Lemieux (1998) and Juhn, Murphy, and Pierce (1993) where the rank of the variable is used in OLS regression.

regressions. But because the residual which is defined by (8.1) is represented by its cumulative distribution, the orthogonality conditions do not form a set of linear equations. Hence, R-regression does not belong to the covariance-based family. However, if one is ready to give up on the linearity requirement then the structure of the multiple regression that is presented in (8.4) can cover other regression methods. In that case we would have to solve a set of nonlinear equations. As a result, we will not have an explicit solution to the unknown regression coefficients. A similar case occurs if one considers minimizing the EG of the residual, which will result in a set of covariances that are equal to zero, namely, $\mathrm{cov}(X_k, [1 - F(\varepsilon)]^{\nu}) = 0, k = 1, \ldots, K,$[2] and similar to the case of the GMD, it will provide a set of nonlinear equations. This set of regressions resembles quantile regressions (Koenker & Bassett, 1978; Koenker & Hallock, 2001) because by increasing v one can stress the lower portion of the distribution of ε.[3] Because the normal equations are not linear we will not discuss them further in this book.

A special case is when the explanatory variables are independent. If the explanatory variables are statistically independent, then the matrix A in (8.6) is a diagonal matrix and the estimate of each β_k will be identical to the estimate one gets in the simple (i.e., one explanatory variable) regression of X_k on Y, based on the same orthogonality conditions. However, in general, because of the correlations among the explanatory variables, a change in one orthogonality condition may affect all β_k's. This is an important property of the mixed regression suggested above because one can create regression methods that enable the investigator to move gradually from one *pure* regression technique to another. This way one can uncover the effects of the correlations among the explanatory variables on the regression coefficients. For example, one can move from OLS to the Gini regression in a step-wise way by changing one T_k at a time in a given order. Unless the explanatory variables are statistically independent, the changes in the estimates will be path dependent because the order in which the explanatory variables are selected can affect the estimated regression coefficients. The effect of changing the regression methodology with respect to one explanatory variable on the sign of the regression coefficient of another explanatory variable will be demonstrated in Chap. 19.

The last property is the following: if the model in the population is truly linear and the residuals are independent of the explanatory variables, as assumed in the classical model of regression, then the regression methodology used will not affect the expected values of the estimators. However it may affect the efficiency of the estimation procedure. Using the term "efficiency" to evaluate the performance of an

[2]Note that when minimizing the EG of the error term there is only one v: the one applied to the residual.

[3]A critical point that distinguishes between the two approaches of the extended Gini regression is the variable to which the weighting scheme is applied: in EG regressions that belong to the covariance based family the application of the weighting scheme is to the explanatory variables, while in the EG minimization, the application is to the residuals. Under the quantile regression regime the application of the weighting scheme is also to the residuals. See Ben Hur, Frantskevich, Schechtman, and Yitzhaki (2010).

estimator may be a bit problematic because the choice of the best estimator may be affected by the variability measure used to measure efficiency. For example, using the GMD (or the square of it) as a measure of variability may bias the selection procedure of the best regression toward a GMD-based regression, while using the variance may cause the OLS to be the most efficient. However, because for large samples the distributions of the estimators tend to converge toward the normal distribution, it seems that defining efficiency in terms of the variance is justified, at least in large samples.

The methodology described above has one restriction. In order to have a solution, the matrix A in (8.6) must be of rank K. This may create a problem in the GMD or the EG regressions whenever one uses several monotonic transformations of a variable as explanatory variables in the regression model. The reason is because monotonic transformations do not change the rankings of the observations; therefore, the columns representing the cumulative distributions of the explanatory variables will be identical, resulting in multicollinearity. In order to overcome this problem one can either use a different ν_k for each explanatory variable or, alternatively, impose a given structure on the relationship between the explanatory variables. For example, assume that we want to include both X and X^2 in the Gini regression model. That is, the estimated model is

$$\hat{Y} = a + bX + cX^2$$

where b and c are the regression coefficients that were obtained by OLS for X and X^2, respectively.

Then, in the Gini regression we define a new explanatory variable, X_1, as $X_1 = X + \frac{c}{b}X^2$ (where b and c are taken from the OLS regression) and run the model

$$\hat{Y} = a' + dX_1.$$

So we have

$$\hat{Y} = a' + dX_1 = a' + d\left(X + \frac{c}{b}X^2\right) = a' + dX + \frac{dc}{b}X^2.$$

Now, if b = d then the meanings of the coefficients are the same as in OLS. However if they are different then the coefficients for both X and X^2 will change.

Alternatively one can use a mixture of methods in the same regression model—do OLS on some variables and do Gini regression on the others. Those issues will be elaborated on in Chap. 21 which presents applications of the suggested methodology.

Having solved for the regression coefficients we move to determine the constant term, α. This term can be selected according to several criteria, depending on through where one wants the linear approximation to the regression curve to pass. In order to see that define the residual without subtracting a constant. That is, let

$$\varepsilon' = Y - \beta_1 X_1 - \cdots - \beta_K X_K = \varepsilon + \alpha. \quad (8.7)$$

Then

$$\text{Min}_\alpha E\{(\varepsilon' - \alpha)^2\} \quad (8.8)$$

yields a constant term that will cause the regression line to pass through the expected values of the random variables, while

$$\text{Min}_\alpha E\{|\varepsilon' - \alpha|\} \quad (8.9)$$

yields a constant term that will cause the regression line to pass through the medians of the variables. In short, one can use one criterion for selecting the slopes and another one in order to select the constant, and there is no a priori reason not to separate those criteria.

8.2 Gini Regression as a Linear Approximation of the Regression Curve

The semi-parametric Gini regression belongs to the covariance-based family of regressions which can be interpreted as having a linear structure *imposed* on the data. In this section we develop an additional interpretation as a linear *approximation* of the regression curve. This interpretation will make it easier to compare and to incorporate the Gini and EG regressions into regressions other than the OLS. In order to save space and repetitions we present the linear approximation approach only for the EG regressions. The Gini regression is a special case.

Let $(Y, X_1, \ldots, X_K)$ be a $(K + 1)$-variate random variable with expected values $(\mu_Y, \mu_1, \ldots, \mu_K)$ and a finite variance–covariance matrix Σ. Let $g(x) = E\{Y|X_1 = x_1, \ldots, X_K = x_K\}$ be the regression curve. The residual at $(Y_i, X_{1i}, X_{2i}, \ldots, X_{Ki})$ is defined as the deviation of Y_i from the linear approximation

$$\alpha + \beta_1 X_{1i} + \beta_2 X_{2i} + \cdots + \beta_K X_{Ki},$$

i.e.,

$$\varepsilon_i = Y_i - \alpha - \beta_1 X_{1i} - \cdots - \beta_K X_{Ki}.$$

Again, no assumptions are imposed on the residual, and the regression curve need not be a linear function of the explanatory variables.

An investigator is interested in estimating a linear approximation of the regression curve. Consider a first-order Taylor expansion around zero of the regression curve. By construction, the expansion is linear.

The slopes of the linear approximation can be written as:

$$\begin{pmatrix} \frac{dy}{dx_1} \\ \frac{dy}{dx_i} \\ \frac{dy}{dx_K} \end{pmatrix} = \begin{pmatrix} \frac{\partial g}{\partial x_1}+ & \frac{\partial g}{\partial x_2}\frac{dx_2}{dx_1}+ & \cdots & +\frac{\partial g}{\partial x_K}\frac{dx_K}{dx_1} \\ \cdots & \cdots & \cdots & \cdots \\ \frac{\partial g}{\partial x_1}\frac{dx_1}{dx_i}+ & \cdots & \cdots & +\frac{\partial g}{\partial x_K}\frac{dx_K}{dx_i} \\ \frac{\partial g}{\partial x_1}\frac{dx_1}{dx_K}+ & \frac{\partial g}{\partial x_2}\frac{dx_2}{dx_K}+ & \cdots & +\frac{\partial g}{\partial x_K} \end{pmatrix} \tag{8.10}$$

Using the simple regression coefficients developed in Chap. 7 to represent the simple slopes in the Taylor expansion presented in (8.10), we now replace $\frac{dx_j}{dx_k}$ by $\beta_{jk}(\nu_k)$,[4] and $\frac{dy}{dx_k}$ by $\beta_{0k}(\nu_k)$, where the subscript 0 refers to the dependent variable and k = 1, …, K indicate the explanatory variables. Note that if g(x) was truly linear then $\frac{dx_j}{dx_k}$ would be exactly equal to $\beta_{jk}(\nu_k)$ (regardless of the value of ν_k) and $\frac{dy}{dx_k}$ would be exactly equal to $\beta_{0k}(\nu_k)$. Having done that, we define $\beta_1, \ldots, \beta_K$ by

$$\begin{pmatrix} \beta_{01}(\nu_1) \\ \cdots \\ \beta_{0i}(\nu_i) \\ \beta_{0K}(\nu_K) \end{pmatrix} = \begin{pmatrix} \beta_1+ & & \cdots & +\beta_K\,\beta_{K1}(\nu_1) \\ \cdots & \cdots & \cdots & \cdots \\ \beta_1\,\beta_{1i}(\nu_i)+ & \cdots & \cdots & +\beta_K\,\beta_{Ki}(\nu_i) \\ \beta_1\,\beta_{1K}(\nu_K)+ & & \cdots & +\beta_K \end{pmatrix}$$
$$= \begin{pmatrix} 1 & \beta_{21}(\nu_1) & \cdots & \beta_{K1}(\nu_1) \\ & & & \\ \beta_{1K}(\nu_K) & \beta_{2K}(\nu_K) & \cdots & 1 \end{pmatrix} \begin{pmatrix} \beta_1 \\ \\ \beta_K \end{pmatrix} \tag{8.11}$$

The rationale is that β_i (the solution to the set of equations given by (8.11)) seems to be a reasonable estimate of $\frac{\partial g}{\partial x_i}$. Using (8.11) one can solve for the estimators of the partial derivatives $\frac{\partial g}{\partial x_k}$:

$$\begin{pmatrix} \beta_1 \\ \cdot \\ \cdot \\ \beta_K \end{pmatrix} = \begin{pmatrix} 1 & \beta_{21}(\nu_1) & \cdots & \beta_{K1}(\nu_1) \\ \cdot & \cdot & \cdot & \cdot \\ \cdot & \cdot & \cdot & \cdot \\ \beta_{1K}(\nu_K) & \beta_{2K}(\nu_K) & \cdots & 1 \end{pmatrix}^{-1} \begin{pmatrix} \beta_{01}(\nu_1) \\ \cdot \\ \cdot \\ \beta_{0K}(\nu_K) \end{pmatrix}. \tag{8.12}$$

[4]The term $\beta_{jk}(\nu_k)$ is intended to allow a different treatment for each explanatory variable, according to ν_k, the parameter of the extended Gini. See Schechtman et al. (2008).

Note that in (8.12), the vector on the right-hand side depends on all the v_i's. Also, the denominator of row k (k = 1, ..., K) in the matrix (before inverting it) is

$$\Delta(v_k, X_k) = -(v_k + 1)\mathrm{cov}(X_k, [1 - F_k(X)]v_k),$$

i.e., the denominator in each row is the extended Gini of the respective variable.

We refer to the parameters β_i as *implied* partial derivatives because we do not argue that they represent the derivatives at a given point, but if one accepts the notion of a linear approximation and accepts simple regression coefficients as representing weighted averages differentials then for consistency it seems reasonable to accept the implied partial derivatives as representing the partial regression coefficients.

If (8.10) represents slopes of a truly linear model, then all the coefficients (i.e., for all values of v_i) at the right-hand side of (8.12) are constants, and the left-hand side must represent the partial derivative of the regression curve (by construction). On the other hand, if the regression curve is not linear, then by changing v_i, one can trace the change in $\partial g/\partial x_k$, other things being equal, by changing the weighting scheme attached to the slopes of variable k. Note that by other things being equal it is meant that all rows except row k in the matrix of the regression coefficients in (8.10) remain unaffected, and all elements in the vector of the simple regression coefficients of the dependent variable on the explanatory variables except element k do not change. This is a unique property of the EG, which is due to the fact that there are two covariances and two correlations between each pair of random variables. Therefore, $\beta_{ij}(v_j)$ can be changed without affecting $\beta_{ji}(v_i)$. (To see this, note that each line is normalized separately.)

Because we are allowed to multiply each row by a constant (in our case the constant is the extended Gini of the explanatory variable), the matrix can be presented in a way which is similar to the variance–covariance matrix in OLS, with Gini's and co-Gini's replacing the variances and covariances, respectively.

Similar to the simple regression case, the multiple regression procedure generates equivalents to the OLS's normal equations although it is not based on an optimization procedure. By defining the residual and substituting for the multiple regression coefficients, it can be shown that

$$\mathrm{cov}(\varepsilon, [1 - F_k(X)]^{v_k}) = 0 \quad \text{for } k = 1, \ldots, K. \tag{8.13}$$

Similar to the GMD regression case, one can use the other EG covariance (between X and (a function of) ε) to see whether it is also equal to zero. This is an additional informal test for the specification of the model (see Chap. 7). However, we do not pursue this line of research in this book.

To sum up: the usual interpretation of the estimates of regression coefficients is that they are derived by applying an optimization to a target function (minimum sum of squared errors, maximum likelihood) subject to a constraint that the true regression model is linear. In this section we added another interpretation: the regression coefficients in a multiple regression framework can be interpreted as being the coefficients of a linear approximation of an unknown regression curve. Those alternative interpretations enable the user to produce mixed regressions. That is, some of the normal equations can be borrowed from the OLS while others can be borrowed from the Gini method. In covariance-based regressions we characterize the solution as based on solving linear equations with covariances serving as the parameters in the linear equations. However if one is ready to give up the linearity of the equations to be solved, and as a result he is ready to lose the ability to get explicit solutions, then one can substitute first-order conditions for optimization by other criteria such as MAD or quantile regressions in some (or all) of the equations solved in (8.10). We will not analyze the implications of such regressions in this book.

8.3 Combining the Two Regression Approaches: The Multiple Regression Case

We have shown in Chap. 7 that the minimization of the GMD of the residuals in the simple regression case leads to R-regression. This methodology was already presented in Hettmansperger (1984) and others in a multiple regression framework and the derivation is similar to the one for the simple regression case, except that the optimization is carried out for K variables. From our point of view the major drawback of this methodology is that (similar to MAD and quantile regressions) there are no explicit expressions for the regression coefficients. This property impairs the ability to have an intuitive explanation of the results of the optimization because one cannot see how the estimators are composed. In addition it makes it hard to analyze the effects of the interconnections between the explanatory variables on the regression coefficients or even to see the roles that different variables play because it is an "all or nothing" approach. Technical optimization without an intuition is a bit dangerous because one has to trust the results of a computer program. However, having presented the multiple regression coefficients as composed of simple regression coefficients enables us to combine the two Gini regression methodologies, the semi-parametric (no model is assumed) and the minimization (a linear model is assumed) regressions, in order to check whether the two Gini regressions yield the same regression coefficients.

In order to do that, the first step is to use the semi-parametric Gini regression of (8.6), case (b). As a result of applying the method on the data one gets the semi-parametric regression coefficients and a vector of residuals ε_N that satisfies (8.5).

Let us concentrate on the numerator of (8.5) and denote it by

$$\mathrm{cov}(\varepsilon_N, F(X_k)) \equiv 0 \quad (k = 1, \ldots, K). \tag{8.14}$$

The K equations in (8.14) are written as identities because they are imposed on the data. We can now use the first-order conditions of the minimization approach in order to construct a specification test for the linear multiple regression model. The first-order conditions of the minimization approach are identical to (8.15) below except that the residual that is being used is different than the one that comes from the semi-parametric approach. That is, the first-order conditions for the minimization approach are $\mathrm{cov}(X_k, F(e_M)) = 0, \quad k = 1, \ldots, K$ where the vector e_M denotes the residuals arising from the minimization approach. We now insert e_N into these first-order conditions and test whether the covariances are equal to zero. That is,

$$\mathrm{cov}(X_k, F(\varepsilon_N)) \stackrel{?}{=} 0, \quad k = 1, \ldots, K. \tag{8.15}$$

If all the covariances in (8.15) are equal to zero, then the residual of the covariance-based GMD regression satisfies the orthogonality conditions for the minimization of the GMD of the residual. This means that the regression coefficients that were derived for the semi-parametric GMD regression can also serve as the solution of the minimization of the GMD of the residuals. If, on the other hand, the covariance between an explanatory variable X_j and the (distribution function of the) residual in (8.15) is different from zero for any explanatory variable X_j, this means that the residual and X_j cannot be considered statistically independent and the two Gini regressions produce different coefficients. In this case we may say that the linear specification of the model failed the built-in specification test offered by the GMD for the variable X_j. Note that this test can be performed for each explanatory variable separately.

The same relationships that were shown in (8.14) and (8.15) hold in the case of the EG as well. In this case the equivalent of (8.14) is (8.16) (below) while the equivalent of (8.15) is (8.17). The explanation for this similarity is that the EG method, just like the GMD method, has two covariances defined between each pair of variables, and a proper specification implies independence between the residuals and each of the explanatory variables.

$$\mathrm{cov}(\varepsilon_N, [1 - F(X_k)]^{\nu_k}) = 0, \quad k = 1, \ldots, K \tag{8.16}$$

$$\mathrm{cov}(X_k, [1 - F(\varepsilon_N)]^{\nu_k}) \stackrel{?}{=} 0, \quad k = 1, \ldots, K. \tag{8.17}$$

However we will not investigate this topic in this book and will restrict ourselves to the GMD only.

8.4 OLS and Gini Instrumental Variables

There are many conceptual issues that are debated in the literature concerning the use of instrumental variables (IV) in order to interpret a statistical relationship as a causal relationship. We do not intend to participate in those debates because the use of the GMD methodology does not affect the validity or invalidity of any conceptual argument that was raised in the debate which relied on the OLS. The advantage of using the GMD in the case of IV is that one can get more information about the relationship in the data, such as whether the relationships between the dependent variable and the explanatory variables and between the explanatory variables themselves are monotonic or not. One can learn about the robustness of the estimates with respect to monotonic transformations or other actions taken by the investigator and more issues concerning the handling of the data. Those issues can be helpful in the debates about the interpretations in specific cases, but have nothing to contribute to the guiding principles such as whether a natural experiment is really exogenous or not, or in issues concerning the identification strategy.

The objective of this section is to present the OLS-IV and Gini-IV in the multiple regression setting and to point out the major practical differences between them. Those differences enable the user to get additional information that may shed some light about the conceptual differences.

From our point of view the difference between OLS (OLS-IV) and Gini (Gini-IV) is caused by the different measures of variability: the variance (covariance) in the first case and the Gini (co-Gini) in the second.

This difference is concerned with whether one can use a two-stage regression in order to derive the IV parameter or not.

It is well known, and it is demonstrated below, that one can derive the OLS-IV estimators (or coefficients) using two alternative ways: a direct application of an IV or by using two-stage least squares. This convenient way does not hold for the Gini-IV. That is, under the Gini methodology the two methods can result in totally different estimators.

We proceed as follows: we first present the equivalence of the two procedures in the OLS setting. Then we show that when using the Gini method the equivalence does not hold in general.

8.4.1 Two-Stage Least Squares and Instrumental Variables

We start with the basic presentation. Let Z be the matrix of instrumental variables correlated with X but not with ε. One can construct an IV estimator that will be a consistent estimator for β:

$$\hat{\beta}_{\mathrm{IV}} = (\mathrm{X}'\mathrm{P}_{\mathrm{Z}}\mathrm{X})^{-1}\mathrm{X}'\mathrm{P}_{\mathrm{Z}}\mathrm{Y} \tag{8.18}$$

where P_{z}, the projection matrix of Z, is defined by

$$\mathrm{P}_{\mathrm{Z}} = \mathrm{Z}(\mathrm{Z}'\mathrm{Z})^{-1}\mathrm{Z}'. \tag{8.19}$$

Note that combing the two equations gives:

$$\hat{\beta}_{IV} = (X'Z(Z'Z)^{-1}Z'X)^{-1}X'Z(Z'Z)^{-1}Z'Y. \quad (8.20)$$

The two-stage least squares is an instrumental variable estimation technique for estimating the regression coefficient. In the first stage one fits the model $X = Z\pi + r$, estimates π just to obtain the predicted value of X, namely, $\hat{X}$ (because there is no interest in π itself). In the second stage one runs a regression of Y on $\hat{X}$ and estimates the regression coefficient of interest from this regression.

Proposition 8.1 *In an OLS setting the direct way of estimating the regression coefficient in an IV regression is identical to the two-stage estimation procedure.*

Proof The direct estimate is $\beta_D^{OLS} = (Z'X)^{-1}Z'Y. \quad (8.21)$

The two-stage procedure is the following: in the first stage one assumes that $X = Z\pi + r$. Then the predicted value of X is

$$\hat{X} = Z\hat{\pi} = Z(Z'Z)^{-1}Z'X. \quad (8.22)$$

In the second stage one uses OLS to regress Y on $\hat{X}$ to get $\hat{\beta}_{IV-2}^{OLS}$

$$\hat{\beta}_{IV-2}^{OLS} = (\hat{X}'\hat{X})^{-1}(\hat{X}'Y) = ((\hat{X}'\hat{X})^{-1}X'Z(Z'Z)^{-1}Z'Y. \quad (8.23)$$

Using basic matrix operations we get that

$$(\hat{X}'\hat{X})^{-1} = (X'Z(Z'Z)^{-1}Z'Z(Z'Z)^{-1}Z'X)^{-1} = (X'Z(Z'Z)^{-1}Z'X)^{-1}. \quad (8.24)$$

Inserting (8.24) into (8.23) proves the equivalence of the two methods.

$$\begin{aligned}\hat{\beta}_{IV-2}^{OLS} &= (\hat{X}'\hat{X})^{-1}(\hat{X}'Y) = ((\hat{X}'\hat{X})^{-1}X'Z(Z'Z)^{-1}Z'Y \\ &= (X'Z(Z'Z)^{-1}Z'X)^{-1}X'Z(Z'Z)^{-1}Z'Y \\ &= (Z'X)^{-1}(Z'Z)(X'Z)^{-1}X'Z(Z'Z)^{-1}Z'Y = (Z'X)^{-1}Z'Y\end{aligned}$$

8.4.2 Two-Stage and IV in Gini Regressions

Unfortunately the convenient way of using two-stage Gini as an equivalent way to identify the Gini-IV estimators is not applicable, as is stated in Proposition 8.2. We differentiate here between the direct way of using a Gini-IV, which is based on (8.6), (e), and the indirect way, which is based on two-stage Gini regression: in the first stage one estimates an IV, while in the second stage the IV participates in a Gini regression.

Proposition 8.2 *In a Gini setting the direct way of estimating an IV is in general NOT identical to the two-stage estimation procedure.*

Proof The regression coefficients in a Gini-IV regression are:

$$\beta_{D}^{G} = (F_{Z}'X)^{-1}F_{Z}'Y$$

where F_Z is the matrix of cumulative distribution functions of the instrumental variables.

Performing two-stage estimation procedure means the following:

The first stage is

$$\hat{X} = Z(F_{z}'Z)^{-1}F_{z}'X$$

while the second stage

$$\hat{\beta}_{IV-2}^{G} = (F_{\hat{X}}'\hat{X})^{-1}F_{\hat{X}}'Y = ((F_{\hat{X}}'Z)(F_{z}'Z)^{-1}F_{z}'X)^{-1}F_{\hat{X}}'Y = (F_{z}'X)^{-1}F_{z}'Z(F_{\hat{X}}''Z)^{-1}F_{\hat{X}}'Y.$$

However, the matrix F in the direct way is F of Z, while two of the F's in the two-stage procedure are F's of $\hat{X}$. Therefore, unless the ranks of all instruments and the original explanatory variable are identical, we should expect different results.

The explanation to this result is that the cumulative distribution is not in general a linear transformation of the variate, and therefore, it is as if one uses a nonlinear transformation and expects the linear relationship to hold.

8.5 Effects of Commonly Used Practices

We have demonstrated in (8.6) and (8.12) that the regression coefficients in OLS and in the semi-parametric Gini regressions are derived by solving linear equations with the simple regression coefficients between all pairs of variables being the parameters of the equations. The decompositions of the simple regression coefficients (Chap. 7) and the NLMA curves (Chap. 5) enable one to check the sensitivity of the estimated regression coefficients to some actions that are commonly taken by the researcher (see an application in Chap. 19). However we do not have a general formula for performing sensitivity analyses and our discussion in this section will be based on classification of the most frequently performed actions and their potential effects on the results.

As in the simple regression case, the sensitivity analysis we are interested in is focused on the following actions:

(a) Throwing extreme observations (a.k.a. outliers).
(b) Throwing irrelevant observations and using a subgroup of the population—e.g., imposing bounds on observations that participate in the regression.
(c) Substituting a continuous variable by a discrete one with the data entries taken to be either mid-points or averages. (Binning according to Wainer, Gessaroli, and Verdi (2006)).
(d) Applying a monotonic non-decreasing transformation to one or more variables.

We have pointed out two complementary methods for sensitivity analysis. One method is based on using the concentration and NLMA curves (Chap. 5), while the other is based on the decomposition of simple regression coefficients (Chap. 7). The approach based on concentration curves fully describes the contributions of different segments of the distribution of the explanatory variable to the regression coefficient but it can only handle actions that affect one explanatory variable at a time. On the other hand the approach that is based on the decomposition of the simple regression coefficients can handle actions that affect several explanatory variables simultaneously, because one can simultaneously decompose all simple regression coefficients. However, it is limited to the sections chosen by the investigator. Hopefully, future research will lead to computer software that will enable more complicated analyses. Because most of the analyses described in this section are actually based on the analysis carried out in the simple regression case (see Chap. 7), we will only focus on the differences between the multiple regression and the simple regression cases.

Note that under cases (a) and (b) above, omitting an observation or several observations means that the coefficients of *all* the variables which participate in the regression are affected, while under cases (c) and (d) only some specific coefficients are affected. Also, case (c) can be viewed as a special case of (d). In this section we are mainly interested in the effect of a transformation on the sign of the estimator. The reason for stressing sign change is that it may reverse the conclusion reached. Instead of positive (negative) effect it may turn the effect into a negative (positive) one.

For convenience we replicate (8.6).

$$\begin{pmatrix} \beta_1 \\ \\ \beta_K \end{pmatrix} \equiv \begin{pmatrix} 1 & \beta_{21} & \beta_{K1} \\ & & \\ \beta_{1K} & & 1 \end{pmatrix}^{-1} \begin{pmatrix} \beta_{01} \\ \\ \beta_{0K} \end{pmatrix} \equiv \mathbf{A}^{-1}\beta_0. \tag{8.6}$$

It is easy to see that throwing extreme or irrelevant observations affects all the coefficients, and therefore it will be the hardest to analyze. With respect to the other two actions, it is worthwhile to distinguish between actions taken with respect to the dependent variable and the ones taken with respect to the explanatory variables.

An action taken on the dependent variable will not affect the matrix A and only the vector β_0 will be affected. In this case, substituting a continuous variable by a discrete variable can be analyzed by decomposing each element in the vector β_0 to the effect of intra- and inter-group coefficients, with the groups being defined as the sections of the variable that were aggregated. The effect of binning will be the elimination of the intra-group component from the equation and being left with the inter-groups component. The effect of applying a monotonic increasing transformation is more complicated to analyze. It is our intuition that it is worth to check the regression coefficients that are based on non-monotonic relationship because then one can turn a positive (negative) component into a negative (positive) one.

The effect of an action taken with respect to an explanatory variable is more complicated to analyze than the effect of an action taken with respect to the dependent variable because in this case both a column and a row are affected in (8.6), the former with a transformation applied to the dependent variable in a simple regression, while the latter with a transformation on the explanatory variable. Again the effect of binning can be seen by decomposing each component into the omitted and non-omitted parts, but the ability to guess the effect on the inverted matrix is pretty limited. A similar case occurs when dealing with a monotonic transformation of the variable. Note, however, that because of the non-symmetrical relationship in the Gini regression the type of concentration curves analyzed to discover non-monotonicity with respect to variables arranged in a row are different than the ACCs analyzed when analyzing the effect of a transformation on the regression coefficient in a column. Because of the complications in dealing with a multiple regression, we will restrict the analysis in the empirical section to the case of two explanatory variables, which is as if one analyzes the effect of a transformation of one variable, with the rest of the model being kept untouched. However, appropriate software is needed in order to enable handling multiple regression with more than two explanatory variables.

8.6 Summary

In this chapter we have introduced the covariance-based family of regressions, which is characterized by having linear normal equations that are composed of covariances between the variables. The fact that OLS, Gini, and extended Gini methods belong to one family enables the user to stress the similarities and differences among the various members and to combine OLS, Gini, and extended Gini methods in the same regression. The advantage of using the family is that it enables the researcher and the reader to see whether (and how) the estimated regression coefficients are sensitive to the decisions made by the researcher such as the regression method used, the actions of omitting observations, transforming a continuous variable into a binary (or grouped) one, and the use of monotonic transformations.

The major advantage of using the extended Gini regression is its ability to stress different regions along the range of each explanatory variable without applying monotonic transformations to the data. The result is that all the properties that the original data possess, such as an aggregation property, are kept intact.

Because the parameters in the multiple regression case are based on simple regression parameters, all the properties which were mentioned when describing the simple regression coefficients can be applied to the multiple regression framework. For example, the decomposition of the regression coefficient into intra- and inter-group components with the weights being derived from the variability measures of the explanatory variable can be applied into a multiple regression framework by substituting the simple regression coefficient by its components.

Additional advantages of those regressions are that: (a) the investigator does not have to impose a linear model on the population, but rather to produce linear approximations to the regression curves. If the linear approximation is not good enough then those deviations will percolate to the properties of the resultant residuals. (b) One can mix methods of regression in order to check the sensitivity of the estimates to the regression methodology used. This is illustrated in Chap. 21. A disadvantage of the suggested methodology is that it is inaccurate to describe the estimates as best or optimal, because they are results of mixed methodologies.

Chapter 9
Inference on Gini-Based Parameters: Estimation

Introduction

The population parameters based on Gini were introduced in previous chapters. One of the objectives in practice is to estimate them from a given data set. This is the main objective of this chapter. When dealing with estimation, several issues come in mind. Is the data based on individual observations or are they grouped? Is the sampling procedure based on equal probability or is it a stratified one? Are the variables of interest coming from a continuous or a discrete distribution? The estimation procedures depend on the answers to the above questions. In addition, the Gini-based parameters have various presentations which lead to different estimators, each one being the natural estimator of a specific definition.

The estimation technique used throughout this chapter is mainly based on U-statistics theory (Hoeffding, 1948). The theory related to U-statistics allows a single theoretical framework to be used in nonparametric statistics to prove results for a wide range of test-statistics and estimators relating to the asymptotic normality and to the variance (in finite samples) of such quantities. In addition the theory has applications to estimators which are not themselves U-statistics, but functions of (dependent) U-statistics, which is the case when the correlation and the overlapping index are estimated. It turns out that most of the estimators of the Gini parameters obtained in this chapter are the sample's representations of the parameters. We start with a short review of the method.

Let *P* be a family of probability measures on an arbitrary measurable space. The problems treated here are nonparametric, which means that *P* will be taken to be a large family of distributions subject only to mild restrictions such as continuity or existence of moments. Let $\theta(P)$ denote a real-valued function defined for P $\in$ *P*. We say that θ(P) is an *estimable parameter* within *P* if for some integer m there exists an unbiased estimator of θ(P) based on m i.i.d. random variables distributed according to P; that is, if there exists a real-valued measurable function h $(X_1, \ldots, X_m)$ such that

S. Yitzhaki and E. Schechtman, *The Gini Methodology: A Primer on a Statistical Methodology*, Springer Series in Statistics 272, DOI 10.1007/978-1-4614-4720-7_9,

$$E_P[h(X_1,\ldots,X_m)] = \theta(P) \quad \text{for all } P \in \mathcal{P} \tag{9.1}$$

where $X_1, \ldots, X_m$ are i.i.d. random variables having a distribution P. The function $h(X_1, \ldots, X_m)$ is called the *kernel* and the smallest integer m with this property is called the *degree* of the kernel. The function h may be assumed to be a symmetric function of its arguments without loss of generality because if h is an unbiased estimator of θ(P), then the average of h applied to all permutations of the variables is still unbiased and is, in addition, symmetric.

For a real-valued measurable function $h(X_1, \ldots, X_m)$ and for a sample $x_1, \ldots, x_n$, of size $n \geq m$ from a distribution P, a *U-statistic* with kernel h is defined as

$$U_n = U_n(h) = \frac{1}{\binom{n}{m}} \sum_{C_{m,n}} h(x_{i_1},\ldots,x_{i_m}) \tag{9.2}$$

where the summation is over the set $C_{m,n}$ of all $\binom{n}{m}$ combinations of m integers $i_1 < i_2 < \cdots < i_m$ chosen from (1, 2, …, n).

If $\theta(P) = E_P[h(X_1, \ldots, X_m)]$ exists for all $P \in \mathcal{P}$, then an obvious property of the U-statistic, U_n is that it is an unbiased estimate of $\theta(P)$. Moreover it has the optimality property of being consistent and being a best unbiased estimate of $\theta(P)$ if $\mathcal{P}$ is large enough. The asymptotic distributions of a U-statistic and of functions of (dependent) U-statistics will be discussed in Chap. 10, the subject of which is testing. The U-statistic discussed above is a one-sample U-statistic. The natural extension to a k-sample U-statistic is called the generalized U-statistic. A k-sample U-statistic will be based on a kernel function of the k samples, with degrees $(r_1, \ldots, r_k)$, where r_i is the degree with respect to the ith sample. Details can be found in Randles and Wolfe (1979).

The structure of the chapter is as follows: Sect. 9.1 deals with estimators based on individual observations coming for a continuous distribution while Sect. 9.2 provides the estimators for the discrete case. In Sect. 9.3 we discuss the case of individual data, weighted, and Sect. 9.4 is devoted to grouped data. Section 9.5 concludes.

9.1 Estimators Based on Individual Observations: The Continuous Case

9.1.1 The Gini Mean Difference and the Gini Coefficient

We start with the basic parameter, the Gini mean difference (GMD). A close relative of the GMD is the Gini coefficient. The GMD is an absolute measure while the Gini coefficient is a relative one. The GMD is mainly used in the areas of finance and econometrics, while the Gini coefficient is mainly used in the areas of income distribution and public policy evaluations. We will give estimates for both versions. We start with the GMD. Let $X_1, \ldots, X_n$ be independent and identically distributed random variables from a distribution with cdf F(x) having a finite mean. The GMD is defined as

$$\Delta_X = E(|X_1 - X_2|) = 4COV(X, F(X))$$

where X_1 and X_2 are two independent draws from a continuous distribution F with a finite mean. There are several estimators for Δ_X, each corresponds to a different way of spelling Gini.

Following the U-statistics theory, the most natural one is to define a symmetric kernel of degree 2 as $h(X_1, X_2) = |X_1 - X_2|$, as detailed in the following proposition:

Proposition 9.1 *Let (X_1, X_2) be a random sample of size 2 from a continuous distribution function with a finite first moment. Let $h(X_1, X_2) = |X_1 - X_2|$. Then $h(X_1, X_2)$ is a symmetric kernel of degree 2 for*

$$\Delta_X = E(|X_1 - X_2|) = 4COV(X, F(X))$$

and the U-statistic is given by

$$U(\Delta_X) = \frac{1}{\binom{n}{2}} \sum_{i<j} |x_i - x_j|. \tag{9.3}$$

Proof $E(h(X_1,X_2)) = E(|X_1 - X_2|) = \Delta_X$ (as shown in Chap. 2 (2.1), hence $U(\Delta_X)$ is a U-statistic for Δ_X. That is, it is a consistent and unbiased estimator of Δ_X (see Randles & Wolfe, 1979; Schechtman & Yitzhaki, 1987).

Several additional ways to estimate the GMD were suggested in the literature. The first is as a linear combination of order statistics

$$U(\Delta_X) = \frac{1}{4\binom{n}{2}} \sum_{i=1}^{n} (2i - 1 - n)x_{(i)} = \frac{1}{4\binom{n}{2}} \sum_{i=1}^{n} ix_{(i)} - \frac{n(n+1)}{2}\bar{x}$$
$$= \frac{1}{8\binom{n}{2}} \sum_{i=1}^{n-1} i(n-i)(x_{(i+1)} - x_{(i)}) \tag{9.4}$$

where $x_{(i)}$ is the ith order statistic of $x_1, \ldots, x_n$ (see Kendall & Stuart, 1969; Yitzhaki & Olkin, 1988 for details). The advantage of a presentation based on a linear combination of order statistics is that one can use the available literature on order statistics in order to establish the properties of the estimator (see David, 1981; Stigler, 1974).

A natural way to estimate the Gini coefficient, which is defined as $G = \frac{\Delta_X}{2\mu}$, is simply by dividing each of the above-mentioned estimators by the natural estimator of 2μ, namely, by $2\bar{x}$. One suggestion along this idea is given by

$$\hat{G} = \frac{1}{n(n-1)\bar{x}} \sum_{i=1}^{n-1} i(n-i)(x_{(i+1)} - x_{(i)})$$

(see Gastwirth, Modarres, & Bura, 2005).

An alternative estimator of the Gini coefficient is based on the following definition:

$$G = \frac{2}{\mu} \int_0^\infty xF(x)dF(x) - 1.$$

Using this presentation, a natural way to estimate G is by the plug-in estimator $\hat{G}_n$

$$\hat{G}_n = \frac{2}{\bar{X}} \int_0^\infty xF_n(x)dF_n(x) - 1.$$

Standard arguments show that $\hat{G}_n$ is a consistent estimate of G under weak regularity conditions (such as the existence of the second moment).

The third direction of estimation is based on the covariance presentation of the GMD (Lerman & Yitzhaki, 1984). From the covariance formula it becomes simple to calculate the Gini from individual observations. First, obtain the rank (R_i) for each observation x_i. Next, calculate the sample covariance between R_i/n (which represents the empirical distribution function) and x_i. Note that unlike standard approaches for calculating the Gini, this method does not require grouping of individual data to economize on computations. In addition, the estimator can be computed by using any standard statistical software.

Finally, the GMD and Gini coefficient can be estimated from the Lorenz curve. Given a Lorenz curve LC(p), the standard way of estimating the Gini coefficient is to approximate the area by choosing k percentiles $0 = p_0 < p_1 < \cdots < p_k < p_{k+1} = 1$ and computing the area of the polygon with vertices (0, 0), (p_1, LC(p_1)), …, (p_k, LC(p_k)), and (1, 1). That is, the points are connected by straight lines. This issue will be further discussed below. This procedure obviously leads to an underestimate of the Gini coefficient because the straight line connecting the adjacent points always lies above the convex curve LC(p).

Several bounds exist for the GMD. Generally, if F is supported on a finite interval [a, b] and has mean μ, then $0 \le GMD \le \frac{2(\mu-a)(b-\mu)}{(b-a)}$. For "open-ended" intervals of the form [a, ∞) the bound is $\text{GMD} \leq 2(\mu - \text{a})$. If F has a finite second moment (denoted by σ^2) then $\text{GMD} \leq \frac{4\sqrt{3}}{3}\sigma$. If F has a unique median, the bound becomes tighter: $\text{GMD} \leq \frac{2\sqrt{3}}{3}\sigma$ (Cerone & Dragomir 2006).

If F is concave, supported on [a, b], then $\frac{2}{3}(\mu - \text{a}) \leq \text{GMD} \leq \frac{2(\mu-\text{a})}{(\text{b}-\text{a})}\left[(\text{b}-\mu) - \frac{1}{3}(\mu - \text{a})\right]$, and if F is convex on [a, b], the bounds are $\frac{2(\text{b}-\mu)}{3(\text{b}-\text{a})} \leq \text{GMD} \leq \frac{2(\text{b}-\mu)}{3(\text{b}-\text{a})}[4(\mu - \text{a}) - (\text{b} - \text{a})]$. (Cerone, 2008; Gastwirth, 1972).

A comprehensive review on bounding and approximating the GMD can be found in Cerone (2008).

Whenever the sample size is small there are two problems that arise, even if the underlying distribution is continuous:

(a) A downward bias in the estimation of the GMD and the Gini coefficient.
(b) An incompatibility between the estimates that are derived by using different formulas of the Gini.

We start with the intuitive explanation of the cause and the size of the bias and a suggestion how to fix it.

Let us start with the calculation of the Gini coefficient through the Lorenz curve. For simplicity assume that all observations are positive and that the sample is of size n. Then we have n linear segments connecting (0, 0) to (1, 1). Maximum inequality in the sample occurs if (n − 1) observations have zero income and one observation is positive. It is easy to see that the maximum value of the Gini coefficient is (n − 1)/n. This value is lower than the upper bound for the population which is equal to one. To correct for this finite population bias one can multiply the estimate of the Gini coefficient by the finite population correction factor n/(n − 1).

An alternative way of interpreting the finite population correction factor is to define the estimate of the Gini coefficient as representing the area between the diagonal and the Lorenz curve, divided by the maximum possible inequality. Because the maximum possible inequality is equal to (n − 1)/n we get the same result as before.

A third interpretation is that the finite sample correction factor is similar to correcting for the degrees of freedom. The bias due to a small sample also occurs in estimating the variance. The correction in the variance world is through the use of degrees of freedom. Using the degrees of freedom correction (i.e., dividing by (n − 1)) solves the small sample bias.

It is worth pointing out that the formula of decomposition of the Gini coefficient (ANOGI, see Sect. 4.2) enables us to evaluate the magnitude of the bias caused by the use of a small sample. Applying the formula for the decomposition of the Gini coefficient of the entire population into the contributions of the subgroups and taking into account that each group, which is actually an observation, occupies a section of size 1/n along the horizontal axis and that the groups (sections) do not overlap, the connection between the Gini coefficient in the population and its estimate is:

$$G = \hat{G} + \sum_{i=1}^{n} p_i s_i G_i$$

where G is the Gini coefficient in the population, $\hat{G}$ is the estimator based on the Lorenz curve (and represents the between-groups Gini), while $p_i = 1/n$, s_i, and G_i are the share of each observation (i.e., section) in the population, its share in the total income, and the Gini of the ith section, respectively. Assuming a specific distribution enables one to estimate the bias. However, in this book we do not assume specific distributions.

We now turn to the incompatibility between the estimates that are derived using different formulas of the Gini.

The covariance formula of the Gini coefficient is based on the cumulative distribution function. In the sample the cumulative distribution is represented by the empirical cumulative distribution function, which is written as a step function. An alternative definition of the Gini is through the Lorenz curve. In the sample, the empirical Lorenz curve is connected by linear segments. This implies an implicit assumption of a continuous distribution function, because the horizontal axis represents the cumulative distribution. This discrepancy between the approaches does not create a wedge between the formulae in the case of an equal probability sampling method because adding (or subtracting) a constant from a variable in the covariance formula does not change the value of the estimate. However it does create a wedge between the formulas when the sample is not based on an equal probability sampling method or when dealing with the extended Gini (see Chotikapanich & Griffiths, 2001; Lerman & Yitzhaki, 1989; Schechtman & Yitzhaki, 2008).[1] Section 9.3 is devoted to this issue.

9.1.2 *The Gini Covariance and Correlation*

Let (X, Y) have a continuous bivariate distribution function with marginal cdfs F and G, respectively. There are two Gini covariances (co-Gini's) between any pair of variables: Gcov(X, Y) = cov(X, G(Y)) and Gcov(Y, X) = cov(Y, F(X)) depending on which variable is taken in its variate value and which one is ranked. Using the U-statistic method, the kernel is given in the next proposition which is a generalization of Proposition 9.1.

Proposition 9.2 *Let* (X_1, Y_1) *and* (X_2, Y_2) *be a random sample of size 2 from a continuous bivariate distribution function with finite first moments. Let*

[1] Gini (1914, reprinted 2005) was well aware of this problem. The way he corrected it was by presenting both the diagonal and the Lorenz curve as step functions. However, because the convexity/concavity of the Lorenz curve carries important information concerning the properties of the random variables it seems that this approach is not very useful because of the properties we are interested in.

$h((X_1, Y_1), (X_2, Y_2)) = (X_1 - X_2)I_{(Y_1>Y_2)} + (X_2 - X_1)I_{(Y_2>Y_1)}$ *where* $I_{a>b}$ *is defined as*

$$I_{a>b} = \begin{cases} 1 & \text{if a>b} \\ 0 & \text{otherwise} \end{cases}.$$

Then $h((X_1, Y_1),(X_2, Y_2))$ *is a symmetric kernel of degree (2, 2) for* $\Delta_{X,Y} = 4\mathrm{cov}(X, G(Y)) = 4G\mathrm{cov}(X, Y)$. *(A degree (2, 2) means that one needs two independent Xs and two independent Ys in order to find an unbiased estimator).*

Proof We need to show that $E\{h((X_1,Y_1),(X_2,Y_2))\} = \Delta_{X,Y}$ In what follows the domain of the integration is $(-\infty, \infty)$ unless stated differently

$$E(h(X_1,Y_1),(X_2,Y_2)) = E[(X_1 - X_2)I_{(Y_1>Y_2)} + (X_2 - X_1)I_{(Y_2>Y_1)}] =$$

$$\iiint\int_{y_2}^{\infty} (x_1 - x_2)f(x_1,y_1)f(x_2,y_2)dy_1dx_1dy_2dx_2$$

$$+\iiint\int_{-\infty}^{y_2} (x_2 - x_1)f(x_1,y_1)f(x_2,y_2)dy_1dx_1dy_2dx_2$$

$$= 2\iint x_1f(x_1,y_1)G(y_1)dy_1dx_1 - 2\iint x_1f(x_1,y_1)(1 - G(y_1))dy_1dx_1$$

$$= 2[E(XG(Y)) - E(X(1 - G(Y)))] = 4E(XG(Y)) - 4 \times \frac{1}{2} \times E(X)$$

$$= \quad 4\mathrm{cov}(X,G(Y)) = \Delta_{X,Y}.$$

Using the kernel above, the U-statistic

$$U(\Delta_{X,Y}) = \frac{1}{\binom{n}{2}} \sum\sum_{i<j} h((x_i,y_i),(x_j,y_j))$$

$$= \frac{1}{\binom{n}{2}} \sum\sum_{i<j} [(x_i - x_j)I_{(y_i>y_j)} + (x_j - x_i)I_{(y_j>y_i)}] \tag{9.5}$$

is a U-statistic for the parameter 4cov(X, G(Y)) and hence an unbiased and consistent estimator.

An alternative definition, based on a linear combination of concomitants of order statistics, is given by

$$U(\Delta_{X,Y}) = \frac{1}{4\binom{n}{2}} \sum_{i=1}^{n} (2i - 1 - n)x_{y_{(i)}}$$

where $x_{y_{(i)}}$ is the x that belongs to $y_{(i)}$, the ith order statistic of $y_1, \ldots, y_n$.

Having estimated both cov(X, F(X)) and cov(X, G(Y)), it is natural to use the two estimators in order to obtain an estimator for the Gini correlation between X and Y.

The Gini correlation between X and Y is defined as

$$\Gamma_{X,Y} = \frac{\mathrm{cov}(X, G(Y))}{\mathrm{cov}(X,F(X))}.$$

See Chap. 3 (3.4) for details. (Recall that this is one of the two correlation coefficients between X and Y).

Hence an estimator of $\Gamma_{X,Y}$, based on a ratio of two dependent U-statistics is $\hat{\Gamma}_{X,Y} = \frac{U(\Delta_{X,Y})}{U(\Delta_X)}$, as stated in the next proposition.

Proposition 9.3 *Let* $U(\Delta_x)$ *and* $U(\Delta_{X,Y})$, *as given in (9.3) and (9.5), be the U-statistics for 4cov(X, F(X)) and 4 cov(X, G(Y)), respectively. Then* $\hat{\Gamma}_{X,Y} = \dfrac{U(\Delta_{X,Y})}{U(\Delta_X)}$ *is a consistent estimator of* $\Gamma_{X,Y}$.

Proof

$$\hat{\Gamma}_{X,Y} = \frac{U(\Delta_{X,Y})}{U(\Delta_X)} = \frac{U(\Delta_{X,Y})}{4\mathrm{cov}(X,G(Y))} \frac{\mathrm{cov}(X, G(Y))}{\mathrm{cov}(X,F(X))} \frac{4\mathrm{cov}(X,F(X))}{U(\Delta_X)}.$$

Because $U(\Delta_x)\,(U(\Delta_{X,Y}))$ is a U-statistic for 4 cov(X, F(X)) (4 cov(X, G(Y))), we get that $U(\Delta_x)\,(U(\Delta_{X,Y}))$ converges in probability to 4 cov(X, F(X)) (4 cov(X, G(Y))) and therefore both $\dfrac{U(\Delta_{X,Y})}{4\mathrm{cov}(X,G(Y))}$ and $\dfrac{4\mathrm{cov}(X,F(X))}{U(\Delta_X)}$ converge in probability to 1, i.e., $\hat{\Gamma}_{X,Y}$ converges in probability to $\Gamma_{X,Y}$.

It is worth pointing out that the Gini correlation and the Gini regression coefficient suffer from the small sample bias. However, they suffer less than the Gini coefficient. The reason is because they are expressed as ratios and both the numerator and the denominator are biased downward. For additional explanation see Wodon and Yitzhaki (2003b).

9.1.3 The Overlapping Index

The overlapping index between two populations denoted by i and j having cumulative distribution functions F_i and F_j, respectively, was defined in Chap. 4 and is given there, below (4.19), as

$$O_{ji} = \frac{\mathrm{cov}_i(Y,F_j(Y))}{\mathrm{cov}_i(Y,F_i(Y))} \tag{9.6}$$

where by cov_i it is meant that the covariance is calculated under the cumulative distribution F_i. The intuitive meaning of the numerator is the covariance between an observation from population i with its rank, had it been ranked within the distribution of population j. Note that (9.6) can be expressed as a function of four parameters as follows:

$$O_{ji} = \frac{cov_i(Y,F_j(Y))}{cov_i(Y,F_i(Y))} = \frac{E_i(YF_j(Y)) - E_i(Y)E_i(F_j(Y))}{cov_i(Y,F_i(Y))} = \frac{\theta_1 - \theta_2\,\theta_3}{\theta_4}$$

where

$$\theta_1 = E_i(YF_j(Y)),\ \ \theta_2 = E_i(Y),\ \ \theta_3 = E_i(F_j(Y))\ \text{ and }\ \ \theta_4 = cov_i(Y,F_i(Y)).$$

Note that the overlapping index is actually similar in structure to a Gini correlation, except that the cumulative distribution in the numerator does not lie between zero and one, but takes the value of the cumulative distribution of the other variable.

We apply the U-statistic technique. Two types of U-statistics will be involved. The U-statistics for θ_1 and θ_3 are two-sample U-statistics (i.e., generalized U-statistics) while the U-statistics for θ_2 and θ_4 are one-sample U-statistics. It is easy to see that the U-statistic for θ_2 is simply the sample mean coming from population i. For θ_4, which is the Gini for distribution i, the U-statistic is given in (9.3). We now find U-statistics for θ_1 and θ_3. For simplicity of notation let X have distribution F_i and let Y have distribution F_j.

Proposition 9.4 *The kernel of degree (1, 1) for estimating* θ_1 *is given by*

$$h(X,Y) = \begin{Bmatrix} X & \text{if } Y \le X \\ 0 & \text{otherwise} \end{Bmatrix}.$$

Proof

$$F_{x,y}(X,Y) - E_X[E_Y(h(X,Y)|X = x)] = E_X[XF_Y(X)] = \theta_1.$$

We use the kernel to define a U-statistic as follows:

$$U_1 = U(x_1, \ldots, x_{n_x}; y_1, \ldots, y_{n_y}) = \frac{1}{n_x n_y}\sum\sum h(x_i,y_j)$$
$$= \frac{1}{n_x n_y}\sum\sum x_i I(y_j \le x_i) = \frac{1}{n_x n_y}\sum x_i(\#y\text{'s} \le x_i).$$

We now turn to the last parameter, θ_3.

Proposition 9.5 *The kernel of degree (1, 1) for estimating* θ_3 *is given by*

$$h(X,Y) = \begin{cases} 1 & \text{if } Y \le X \\ 0 & \text{otherwise} \end{cases}.$$

The proof is trivial and the resulting U-statistic is

$$U_3 = \frac{1}{n_x n_y} \sum (\#y's \le x_i).$$

Combining these results, the estimator of O_{ji} is given by

$$U = \frac{U_1 - U_2 U_3}{U_4}.$$

We note that each of the four U-statistics is a consistent estimator of the corresponding parameter. Therefore by applying Slutzky's theorem (see, for example, Randles & Wolfe, 1979) we get that U is a consistent estimator for the overlapping parameter.

9.1.4 The Extended Gini, Extended Gini Covariance, and Extended Gini Correlation

The extended Gini (EG) was introduced in Chap. 6. It is defined as

$$\Delta(v, X) = -(v + 1)\ \mathrm{cov}(X, [1 - F(X)]^v). \tag{9.7}$$

Similarly, the extended Gini covariance between Y and X is given by

$$\Delta(v, Y, X) = -(v + 1)\ \mathrm{cov}(Y, [1 - F(X)]^v). \tag{9.8}$$

We proceed by finding a kernel and defining a U-statistic for the extended Gini covariance, and then we obtain the estimate of the EG of X as a special case with X = Y.

Proposition 9.6 *Let $(X_1, Y_1), \ldots, (X_{v+1}, Y_{v+1})$ be a random sample of size $(v + 1)$ from a continuous bivariate distribution $F_{X,Y}$ with finite second moments. Let*

$$h((X_1,Y_1),\ldots,(X_{v+1},Y_{v+1})) = \overline{Y}_{v+1} - Y_{X_{(1)}}, \tag{9.9}$$

where $\bar{Y}_{v+1}$ is the average of $Y_1, \ldots, Y_{v+1}$ and $Y_{X_{(1)}}$ is the Y that belongs to $X_{(1)}$, the minimum of $X_1, \ldots, X_{v+1}$. Then $h((X_1, Y_1), \ldots, (X_{v+1}, Y_{v+1}))$ is a symmetric kernel of degree $v + 1$ for the parameter $\Delta(v, Y, X)$ of (9.8). That is, $h((X_1, Y_1), \ldots, (X_{v+1}, Y_{v+1}))$ is an unbiased estimator of the parameter $\Delta(v, Y, X)$ based on $v+1$ bivariate observations.

Proof The parameter $\Delta(v, \mathrm{Y}, \mathrm{X})$ of (9.8) can be expressed as follows:

$$\begin{aligned}\Delta(v,Y,X) &= -(v+1)\ \mathrm{COV}(\mathrm{Y},[1-\mathrm{F(X)}]^{v}) \\ &= (v+1)\mathrm{E}\{\mathrm{Y}\}\mathrm{E}\{[1-\mathrm{F(X)}]^{v}\} - (v+1)\ \mathrm{E}\{\mathrm{Y}\ [1-\mathrm{F(X)}]^{v}\} \\ &= \ \mu_{\mathrm{Y}} - (v+1)\ \mathrm{E}\{\mathrm{Y}[1-\mathrm{F(X)}]^{v}\}.\end{aligned}$$

Therefore we only need to show that $\mathrm{E}\{\mathrm{Y}_{\mathrm{X}(1)}\} = (\nu + 1)\mathrm{E}\{\mathrm{Y}[1 - \mathrm{F(X)}]^{\nu}\}$.

Claim $\mathrm{E}\ \left\{\mathrm{Y}_{\mathrm{X}(1)}|\mathrm{X}_{(1)} = \mathrm{x}\right\} = \mathrm{E}\{\mathrm{Y}|\mathrm{X} = \mathrm{x}\}$.

The proof of the claim is restricted to the discrete case for simplicity.

Proof of the claim

$$\begin{aligned}\mathrm{E}\{\mathrm{Y}_{\mathrm{X}(1)}|\mathrm{X}_{(1)} = \mathrm{x}\} &= \sum_{i=1}^{v+1} \mathrm{y_i P}\left(\mathrm{Y}_{\mathrm{X}_{(1)}} = \mathrm{y_i}|\mathrm{X}_{(1)} = \mathrm{x}\right) \\ &= \sum_{j=1}^{v+1}\sum_{i=1}^{v+1} \mathrm{y_i P}\left(\mathrm{Y}_{\mathrm{X}_{(1)}} = \mathrm{y_i}|\mathrm{X}_{(1)} = \mathrm{x}, \mathrm{X_j} = \mathrm{X}_{(1)}\right)\mathrm{P}\left(\mathrm{X_j} = \mathrm{X}_{(1)}\right) \\ &= \sum_{j=1}^{v+1}\sum_{i=1}^{v+1} \mathrm{y_i P}\left(\mathrm{Y_j} = \mathrm{y_i}|\mathrm{X_j} = \mathrm{x}\right)1/(v+1) \\ &= \sum_{j=1}^{v+1} \mathrm{E}\left(\mathrm{Y_j}|\mathrm{X_j} = \mathrm{x}\right)1/(v+1) = \mathrm{E}(\mathrm{Y}|\mathrm{X} = \mathrm{x}).\end{aligned}$$

Using the claim,

$$\begin{aligned}\mathrm{E}\left\{\mathrm{Y}_{\mathrm{X}_{(1)}}\right\} &= \mathrm{E}_{\mathrm{X}_{(1)}}\left\{\mathrm{E}(\mathrm{Y}_{\mathrm{X}_{(1)}}|\mathrm{X}_{(1)} = \mathrm{x}\right\} = \int \mathrm{E}\left(\mathrm{Y}_{\mathrm{X}_{(1)}}|\mathrm{X}_{(1)} = \mathrm{x}\right)\mathrm{f}_{\mathrm{X}_{(1)}}(\mathrm{x})\mathrm{dx} \\ &- (v+1)\int \mathrm{E}\left(\mathrm{Y}_{\mathrm{X}_{(1)}}|\mathrm{X}_{(1)} = \mathrm{x}\right)[1-\mathrm{F(x)}]^{v}\mathrm{f(x)dx} \\ &= (v+1)\int \mathrm{E}(\mathrm{Y}|\mathrm{X}=\mathrm{x})[1-\mathrm{F(x)}]^{v}\mathrm{f(x)dx} \\ &= (v+1)\iint \mathrm{yf(y|x)}[1-\mathrm{F(x)}]^{v}\mathrm{f(x)dydx} \\ &= (v+1)\iint \mathrm{y}[1-\mathrm{F(x)}]^{v}\mathrm{f(x,y)dydx} = (v+1)\mathrm{E}\{\mathrm{Y}[1-\mathrm{F(X)}]^{v}\}.\end{aligned}$$

The symmetry of $\mathrm{h}((\mathrm{X}_1, \mathrm{Y}_1), \ldots, (\mathrm{X}_{\nu+1}, \mathrm{Y}_{\nu+1}))$ is obvious.

Let $\mathrm{h}((\mathrm{X}_1,\mathrm{Y}_1),\ldots,(\mathrm{X}_{v+1},\mathrm{Y}_{v+1})) = \bar{Y}_{v+1} - \mathrm{Y}_{\mathrm{X}(1)}$, as in (9.9) and let

$$\begin{aligned}\mathrm{U}(\Delta(v,\mathrm{X},\mathrm{Y})) &= \frac{1}{\binom{\mathrm{n}}{\mathrm{v}+1}}\sum_{\mathrm{i}_1<}\sum_{\cdots<}\sum_{\mathrm{i}_{v+1}} \mathrm{h}((\mathrm{x}_{\mathrm{i}_1},\mathrm{y}_{\mathrm{i}_1}),\ldots,(\mathrm{x}_{\mathrm{i}_{v+1}},\mathrm{y}_{\mathrm{i}_{v+1}})) \\ &= \frac{1}{\binom{\mathrm{n}}{v+1}}\sum_{\mathrm{i}_1<\cdots<\mathrm{i}_{v+1}}\cdots\sum\left(\frac{\sum_{\mathrm{j}=1}^{v+1}\mathrm{y}_{\mathrm{i_j}}}{v+1} - \mathrm{y}_{\min(\mathrm{x}_{\mathrm{i}_1},\ldots,\mathrm{x}_{\mathrm{i}_{v+1}})}\right),\end{aligned}$$

where $(i_1, \ldots, i_{v+1})$ is a permutation of $(v + 1)$ indices chosen from $(1, \ldots, n)$. Then $U(\Delta\ (v, Y, X))$ is a U-statistic for the parameter $\Delta(v, Y, X)$ and is therefore an unbiased and consistent estimator of $\Delta(v, Y, X)$. Using combinatorial arguments $U(\Delta\ (v,Y, X))$ can be simplified and written as a linear combination of concomitants of the order statistics as follows:

$$U(\Delta(v, Y, X)) = \frac{1}{\binom{n}{v+1}} \sum_{i=1}^{n} \left[\frac{1}{v+1} \binom{n-1}{v} - \binom{n-i}{v} \right] y_{X_{(i)}}. \tag{9.10}$$

(Note that if $v > (n - i)$ then $\binom{n\text{ - }i}{v} = 0$).

A special case with $X = Y$ gives the U-statistic for estimating the EG,

$$U(\Delta(v,X)) = \frac{1}{\binom{n}{v+1}} \sum_{i=1}^{n} \left[\frac{1}{v+1} \binom{n-1}{v} - \binom{n-i}{v} \right] x_i. \tag{9.11}$$

9.1.5 Gini Regression and Extended Gini Regression Parameters

We start the presentation with Gini simple and multiple regression coefficients and then move to the extended Gini regressions. Let (X, Y) have a continuous bivariate distribution function with final second moments and denote the marginal cdfs by F and G, respectively. In the simple GMD regression the investigator is interested in constructing a linear predictor of Y that is based on X. The linear predictor is (see (7.1)):

$$\hat{Y} = \alpha + \beta X.$$

where (α, β) are the intercept and the slope of the linear predictor, respectively.

The Gini simple regression coefficient (i.e., the slope) was defined in Chap. 7 (see (7.5)) by

$$\beta = \beta^N = \frac{cov(Y, F(X))}{cov(X,F(X))}. \tag{9.12}$$

The superscript N will be ignored when it is not confusing. Using the U-statistic method, it is easy to see that β can be estimated by

$$\hat{\beta} = \frac{U(\Delta_{Y,X})}{U(\Delta_X)}$$

where

$$\begin{aligned}U(\Delta_{Y,X}) &= \frac{1}{\binom{n}{2}}\sum\sum_{i<j} h((y_i,x_i),(y_j,x_j))\\ &= \frac{1}{\binom{n}{2}}\sum\sum_{i<j}[(y_i - y_j)I_{(x_i>x_j)} + (y_j - y_i)I_{(x_j>x_i)}]\end{aligned}$$

(see (9.5) for details), and $U(\Delta_X)$ is the U-statistic for estimating the Gini of X (see Sect. 9.1.1).

Because the GMD is a member of the extended Gini (EG) family, one can replicate the previous part of this section with the extended Gini. All one has to do is substitute F(X) by $-[1 - F(X)]^v$ or $F(X)^v$ or a combination of the two, depending on whether one is interested in applications that fit concave, convex, or symmetric weighting schemes.

For example, similar to the definition in (9.12) the EG regression coefficient is defined by

$$\beta_N(\nu) = \frac{\text{cov}(Y,[1 - F_x(X)]^{\nu-1})}{\text{cov}(X,[1 - F_x(X)]^{\nu-1})}.$$

The natural estimator is based on the ratio of two U-statistics. The U-statistic for $\beta_N(\nu)$ is then given by

$$\hat{\beta}_N(\nu) = \frac{U(\Delta(\nu,Y,X))}{U(\Delta(\nu,X))}$$

where $U(\Delta(\nu, Y,X))$ and $U(\Delta(\nu, X))$ are given in (9.10) and (9.11), respectively.

9.1.6 *Lorenz Curve and Concentration Curves*

There are several definitions of the absolute Lorenz curve (ALC). We follow Gastwirth's (1971, 1972) definition (see Sect. 2.1), which is based on the inverse of the cumulative distribution X(p): p is plotted on the horizontal axis while the vertical axis represents the cumulative value of the variate, $\int_{-\infty}^{p} X(t)dt$.

The empirical ALC is generated from the data in the following way: given a set of n ordered numbers $x_1 \le x_2 \le \cdots \le x_n$, the empirical ALC is defined at the points $\frac{i}{n}$, i = 0, ..., n by ALC(0) = 0 and $\text{ALC}\left(\frac{i}{n}\right) = s_i$ where $s_i = (x_1 + \cdots + x_i)/n$.

The empirical ALC, ALC(p), is then defined for all p in the interval (0, 1) by linear interpolation. That is, the points are connected by linear segments.

The familiar Lorenz curve (LC) is derived from the ALC by dividing the cumulative value of the variate by its mean. The vertical axis is then $\frac{1}{\mu}\int_{-\infty}^{p} X(t)dt$. The Lorenz curve is estimated in a similar way to that of the ALC, where the only difference is that now $LC\left(\frac{i}{n}\right) = \frac{S_i}{S_n}$. The absolute concentration curve (ACC) is similar to the ALC, with one difference: the vertical axis represents the cumulative value of Y, while the horizontal axis represents the cumulative distribution of X. Therefore plotting an empirical absolute concentration curve is identical to plotting the ALC except that each point on the vertical axis represents the cumulative value of Y, arranged according to a non-decreasing value of X, divided by n. Again, the points are connected by linear segments.

9.2 Estimators Based on Individual Observations: The Discrete Case

As elaborated in Chap. 5 and in Sect. 9.1, there is a critical difference between the definitions of the cumulative distribution function and the one which is used in the Lorenz curve in the case of discrete distributions. While the cumulative distribution is portrayed as a step function so that it is discontinuous in F, the points of the Lorenz curve are connected by linear segments that "make" the distribution function a continuous one. On top of that there is a problem of a small sample bias that was elaborated upon earlier in this chapter. Ignoring the small sample bias, the use of the different formulas of the Gini in the case of equal probability sampling does not cause any difference between the alternative definitions. However, the different formulas result in different estimates of the GMD, the Gini coefficient, and the Gini correlations whenever the sampling is based on an unequal probability sampling.

The intuitive explanation to the discrepancy between the different formulations of the Gini is the following: connecting the points of the Lorenz curve by straight lines and calculating areas enclosed between the Lorenz and the diagonals is as if one uses the mid-point of the cumulative distribution. On the other hand, using the covariance formula of the Gini uses the cumulative distribution (represented by a step function), which is as if one uses the extreme left point of the distribution in each step. In an equal probability sampling scheme it is as if one adds a constant (which is equal to 1/(2n)) to each value of the cumulative distribution. Because adding a constant to a variable does not change the value of the covariance, no bias is caused by this discrepancy. On the other hand, when the sample is not an equal probability sample, or if the observations are grouped with a different proportion of the population in at least one group, then there is a discrepancy between the two methods of calculations.

The covariance method of estimation has several advantages over the one based on the Lorenz: it can be calculated by any standard statistical software, and it allows for a convenient decomposition of the Gini coefficient of a sum of variables (or populations). Lerman and Yitzhaki (1989) suggested a correction to the covariance formula of the Gini coefficient so that it yields the same estimate as the one that is based on the Lorenz curve. They suggested substituting the cumulative distribution in the covariance formula by the mid-point of the cumulative distribution in each section. The description of their suggestion is detailed in Sect. 9.3.1. However, this correction is not applicable for the extended Gini because it is not a linear function of the cumulative distribution. The correction required for the EG is dealt with in Sect. 9.3.2.

9.3 Individual Data, Weighted

In general, the weighting schemes in samples produced by official statistical offices can be the results of three alternative processes:

1. Weights that are derived from an unequal probability sampling.
2. Grouping of the original observations aimed to maintain confidentiality of respondents.
3. Adjustment of the data to fit the given marginal demographic distributions in order to correct for nonresponse.

It is not always clear which process has actually led to the use of the chosen weights. Moreover, the process that led to the weighting scheme is not always documented. The first process can be a result of two alternative cases: (a) prior knowledge by the producer of the data about differences in the subpopulations with respect to the variance in the population. (b) Differences in the cost of getting the data so that although the variance in the population is assumed to be given, the cost of interviewing causes the division of the data into different strata. The second process can be a result of tax or statistical authorities intending to protect the confidentiality of the population. The third process can be a result of the tendency of official statistical agencies to reduce the bias caused by differential nonresponse and to decrease the variance in the sample by adjusting the demographic properties of the sample to the properties of the population.

We do not cover the issues of handling heteroscedasticity and aggregation in this book. Note that aggregation can also be considered as related to heteroscedasticity because by aggregating observations one creates differences in the variance. We assume that the handling of heteroscedasticity in our context is not different than in a regular case.

9.3.1 *Estimating the Gini Coefficient from Weighted Data*

As mentioned above (Sect. 9.2), in the case of unequal weights there is a difference between the estimators obtained from the Lorenz and the ACCs and those which are obtained from the covariance formulas. The discrepancy between the two is due to the fact that while in the Lorenz curve approach one uses the mid-point of the cumulative distribution, in using the covariance formula of the Gini one uses the cumulative distribution represented by a step function, which is as if one uses the extreme left point in each step. The suggested solution (Lerman & Yitzhaki, 1989) is to estimate the cumulative distribution function by a quantity that will reflect the "mid-point" idea and use this estimate in the covariance-based formula.

Formally, the correction of the covariance method to get the Lorenz-based estimate is the following:

Let $w_i, y_i, i = 1, \ldots, n$ be the weights and incomes in a sample of n observations, respectively. Assume that $\sum_{i=1}^{n} w_i = 1$, and order the observations so that the y_i are in a non-decreasing order ($j < k$ implies $y_j < y_k$). The Gini coefficient of Y is

$$G = 2\text{cov}(Y, F(Y))/\mu.$$

The suggested estimator of $F_i = F(y_i)$ is

$$\hat{F}_i = \hat{F}(y_i) = \sum_{j=0}^{i-1} w_j + w_i/2, \text{ where } w_0 = 0.$$

Note that $\hat{F}$ is *not* a cumulative distribution function.

The formula for the covariance-based estimator of the Gini coefficient in a weighted sample is

$$\hat{G} = \sum_{i=1}^{n} w_i(y_i - \bar{y})\hat{F}_i,$$

which results in the same estimate as the one obtained by the Lorenz curve (Lerman & Yitzhaki, 1989).

Similar corrections have to be performed for the Gini covariance and correlation in order to adjust the covariance formula to yield the definitions of the same parameters obtained by the ACCs. Further research is needed to select the preferred estimation method. We note that the proposed covariance-based estimator, when extended to the extended Gini case, is not equal to the estimator obtained by the Lorenz curve approach. Furthermore, the covariance-based estimator is more biased and has a larger MSE (Chotikapanich & Griffiths, 2001). This issue is dealt with in the next section.

9.3.2 *Estimating the Extended Gini Coefficient from Weighted Data*

As mentioned above, when dealing with the estimation of the extended Gini and the parameters that are related to it (extended Gini correlation, extended Gini regression coefficient) in the case of unequal weights there is a difference between the estimator obtained from the Lorenz curve and the one obtained from the covariance-based formula. The discrepancy between the two is caused by the different definitions used for the cumulative distribution function. Unfortunately the solution suggested in Sect. 9.3.1 cannot be applied in the extended Gini. Furthermore, according to Chotikapanich and Griffiths (2001) the covariance-based estimator obtained this way is more biased and has a larger MSE. However, if one replaces the cdf by a proper function, the two estimates will produce algebraically identical estimators.

Formally, let X be a random variable and let F(x) and μ (>0) be the cumulative distribution function and the mean of X, respectively, and let w_i, x_i, $i = 1, \ldots, n$, be the weights and incomes in a sample of n observations, respectively. Assume that $\sum_{i=1}^{n} w_i = 1$.

The extended Gini coefficient can be expressed as

$$G = \nu(\nu - 1)\int_0^1 (1 - p)^{(\nu-2)}(p - LC(p))dp, \tag{9.13}$$

where $\nu > 1$ is the extended Gini parameter and LC(p) is the Lorenz curve defined by

$$LC(p) = \frac{1}{\mu}\int_0^p F^{-1}(t)dt, \tag{9.14}$$

with $0 < p < 1$. (See (6.15) for the definition of the absolute extended Gini.)

Alternatively,

$$G = -\frac{\nu}{\mu}\mathrm{cov}(X,[1 - F(X)]^{(\nu-1)}). \tag{9.15}$$

As mentioned above, when F is represented by a step function the two definitions yield different estimators. Using (9.14) and integration by parts, (9.13) can be written as

$$G = 1 - \frac{\nu}{\mu}\int_0^1 F^{-1}(t)(1 - t)^{(\nu-1)}dt,$$

which in the continuous case becomes $-\frac{\nu}{\mu}\mathrm{cov}(X,(1 - F(X))^{(\nu-1)})$. (Chotikapanich & Griffiths, 2001; Schechtman & Yitzhaki, 2008).

In the discrete case Chotikapanich and Griffiths (2001) show that the extended Gini coefficient, based on the Lorenz curve, is given by

$$\hat{G} = 1 + \sum_{i=1}^{n}\left(\frac{\varphi_i}{w_i}\right)[(1 - \pi_i)^{\nu} - (1 - \pi_{i-1})^{\nu}], \tag{9.16}$$

where $\pi_i = w_1 + \cdots + w_i$ and $\varphi_i = \frac{w_i x_i}{\sum_{j=1}^{n} w_j x_j} = \frac{w_i x_i}{\mu}$.

Schechtman and Yitzhaki (2008) show that (9.16) can be written as

$$\hat{G} = -\frac{\nu}{\mu}\text{cov}(X,(1 - F^*(X))^{(\nu-1)}),$$

where F^* is implicitly defined by

$$(1 - F_i^*)^{(\nu-1)} = \frac{(1 - w_1 - \cdots - w_{(i-1)})^\nu - (1 - w_1 - \cdots - w_i)^\nu}{\nu w_i}. \tag{9.17}$$

We note in passing that there is a relationship between F and F^*, as we describe next. From (9.17) one can get

$$\begin{aligned} \nu(1 - F_i^*)^{(\nu-1)} w_i &= (1 - w_1 - \cdots - w_{(i-1)})^\nu - (1 - w_1 - \cdots - w_i)^\nu \\ &= (1 - F_{(i-1)})^\nu - (1 - F_i)^\nu. \end{aligned}$$

Therefore

$$\begin{aligned} \mu\hat{G} &= -\nu\text{cov}\left(X, (1 - F^*(X))^{(\nu-1)}\right) = -\nu\sum_{i=1}^{n} w_i(x_i - \bar{x})(1 - F^*(x_i))^{(\nu-1)} \\ &= \nu(\text{cov}(X,(1 - F_i)^\nu) - \text{cov}(X,(1 - F_{(i-1)})^\nu)). \end{aligned}$$

It is worth noting that those differences occur because the EG, unlike the Gini, is not a linear function of the cumulative distribution.

9.4 Estimators Based on Grouped Data

The method of interpolation between the points of the Lorenz curve by straight lines is a commonly used technique. However other options exist in the literature for the case of grouped data. Gastwirth and Glauberman (1976) suggest the use of Hermite interpolation which provides a cubic polynomial in each interval, with some modifications at the first and the last intervals. They assess the accuracy of the piecewise Hermite interpolator by comparing it to known Lorenz curves. By "known Lorenz curves" it is meant that the Lorenz curves can be theoretically calculated. The two examples are the Lorenz curves for the Pareto case ($\alpha = 2$), $LC(p) = 1 - (1 - p)^{0.5}$ and for the exponential case, $LC(p) = p + (1 - p)\ln(1 - p)$. They find that the method works well in all intervals except for the extrapolation beyond the range of the given data (the last interval). The reason is

(Gastwirth & Glauberman, 1976) that the Lorenz curve near 1 depends on the distribution of income within the largest income class and when the data is grouped, only the average of this group is given. Gastwirth and Glauberman (1976) state necessary conditions to guarantee that the Hermite interpolator will possess the desired properties: convexity and being strictly increasing at the end points. These conditions are beyond the scope of this book.

The Gini coefficient derived from this interpolator is given by

$$\hat{G} = 1 - 2\sum_{i=1}^{k}\left(\Delta_i LC_i' + \Delta_i^2\frac{LC_i'}{2} + \Delta_i^3\frac{a_2}{3} - \frac{a_3}{12}\Delta_i^4\right)$$

where LC' is the derivative of LC, $\Delta_i = p_{i+1} - p_i$ and a_2 and a_3 are functions of LC, LC' and Δ.

Many authors demonstrated that there is a need to find lower and upper bounds to the Gini rather than a point estimator, because the estimator of the Gini coefficient is sensitive to the specification of the underlying Lorenz curve (see, for example, Cowell, 1991; Fuller, 1979; Gastwirth, 1972; Giorgi & Pallini, 1987; Mehran, 1975; Murray, 1978; Ogwang, 2003; Ogwang & Rao, 1996; Ryu & Slottje, 1999; Silber, 1990). These bounds were obtained regardless of the functional form of the underlying distribution of income—the approach was nonparametric. The basic assumptions were: in each income bracket, all units receive the average income of that income bracket for the lower bound and the maximum inequality in each bracket for the upper bound. The papers differ in their assumptions about the available information on the income brackets.

Most of the bounds are similar and take the following geometric approach: for the lower bound they draw a piecewise linear Lorenz curve by connecting the observed points by line segments and take the lower bound to be one minus twice the area below it. For the upper bound they construct tangents to the Lorenz curve at the observed points and use one minus twice the area below these tangents as the upper bound. The disadvantage of this approach is that it requires information on the limits of the income brackets and group mean incomes or the overall mean income. Silber (1990) derived the coordinates of the points of intersection of the tangents to the Lorenz curve at the observed points without using information on the limits of the income brackets, the groups' mean incomes, or the overall mean income. His points of intersection are based purely on population shares and income shares, without information on income brackets. A modified way to find the coordinates of the points of intersection of the tangents to the Lorenz curve at the observed points, assuming that there is information on the limits of the income brackets and full or sparse information on mean incomes, is suggested by Ogwang and Wang (2004). Their suggestion is equivalent to Gastwirth's, Fuller's and Ogwang's, but much simpler to implement empirically. The formulae are given in Ogwang and Wang (2004).

9.5 Summary

In this chapter we deal with estimating the parameters that are based on the Gini measure. Most of the parameters can be estimated by U-statistics or functions of (dependent) U-statistic as we have shown. There are two advantages to this fact. First, we use common language, which makes it easier to follow, and second, it turns out that estimates that are U-statistics or functions of U-statistics have a desirable asymptotic theory. We divide the discussion to cases where the data come as individual observations or as grouped data, data with equal or unequal sampling probabilities (weighted vs. unweighted data), and discrete vs. continuous distributions. Each of the above-mentioned cases is treated separately. However, our preferred correction is to use a correction for degrees of freedom, in the same way that those problems are dealt within the classical case of the variance world.

Chapter 10
Inference on Gini-Based Parameters: Testing

Introduction

Chapter 9 dealt with the estimation of the parameters based on the Gini. In this chapter we introduce methods of testing for the parameters that are based on the Gini. Most of the estimators that were derived in Chap. 9 are based on U-statistics or functions of (dependent) U-statistics. The advantage is that we can use known facts about the limiting distributions of U-statistics and of functions of them in order to obtain statistical tests. In what follows we concentrate on the asymptotic normality but do not give explicit formulas for the variances. Instead we suggest estimating the variances using the jackknife method (to be explained below). Therefore, the explicit variances which sometimes have complicated expressions are not needed for the applications.

Using the notation and the definitions of Chap. 9, let $h(X_1, \ldots, X_m)$ be a kernel of degree m for θ, based on a sample $x_1, \ldots, x_n$, of size $n \geq m$ from a distribution F, and let the U-statistic with kernel h be defined as

$$U_n = U_n(h) = \frac{1}{\binom{n}{m}} \sum_{C_{m,n}} h(x_{i_1}, \ldots, x_{i_m}),$$

where the summation is over the set $C_{m,n}$ of all $\binom{n}{m}$ combinations of m integers $i_1 < i_2 < \cdots < i_m$ chosen from $(1, 2, \ldots, n)$.

The following results are used throughout the chapter. They are taken from Hoeffding (1948), Arvesen (1969), Randles and Wolfe (1979), Serfling (1980), Lee (1990), and Lehmann (1999).

Theorem 10.1 *The limiting distribution of a one-sample U-statistic.*

Let $X_1, \ldots, X_n$ be independent random variables having a distribution function F with finite second moment. The U-statistic for the parameter θ, with a symmetric kernel $h(X_1, \ldots, X_m)$ of degree m, is an unbiased estimator for θ and the

S. Yitzhaki and E. Schechtman, *The Gini Methodology: A Primer on a Statistical Methodology*, Springer Series in Statistics 272, DOI 10.1007/978-1-4614-4720-7_10,

distribution of $\sqrt{n}(U - \theta)$ *tends to a normal distribution as* $n \to \infty$ *under the condition that* $E(h^2(X_1, \ldots, X_m))$ *exists.*

Theorem 10.2 *The asymptotic distribution of a two-sample (generalized) U-statistic.*

Let $U(X_1, \ldots, X_n; Y_1, \ldots, Y_m)$ *be a two-sample U-statistic for the parameter* θ, *with a symmetric kernel* $h(X_1, \ldots, X_{r1}; Y_1, \ldots, Y_{r2})$ *of degree* (r_1, r_2). *Let* $N = n + m$. *If* $\lim \frac{n}{N} = \lambda$ *and* $\lim \frac{m}{N} = 1 - \lambda$ *with* $0<\lambda<1$, *and if* $E(h^2(X_1, \ldots, X_{r1}; Y_1, \ldots, Y_{r2}))<\infty$, *then* $\sqrt{N}(U - \theta)$ *has a limiting normal distribution.*

Theorem 10.2 can be easily extended to the case of k independent samples.

Theorem 10.3 *The asymptotic joint distribution of several (dependent) generalized U-statistics.*

Let $U_1, \ldots, U_t$ *be k-sample* $(k > 1)$ *U-statistics, with* U_a *corresponding to a parameter* θ_a *and a symmetric kernel* h_a *of degree* $(r_1^{(a)}, \ldots, r_k^{(a)})$, *for* $a = 1, \ldots, t$. *Under the assumptions similar to those of Theorem 10.2, the joint limiting distribution of* $\sqrt{N}(U_1 - \theta_1), \ldots, \sqrt{N}(U_t - \theta_t)$, *with* $N = n_1 + \cdots + n_k$ *is t-variate normal.*

Theorem 10.4 *The asymptotic distribution of a function of several (dependent) U-statistics.*

Let $(U') = U_1, \ldots, U_t$ *be t U-statistics based on a sample* $x_1, \ldots, x_n$ *of size n, with* U_i *corresponding to* θ_i *(with kernel* h_i*),* $i = 1, \ldots, t$. *If the function* $g(y) = g(y_1, \ldots, y_t)$ *does not involve n and is continuous together with its partial derivatives in some neighborhood of the point* $(y) = (\theta) = (\theta_1, \ldots, \theta_t)$ *and if* $E(h_i^2(X_1, \ldots, X_{m_i}))$ *exist for all i then the distribution of* $\sqrt{n}(g(U') - g(\theta))$ *tends to the normal distribution as* $n \to \infty$.

Theorem 10.4 can be easily extended to several generalized U-statistics.

Given that the limiting distributions of the estimators are normal (under some regularity conditions), the remaining issue is to find their asymptotic variances. The references mentioned above provide the needed formulations for the variances (and for the variance–covariance matrices for the multivariate normal case). These expressions involve relatively complicated functions of the parameters, and hence, they are hard to estimate and to use in practice. An alternative method for estimating the variances is the jackknife method. The jackknife method is a computer intensive method, first introduced by Quenouille (1949), which is used in statistical inference in order to estimate the bias and the variance of an estimator. The basic idea of the jackknife method is as follows: given a sample of size n and an estimator T_n, the jackknife method creates a series of n estimators T_{-i}, $i = 1, \ldots, n$, where T_{-i} is the estimator T_n computed from a sample of size $(n - 1)$, after deleting the ith observation from the original sample. The bias and the variance are computed from these n estimators. The disadvantage of the jackknife method is that when the data set is large, the jackknife is very computer intensive. Yitzhaki (1991) suggests an efficient algorithm for estimating the variances of estimators of the Gini-based parameters for the one sample case using jackknife. The advantage of his method is that one needs to go over the data only twice (whereas in the ordinary jackknife one has to do so n times, deleting one observation at a time).

We now give a brief description of the algorithm. The algorithm is based on the fact that the Gini parameters are composed of variations of two parameters: the mean and cov(X, F(Y)). It was shown in Chap. 9 that cov(X, F(Y)) can be estimated by a U-statistic. One of the presentations of the estimator (not shown in Chap. 9 but can be easily derived) is

$$U = \frac{1}{n(n-1)} \sum i(x_{y_{(i)}} - \bar{x})$$

where $x_{y_{(i)}}$ is the value of the x that corresponds to the ith order statistic of $y_1, \ldots, y_n$.

Generally, the jackknife estimator of the variance of an estimator U(n, k) is given by

$$V_J = \frac{(n-1)}{n} \sum_{k=1}^{n} [U(n,k) - U(n,.)]^2$$

where U(n, k) is the estimator based on a sample of size n after deleting the kth observation, and U(n,.) is the average of the U(n, k)'s. Denote the estimator based on all observations by U(n, 0). Assume that the data is ordered in an increasing order of Y. The first run on the data calculates U(n, 0), $\sum_{i=1}^{n} x_{y_{(i)}}$, and $\bar{x}(n, 0)$. The second run calculates U(n, k) from U(n, 0) and calculates summary statistics accumulated till the kth observation so that all estimators can be calculated. The derivation of $\bar{x}(n, k)$ is trivial:

$$\bar{x}(n, k) = \frac{1}{n-1} \left[n\bar{x}(n, 0) - x_{y_{(k)}}\right].$$

The derivation of U(n, k) is detailed in Yitzhaki (1991).

Some of the Gini parameters involve two independent samples (for example, the overlapping index). The jackknifing for a two-sample problem is discussed in Arvesen (1969). The suggestion is to compute two pseudo values: One by leaving one x out (and keeping all the ys) and the other by reversing the roles of x and y. Then the n + m pseudo values are averaged (by taking their sum and dividing by n + m) in order to obtain the jackknife estimate of the parameter. The jackknife variance is obtained by simply adding two sums of squares (each divided by n(n − 1), where n is the respective sample size): the squared deviation of pseudo values based on eliminating one x at a time around their mean, and the same for the ys. Furthermore, provided that the function of the U-statistics has bounded second partial derivatives in the neighborhood of the parameter of interest, the variance of the jackknife estimator converges in probability to the asymptotic variance, and the limiting distribution of the function of several U-statistics, properly standardized, is approximately normal. Using the above results, one can estimate the parameter (point and interval estimation) and perform tests using the limiting normal distribution in the obvious way.

An alternative option to estimate the variance is by the bootstrap method. The bootstrap method is a computer intensive method for obtaining a variance of an estimator. The procedure is the following: given a sample $x_1, \ldots, x_n$ of size n and an estimator T_n, draw B bootstrap samples of size n with replacement from $x_1, \ldots, x_n$, calculate the estimator for each one of them and obtain B values of the estimator, denoted by $T_1^*, \ldots, T_B^*$. Now, these values are used in order to estimate the variance of the original estimator T_n. Namely, the sample variance of $T_1^*, \ldots, T_B^*$ is used as the bootstrap variance estimator of the variance of the original statistic T_n.

We choose to use the jackknife method over the bootstrap in our applications sections because in the current technology of computing one can replicate the jackknife, while replicating the bootstrap results requires knowing the algorithm for the random number generator and knowing the seed used, so in practice it is impossible to replicate the results. Therefore, one cannot verify the estimates. Generally the two methods provide similar results.

As was mentioned above, our approach is to estimate the Gini-related parameters by U-statistics or functions of several (dependent) U-statistics. Therefore, the inference procedures for them are similar in nature. In what follows we describe the procedures for several types of parameters: (1) A single parameter, based on one univariate distribution. (2) A single parameter based on one bivariate distribution. (3) The difference between two parameters, based on one bivariate distribution. (4) A single parameter based on two univariate distributions. (5) The difference between two parameters based on two univariate distributions.

The extensions of these procedures to inferences on other parameters (for example, testing on the extended Gini multiple regression coefficient) are straightforward. The structure of the chapter is as follows: In Sect. 10.1 we discuss one sample problems related to univariate or bivariate distributions: tests for GMD, the Gini coefficient, the extended Gini, the Gini correlation, the Gini regression coefficient, and a test for the equality of two Gini correlations. In Sect. 10.2 we discuss two-sample problems: inference on the overlapping index and the comparison of two Gini coefficients. Section 10.3 concludes.

10.1 The One Sample Problem

10.1.1 Inference on the GMD and the Gini Coefficient

The GMD has several presentations, each one resulting in its natural point estimator. This fact implies that there are different approaches to drawing inference. In this section we use the point estimator for the Gini Mean Difference (GMD), based on a U-statistic. The estimate is given in Chap. 9 ((9.3) and (9.4)) by

$$U(\Delta_X) = \frac{1}{\binom{n}{2}} \sum_{i<j}\sum |x_i - x_j| = \frac{1}{4\binom{n}{2}} \sum_{i=1}^{n} (2i - 1 - n)x_{(i)}$$

where $x_{(i)}$ is the ith order statistic of $x_1, \ldots, x_n$. The variance of U under the normal distribution is given by

$$\mathrm{var}(U(\Delta_X)) = \frac{\sigma^2}{\pi} \frac{\left[n\left(\frac{\pi}{3} + 2 \times \sqrt{3} - 4\right) + \left(6 - 4 \times \sqrt{3} + \frac{\pi}{3}\right)\right]}{n(n-1)}$$

where σ^2 is the variance of X (Budescu, 1980). However, one use of GMD is when the sample comes from a distribution with tails heavier than the normal. Therefore, the preferred procedures should be applicable under a large class of distributions, including the normal, and should be essentially "distribution-free," or at least asymptotically distribution free. In order to draw inference on the population GMD we take advantage of the fact that its estimator is a U-statistic of degree 2. Hence, we can use U-statistics theory (Theorem 10.1 above) to develop hypotheses tests and interval estimation based on the asymptotic normality of U. The main issue is to find the variance of U. Arvesen (1969) has shown that one can obtain consistent estimators of the standard error of U-statistics by jackknifing. Following this suggestion, let $U(\Delta_X)_{-k}$ be the estimator $U(\Delta_X)$ based on (n − 1) observations only (after deleting the kth observation). It can be written as

$$U(\Delta_X)_{-k} = \left(\frac{n}{n-2}\right)U(\Delta_X) - \frac{2\sum_{j=1}^{n} |x_k - x_j|}{(n-1)(n-2)}.$$

The sample variance of $U(\Delta_X)_{-1}, \ldots, U(\Delta_X)_{-n}$ is given by

$$S^2(U) = \frac{4}{(n-2)^2}\left[\frac{\sum_{k=1}^{n}\left(\sum_{j=1}^{n} |x_k - x_j|\right)^2}{(n-1)} - n(n-1)U^2(\Delta_X)\right].$$

Finally, a $100(1 - \alpha)\%$ symmetric confidence interval for Δ_X, for large n, is given by

$$\left[U(\Delta_X) - t_{(n-1),\frac{\alpha}{2}}\frac{S(U_X)}{\sqrt{n}} \leq \Delta_X \leq U(\Delta_X) + t_{(n-1),\frac{\alpha}{2}}\frac{S(U_X)}{\sqrt{n}}\right]$$

where $t_{(n-1),\frac{\alpha}{2}}$ is the upper $\frac{\alpha}{2}$ value from a t-distribution with (n − 1)df (Budescu, 1980). The confidence interval can serve for hypotheses tests in the ordinary way. Note that the parameter Δ_X is nonnegative. Therefore, if the lower bound of the interval comes out negative it can be replaced by 0. Another option is to use the efficient jackknife algorithm that was mentioned in the introduction above.

The Gini coefficient can be estimated by a ratio of two dependent U-statistics as

$$\hat{G} = \frac{U(\Delta_X)}{2\bar{x}}$$

where $\bar{x}$ is the average of $x_1, \ldots, x_n$. Therefore, by Theorem 10.4 the asymptotic distribution of $\hat{G}$ is normal, under some regularity conditions. Hence, inference can be based on the standard normal critical values.

A different approach is to base inference on the following estimator of the Gini coefficient:

$$\hat{G}_n = \frac{2}{\bar{x}} \int_0^{\infty} xF_n(x)dF_n(x) - 1 \tag{10.1}$$

as given Sect. 9.1.1. In what follows it is shown that $\sqrt{n}(\hat{G}_n - G)$ is asymptotically normal under regularity conditions, and a relatively simple way to compute its asymptotic variance is obtained (see Davidson, 2009). Let

$$I = \int_0^{\infty} xF(x)dF(x) \quad \text{and} \quad \hat{I} = \int_0^{\infty} xF_n(x)dF_n(x).$$

Then $\sqrt{n}(\hat{G}_n - G) = \sqrt{n}\left(\frac{2\hat{I}}{\bar{x}} - \frac{2I}{\mu}\right) = \frac{2}{\mu\bar{x}}\left(\mu\sqrt{n}(\hat{I} - I) - I\sqrt{n}(\bar{x} - \mu)\right)$. Because both $\sqrt{n}(\bar{x} - \mu)$ and $\sqrt{n}(\hat{I} - I)$ are of order 1 in probability, $\mu\bar{x}$ can be replaced in the above equation (to a leading order) by μ^2.

Note that $\sqrt{n}(\bar{x} - \mu) = \frac{1}{\sqrt{n}} \sum_{i=1}^{n} (x_i - \mu)$, which is an asymptotically normal random variable. The second term, $\sqrt{n}(\hat{I} - I)$, is also asymptotically normal (the proof is beyond the scope of the book and can be found in Davidson, 2009). $\sqrt{n}(\hat{I} - I)$ can be expressed as

$$\frac{1}{\sqrt{n}} \sum_{i=1}^{n} (x_i F(x_i) - m(x_i) - (2I - \mu))$$

where $m(x) = \int_0^x t dF(t)$.

Combining these pieces together,

$$\sqrt{n}(\hat{G}_n - G) \approx \frac{1}{\sqrt{n}} \frac{2}{\mu} \sum_{i=1}^{n} [-\frac{I}{\mu}(x_i - \mu) + x_i F(x_i) - m(x_i) - (2I - \mu)].$$

This presentation is (approximately) a sum of normalized i.i.d. random variables, each having mean zero. Therefore, the asymptotic normality is obtained.

Using the fact that $G = \frac{2I}{\mu} - 1$, the theoretical asymptotic variance of $\sqrt{n}(\hat{G}_n - G)$ is

$$\frac{1}{n\mu^2}\sum_{i=1}^{n} \mathrm{var}(-(G+1)x_i + 2(x_i F(x_i) - m(x_i)).$$

Let $T_i = -(G+1)x_{(i)} + 2(x_{(i)}F(x_{(i)}) - m(x_{(i)}))$ where $x_{(i)}$ is the ith order statistic of $x_1, \ldots, x_n$. Then

$$\hat{T}_i = -(\hat{G}_n + 1)x_{(i)} + \frac{2i-1}{n}x_{(i)} - \frac{2}{n}\sum_{i=1}^{n} x_{(i)} \tag{10.2}$$

and the mean and variance of the distribution of T_i are estimated by $\bar{\hat{T}} = \frac{1}{n}\sum \hat{T}_i$ and $\frac{1}{n}\sum(\hat{T}_i - \bar{\hat{T}})^2$, respectively. Before inference can be carried out, it is recommended to correct for the bias in $\hat{G}_n$. It turns out that $E(\hat{G}_n - G) \approx -\frac{G}{n}$, and hence, a bias-corrected estimator (still biased, but of a smaller order) of G is $\tilde{G} - \frac{n\hat{G}_n}{(n-1)}$.

We now summarize the steps needed in order to draw inference on the Gini coefficient (Davidson, 2009).

1. Calculate the sample mean.
2. Sort the sample in increasing order. Obtain $x_{(1)} \leq \cdots \leq x_{(n)}$.
3. Form two series of random variables $w_i = \frac{(2i-1)x_{(i)}}{2n}$ and $v_i = \frac{1}{n}\sum x_{(i)}$. Then $\hat{I} = \overline{w}$, the mean of the w_i.
4. Compute the bias-corrected estimator of the Gini coefficient $\tilde{G} = \frac{n}{(n-1)} \times \left(\frac{2\hat{I}}{(\overline{X}-1)}\right)$.
5. Form the series $\hat{T}_i = -(\tilde{G}+1)x_{(i)} + 2(w_i - v_i)$ and compute $\bar{\hat{T}}$. The estimated variance of G is the sum of the squared deviations of $\hat{T}_i$ from their mean, divided by $(n\overline{x})^2$.

The above discussion is applicable in general. However, when the underlying distribution is known to be exponential or Pareto, the sampling distribution of the estimator of the Gini coefficient can be obtained theoretically. The exact distribution of the estimator of the Gini coefficient in the case of the *exponential* underlying distribution has a cumbersome presentation. However, for practical use, the normal approximation of the distribution can be used and it is accurate even for $n \leq 10$. The asymptotic distribution of the Gini coefficient under a *Pareto* law with parameter λ is normal with mean $(2\lambda - 1)$ and variance $4\lambda(\lambda - 1)/[n(\lambda - 2)(2\lambda - 1)2(3\lambda - 2)]$ (Gail & Gastwirth, 1978; Gastwirth, Modarres, & Bura, 2005; Moothathu, 1985. See also Giorgi & Madarajah, 2010 for various Gini indices for various parametric distributions).

For completeness we note that the variance of the Gini coefficient can be obtained by applying the U-statistic theory. Several researchers take this route

(Bishop, Formby & Zheng, 1997; Xu, 2007, among others). A somewhat controversial approach is taken by Giles (2004, 2006) and Ogwang (2000, 2004, 2006). They provide a method of estimating the Gini coefficient by an OLS regression and show how to use the regression in order to simplify the computation of the jackknife standard deviation of the estimator of the Gini coefficient. A similar method of estimation is suggested by Deltas (2003). This approach is criticized by Modarres and Gastwirth (2006) who show that the standard deviations obtained are inaccurate.

Inference can be based on confidence intervals. This line of research was taken by Gastwirth et al. (2005) who carry out a simulation study on evaluating the usefulness of the percentile bootstrap in forming confidence intervals for the Gini coefficient (G) and the coefficient of dispersion (CD) for small samples.

Formally, given a random sample $y_1, \ldots, y_n$ from an unknown distribution function F, the interval estimates of G and CD are computed using the bootstrap percentile technique. The procedure is the following: first one calculates B values of G (or CD) from B bootstrap resamples from the data. Denote the B values by, say, $\hat{G}_1^*, \ldots, \hat{G}_B^*$. Next, order these B values from smallest to largest. The bootstrap percentile confidence interval of size $(1 - \alpha)100\%$ is $(\hat{G}_{\alpha/2}^*, \hat{G}_{(1-\alpha/2)}^*)$ where $\hat{G}_p^*$ is the [pB]th order statistic of the bootstrap distribution of the $\hat{G}^*$ or (CD). For example, if $B = 1{,}000$ and $p = 0.025$ then $\hat{G}_{\alpha/2}^* = \hat{G}_{0.025}^*$ is the $1{,}000 \times 0.025 = 25$th order statistic of the G_i^*'s. Gastwirth et al. (2005) construct confidence intervals for the Gini coefficient and the coefficient of dispersion when the underlying random variable is positive with a parent distribution that is right-skewed. The exponential, Pareto and lognormal distributions are used for data generation in a Monte Carlo study. The sampling distribution of G depends on the underlying distribution of the observations. The sampling distribution of G on data from the exponential distribution has been studied by Gail and Gastwirth (1978) and Giorgi (1990) and converges rapidly to its asymptotic normal approximation. However, the result is not true in general.

How well do the bootstrap estimates of the G and the CD perform in terms of coverage probabilities? These coverage probabilities are estimated by the fraction of the percentile bootstrap confidence intervals in 1,000 Monte Carlo simulations that contain the true parameter G or CD. It is shown that with moderate sample sizes, the 95% coverage for the G using the percentile bootstrap procedure is somewhat below the nominal value. This is more pronounced for the Pareto and the log normal distributions. The coverage improves with increasing sample size. In the case of the lognormal distribution the confidence intervals do not provide adequate coverages even in samples of size 500. This result is due to the "heavy-tailed" nature of the underlying data (Gastwirth et al. 2005).

10.1.2 *Inference on Gini Correlation and Gini Regression*

The point estimators of the co-Gini and of the Gini correlation coefficient are given in Chap. 9. The Gini correlation can be estimated by the ratio of two dependent U-statistics, one of which is the U-statistic for the co-Gini. Therefore, we only discuss inference on the Gini correlation. The test for the co-Gini can be obtained from the discussion below. Let $U(\Delta_X) = \frac{1}{4\binom{n}{2}} \sum_{i=1}^{n} (2i-1-n)x_{(i)}$ be the point estimator of 4 cov(X, F(X)), and let

$$U(\Delta_{X,Y}) = \frac{1}{\binom{n}{2}} \sum_{i<j}\sum [(x_i - x_j)I_{(y_i>y_j)} + (x_j - x_i)I_{(y_j>y_i)}]$$

$$= \frac{1}{4\binom{n}{2}} \sum_{i=1}^{n} (2i-1-n)x_{y_{(i)}}$$

where I is the indicator function and $x_{y_{(i)}}$ is the x that belongs to $y_{(i)}$, (usually, referred to as the concomitant) the ith order statistic of $y_1, \ldots, y_n$, be the U-statistic for estimating 4 cov(X, G(Y)).

Then the Gini correlation coefficient between X and Y, defined as

$$\Gamma_{X,Y} = \frac{\text{cov}(X,G(Y))}{\text{cov}(X,F(X))}$$

is estimated by $\hat{\Gamma}_{X,Y} = \frac{U(\Delta_{X,Y})}{U(\Delta_X)}$.

The limiting distribution of $\sqrt{n}(\hat{\Gamma}_{X,Y} - \Gamma_{X,Y})$ is normal by Theorem 10.4 above. Schechtman and Yitzhaki (1987) provide an estimate for the variance of $\hat{\Gamma}_{X,Y}$ which is based on five dependent U-statistics. However, as stated above, a more practical approach is to use the jackknife method in order to estimate the variance of $\hat{\Gamma}_{X,Y}$.

Likewise, the Gini simple regression coefficient, which is defined (Chap. 3 (3.7)) as $\beta = \frac{\text{cov}(Y,F(X))}{\text{cov}(X,F(X))}$ is estimated by $\hat{\beta} = \frac{U(\Delta_{Y,X})}{U(\Delta_X)}$, which is a ratio of U-statistics. Hence, inference can be made, based on Theorem 10.4.

10.1.3 Testing for the Symmetry of the Gini Correlation

Due to the asymmetrical nature of $\Gamma_{X,Y}$ there are two Gini correlations between each pair of random variables, depending on which variable is taken in its variate values and which one is ranked. The two Gini correlations are generally not equal. We have shown earlier (Chap. 4) that the question of whether the decomposition of GMD of a linear combination of random variables follows the same structure as the decomposition of the variance depends on whether the Gini correlation coefficients, especially between each of the individual variables and the linear combination of them, are equal.

In this section we describe a test for the equality of the two Gini correlations between X and Y (Schechtman, Yitzhaki, & Artsev, 2007). The hypothesis of interest is

$H_0 : \Gamma_{X,Y} = \Gamma_{Y,X}$, vs. the alternative
$H_1 : \Gamma_{X,Y} \neq \Gamma_{Y,X}$.

The test is based on the estimator of $\delta = \Gamma_{X,Y} - \Gamma_{Y,X}$. The point estimator of each of the two correlation coefficients $\Gamma_{X,Y}$ and $\Gamma_{Y,X}$ is given in Chap. 9 (9.5). It is based on a ratio of the two relevant dependent U-statistics. For example,

$$\hat{\Gamma}_{X,Y} = \frac{U(\Delta_{X,Y})}{U(\Delta_X)}$$

where $U(\Delta_X)$ and $U(\Delta_{X,Y})$ are the U-statistics for the parameters 4 cov(X, F(X)) and 4cov(X, G(Y)), respectively.

A similar derivation gives an estimator of $\Gamma_{Y,X}$. Finally, we can express the estimator of $\delta = \Gamma_{X,Y} - \Gamma_{Y,X}$ as a function of four dependent U-statistics as follows:

$$\hat{\delta} = \hat{\Gamma}_{X,Y} - \hat{\Gamma}_{Y,X} = \frac{U(\Delta_{X,Y})}{U(\Delta_X)} - \frac{U(\Delta_{Y,X})}{U(\Delta_Y)}.$$

By Theorem 10.4 above (Hoeffding, 1948), it is known that a function of several dependent U-statistics has a limiting normal distribution after appropriate normalization. Hoeffding (1948) provides a way to calculate the variance as well. However, the formulas are complicated. We suggest estimating the variance using the jackknife method.

The test statistic for testing H_0: $\delta = 0$ is based on

$$Z = \frac{\hat{\delta}}{\sqrt{\hat{V}(\hat{\delta})}},$$

and the rejection region is the standard one, namely: reject H_0 in favor of H_1 if $|Z| \geq Z_{\alpha/2}$ where $Z_{\alpha/2}$ is the upper $\alpha/2$th percentile of the standard normal distribution.

Another way to test for the equality of the two Gini correlations is by using the fact that a sufficient condition for the equality of the two Gini correlations is that X and Y are exchangeable (see property 7 in Sect. 3.4 and the proof that follows). Tests for bivariate exchangeability were suggested by Modarres (2008).

As mentioned above, one of the uses of the test of equality of two Gini correlations is to check whether the decomposition of the GMD of a linear combination of random variables follows the same structure as the decomposition of the variance. It is shown in Chap. 4 that the answer depends on whether the Gini correlation coefficients, especially between each of the individual variables and the linear combination of them, are symmetric. That is, the hypotheses of interest are simultaneous hypotheses. For simplicity, we formalize the hypotheses for the case of two variables X_1 and X_2. Let $Y = \beta_0 + \beta_1X_1 + \beta_2X_2$. The following identities hold (proposition 4.2 of Chap. 4):

(a)

$$\begin{aligned}\Delta_Y^2 - [\,\beta_1 D_{1Y}\Delta_1 + \beta_2 D_{2Y}\Delta_2]\Delta_Y \\ = \beta_1^2\Delta_1^2 + \beta_2^2\Delta_2^2 + \beta_1\beta_2\Delta_1\Delta_2(\Gamma_{12}+\Gamma_{21})\end{aligned} \tag{10.3}$$

where $D_{iY} = \Gamma_{iY} - \Gamma_{Yi}$, $i = 1, 2$ and $\Gamma_{ij} = \Gamma_{X_iX_j}$.

(b) Provided that $D_{iY} = 0$, for $i = 1, 2$, and $\Gamma_{12} = \Gamma_{21} = \Gamma$, then:

$$\Delta_Y^2 = \beta_1{}^2\Delta_1{}^2 + \beta_2{}^2\Delta_2{}^2 + 2\,\beta_1\beta_2\Delta_1\Delta_2\Gamma. \tag{10.4}$$

Note that (10.4) is identical in structure to the decomposition of the variance. Therefore, we are interested to test

(a) $D_{iY} = \Gamma_{iY} - \Gamma_{Yi} = 0, \quad i = 1, 2$
(b) $\Gamma_{12} = \Gamma_{21} = \Gamma$.

In General, the testing procedure involves three steps. The first step is a test for the equality of each pair of correlations. Given n variables, (a) involves n hypotheses - between each X variable and Y, while (b) involves $n(n-1)/2$ hypotheses—between each pair of X variables (for our special case n = 2). The second step is to test two intersection hypotheses. For (a) the hypothesis is $\bigcap_{i=1}^{n} (D_{iY} = 0)$. For (b) the hypothesis is $\bigcap_{i,j-1}^{n} (D_{ij} = 0)$. Finally, the third step is the simultaneous test for the last two intersection hypotheses.

The first step, the test of symmetry for each individual pair of variables was described above. The second step, a test for the intersection hypothesis uses the outcomes of the first step, namely the p-values, as inputs. Following Simes (1986), let $p_{(1)} < p_{(2)} < \cdots < p_{(n)}$ be the ordered *p*-values for the first intersection hypothesis, involving n comparisons.

Define

$$\text{adj - } p \text{ - value} = \min\left(\frac{np_{(j)}}{j}\right).$$

Reject H_0 if the level of significance exceeds the adj-p-value. The same procedure is repeated for the second intersection hypothesis, with n replaced by n(n − 1)/2. Because the second step involves two simultaneous (intersection) hypotheses, the third step is to use the Bonferroni correction, that is: replace the levels used in step 2 by $\alpha/2$ per comparison, to keep the overall level at (approximately) α.

It is worth mentioning that there is an additional use of the test for equality of two Gini correlations between Y and X, as described in (10.3) and (10.4). It is related to a check when (that is, for what value of n) the distribution of a sequence of averages of i.i.d. random variables convergences to the normal distribution. Details are given in Chap. 23.

10.1.4 The Extended Gini and the Extended Gini Regression Coefficients

Additional parameters related to the GMD are the extended Gini (EG) and its variants: the EG coefficient, the EG covariance, the EG correlation and the EG regression coefficients, as detailed in Chaps. 6–8. We start with the extended Gini and the extended Gini covariance. There are several alternative presentations of EG and each one results in its natural estimator. We start with the following presentations: $\Delta(v, X) = -vCOV(X, [1 - F(X)]^{v-1})$ and $\Delta(v, Y, X) = -vCOV(Y, [1 - F(X)]^{v-1})$ for EG and EG covariance, respectively.

These parameters are estimated by the following U-statistics (as shown in Chap. 9)

$$U(\Delta(v, X)) = \frac{1}{\binom{n}{v}} \sum_{i=1}^{n} \left[\frac{1}{v} \binom{n-1}{v-1} - \binom{n-i}{v-1} \right] x_i \tag{10.5}$$

and

$$U(\Delta(v, Y, X)) = \frac{1}{\binom{n}{v}} \sum_{i=1}^{n} \left[\frac{1}{v} \binom{n-1}{v-1} - \binom{n-i}{v-1} \right] y_{x(i)}, \tag{10.6}$$

respectively. Hence, by Theorem 10.1, inference is based on the approximate normal distribution, after proper standardization.

However, it turns out that the convergence to normality is not uniform. Giorgi, Palmitesta, and Provasi (2006) develop inference on a *variant* of the extended Gini coefficient and conclude that the convergence depends on the sampling distribution of the variable (usually the variable is the income) and on their extended Gini parameter δ.[1]

The form of their EG index is

$$I_{F,\delta} = 2\left(\int_0^1 (t - LC_F(t))^{\delta} dt\right)^{\frac{1}{\delta}}$$

where $LC_F(t) = \frac{1}{\mu}\int_0^1 F^{-1}(s)ds$ is the Lorenz curve.

A Monte Carlo study is conducted to evaluate the convergence, where the distribution of the income is taken to be the generalized Beta of the second kind (GB2) (details can be found in Giorgi et al., 2006). This distribution is flexible and includes cases with a wide variety of shapes. Special or limiting cases include the lognormal, Weibull, gamma and loglogistic distributions. It is shown that the larger the value of δ the better the approximation of the normal distribution. In addition, the nominal confidence levels for small to moderate sample sizes (smaller than 100) are not achieved when the normal approximation is applied.

An alternative method for obtaining confidence intervals for the EG is described in Giorgi et al. (2006) for a given distribution F. It is based on the bootstrap method. The bootstrap method is presented here for completeness, as it is a general methodology, although the version of the EG index used below is not discussed in the book.

The estimate of the EG index of Giorgi et al. (2006) is given by

$$I_{n,\delta} = 2\left[\sum_{i-1}^{n}\left(\frac{i}{n} - LC_n\left(\frac{i}{n}\right)\right)^{\delta}\right]^{\frac{1}{\delta}}$$

where $LC_n(t) = \frac{1}{X}\int_0^t F_n^{-1}(s)ds$ with $0 \le t \le 1$

and its variance $\sigma_{F,\delta}$ for a general F distribution (having a finite second moment) is derived explicitly (Zitikis, 2003). The variance is complex (and depends on F and δ) but can be estimated from the data (by $\sigma_{n,\delta}$) in the natural way, by estimating F by the empirical distribution. The bootstrap method is used here both to estimate the variance and to find the confidence interval. It was described in general in the introduction above. Let the B bootstrap values, denoted by $T^{*1}_{n,\delta}, \ldots, T^{*B}_{n,\delta}$, be $T^{*b}_{n,\delta} = \sqrt{n}\frac{I^*_{n,\delta} - I_{n,\delta}}{\sigma^*_{n,\delta}}$, b = 1, …, B, where $I^*_{n,\delta}$ and $\sigma^*_{n,\delta}$ are the bootstrap versions of $I_{n,\delta}$ and $\sigma_{n,\delta}$, respectively. Then a $(1 - \alpha)100\%$ bootstrap-t confidence interval for $I_{F,\delta}$ is $\left(I_{n,\delta} - C^{1-\frac{\alpha}{2}}_{n,\delta}\frac{\sigma_{n,\delta}}{\sqrt{n}},\ I_{n,\delta} - C^{\frac{\alpha}{2}}_{n,\delta}\frac{\sigma_{n,\delta}}{\sqrt{n}}\right)$ where the bootstrap quantile estimates of $T^*_{n,\delta}$ are given by $C^{\alpha}_{n,\delta}$ such that $\sum_{b=1}^{B} I(T^{*b}_{n,\delta} \le C^{\alpha}_{n,\delta}) = \alpha$ and I is the indicator function.

[1] We do not cover this version of the extended Gini here because it is not linear in the variable, nor is it linear in the distribution function; hence, it is not relevant to our discussion. We describe it here for completeness.

We note that inference about the EG correlation and regression coefficients is straightforward. As shown in Chap. 9, the EG correlation between X and Y, which is given by

$$\xi(X,Y,\nu) = \frac{-\ \nu \mathrm{cov}(X,(1-F_Y(Y))^{\nu-1})}{-\ \nu \mathrm{cov}(X,(1-F_X(X))^{\nu-1})}$$

can be estimated by a ratio of two (dependent) U-statistics—the U-statistic defined in (10.6) divided by the U-statistic defined in (10.5). Therefore, inference can be drawn as detailed in Theorem 10.4. Inference about an EG regression coefficient can be drawn in a similar way.

10.2 The Two Sample Problem

10.2.1 The Overlapping Index

The overlapping index between two distributions denoted by i and j having cumulative distribution functions F_i and F_j, respectively, was defined in Chap. 4 and is given there, below (4.19), as

$O_{ji} = \frac{\mathrm{cov}_i(Y,F_j(Y))}{\mathrm{cov}_i(Y,F_i(Y))}$. It is estimated (see Chap. 9) by a function of four dependent U-statistics as follows: $\hat{O}_{ji} = \frac{U_1-U_2U_3}{U_4}$ where U_2 and U_4 are the sample mean and the estimator of GMD, respectively, and

$$\begin{aligned} U_1 &= U(X_1,\ldots,X_{n_x};Y_1,\ldots,Y_{n_y}) = \frac{1}{n_x n_y}\sum\sum h(x_i,y_j) \\ &= \frac{1}{n_x n_y}\sum\sum x_i I(y_j \le x_i) = \frac{1}{n_x n_y}\sum x_i(\#\ y\text{'s} \le x_i) \end{aligned}$$

$$U_3 = \frac{1}{n_x n_y}\sum(\#\ y\text{'s} \le x_i).$$

By Theorem 10.4, the limiting distribution of $\hat{O}_{ji}$ is normal. Therefore, inference about the overlapping index O_{ji} will be based on $OL = \frac{\hat{O}_{ji}-O_{ji}}{\sqrt{\mathrm{var}(\hat{O}_{ji})}}$ where the variance is estimated by the jackknife method as mentioned above. Schechtman (2005) reported on a simulation study to assess the convergence to normality. Three cases were considered: (a) X and Y are equally distributed, in which case the theoretical value of the overlapping index is equal to 1. (b) Y is contained in X, near its mean (relatively large overlapping index), and (c) X and Y occupy different ranges on the X-axis with relatively small intersection (relatively small overlapping index). Two underlying distributions were used: the normal (symmetric) and the lognormal (asymmetric). The theoretic value of the overlapping index was calculated (in the case of lognormal distribution MAPLE software was used). The reported figures were (among others) the average length of a 95% confidence interval based on the normal cutoff points and the percent coverage (should be 95%). The findings were that as long as the overlapping was substantial the coverage rate was within two standard errors of the theoretical one (of 95%) even for samples of size 50. When the overlapping index was small, the coverage was lower and a bigger sample size was needed ($n = 150$) and when the overlapping index was around 1.5, $n = 100$ was found to be good enough.

A natural extension of the above inference is the comparison of two independent overlapping indices. For example, the comparison between the overlapping index of men and women in the Israeli and the American labor markets. The inference is based on the difference between their point estimators. This difference is based on eight dependent U-statistics. Using Theorem 10.4 and the jackknife method for variance estimation, inference can be drawn on the difference between two independent overlapping measures.

10.2.2 Comparing Two GMDs and Two Gini Coefficients

The hypothesis of equality of two independent GMDs can be tested in a way which is similar to the test for one GMD parameter. The test statistic is based on the difference between the two point estimators (namely the difference between two independent U-statistics). By Theorem 10.2 above, the limiting distribution of the difference is normal. Again, the main issue is to find the asymptotic variance of the difference. The simplest way is to use the fact that

$$\mathrm{Var}\{\mathrm{U}(\Delta_1) - \mathrm{U}(\Delta_2)\} = \mathrm{Var}(\mathrm{U}(\Delta_1)) + \mathrm{Var}(\mathrm{U}(\Delta_2))$$

where $\mathrm{Var}(\mathrm{U}(\Delta_i))$ is the variance of $\mathrm{U}(\Delta_i)$, and estimate the individual variances as in Sect. 10.2.1 above. Budescu (1980) suggests using a two-sample jackknife procedure, following Arvesen (1969). This procedure suggests to compute the statistic and then jackknife it by successively deleting observations from the first sample and then from the second sample. The detailed calculations can be found in Budescu (1980).

A test for the equality of two independent Gini coefficients can be based on

$$\mathrm{Z} = \frac{\hat{\mathrm{G}}_1 - \hat{\mathrm{G}}_2}{\sqrt{\hat{\sigma}^2_{\mathrm{G}_1} + \hat{\sigma}^2_{\mathrm{G}_2}}}$$

where $\hat{G}_1$ and $\hat{G}_2$ are the point estimates of G_1 and G_2 as in (10.1), and the asymptotic variances are estimated as in Sect. 10.1.1 above. Inference is drawn based on the asymptotic standard normal distribution of Z. If the two samples are correlated then the covariance between the two estimates should be taken into account. In order to do that, two series should be formed: $\hat{T}_{1i}$ and $\hat{T}_{2i}$, $i = 1, \ldots, n$, as in (10.2). Then, after ordering the two series in the same way, the covariance between $\hat{G}_1$ and $\hat{G}_2$ is estimated by

$$\hat{cov}(\hat{G}_1, \hat{G}_2) = \frac{1}{n^2 \bar{x}_1 \bar{x}_2} \sum (\hat{T}_{1i} - \bar{\hat{T}}_1)(\hat{T}_{2i} - \bar{\hat{T}}_2),$$

where n is the size of each sample (Davidson, 2009). The denominator of Z is adjusted to the dependence by subtracting twice the estimated covariance under the square root sign. See also Davidson and Duclos (1997, 2000) for related applications.

10.3 Summary

The objective of the chapter was to introduce methods of inference for the parameters that are based on the Gini. There are several alternative ways of estimation and inference in the literature. We chose to use a common technique, based on U-statistics theory. One advantage of using a unified method is the ease of adjustment to other parameters not covered here. In addition, we can use known facts about the limiting distributions of U-statistics and of functions of them in order to obtain statistical tests.

Because all of our estimators are based on (functions of) U-statistics, the limiting distributions are normal. However, the asymptotic variances are usually hard to write explicitly. Therefore, we recommend using the jackknife method instead.

Chapter 11
Inference on Lorenz and on Concentration Curves

Introduction

In a pathbreaking paper, Atkinson (1970) proved several results concerning the ranking of income distributions according to expected values of all concave social welfare functions. One of the important results is that for distributions with equal means, all social welfare functions show the same order of average social welfares (i.e., the same ordering of inequality) if and only if the appropriate Lorenz curves do not intersect. If, on the other hand, the Lorenz curves intersect then it is possible to find two alternative social welfare functions which rank average social welfares differently (to be discussed in Chaps. 13 and 14). This finding by Atkinson has opened the way for using the Lorenz curve as a basic tool in the application of the concept of second-degree stochastic dominance (SSD, to be defined below). This tool allows the analyses of the effects of tax reforms and decision under risk to be applied to a wide group of utility functions, freeing the analysis from the need to specify the utility function. Shorrocks (1983) proved that X dominates Y according to SSD if and only if the absolute Lorenz curve ALC of X is not lower than the ALC of Y. This result enables to extend the possible applications to distributions with different expected values. There are three possible outcomes when comparing two absolute (and relative) Lorenz curves: Lorenz dominance, equivalence, and crossing. Bishop, Chakravarty, and Thistle (1989) extend the works by Gail and Gastwirth (1978), Beach and Davidson (1983), and Gastwirth and Gail (1985) who deal with relative Lorenz curves and suggest a pair-wise multiple comparisons method of sample absolute (generalized) Lorenz ordinates to test for differences.

Lorenz and absolute concentration curves are used as descriptive devices for more than a century. Following Atkinson (1970) and Shorrocks (1983) they also became tools for identifying second-degree stochastic dominance. Recent developments use those tools for identifying monotonic relationship among variables and the possibility of changing the sign of a regression coefficient by monotonic transformations. Statistical inference may turn these descriptive tools into analytical tools.

S. Yitzhaki and E. Schechtman, *The Gini Methodology: A Primer on a Statistical Methodology*, Springer Series in Statistics 272, DOI 10.1007/978-1-4614-4720-7_11,

The objective of this chapter is to introduce tools for formal statistical inference. However, we must apologize up front that the statistical theory is not well developed and further research is needed in order to establish a well accepted theory. A fundamental result is derived by Goldie (1977) who developed convergence theorems concerning the Lorenz curve. Those theorems were followed by Beach and Davidson (1983) and Beach and Richmond (1985).

The possible tests can be classified into two classes: tests that are based on violation (or fulfillment) of necessary conditions only, and those that are based on necessary and sufficient conditions. The tests that are based on necessary conditions are well established because they are based on testing whether a set of parameters fulfills (or violates) a specific condition that is based on parameters of the distribution. Those tests are actually covered in Chap. 10. However, it is clear that they cannot substitute for those that are based on necessary and sufficient conditions, which are required whenever the question posed is whether two curves intersect. Further research is needed in order to see whether one can improve upon those tests.

We introduce a test on the ordinates of the Lorenz curve and provide two tests for the intersection of concentration curves. The first test is for a necessary condition for second order stochastic dominance (SSD). The test is based on a specific parameter—the area below the Lorenz or the concentration curve. The second test is for necessary and sufficient conditions for SSD which are equivalent to the necessary and sufficient conditions for the intersection of two absolute concentration curves (ACC).

The structure of the chapter is the following: in Sect. 11.1 we deal with inference on Lorenz curves. Section 11.2 is devoted to necessary conditions for second order stochastic dominance, while in Sect. 11.3 tests for intersection of two ACCs are detailed. Section 11.4 concludes.

11.1 Inference on the Ordinates of the Lorenz Curves

As mentioned in Bishop, Formby, and Smith (1991) simple comparison of Lorenz ordinates may indicate crossings of the curves, which in fact are not statistically significant. It is important to take the level of dispersion of incomes into account.

The objective of this section is to derive the asymptotic joint variance–covariance matrix for Lorenz curve ordinates, to provide consistent estimates for them and to derive the asymptotic distribution of the Lorenz curve ordinates. Inference procedures follow.

Let Y be a random variable with a cdf F(y), assumed to be continuous and second order differentiable. Let the set of K ordinates be denoted by $\{\Phi_i | i = 1,..., K\}$, corresponding to the abscissa points $\{p_i | i = 1,..., K\}$. For example, in the case of deciles, $K = 9$ and $p_i = \frac{i}{10}$, $i = 1, \ldots, 9$. The ordinates on the Lorenz curve are $\Phi(\xi_{p1}) < \Phi(\xi_{p2}) < \cdots < \Phi(\xi_{pK})$ where the population income quantile ξ_{pi} is defined by $F(\xi_{pi}) = p_i$, and

$$\Phi(\xi_{\mathrm{pi}}) = \frac{1}{\mu}\int\limits_{0}^{\xi_{\mathrm{pi}}} \mathrm{udF(u)} = \frac{\mathrm{F}(\xi_{\mathrm{pi}})}{\mu}\int\limits_{0}^{\xi_{\mathrm{pi}}} \frac{\mathrm{udF(u)}}{\mathrm{F}(\xi_{\mathrm{pi}})} = \mathrm{p_i}\frac{\gamma_{\mathrm{i}}}{\mu} \tag{11. 1}$$

with $\gamma_{\mathrm{i}} = \mathrm{E}(\mathrm{Y}|\mathrm{Y} \le \xi_{\mathrm{pi}})$ being the conditional mean of income less than or equal ξ_{pi}.

Let $\mathrm{y}_{(1)} < \mathrm{y}_{(2)} < \cdots < \mathrm{y}_{(\mathrm{n})}$ be the ordered sample of size n. Then the sample quantile $\hat{\xi}_r$ is simply the rth order statistic, where r = [np], and [np] means the greatest integer less than or equal to np. Under some regularity conditions (i.e., F strictly monotonic and differentiable for any finite set $\{\mathrm{p}_1, \ldots, \mathrm{p}_\mathrm{K}\}$), the $\hat{\xi}'_{pi}$s have an asymptotic multivariate normal distribution (Beach & Davidson, 1983; Wilks, 1962). More precisely, let $\hat{\xi} = (\hat{\xi}_{p1}, \ldots, \hat{\xi}_{pK})$ be the vector of the K sample quantiles, then $\sqrt{n}(\hat{\xi} - \xi)$ converges to a K-variate normal distribution with mean zero and covariance matrix Λ where

$$\Lambda = \begin{pmatrix} \frac{p_1(1-p_1)}{f_1^2} & \cdot & \cdot & \cdot & \frac{p_1(1-p_K)}{f_1 f_K} \\ \cdot & \cdot & \cdot & \cdot & \cdot \\ \cdot & \cdot & \cdot & \cdot & \cdot \\ \cdot & \cdot & \cdot & \cdot & \cdot \\ \frac{p_1(1-p_K)}{f_1 f_K} & \cdot & \cdot & \cdot & \frac{p_K(1-p_K)}{f_K^2} \end{pmatrix}$$

and $\mathrm{f_i}$, i = 1, …, K are the density functions (assumed to be positive).

In order to draw inference on the population ordinates, several more steps are needed. First, the unknown parameters have to be estimated. The sample estimates of the Lorenz curve ordinates are

$$\hat{\Phi}_{\mathrm{i}} = \frac{\sum_{\mathrm{j}=1}^{\mathrm{r_i}} \mathrm{y}_{(\mathrm{i})}}{\sum_{\mathrm{j}=1}^{\mathrm{n}} \mathrm{y}_{(\mathrm{j})}} \doteq \mathrm{p_i}\frac{\hat{\gamma}_{\mathrm{i}}}{\bar{\mathrm{y}}}$$

where $\mathrm{r_i} = [\mathrm{np_i}]$ and

$$\hat{\gamma}_{\mathrm{i}} = \frac{\sum_{\mathrm{j}=1}^{\mathrm{r_i}} \mathrm{y}_{(\mathrm{j})}}{\mathrm{r_i}}.$$

The second step is to obtain the sampling distribution of the vector of sample ordinates $\hat{\Phi} = (\hat{\Phi}_1, \ldots, \hat{\Phi}_K)$. It has been shown that $\sqrt{n}(\hat{\Phi} - \Phi)$ has an asymptotic K-variate normal distribution with mean zero and a covariance matrix which can be calculated, but has a complex form (Beach & Davidson, 1983). It is important to note that the covariance matrix does not require knowledge of the underlying distribution from which the data was drawn, but only of the proportions $\mathrm{p_i}$, of the unconditional mean and variance, of the income quantiles ξ_{pi}, and of the conditional means and variances. All these quantities can be estimated consistently from the sample. Hence, the inference is distribution free. To sum up, the procedure involves three steps. First, sort the raw data set and determine the sample quantiles. Second, compute the conditional and unconditional means and variances from the sorted data. Last, compute the sample Lorenz curve variances and covariance and calculate the test statistics from them as follows: to test H_0: $\Phi = \Phi^0$ where $\Phi_0 = (\Phi^0_1, \ldots, \Phi^0_K)$ is an hypothesized Lorenz curve, one can use an asymptotic chi-square test statistic $C_1 = n(\hat{\Phi} - \Phi^0)'\hat{V}_L^{-1}(\hat{\Phi} - \Phi^0)$ with K degrees of freedom, where $\hat{V}_L^{-1}$ is the inverse of the sample variance–covariance matrix.

To compare two independent Lorenz curves Φ_1 and Φ_2 from two independent samples of sizes n_1 and n_2, respectively, that is, to test H_0: $\Phi_1 = \Phi_2$ one can use the statistic

$$C_2 = (\hat{\Phi}_1 - \hat{\Phi}_2)'\left(\frac{\hat{V}_{L1}}{n_1} + \frac{\hat{V}_{L2}}{n_2}\right)^{-1}(\hat{\Phi}_1 - \hat{\Phi}_2)$$

which is asymptotically chi-square distributed with K degrees of freedom, where $\hat{V}_{Li}$, $i = 1, 2$ are the sample variances based on the two samples.

Once the null hypothesis that the two curves are the same is rejected one may be interested to know which particular differences in the ordinates are significantly different from zero. The natural way to proceed is by performing multiple comparisons. These methods are beyond the scope of this book and can be found in Beach and Richmond (1985).

The inference about the absolute Lorenz curve is derived as follows. Using similar notation, let $\varphi(\xi_{pi}) = \int_0^{\xi_{pi}} u\,dF(u)$, then the ordinates of the absolute Lorenz curve are given by $\varphi(\xi_{p1}) \leq \cdots \leq \varphi(\xi_{pK})$. The estimate of the p_ith ordinate is $\hat{\varphi}(\xi_{pi}) = \frac{\sum_{j=1}^{r_i} y_{(j)}}{n}$ where $r_i = [np_i]$. It can be shown that $\hat{\varphi}(\xi_{pi})$ can be approximated by

$$\hat{\varphi}(\xi_{pi}) \sim \frac{1}{n}\sum_{j=1}^{n}(y_j - \xi_{pi})I(y_j \leq \xi_{pi}) + \xi_{pi}p_i. \tag{11.2}$$

Using the presentation in (11.2) it is easy to see that the asymptotic distribution of $\hat{\varphi} = (\hat{\varphi}_{p1}, \ldots, \hat{\varphi}_{pK})$ is K-variate normal. The covariance matrix of size $K \times K$ is given by $A = [a_{k,j}]$, $k, j = 1, \ldots, K$, where for $k \leq j$

$$a_{k,j} = \frac{1}{n}\left\{\int_0^{\xi_k}(x - \xi_k)(x - \xi_j)dF(x) - \int_0^{\xi_k}(x - \xi_k)dF(x)\int_0^{\xi_j}(x - \xi_j)dF(x)\right\}$$

(Zheng, 1996, 2002).

The procedures described above are increasingly implemented in empirical studies. The underlying assumption is that the data come from a simple random sample. Unfortunately, most economics data do not come from simple random samples. They come from stratified, cluster or multistage samples. Therefore, the theory described above will tend to yield biased estimators, which will imply inaccurate inference. However, at least in the area of income distributions, there are several alternative reasons for attaching weights to observations, such as reducing bias due to nonresponse and different probabilities of sampling. Therefore, an analysis of the causes for the different weights is called for because different causes may require different treatments of the data. This issue, which is general, is beyond the scope of this book.

The estimates and the variance–covariance matrices for testing Lorenz and absolute Lorenz curves when the samples are not simple random samples appear in the literature (Zheng, 2002). It turns out that each sampling method results in its own way of estimating the Lorenz curve and the variance covariance matrix.

11.2 Necessary Conditions for Second Order Stochastic Dominance

The formal definition of Second order Stochastic Dominance (SSD) is as follows. Let U denote the utility function and let X and Y be two random variables having cdf's F and G, respectively.

We say that F dominates G in the SSD meaning if

$$E_F(U(X)) \geq E_G(U(Y)). \tag{11.3}$$

for all U for which $U' > 0$ and $U'' < 0$, where U' and U'' are the first and second derivatives, respectively, with a strict inequality for at least one U. Condition (11.3) holds if and only if

$$\int_{-\infty}^{z} (F_X(t) - G_Y(t))dt \leq 0 \quad \text{for all z, and} \tag{11.4}$$

$$\int_{-\infty}^{z} F_X(t) - G_Y(t))dt < 0 \quad \text{for some z.} \tag{11.5}$$

This is the most common way of defining the conditions for second degree stochastic dominance. For a survey on stochastic dominance the reader is referred to Levy (1992, 2006).

An alternative and equivalent way of presenting (11.4) is through the ALC.

That is (11.3) holds if

$$\mathrm{ALC(X)} \geq \mathrm{ALC(Y)} \tag{11.6}$$

everywhere and the inequality strictly holds at some points.

In this book we naturally deal with statistical inference that is based on the ALC. We note in passing that other approaches exist as well. For example, Schmid and Trede (1996) develop nonparametric inference for second order stochastic dominance of two random variables when their distribution functions are unknown and have to be inferred from observed realizations. They establish two methods to take the sampling error into account. The first one is based on the asymptotic normality of the point estimators, while the second one relies on resampling techniques. Both methods are used to develop statistical tests for second order stochastic dominance. They show that tests based on resampling techniques are more useful in practical applications and recommend the application of the permutation principle for the determination of the critical values of the test statistics. They further show that these tests can also be used for testing for first order stochastic dominance. Kaur, Rao, and Singh (1994) and Xu, Fisher, and Wilson (1995) present large sample tests which are based on the asymptotic distributions of various test criteria. A different approach was taken by McFadden (1989) and Klecan, McFadden, and McFadden (1991) who suggested tests based on the bootstrap principle, which are applicable even for small samples.

In this section we focus on inference that is based on necessary conditions. It is based on the following logic: condition (11.6) implies that the ALC of X is not lower than the ALC of Y everywhere. This means that the area below the ALC of X is greater than the area below the ALC of Y. In other words, the necessary condition can be translated to a condition on one parameter—the area below the ALC.

The area below the ALC can be written as the area below the LOI, which is the area of a triangle and is equal to $\mu/2$, minus the area enclosed between the LOI and the Lorenz curve, which is cov(X, F(X)).

This condition can be translated into the condition $\theta < 0$, where

$$\theta = \mu_{\mathrm{Y}} - \mu_{\mathrm{X}} - 2\ \mathrm{cov(Y, G(Y))} + 2\ \mathrm{cov(X, F(X))}. \tag{11.7}$$

Then if X stochastically dominates Y in the second order meaning then $\theta < 0$ (Yitzhaki 1982a, 1983). It is convenient to add another condition which is: $\mu_{\mathrm{X}} \geq \mu_{\mathrm{Y}}$ which means that the last point (from the right) of the ALC of X is not lower than the equivalent point of Y.

Following the same logic, one can add additional necessary conditions by requiring that specific points on the Lorenz of X will not be lower than those of the Lorenz curve of Y, and/or to base the additional conditions on the EG. For example we may require that

$$\begin{aligned} &\mu_{\mathrm{X}} \geq \mu_{\mathrm{Y}} \text{ and.} \\ &\mu_{\mathrm{X}} - \ \nu\mathrm{cov}(\mathrm{X}, -[1-\mathrm{F(X)}]^{\nu-1}) \\ &\geq \mu_{\mathrm{Y}} - \ \nu\mathrm{cov}(\mathrm{Y}, -[1-\mathrm{G(Y)}]^{\nu-1}). \end{aligned} \tag{11.8}$$

for several values of ν. However, a counterexample (Yitzhaki, 1983) can be constructed to show that even if distribution X fulfills (11.8) for all possible values of ν, it need not provide a sufficient condition for SSD.[1] On the other hand, the advantage of relying on necessary conditions only is that the inference is relatively simple.

Next we introduce a test for the necessary conditions for SSD. Let F and G be two absolutely continuous distribution functions having finite second moments. Formally we are interested to test

H_0: F = G
H_1: F stochastically dominates G in the second order meaning.

An alternative formulation, based on θ, is

$$H_0 : \theta = 0$$
$$H_1 : \theta < 0.$$

Let

$$d_{F,G}(x) = \int_{-\infty}^{x} (F(t) - G(t))dt$$

and

$$D_{F,G} = 0.5\left[\int_{-\infty}^{\infty} d_{F,G}(x)dG(x) + \int_{-\infty}^{\infty} d_{F,G}(y)dF(y)\right].$$

Based on the definitions above we can reformulate H_0 and H_1 as follows:

$$H_0 : D_{F,G} = 0$$
$$H_1 : D_{F,G} < 0$$

(see Eubank, Schechtman, & Yitzhaki, 1993). Let $x_1, \ldots, x_n$ and $y_1, \ldots, y_m$ be two independent random samples of sizes n and m, from distribution functions F and G respectively. The test is based on $\hat{D}_{n,m}$, the sample version of $D_{F,G}$. Let

[1] The explanation to the counterexample is that in each condition (11.8) relies on a constant ν. Therefore, a small crossing of curves can be hidden by a large deviation elsewhere.

$$d_{n.m}(x) = \int_{-\infty}^{x} (F_n(t) - G_m(t))dt \text{ and}$$

$$\hat{D}_{n,m} = \frac{1}{2}\left[\int_{-\infty}^{\infty} d_{n,m}(x)dG_m(x) + \int_{-\infty}^{\infty} d_{n,m}(y)dF_n(y)\right]$$

then $\hat{D}_{n,m}$ estimates $D_{F,G}$ and inference can naturally be based on $\hat{D}_{n,m}$.

Next, we show that $\hat{D}_{n,m}$ can be presented as

$$\hat{D}_{n,m} = \frac{1}{2}\left[\bar{Y} - \bar{X} + \frac{1}{2\binom{n}{2}}\sum_{i<j}|x_i - x_j| - \frac{1}{2\binom{m}{2}}\sum_{i<j}|y_i - y_j|\right]. \tag{11.9}$$

Using presentation (11.9) it is easy to see that $\hat{D}_{n,m}$ is an unbiased estimator of 0.5 θ of (11.7). The connection to the alternative formulation of the hypotheses is based on θ.

Claim 1 Let $x_1, \ldots, x_n$ and $y_1, \ldots, y_m$ be two independent random samples of sizes n and m from distribution functions F and G, respectively. Then

$$\hat{D}_{n,m} = \frac{1}{2}\left[\frac{1}{n}\sum_{i=1}^{n} W_i - \frac{1}{m}\sum_{j=1}^{m} T_j\right] + \frac{1}{2}\left[\int_{-\infty}^{\infty} F_n(x)(1 - F_n(x))dx - \int_{-\infty}^{\infty} G_m(y)(1 - G_m(y))dy\right]$$

where

$$W_i = \int_{x_i}^{\infty} (t - x_i)dG_m(t) \quad \text{and } T_j = \int_{y_j}^{\infty} (t - y_j)dF_n(t).$$

The proof of claim 1 is technical and is detailed in Eubank et al. (1993).

Claim 2 Let $x_1, \ldots, x_n$ and $y_1, \ldots, y_m$ be two independent random samples of sizes n and m from distribution functions F and G respectively. Then

$$\hat{D}_{n,m} = \frac{1}{2}\left[\bar{Y} - \bar{X} + \frac{1}{2\binom{n}{2}}\sum_{i<j}|x_i - x_j| - \frac{1}{2\binom{m}{2}}\sum_{i<j}|y_i - y_j|\right].$$

Proof of claim 2 We start by proving that

$$\frac{1}{n}\sum_{i=1}^{n} W_i = \int_{-\infty}^{\infty} (1 - G_m(t))F_n(t)dt. \tag{11.10}$$

To see that, let $x_{(1)} < x_{(2)} < \cdots < x_{(n)}$ be the ordered x's, and let $I_{[a,b]}$ be the indicator function of [a, b]. Then

$$\int_{-\infty}^{\infty} (1 - G_m(t))F_n(t)dt = \frac{1}{n}\sum_{i=1}^{n} \int_{-\infty}^{\infty} (1 - G_m(t))I_{[x_{(i)},\infty]}(t)dt = \frac{1}{n}\sum_{i=1}^{n} \int_{x_i}^{\infty} (1 - G_m(t))dt$$

and integrating by parts gives

$$= \frac{1}{n}\sum_{i=1}^{n}\left(-x_i(1 - G_m(x_i)) + \int_{x_i}^{\infty} tdG_m(t)\right) = \frac{1}{n}\sum_{i=1}^{n}\int_{x_i}^{\infty} (t - x_i)dG_m(t) = \frac{1}{n}\sum_{i=1}^{n} W_i.$$

Notice that $\int_{-\infty}^{\infty} G_m(y)(1 - G_m(y))dy$ is one half of GMD of Y (see (2.9)), which can also be written as $\frac{1}{2\binom{m}{2}}\sum_{i<j} |y_i - y_j|$. Therefore, we only need to show that $\frac{1}{n}\sum_{i=1}^{n} W_i - \frac{1}{m}\sum_{j-1}^{m} T_j = \bar{X} - \bar{Y}$.

We have (by (11.10))

$$\frac{1}{n}\sum_{i=1}^{n} W_i = \int_{-\infty}^{\infty} (1 - G_m(t))F_n(t)dt = \frac{1}{m}\sum_{j=1}^{m}\int_{-\infty}^{y_i} F_n(t)dt.$$

Integrating by parts gives

$$\frac{1}{m}\sum_{j=1}^{m}\left[y_jF_n(y_j) - \int_{-\infty}^{y_j} tdF_n(t)\right] = \frac{1}{m}\sum_{j=1}^{m}\int_{-\infty}^{y_j} (y_j - t)dF_n(t).$$

Thus,

$$\frac{1}{n}\sum_{i=1}^{n} W_i - \frac{1}{m}\sum_{j=1}^{m} T_j = \frac{1}{m}\sum_{j=1}^{m}\left(\int_{-\infty}^{y_j}(y_j - t)dF_n(t) - \int_{y_j}^{\infty}(t - y_j)dF_n(t)\right)$$

$$= \frac{1}{m}\sum_{j=1}^{m}\left(\int_{-\infty}^{y_j} y_j dF_n(t) - \int_{-\infty}^{y_j} t dF_n(t) - \int_{y_j}^{\infty} t dF_n(t) + \int_{y_j}^{\infty} y_j dF_n(t)\right)$$

$$= \frac{1}{m}\sum_{j=1}^{m}\left(y_j - \int_{-\infty}^{\infty} t dF_n(t)\right) = \bar{Y} - \bar{X}.$$

The final step is deriving the asymptotic distribution of $\hat{D}_{n,m}$. $\hat{D}_{n,m}$ can be expressed as a two-sample U-statistic with kernel of degree (2, 2): $h(x_1, x_2; y_1, y_2) = g(y_1, y_2) - g(x_1, x_2)$

where

$$g(t_1, t_2) = \frac{(t_1 + t_2 - |t_1 - t_2|)}{2}.$$

Therefore, under some regularity conditions the limiting distribution of $\hat{D}_{n,m}$, after proper standardization is asymptotically normal (Eubank et al., 1993) and inference can be based on cutoffs from the normal distribution.

11.3 Testing for Intersection of Two ACCs

As was mentioned above, a necessary and sufficient condition for X to dominate Y according to SSD is that the absolute Lorenz curve of X will not be lower than the ALC of Y. If the ALCs intersect then there exist two legitimate utility functions that will rank X and Y in reverse order. Formally, $ALC(X) \geq ALC(Y)$ if and only if X dominates Y in the SSD meaning (Shorrocks, 1983; Yitzhaki & Olkin, 1988, 1991). In a portfolio context or in welfare economics we are also interested in Marginal Conditional Stochastic Dominance (MCSD), which is an extension of the stochastic dominance rules to multivariate data. In those cases the rules that apply to the ALC are substituted by the ACC curves.

The objective of this section is to introduce a test for the intersection of two absolute concentration curves (ACCs). The ALC is a subclass of all ACCs and hence the test is applicable to ALC as well. We consider statistical tests concerning various relationships between two ACCs. In particular, we consider tests for determining if the two ACCs coincide, if one is above another in a specified order, or if they do not intersect without specifying which one is above/below the other.

The procedures presented in this section can easily be adapted for testing analogous hypotheses about two arbitrary curves for which empirical estimators can be constructed and weak convergence of the appropriately normalized differences between the empirical and the theoretical curves established. In this sense, it can be put into a broad context of testing for stochastic dominance and monotonic relationship in econometrics and actuarial science.

Let (X, Y, Z) be three random variables coming from a continuous cumulative distribution functions F, G, and H, respectively. We are interested in testing hypotheses about various relationships between the two ACCs corresponding to the pairs (Y, X) and (Z, X). The ACC was introduced in Chap. 5 and is briefly given below.

Let μ_X and μ_Y denote the means of X and Y respectively, and let $f_{Y|X}$ denote the conditional density function of Y given X. The conditional expectation is $g(x) = \mu_{Y.X} \equiv E\{Y|X = x\}$. It is assumed that all densities are continuous and differentiable, and all second moments exist. The *absolute concentration curve* (ACC) of Y with respect to X, $ACC_{Y.X}(p)$, is implicitly defined by the relationship

$$ACC_{Y.X}(p) = \int_{-\infty}^{X_p} g(t) dF_X(t)$$

where X_p is defined by $p = \int_{-\infty}^{X(p)} dF_X(t)$.

In words, X(p) is the pth percentile of the distribution of X. The special case $ACC_{X.X}(p)$ is referred to as the *absolute Lorenz curve* (ALC) (see Chap. 5).

An alternative way to present ACC is in terms of the quantile function. Let F^{-1} denote the quantile function of F, defined on [0, 1] by the equation $F^{-1}(t) = \inf\{x|F(x) \geq t\}$, with $F^{-1}(0) = -\infty$ by definition. Then the two ACCs are defined on [0, 1] by $A(t) = E[YI_{X \leq F^{-1}(t)}]$ and $B(t) = E[ZI_{X \leq F^{-1}(t)}]$ where $I_{X \leq F^{-1}(t)}$ is equal 1 if $X \leq F^{-1}(t)$ and 0 otherwise.

The objective is to test whether two curves coincide i.e., $A = B$ or if there is dominance in a particular direction, say, $A < B$. More generally, it might be of interest to test whether there is dominance in any direction, without specifying the direction explicitly. Note that rejecting the latter (two-sided) null hypothesis means accepting the claim that the ACCs A and B intersect. By their nature, such hypotheses are complex, because they involve comparisons of A(t) and B(t) over the values $t \in [0, 1]$. It is therefore natural to aim at reformulating the hypotheses in a simpler way, using a single parameter τ such that the null hypothesis would become $\tau = 0$ and the alternative $\tau > 0$. The construction of τ naturally depends on the hypotheses considered, and it may not be unique. With the help of one parameter $s(h) = \sup_{0 \leq t \leq 1} h(t)$, it is possible to identify four mutually exclusive cases, written as (i)–(iv) in Table 11.1, that are used for constructing the aforementioned parameter τ, depending on the hypotheses of interest.

For example, suppose that we are interested in testing the hypotheses

$$H_{01}: A(t) = B(t) \forall t \in [0, 1]$$

$$H_{11}: \exists t \in [0,1] \text{s.t.} A(t) \neq B(t).$$

Table 11.1 Analyzing hypotheses with the help of s(h)

Case	S(A − B)	S(B − A)
(i) $A(t) = B(t) \forall t \in [0, 1]$	=0	=0
(ii) $A(t) \leq B(t) \forall t \in [0, 1]$ and $\exists t_1 : A(t_1) < B(t_1)$	=0	>0
(iii) $A(t) \geq B(t) \forall t \in [0, 1]$ and $\exists t_2 : A(t_2) > B(t_2)$	>0	=0
(iv) $\exists t_1, t_2 \in [0,1] : A(t_1) < B(t_1)$ and $A(t_2) > B(t_2)$	>0	>0

Source: Schechtman, Shelef, Yitzhaki, and Zitikis (2008a), Table 1, p. 1048

Define the parameter $\tau_1 = s(A - B) V s(B - A)$ where $x V y$ denotes the maximum between x and y. Then the hypotheses above can be translated into

$$H_{01} : \tau_1 = 0$$

$$H_{11} : \tau_1 > 0.$$

The idea behind this translation is that under H_{01}, both $s(A - B)$ and $s(B - A)$ are equal to 0, while under H_{11} at least one of $s(A - B)$ and $s(B - A)$ is positive, and hence the maximum is positive. Following the same logic, the test for one-sided dominance is formally stated as:

$$H_{02} : A(t) \leq B(t) \forall t \in [0, 1]$$

$$H_{12} : \exists t_0 \in [0,1] \text{ s.t. } A(t_0) > B(t_0)$$

is based on $\tau_2 = s(A - B)$ which leads to the following reformulation of the hypotheses:

$$H_{02} : \tau_2 = 0$$

$$H_{12} : \tau_2 > 0.$$

Lastly, the hypotheses

$$H_{03}\text{: either } A(t) \leq B(t) \forall t \in [0, 1] \text{ or } A(t) \geq B(t) \forall t \in [0, 1]$$

$$H_{13}\text{: } \exists t_1, t_2 \in [0,1] \text{ s.t. } A(t_1) < B(t_1) \text{ and } A(t_2) > B(t_2)$$

are tested based on $\tau_3 = s(A - B) \wedge s(B - A)$ where $x \wedge y$ is the minimum between x and y, and translated to

$$H_{03} : \tau_3 = 0$$

$$H_{13} : \tau_3 > 0.$$

The last hypotheses test whether the two curves intersect or not.

Once the hypotheses are formulated in terms of the parameters, one needs to estimate them empirically. The idea in Schechtman, Soffer, and Yitzhaki (2008) is to construct two empirical ACCs and then estimate τ_1,τ_2 and τ_3 from those curves. The final step is to construct tests based on the estimated parameters. The interested reader is referred to Schechtman et al. (2008) for details.

11.4 Summary

Graphical tools become increasingly popular. Lorenz curve, concentration curves and plots that are derived from them such as LMA are mostly used as descriptive tools.

They can be used to check for second degree stochastic dominance, for identifying monotonic relationship among variables and the possibility of changing the sign of a regression coefficient by monotonic transformations and more.

In this chapter we offer several formal tests that are related to those curves: a test on the ordinates of the Lorenz curve, a test on necessary conditions for SSD (which turns out to be a test on a parameter—the area below the ALC curve—rather than a test on the entire curve, and a test for the intersection of two ACCs).

A lot of research is still needed in the area of formal testing. For example, as far as we know there are no tests for intersection of LMA curves and no tests for the concavity or the convexity of such curves. Therefore, at this stage most of the use of the curves is for visual evaluation only.

Part II
Applications

Chapter 12
Introduction to Applications

The main properties of the Gini mean difference (GMD) and the extended Gini (EG) were presented in the first part of the book. We have concentrated on those properties that enable the user to replicate almost everything that can be done when relying on the variance. In some sense we can claim that (almost) every analysis that is performed when using the variance can be done with the GMD, and sometimes with the EG as well. This means more than doubling the number of possible models that can be used because every variance-based model can be replicated by a Gini-based model. This fact raises the question whether it is worth to pursue this direction of research or not and what are the pros and cons of using the Gini methodology. We note that generally speaking when the underlying distribution is multivariate normal then there is nothing to be gained from using the GMD method. The reason is simple: when the underlying distribution is multivariate normal then the estimates of the means, the variances, and the correlations (by Pearson) are sufficient statistics for describing the data, and therefore nothing is gained by using an alternative system for describing the data on one hand, while a loss of efficiency follows because the parameters of the normal distribution are estimated in a circumvent way. However, as pointed out by Huber (1981) and Gorard (2005) even a small deviation from the ideal world of the normal distribution can lead to an advantage of using other measures of variability.

The applications part of the book intends to illustrate the use of the Gini methodology in various areas of economic and statistics research and to emphasize its advantages. Generally speaking, almost all the issues illustrated in the empirical chapters can be and have been analyzed by alternative methods. This should not surprise the reader because we are imitating what was done by other methods. The difference (and advantage) is that we are offering a unified method that can encompass what is done by several alternative methods. The problem with using different methodologies for answering different research questions is that the investigator may unintentionally contradict himself.

To illustrate this argument, let us present an example. Assume that in analyzing the robustness of a regression coefficient an econometrician finds that a specific observation influences the estimates. To increase robustness the econometrician

S. Yitzhaki and E. Schechtman, *The Gini Methodology: A Primer on a Statistical Methodology*, Springer Series in Statistics 272, DOI 10.1007/978-1-4614-4720-7_12,

may omit this observation from the regression. But it may turn out that this observation is the most important observation from the point of view of the financial economist, e.g., representing an extreme crisis in a financial market. The use of the extended Gini, which includes a risk aversion parameter in the estimation process, enables to introduce the financial economist's point of view into the estimation process. (See an illustration of this point in Chap. 18.) In other words, the use of the EG enables one to impose economic considerations on the statistical analysis.

There is a major problem in the presentation and in the applications of those ideas. Although the basic theory is identical for all the areas of applications in the sense that the basic ideas and the mathematical models are the same, there are major differences between the questions asked in the various fields of research, between the empirical data that exist, and even in the terminologies used. Because there is no point in replicating the proof of each proposition using a different terminology, we presented the proofs in the first part of the book (the first eight chapters) while in the applications part we will present the propositions in each field using the terminology used in the specific field. We will refer the interested reader to the relevant chapters for the general theory and proofs. The reason for this kind of presentation is that in some cases there is a huge difference between the skills and the expertise that are needed to apply the methodology and the skills required to follow the proofs.

We start our applications part with Chap. 13 which describes the role of the Gini in representing Runciman's (1966) theory of relative deprivation. This theory was developed as a sociological theory. The theory of relative deprivation was developed without a reference to the Gini. This gives some credibility that the contribution of the Gini in this case can be characterized as supplying a genuine need. The Gini enables one to interpret it as an economic theory in the sense that it is based on accepted economic paradigms. It presents an alternative to the Bergson's type social welfare function approach which dominated the area of formation of theoretical policy recommendations by economists, especially in public economics, welfare, poverty, and income distributions. This application of the Gini does not fall into the category of an empirical application yet, but it may resolve some of the paradoxes we observe in social behavior and it suggests some policy recommendations that are totally different from those recommended by the current social welfare approach to policy recommendations.

Following are some of the advantages of using the Gini:

(a) Unlike the variance, the Gini enables the user to form necessary conditions for stochastic dominance, so that an investigator who uses it can safely summarize the expected utility or social welfare with the mean and the Gini of the distribution, without violating the basic principles of expected utility theory or Yaari's (1987, 1988) dual theory, or Runciman's relative deprivation theory. On the other hand, it serves as a measure of variability, just like the variance. This advantage plays an important role in constructing portfolios that on one hand belong to the efficient set of the second-order stochastic dominance efficient set, and at the same time enables the user to form a complete order of portfolios.

In other words, it enables the user to have an efficient set of portfolios so that no one can find a portfolio in the efficient set that violates economic theory. This point is illustrated in Chaps. 17 and 18.

The same point, based on different terminology and illustrated under totally different circumstances, is replicated in Chaps. 14 and 16 in the areas of welfare economics, tax reforms, and income distribution.

It is also replicated in regression analysis. To see that, assume that the research area is public economics or finance which are characterized by a theory that requests an asymmetric attitude to the variability of income due to the declining marginal utility of income. The researcher runs a linear regression with respect to income but she is unaware of the fact that the data do not support a linear relationship with respect to income. That is, we focus on areas in which the researcher does not care about the linearity of the model because all she is interested in is getting a weighted average of slopes of the regression curve, weighted by the marginal utility of income. The extended Gini methodology of regression enables the user to impose her risk aversion or social welfare attitude on the regression. This way economic theory is imposed on the statistical analysis so that there is no contradiction between the econometrician and the economist in the case where the linearity of the model with respect to income does not hold. (In the case of linearity, the weighting scheme is irrelevant.) This point is the subject of Chaps. 19 and 20.

(b) The relationship between the Gini and the Lorenz curve which can be used to form sufficient conditions for second-order stochastic dominance enables the user of the methodology to perform the investigation in a step-wise manner. First one searches for large deviations from an optimal policy. That is, deviation from stochastic dominance efficiency. Then, if no such deviation is found, one uses necessary conditions only, i.e., uses the Gini. This way of research implies imposing assumptions in a step-wise manner. The more assumptions imposed the less robust the conclusions derived are. This issue is covered in Chaps. 14 and 15 in the area of tax reforms, in Chaps. 17 and 18 in finance, and in Chap. 20 for regression analysis.

(c) The decomposition of the Gini offers more parameters than the decomposition of the variance. Those additional parameters offer additional aspects of analyzing the data. Chapter 15 uses this property in the area of income distribution, Chap. 18 uses it in the area of finance, Chaps. 19–21 use it in regression analysis while Chap. 22 replicates ANOVA. Chapter 23, dealing with further research, points out several additional uses of these parameters.

To summarize, the structure of the empirical part is the following: In Chap. 13 we demonstrate the role of the Gini coefficient in two major competing theories that dominate the theoretical considerations in the area of income distribution, namely: the social welfare function approach and the theory of relative deprivation. In Chap. 14 we illustrate the use of the concentration curves and the Gini methodology in the areas of taxation and progressivity of public expenditure. Chapter 15 deals with the usefulness of several decompositions of the Gini and the extended Gini in analyzing government policies by non-marginal analyses, while in Chap. 16 the marginal analysis is illustrated. The applications in finance are the topics of Chaps. 17 and 18. These applications are relevant whenever one is interested in decision making under risk. Chapters 19–21 are devoted to applications of the Gini regression, in Chap. 19 we apply the simple Gini and extended Gini regressions, in Chap. 20 the multiple regression is applied, and in Chap. 21 we apply the mixed OLS, Gini, and extended Gini regressions. Chapter 22 deals with one application of the GMD and the Gini coefficient in statistics—an application that replicates the commonly used ANOVA and is denoted by ANOGI (ANalysis Of GIni). The last chapter (Chap. 23) concludes and lists several topics for further research.

Chapter 13
Social Welfare, Relative Deprivation, and the Gini Coefficient

Introduction

The aim of this chapter is to elaborate on the role of the Gini coefficient in two major competing theories that dominate the theoretical considerations in the area of income distribution. The major theory that dominates the economic thinking with respect to the role of the government in the area of income distribution is the Bergson's (1938) social welfare function approach (hereafter SWF). The other theory that is gaining acceptability among economists is the theory of relative deprivation (hereafter RD).

The SWF approach is the corner stone in normative public finance and optimal taxation. In this area the Gini plays an interesting role. Although one cannot construct a specific SWF which is based on the Gini coefficient (Lambert, 1985; Newberry, 1970; Sheshinski, 1972), one can use the Gini coefficient to form necessary conditions for "welfare dominance." That is, by using the Gini one can form necessary conditions for detecting improvements in all possible increasing concave SWFs. The combination of the absolute Lorenz curve (ALC) and the Gini enables the researcher to separate the decisions that should be taken by the society into two types: those that can be agreeable by all users of the concave SWF without any additional information and those which require knowing (or assuming) the exact social views of the decision-maker. The former is referred to as "welfare dominance," which means finding the decisions that are agreeable by everyone who agrees that the social evaluation of the marginal utility of income is positive and declining. Those decisions can be found by using the ALC.

However whenever additional specification of the social evaluation of the marginal utility of income is required in order to reach a decision, then the use of the Gini is called for. The advantage of using the criterion of welfare dominance through sufficient and necessary conditions is in allowing the user to introduce the normative assumptions in a sequential order. The more assumptions needed, the closer we are to the fine-tuning of the policy. In this chapter we consider mainly the necessary conditions. The necessary and sufficient conditions can be formed by using the ALCs.

S. Yitzhaki and E. Schechtman, *The Gini Methodology: A Primer on a Statistical Methodology*, Springer Series in Statistics 272, DOI 10.1007/978-1-4614-4720-7_13,

The second theory is the theory of relative deprivation (RD) which was developed as a sociological theory (Runciman, 1966; Stouffer, Suchmam, DeVinney, Star, & Williams, 1949). Unlike the social welfare approach which is intended to construct a decision function for the society this theory is intended to explain the feelings of the members of the society. The Gini coefficient can be used in order to define a special quantitative measure of relative deprivation (Yitzhaki, 1979, 1982a, 1982b, 1982c). In addition the Gini coefficient enables us to reinterpret RD theory as an economic theory which is competing with the SWF approach. The recent use of the theory is mainly in explaining paradoxes that cannot be explained by the SWF approach. For example, this theory can be used to explain the rationale behind a decision of a deputy director in a large and successful company to start his own company while suffering a reduction in his salary, or the phenomenon of "return migration" which is the decision of a migrant who was financially successful in his new country to return to his homeland, while suffering an income loss (Stark & Yitzhaki, 1988).

An additional advantage of the Gini is that it is a statistical measure of variability that is compatible with the SWF and RD theories. Therefore by using it one can perform statistical analyses without violating the principles advocated by those theories (Yitzhaki, 1996).

We start this chapter with a simplistic example of an economy. The example illustrates the difference between the approaches and clarifies the three stages that play roles: allocation of resources by the agents, the equilibrium process, and the satisfaction from the income distribution. In addition the example will help in showing the roles of the Gini coefficient in both theories. Consider the decathlon, an athletic 2-day event with ten different disciplines (100 m sprint, long jump, shot put, high jump, and 400 m run in that order on the first day and 110 m hurdle, discus, pole vaults, javelin, and 1,500 m run on the second day). A scoring table is used to award points for the performance in each discipline, and the winner is the athlete with the highest total score after completing the ten disciplines. To draw the analogy with a market economy, the ten disciplines can be considered as commodities and the scoring system can be considered as their prices. The utility function of each athlete may be defined over the physical units of achievements in each field.

The first stage in constructing the analogy can be referred to as the microeconomic problem: each athlete allocates his/her practicing time in order to maximize his/her utility subject to a time constraint and to the scoring structure (prices). A proper solution (in a competitive environment) is to allocate time so that for each athlete the marginal cost of achieving an additional point is the same for the various disciplines.[1]

The allocation of time and effort may also be interpreted as if each athlete were maximizing his/her points (income) subject to the time constraint. The overall score that an athlete has achieved can be viewed as if we have defined the indirect utility function as a function of income and prices.

[1] A monopolistic behavior implies that an athlete takes into account that his achievements may affect the scoring system. Note, however, that the arguments raised continue to hold under monopolistic behavior.

The second stage is the general equilibrium process which determines the prices. To reach an "equilibrium," the scoring structure is adjusted by a committee to avoid the possibility that any one discipline overshadows all others (in our analogy, this is akin to the demand and supply mechanisms in markets).

The result of applying the prices (scores) to individual achievements in each field in order to obtain the total score of the individual athlete is that we end up with points of equal value from the point of view of production (given the pricing system, each point requires an equal marginal cost or effort to be produced).

The two stages described above generate the distributions of scores, or in an economic context, the distribution of income in a simplistic model, with one factor of production (labor) and no capital accumulation.

We now turn to the third stage, which is the roles of the two theories in explaining the satisfaction from the income distribution. The SWF approach is mainly concerned with the kinds of actions that are needed in order to improve the SWF, defined over the utilities (or indirect utilities) of the athletes. The relative deprivation approach is concerned with hypothesizing about the satisfaction in the society of the individual athletes.

The two theories are related in the sense that the SWF is a drill down approach, starting from the society and from the society it sets the welfares of the individuals, while the relative deprivation approach starts from the individual and its intention is to describe the distribution of the satisfactions in the society. The question of what to do with the distribution of the satisfactions of the individuals and how to improve them is not tackled by the RD theory.

The decathlon example will be used when we describe the roles of the Gini coefficient in representing the two theories. Because both theories are concerned with the society we will assume that the setting of the microeconomic problem is the same: in a market economy, each member of the society is faced with an income constraint and with given prices. It is assumed that the member of the society maximizes a well-behaved utility function defined over a commodity space subject to a given budget constraint. Formally:

$$\begin{aligned} &\mathrm{Max}_{x_1^h,\ldots,x_n^h} U_h(x_1^h,\ldots,x_n^h) \\ &\text{s. t. } y^h = \sum\nolimits_{i=1}^{n} p_i x_i^h \end{aligned} \tag{13.1}$$

where $h = 1,\ldots,H$ represent the H members of the society, x_i^h, $i = 1,\ldots,n$ is the quantity of commodity i consumed by the h-th member, p_i, $i = 1,\ldots,n$ are the given prices of the commodities, and y^h is the given income of the h-th member.[2]

Solving for the first-order conditions for (13.1) we can derive the indirect utility function, $V^h(y, p_1,\ldots,p_n)$, which defines the utility function of the h-th member of the society as a function of the parameters of the problem. Formally

[2] One can complicate the presentation by introducing public goods, leisure, monopolistic behavior, etc. Introducing those issues will complicate the presentation without adding relevant content.

$$V^h\left(y^h, p_1, \ldots, p_n\right) = \begin{array}{c} \mathrm{Max}_{x_1^h,\ldots,x_n^h} U_h(x_1^h,\ldots,x_n^h) \\ \text{s. t.. } y^h = \sum_{i=1}^n p_i x_i^h \end{array}, \tag{13.2}$$

where the properties of the indirect utility function that are relevant to us are:

$$\frac{\partial V^h}{\partial y} = V_y^h > 0 \tag{13.3}$$

that is, the marginal utility of income is positive, and

$$\frac{\partial V^h}{\partial p_i} = -V_y^h\, x_i^h < 0 \tag{13.4}$$

that is, the effect of an increase in the price of commodity i on the indirect utility of the h-th member is negative.

The structure of the chapter is the following: Sect. 13.1 surveys the main issues in the development of the SWF approach. Section 13.2 describes the theory of deprivation, while Sect. 13.3 dwells on the relativity of the concept, and Sect. 13.4 concludes and offers some suggestions for further research.

13.1 The SWF Approach

The Bergson-type SWF approach includes efficiency considerations together with normative and ethical issues. It can be applied regardless of whether markets exist or do not exist. However, because we are interested only in the role of the Gini coefficient we will limit ourselves to the case of existence of market economy and ignore detailed efficiency considerations that are mainly relevant to taxation policy. We will touch upon detailed efficiency considerations only when we deal with policy issues, for completeness. Efficiency considerations are defined by the Pareto criterion. Atkinson and Stiglitz (1980) define Pareto-efficiency in the following way:

"A Pareto-efficient allocation is one where no Pareto-improving move can be made." (p. 337, footnote 2). A Pareto-improving move is defined as a move which improves the utility of at least one member of the society without deteriorating the utility of any member of the society.

Although complicated and abstract, the Pareto-improving principle is the only principle mentioned in this chapter that is actually applied in practice. It is applied whenever decisions are implemented as long as no one objects to them.

The Pareto-improving principle is the concept that divides the social decisions into efficiency considerations and income distribution issues. It implies that increasing the welfare of a member of the society is desirable if it is not hurting other members of the society. If we allow envy then it is clear that every allocation is Pareto efficient. Therefore envy is ignored in efficiency considerations. It is important to mention that our RD theory is not based on assuming envy (Podder, 1996). It is based on extending the principle of declining marginal utility to the area of the satisfaction of the individual from his status in his reference group. In the social context this principle can be better described as the increasing marginal utility from scarcity.

The SWF is the normative decision guide for the society. It should include all the elements that are relevant for the proper decision-making process. Assuming a market economy for which (a) each member of the society has a well-behaved utility function[3] and (b) all individuals face the same prices, we can define the SWF as a function of the indirect utility functions of the members of the society. Formally,

$$W\left(V^1\left(y^1\right),\ldots,V^H\left(y^H\right);p_1,\ldots,p_n\right). \tag{13.5}$$

is the SWF. Applying the Pareto principle we can assume

$$\frac{\partial W}{\partial y^h}=\frac{\partial W}{\partial V^h}\frac{\partial V^h}{\partial y^h}>0 \quad \text{for } h=1,\ldots,H. \tag{13.6}$$

Equation (13.6) says that the social evaluation of the marginal utility of income is positive for all individuals. For reasons of egalitarianism and/or ability of the function to have a maximum it is also assumed that the social evaluation of the marginal utility of income is declining. Formally, it is assumed that

$$\frac{\partial^2 W}{\partial (y^h)^2}<0. \tag{13.7}$$

Additional assumptions on the SWF are the anonymity axiom and the additivity of the SWF. We will make these assumptions, although they are not required for the analysis that concerns the applications of the Gini coefficient.

Adding those assumptions and assuming a continuous distribution function of income, we have the following SWF:

$$W(y^1,\ldots,y^H)=\int w(y)f(y)dy, \tag{13.8}$$

with $w'(y)=\frac{\partial W}{\partial y}=\frac{\partial W}{\partial V}\frac{\partial V}{\partial y}>0$ being the social evaluation of the marginal utility of income, while $w''(y)=\frac{\partial^2 W}{\partial y^2}<0$, which means that the social evaluation of the marginal utility of income is positive and declining with income.

---

[3] A well-behaved utility function is defined only over the commodity owned and consumed by the individual, the marginal utilities are positive and declining, and for every set of prices, there is one optimal allocation of commodities that maximizes it.

It should be mentioned that the assumptions leading to the SWF are very strong and some of the empirical analyses that are performed using the Gini coefficient in the empirical section on statistical analysis of income distribution are redundant, would one accept all the assumptions concerning the SWF. To bring one example, the issue of horizontal inequity in taxation is redundant if one accepts (13.8) because (13.8) implies that all those with identical y have the same SWFs while in the empirical sections one must take into account households with additional different characteristics. (Lambert & Yitzhaki, 1995).

13.1.1 Welfare Dominance: The Role of the Gini Coefficient

We are now in a position to introduce the connection with the Gini coefficient. Our starting point is the concept of welfare dominance (Yitzhaki & Semrod, 1991), which is the application of the concept of stochastic dominance to welfare economics.

Welfare dominance accepts (13.8) as the normative social decision rule but avoids the assumption of a specific SWF. (Note that if we will be ready to assume a specific SWF in (13.8) then this section will be redundant). The concept of welfare dominance is weaker than the Pareto-improving criterion because it allows the acceptance of a policy recommendation even if some members of the society object to it, provided that other individuals benefit from it. The researcher is trying to reach a conclusion on a set of SWFs. The sets of permissible functions are defined by the signs of the agreeable properties of the social evaluation of the marginal utility of income. The largest set of functions is the one that only requires that $w' > 0$. The second largest set, which is a subset of the first one, requires $w' > 0$, $w'' < 0$, and so on. We will limit ourselves to first- and second-degree welfare dominance rules.

The set of functions $\{w(y)\colon w' \geq 0\}$ is the set of non-decreasing SWF, which means that all we are ready to assume is that the social evaluation of the marginal utility of income is non-decreasing. The set of functions $\{w(y)\colon w' \geq 0;\ w'' \leq 0\}$ is the set of permissible SWF that is restricted to functions with the social evaluation of the marginal utility of income being non-negative and non-increasing.

Proposition 13.1 *Let y_1 and y_2 be two continuous income distributions defined over [a,b]. Then $E\{w(y_1)\} \geq E\{w(y_2)\}$ for all functions w in $\{w(y)\colon w' \geq 0\}$ iff*

$$F_2(y) \geq F_1(y) \quad \text{for all y,} \tag{13.9}$$

where F_i is the cumulative distribution function of y_i.

Proof See Chap. 5, Sect. 5.4.

Proposition 13.1 implies that if the cumulative distributions intersect, then one can find two legitimate SWFs which belong to $\{w(y)\colon w' \geq 0\}$ that will rank the distributions in a contradicting order of expected social welfare. This means that the ranking of SWF provided by Proposition 13.1 is incomplete, because the ranking is restricted to non-intersecting cumulative distributions.

It is worth pointing out that Pareto improvement is a stricter requirement than first-order welfare dominance because while Pareto improvement requires that no individual is worse off, welfare dominance allows for making a person worse off provided that another person is better off. Because of the anonymity requirement, it may happen that the rich becomes poor and the poor becomes rich, provided that the income added to the poor is greater than the income taken from the rich.

Proposition 13.2 *Let y_1 and y_2 be two continuous income distributions defined over [a,b]. Then $E\{w(y_1)\} \geq E\{w(y_2)\}$ for all functions w in $\{w(y)$: $w' \geq 0$ and $w'' \leq 0\}$ iff*

$$\mathrm{ALC}_1(\mathrm{p}) \geq \mathrm{ALC}_2(\mathrm{p}) \quad \text{for all p,} \tag{13.10}$$

where $\mathrm{ALC_i}$ *is the absolute Lorenz curve of* $\mathrm{y_i}$.

Proof See Chap. 5, Sect. 5.4.

Proposition 13.2 implies that if the ALCs intersect then one can find two legitimate SWFs in $\{\mathrm{w(y)}: \mathrm{w}' \geq 0 \text{ and } \mathrm{w}'' \leq 0\}$ that will rank the expected SWFs in a contradicting order.

Definition 13.1 The efficient set of distributions according to a given rule is defined as those distributions that are not dominated by any other distribution according to that rule.

A useful property of the efficient set is that the more general the rule that is used to derive it, the larger the efficient set is. Following this property, we note that the Pareto-improving efficient set includes as a subset the first degree welfare dominance efficient set, which includes the second-degree welfare dominance efficient set, etc.

However, unless the set of distributions is constrained, the welfare dominance rules cannot produce a complete ordering of distributions.

Proposition 13.3 *The following are necessary conditions for second-degree welfare dominance:*

(a)
$$\mu_1 \geq \mu_2 \tag{13.11}$$

and

(b)
$$\mu_1 - \Delta_1(v) \geq \mu_2 - \Delta_2(v) \quad \text{for all } v$$

where μ_i *and* $\Delta_i(v)$ *are the mean and the extended Gini coefficient (with extended Gini parameter* v*) of* $\mathrm{Y_i}$*, respectively.*

Proof The proof is immediate if we recall that $\mu_1 - \Delta_1(v)$ is the area between the ALC and the horizontal axis. See Chap. 5.

The usefulness of Proposition 13.3 is the following: because welfare dominance rules produce only an incomplete ordering of distributions, they are not capable of producing decision rules that can be applied in all cases. Actually, they are unable to rank distributions unless an extreme deviation from optimality occurs. The advantage of forming efficient sets according to necessary conditions is that they enable producing a complete ordering of distributions with the created efficient set being a subset of the efficient set of the welfare dominance (WD) rule.

For example, consider the following two-necessary-conditions efficient set: "the mean and the mean minus the GMD" rule (conditions (a) and (b) of Proposition 13.3). That is, a distribution Y_1 belongs to the efficient set if there is no other distribution, Y_2, such that: $\mu_2 \geq \mu_1$ and $\mu_2 - \Delta_2 \geq \mu_1 - \Delta_1$. Each condition by itself forms a complete ordering, and all distributions that belong to the efficient set formed by the two conditions belong to the efficient set of second-degree welfare dominance as well. The distributions that belong to the efficient set have either high average income or their inequality is not "too high" relative to other distributions in the set. The advantage of using the necessary condition for WD is that it enables the researcher to reduce the size of the efficient set, and it cannot be shown that by using it one violates all possible SWFs.

It should be noted that even if all the conditions in (13.11) are met, it does not necessarily mean that distribution 2 does not belong to the second-degree welfare dominance set, because not all possible legitimate SWFs are included in (13.11). (A counter example is provided in Chap. 6.) In other words, fulfilling the infinite number of necessary conditions in (13.11) does not form a sufficient condition for second-degree welfare dominance.

An interesting property of the Gini coefficient is given in proposition 13.4.

Proposition 13.4 *There exists no additive utility function which ranks income distributions in the same order as the Gini coefficient.*

Proof See Newberry (1970).

On the other hand, the Gini and the extended Gini are special cases of Yaari's decision rule (1987, 1988).

To see this, note that Yaari's decision rule is

$$\mathrm{YD} = \int \Psi(1 - \mathrm{F(x)})\mathrm{dx} \tag{13.12}$$

where Ψ is a decreasing function. Substituting $\Psi(1 - F(X))$ by $[1 - F(X)]^{\upsilon}$ we get a special case of Yaari's decision function to be $\mu - \Delta(\upsilon)$ (see Chap. 6). Therefore we can claim that the "mean minus the EG" rules also form necessary conditions for Yaari's dominance rules. We will not elaborate upon Yaari's decision rule here. All we say is that the "mean minus the EG" forms a special case to Yaari's index, similar to the relationship between Atkinson's (1970) family of SWFs and a general SWF.

To sum up: the necessary conditions according to Proposition 13.3 are necessary condition for second-degree welfare dominance and Yaari's theory. We will return to this topic in Chap. 14 when we add marginal analysis to welfare dominance rules.

13.2 The Theory of Deprivation

The purpose of the next two sections is to interpret the sociological theory that is referred to as "relative deprivation" (hereafter RD) as an economic theory that is competing with the SWF approach. Our objective is to illustrate the role of the Gini in quantifying deprivation. In doing so it is useful to separate the "relative" concept from the "deprivation" concept. However, because most of the literature deals with RD theory we will separate the two concepts at a later stage.

A major difference between the RD and SWF approaches is that deprivation is a theory about satisfaction while SWF is a normative intended to find the appropriate decision. Relative deprivation is an early version of what economists refer today as the "happiness literature" (Van Praag & Ferrer-i-Carbonell, 2008) and some of this literature uses the concept of deprivation. It is not intended to be a normative decision function, and there is no requirement for consistency with optimization. However, it can serve as a decision function, because in a democratic country we expect the SWF to be based on and to take into account the utilities (satisfaction) of the members of the society.

The original conceptualization of the theory appears in the famous three-volume research monograph *The American Soldier: Adjustment During Army Life* (Stouffer et al., 1949). The theory has been applied to several fields in order to model social behavior. Crosby (1979) presents an excellent review of the early stages of the theory. However, as pointed out by Merton and Kitt (1950), the concept of relative deprivation is not formally defined in *The American Soldier*. Therefore it is not surprising that Crosby (1979) counts four versions of the theory and that in general there is no agreement on what the exact meaning of the term "relative deprivation" is. By now one can count at least a dozen versions, and there is no advantage in describing them here.

A major advantage of the RD theory from our point of view is that it gives rational explanations to what at first sight might be viewed as a paradox or irrational behavior. For example, how come there is more deprivation concerning promotion among pilots than among the military police personnel, although promotions at the military police is a rare event while among pilots it is abundant? Is it possible that the closer the society is to egalitarian distribution the higher the feelings of deprivation? We will show several applications of the theory at a later stage.

What make the RD theory richer than the classical economic theory are the dimensions it deals with. RD theory is based on three dimensions: income, power, and prestige, and within those dimensions it is based on different reference groups. Traditional economists tend to focus on income only. Prestige, if not ignored at all, is usually considered as another dimension that can be translated into forgone income, while power is mainly considered in game theory. Reference groups are ignored. Whenever RD is used as a theory of revolution, it is based on the conflict between power and income or prestige. However we will follow traditional economics and concentrate on income only. The reason is because in order to show the relationship with economic theory, power and prestige are not needed.

Runciman (1966) described the theory in words, but it can be interpreted as posing exact axioms that enable one to describe the theory as an extension of traditional economic theory into social collective feelings. His book is intended to explain a paradox: why is there a higher feeling of deprivation whenever inequality is relatively low? That is, he wanted to explain the difference between an objective measurement of inequality and subjective feelings.

Runciman defines the conditions for an individual to feel relatively deprived: "We can roughly say that a person is relatively deprived of X when (1) he does not have X, (2) he sees some other person or persons (possibly including himself at some previous or future time) as having X whether or not that is or will be in fact the case, and (3) he sees it as feasible that he should have X." (Runciman, 1966, p. 10).

Runciman rightly suggests that people compare themselves with some reference group within the society rather than with the whole society. The reference group forms the base for the yardstick they use. Because individuals may have several reference groups and different individuals may have different reference groups, we argue that the reference group is responsible for the "relative" part in the RD concept. The concept of relativity will be discussed in the next section. At this stage we will assume that the whole society is the sole reference group for all members of the society so that we are left with deprivation theory. The deprivation theory is based on (1) and (3). However, it is possible to argue that (3) refers to the power dimension, which we ignore.

We shall consider income as the object of deprivation. Even when restricting ourselves to income, there are still numerous versions of the theory of relative deprivation and it is not the purpose of this chapter to review them. We will follow only one version of the theory, the one presented by Runciman (1966) and its quantification as presented in Yitzhaki (1979, 1982b).

Runciman's (1966) theory is based on three dimensions: deprivation, power, and status (prestige). With respect to deprivation, Runciman distinguishes between deprivation of an individual due to the position of his group (hereafter, between-group deprivation, in Runciman's term, "relatively deprived because of group's position in the society," p. 33), and relatively deprived because of his own position in his reference group (to be called "within-group" deprivation).

Runciman mentioned power in his theory, but he did not analyze its implications. It seems that the main reason for that is that power is used in analyzing social conflicts while Runciman was not interested in that part of the theory. One possible explanation for this approach is that social unrest did not seem a very important factor at that period in Britain. Runciman also mentioned status as a dimension relevant to RD theory. Because we are interested in presenting RD as an economic theory we do not take status into considerations.

Yitzhaki (1979) quantified the RD theory while Yitzhaki (1982b) dealt with the implications of the existence of reference groups. However, mathematical difficulties prevented Yitzhaki from having a better analysis of the latter. Meanwhile, Frick, Goebel, Schechtman, Wagner, and Yitzhaki (2006) managed to offer a decomposition of the Gini coefficient according to population subgroups. The decomposition enables the analysis of the effects of reference groups on deprivation. The rest of this section will set the ground for deprivation theory, with the whole society serving as one reference group, while the effect of different reference groups for different individuals, which is responsible for the relativity of the concept, will be dealt with only in the next section.[4]

In order to interpret the theory we have to analyze the stages that lead to social welfare and to analyze the roles they are playing. The description of the considerations leading to the theory is based on three stages (recall the three stages in the decathlon example): the microeconomic problem, the general equilibrium reached through market activity, and the effect of the resultant income distribution on deprivation.

(a) The microeconomic problem: Consider an individual who maximizes her utility function subject to a given budget constraint. That is:

$$\text{Max } U(x_1, \ldots, x_n) \tag{13.13}$$

$$\text{s.t. } \Sigma_i p_i x_i = y.$$

As a result of this optimization we can write the indirect utility function as:

$$V(y, p_1, \ldots, p_n). \tag{13.14}$$

The indirect utility function states that the utility of the individual is a function of her income and the prices that she faces.

Would we label the indifference curves (i.e., equal utility curves) as indifference of deprivation curves, then the consumer problem could have been defined as minimizing deprivation subject to the same income and prices constraint. Formally, let $D(x_1,\ldots,x_n) = B - U(x_1,\ldots,x_n)$ be the deprivation function, where B is the non-reachable Bliss point. Then

$$\text{Min } D(x_1, \ldots, x_n) \tag{13.15}$$

$$\text{s.t. } \Sigma_i p_i x_i = y.$$

[4] Hopkins (2008) surveys the different approaches toward relativity and the implications of the connection between happiness and inequality. Note, however, that the relativity in Runciman's approach arises because deprivation depends on the reference group, and reference groups among individuals may differ. Would we use the same reference group for the whole population then deprivation ceases to be relative.

results in an identical solution to the (13.13) problem. Hence, the only difference between Runciman's deprivation and the utility function is that the utility function is defined over what the individual has, while deprivation is defined over what one does not have. Because the same basket of commodities defines what the individual has or does not have, the two theories seem to be representing the same idea. The only difference is whether one considers the glass as half empty or half full.

(b) The general equilibrium: We now move into the general equilibrium stage which describes the role of markets and the resultant national income.[5] At this stage prices and incomes are *simultaneously* determined. Because it is assumed that $p_1,\ldots,p_n$ are equal for all individuals, the only differences among individuals are the utility functions and incomes. This gives the following meaning to prices: prices enable the society to create units of equal production value. In other words, the n-dimensional commodity space is converted into units of equal purchasing power value. Each unit of income represents a basket of commodities with equal purchasing power. It is important to note that the prices and the income distribution are determined simultaneously. This observation is important because the size of the cake and its distribution are simultaneously determined. It implies that the term "real income" is meaningless at the national level unless a set of equilibrium prices is attached to it (Yitzhaki, 1982b).

(c) Distributional values: the effect of the income distribution on deprivation.

Till now we applied only standard economic theory without referring to deprivation or welfare economics. We now turn to the only assumption needed to apply deprivation theory.

A critical assumption: In evaluating the units of equal purchasing power (i.e., units of income) that she possesses, an individual applies the law of declining marginal utility. That is, the distributional value of a unit of income y (i.e., the unit defined between [y, y + dy]) depends on the scarcity of that unit in the population. Scarcity is the excess demand over supply. To attest that this assumption is based on Runciman's approach note the following: "The more people a man sees promoted when he is not promoted himself, the more people he may compare himself with in a situation where the comparison will make him feel relatively deprived" (1966, p. 19). Note that our addition to Runciman's theory is by applying Runciman's X to each unit of income, and determining that the value attached to X is the proportion of individuals who have X.

In what follows we will use two types of values: production value and distributional value. Production value is the value of the inputs required to produce a dollar value of commodities. The distributional value is the value attached by the individual (or society) to a unit consumed of a product, the production value of which is one

[5] Sen (1976b) describes the assumptions that lead to national income comparisons.

dollar. Each unit of income represents a different basket of commodities available for consumption.

Let y and F(y) represent the income and the cumulative distribution of income, respectively. Each dollar of income represents a different bundle of commodities and its distributional value is determined by its scarcity. The cumulative distribution F(y) represents the scarcity of the y-th unit in the population while 1 – F(y) represents its abundance. 1 – F(y) = P(Y > y) is the share of individuals who do have the y-th unit of income (which represents a bundle of commodities consumed by someone with y units of income). We will refer to 1 – F(y) as the distributional value of the y-th unit. The individual can be deprived of the y-th unit (if he does not have it), or satisfied with having the y-th unit if he has it. Note that the distributional value is determined independently of whether the individual has or does not have that unit.[6]

The total value of the units an individual with income y is deprived of is

$$d(y) = \int_{y}^{\infty} (1 - F(t))dt, \tag{13.16}$$

while the total value of the units of y the individual has is:

$$s(y) = \int_{0}^{y} [1 - F(t)]dt. \tag{13.17}$$

The sum of (13.16) and (13.17) is equal (by definition) to μ, the mean income in the society (assuming that income is non-negative). It is shown in Yitzhaki (1979) that the average deprivation in the society, which is the average of (13.16) over all individuals, is

$$D = \mu G, \tag{13.18}$$

where G is the Gini coefficient. The total satisfaction in the society is

$$S = \mu(1 - G). \tag{13.19}$$

The proof is based on (2.13).

To give a concrete example, imagine the market for stamps. Consider a group of collectors, each one of them is interested in maximizing the value of his collection. The price of each type of stamps is determined in a market that takes into account

[6] See Chakravarty and Mukherjee (1999), Bossert and D'Ambrosio (2006), Bossert, D'Ambrosio, and Peragine (2007), and D'Ambrosio and Frick (2007) for different views. Deaton (2001) uses the theory to explain differential mortality.

tastes and scarcity of stamps. The result of the market activity is a set of prices and a distribution of the values of stamp collections. Real income can be defined in terms of a specific stamp with each unit of income representing a bundle of stamps of equal exchange value. The distributional value attached to each unit of real income depends on the scarcity of that unit. According to deprivation theory each collector feels deprived of units that he does not have and feels satisfied with each unit he possesses. The units of income he is deprived of are those units that would have enabled him to have a larger stamp collection.

An additional example is the decathlon, where athletes compete in several fields.

The microeconomic problem: Given prices, which in this case means the conversion of seconds or meters in each field into scoring points, we can present the athlete's problem as maximizing his utility over the achievements in different fields subject to time constraint and to given prices (points per unit). That is, each athlete allocates his practicing time in a way that will ensure that the marginal cost of time in terms of scores is equal across fields. This problem can be translated into an indirect utility presentation so that the utility of the athlete is a function of his total score and prices. Note that in order to obtain the RD theory there is no need to assume envy or altruism. The externality is imposed on the social evaluation of the marginal utility of income, and individuals evaluate what they have and what they don't have by the same criterion, its scarcity. In this sense, one can view RD as an extension of the economic assumption of declining marginal utility to the feelings of the individuals in the society.[7]

Finally, everything in this section could have been developed using the extended Gini coefficient. Therefore, all EGs can represent RD theory. The difference between using Gini and using EG is that in using Gini we assume a linear relationship between distributional price and scarcity, while in using the EG the relationship is not linear.

13.3 Relative Deprivation

13.3.1 Concepts of Relativity

The term "relative" is used in economics in several different contradicting ways. When we say "relative price" we actually mean "real" price, which is the price of a good in terms of another good. In other cases the interpretation of "relative" is that the units are normalized, as in the case of a relative measure of inequality where we normalize the units of income by the mean income. A third use of the term "relative" occurs when we use the rank of an individual according to some property.

[7] Hey and Lambert (1980) consider a different interpretation. The individual compares his income to the income of each individual in the society. As far as we can see this is a crucial difference in interpreting the implication of the theory.

A fourth type of "relative" occurs when one changes the reference group. Our first step is to show that the last two interpretations of "relativity" occur in relative deprivation theory.

Proposition 13.5 *The individual's satisfaction function is "relative" in the sense that it is an increasing function of the rank of the individual in the society.*

Proof The satisfaction of an individual with income y is:

$$s(y) = \int_0^y (1 - F(t))dt \tag{13.20}$$

which is the same as the total value of the units of y that the individual has (see 13.17). Using integration by parts, with $v' = dt$ and $u = (1 - F(t))$, we get

$$s(y) = y[1 - F(y)] + \int_0^y t\, f(t)dt.$$

Using the notation $p = F(y_p)$, i.e., p is the rank of the individual with income y_p in the society we get

$$s(y_p) = y_p[1 - F(y_p)] + \int_0^{y_p} t\, f(t)dt = (1 - p)\frac{\partial ALC(p)}{\partial p} + ALC(p) \tag{13.21}$$

where ALC is the absolute Lorenz curve. (Note that geometrically, $s(y_p)$ is a linear approximation of the value of ALC(1) evaluated at p).

By looking at the derivative of $s(y_p)$ with respect to p we find:

$$\frac{\partial s(y_p)}{\partial p} = (1 - p)\frac{\partial^2 ALC(p)}{\partial p^2} - ALC'(p) + ALC'(p) = (1 - p)\frac{\partial^2 ALC(p)}{\partial p^2} \geq 0. \tag{13.22}$$

The last inequality is based on the convexity of the ALC.

Hence we can argue that deprivation theory is "relative" in the sense that social evaluation of the utility of the individual is an increasing function of the absolute level of the income of the individual (which is equal to the derivative of ALC) and an increasing function of his rank in the society (13.21). This is known by every sports fan who uses the term "Numero Uno" to convey the message that his team is the best. Equation (13.22) states that the social evaluation of the marginal utility of income declines with the rank of the individual in the society. We now turn to the fourth use of "relative."

13.3.2 Relative Deprivation

In the discussion above the society was viewed as one reference group. As we have demonstrated the deprivation/satisfaction theory differs from the Bergson-type SWF only in one point: the way the social evaluation of the marginal utility of income is determined. While in Bergson-type SWF the social evaluation of the marginal utility of income is determined without reference to incomes of other persons, in the deprivation/satisfaction theory the social evaluation of the marginal utility of income is a function of the income distribution. Specifically, it is determined by the law of diminishing marginal utility applied to each unit of income. Note that deprivation theory does not have an element of envy or altruism. The way an individual determines the distributional value of the dollars he has is identical to the way a stamp collector and the market determine its value—as an inverse function of its scarcity.

We turn now to the relativity dimension of the concept. According to Runciman, the individual determines the distributional value of a dollar with reference to a reference group. The reference group may be composed of the whole society or any subgroup of members of the society.

In a general framework, reference groups should be determined endogenously by the individuals. In a dynamic model, especially when the society is changing and the individual changes his position, we should expect movement from one reference group to another. Even if we ignore the time dimension, we still face complications that arise from the individual having several reference groups, and the reference groups being open rather than closed. By closed groups we mean that if individual A belongs to the reference group of B, then B also belongs to the reference group of A. The fact that the individual is influenced by subpopulations is well recognized by Runciman. This is the basis of the relativity of the concept of deprivation. The idea of relativity is not unfamiliar to economists. Ben-Porath (1980) has coined the F-connection as the natural base of reference groups: Families, Friends, and Firms. Ethnic origin, common language, common religion, neighborhoods, and nationalities are also the bases of group identity. However technical difficulties prevent the analysis of a complicated division of the society into reference groups.[8]

An additional point worth mentioning is Runciman's attitude toward the role of reference groups. Pedersen (2004, p. 39) uses the following quote to describe Runciman's theory:

> Most people lives are governed more by the resentment of narrow inequalities, the cultivation of modest ambitions and the preservation of small differentials than by attitudes to public policy or the social structure as such (Runciman, 1966, p. 285).

In this respect, the reference group in Runciman's approach seems to be the group that causes the feelings of deprivation rather than the group with which an

[8] The classification of a society into classes is the corner stone of the Marxist theory but it is rarely done in the measurement of inequality. For a recent example see Wolff and Zacharias (2007b).

individual identifies himself. This is different than the polarization approach, where the group is the group one identifies himself with. (On polarization, see Duclus, Esteban, and Ray (2004); Esteban and Ray (1994, 1999, 2001, 2005).) It hints that the reference group for comparisons and the group with which one identifies himself may be different. In this sense it seems that Zhang and Kanbur's (2001) reasoning for the alternative polarization index they suggest agrees with Runciman. Note the following: "The three polarization measures discussed so far aim to capture the 'clustering' along the income dimension into high and low income groups. However, debates on polarizations are often conducted in the framework of recognized and accepted non-income grouping. In the U.S., for example, clustering of black and white income levels is of as much concern as 'the disappearing middle class.' In China, as discussed in the introduction, geographical clustering of income is a major policy concern." (Zhang & Kanbur, 2001, p. 93).

Dividing the society into reference groups, especially if one allows individuals to belong to several reference groups and also allows the reference groups to be open makes the analysis complicated and not tractable. Therefore we limit the analysis to closed reference groups, that is if A is in the reference group of B then B is in the reference of A. We also assume that each member of the society belongs to one and only one reference group. This is similar to dividing sports leagues according to ability or regions, while assuming that each team restricts its reference group to its own league.

13.3.3 The Effect of Reference Groups on Deprivation

This section relies on the decomposition of the Gini coefficient with respect to population subgroups in order to analyze the effects of different structures of subgroups on the average deprivation in the society (the reader is referred to Chap. 4 for more details).[9] Let Y_i, $F_i(y)$, $f_i(y)$, μ_i, and p_i represent the income, the cumulative distribution, the density function, the expected value, and the share of subpopulation i in the overall population, respectively. Let $s_i = p_i\mu_i/\mu_u$ denote the share of group i in the overall income. The overall population is composed of the union of the subpopulations. That is: $Y_u = Y_1 \bigcup Y_2 \bigcup \ldots \bigcup Y_n$, where Y_u is the union of subpopulations Y_i, $i = 1, \ldots, n$.

Note that

$$F_u(y) = \sum_i p_i F_i(y). \qquad (13.23)$$

[9] The decomposition is applied to the Gini coefficient only. Although the extended Gini can also represent deprivation and other theories (Ebert & Moyes, 2000; Moyes, 2007), it is not decomposable in the same way as the Gini.

That is, the cumulative distribution of the overall population is the weighted average of the cumulative distributions of the subpopulations, weighted by the relative sizes of the subpopulations.

Following (13.18) deprivation is modeled as $D = \mu G$ where μ is the mean income while G is the Gini coefficient. Assuming that the society is composed of n subgroups, then by using (4.27) we get:

$$D_u = \mu_u G_u = \sum_{i=1}^{n} p_i \mu_i G_i + \sum_{i=1}^{n} p_i \mu_i G_i (O_i - 1) + \mu_u G_{bp} + \mu_u (G_b - G_{bp}). \tag{13.24}$$

In order to be able to illustrate the implications of (13.24) let us distinguish between intra- and inter-group deprivations and the role played by stratification.

Intra-group deprivation (within-group)—the first two components in the right-hand side of the equation, i.e., $\sum_{i=1}^{n} p_i \mu_i G_i$ (denoted by IG) and $\sum_{i=1}^{n} p_i \mu_i G_i (O_i - 1)$, are the intra-group components of deprivation. Note that in the first term the contribution of each group to the overall deprivation is a function of its size, average income, and inequality among its members. IG is a weighted average of the intra-group Gini coefficients. The second term will be discused below, under "stratification".

Between-groups Gini: The last two components of (13.24) represent the between-groups Gini. $\mu_u G_{bp}$ is the between-groups Gini (BG) as defined by Pyatt (1976). It is calculated as if all members of the group received the same income which is equal to the mean income of the group. This term is the "closest" to the polarization index and the main difference is that while BG is homogeneous of degree zero in the share of the population in each group, the polarization index is not.

The second term, $\mu_u (G_b - G_{bp})$, is related to the overlapping between the groups and is explained next.

Stratification: Stratification, which is the inverse of overlapping, has two effects. The first one is the effect on the "within" component and the other is the effect on the "between" component. Assuming that the society is completely stratified, both terms of stratification vanish and the Gini (hence the deprivation) decomposes neatly into purely intra- and inter-group components. Existence of non-stratification, i.e., overlapping between distributions reduces the between-groups component because $\mu_u (G_b - G_{bp})$ is always nonpositive. The other effect is on the intra-group deprivation. It is equal to $\sum_{i=1}^{n} p_i \mu_i G_i (O_i - 1)$, and its components can be positive, zero, or negative depending on the overlappings between the distributions. However, because large values of overlapping tend to be associated with large values of Gini the overall term tends to be positive. An alternative way to see it is that because the overall and Pyatt's BG are given, and $\mu_u (G_b - G_{bp})$ is negative, by definition the sum of all overlapping terms is positive. Hence, overlapping increases the intra-group deprivation component and decreases the inter-group component.

By now we are ready to explore the effects of different structures of reference groups on relative deprivation.

Runciman's approach: Consider first the case where individuals' deprivations arise only from the intra-group component. We can analyze the implications of different scenarios depending on stratification.

Runciman (1966) mentions between-groups deprivation, but seems to stress the role of intra-group deprivation. That is, it is assumed that deprivation arises from comparisons within the reference group of the individual. This is the easiest to handle with clear-cut results. We divide the discussion to two cases: a case with perfect stratification and a case with an imperfect one.

Consider a perfect stratificaton: the society is divided into "leagues" with the aspirations of members of each "league" limited to that league. In this case, it is easy to see that deprivation is low even if inequality is high (because between-groups inequality is ignored). Also, the larger the number of groups, the larger the BG term, the lower the relative deprivation, even if inequality and mean income are kept constant. In the extreme case where the number of groups approaches the number of members in the society, high inequality can prevail with zero deprivation. This case is analyzed in detail in Yitzhaki (1982b). As far as we can see, this classification, if accepted by the members of the society, can allow extreme inequalities to exist without deprivation. Therefore we should expect the upper class in the society to convince others to restrict their aspirations to their reference groups. Examples of such behavior include the separation in the army to soldiers and officers, with officers eating in a separate dining room, the tendency to have ranks at work and the existence of different ranks in universities. In an extreme and unacceptable case, this theory can supply the rationale for an Apartheid policy. If a policy designer can convince each group to stick to its own folks, so no cross-group comparison is done—society can tolerate large inequalities with low level of deprivation.

The extreme case of stratification can be described by the following lines:

The rich man in his castle
The poor man at his gate
God made them, high or lowly,
And ordered their estate.[10]

If, on the other hand, stratification is not perfect, then it is possible that deprivation in one group can be higher than deprivation in the society. To see this, consider the case where the poor and the rich form one group, while the middle class forms another group. Then the mixed group of the rich and poor may have higher intra-group inequality than the inequality in the overall society, leading to higher deprivation in this group.

[10] Nineteenth Century hymn (Mrs. C. F. Alexander) brought to our attention by Susanne Freund.

Another interpretation of the same case to be considered is when individuals' deprivations arise only from the intra-group components, but stratification is not perfect. The overlapping component means that although there are no between-group deprivations, members of each group can see members of other groups and mingle with them. As a result it is possible that deprivation in one group is greater than deprivation in the whole society and in extreme cases the sum of the deprivations in all subgroups may be greater than the deprivation in a society that is not divided into reference groups.[11] In this case the "revolution" may start among the richest class. In some sense, deprivation arises not from an increase in inequality but from the collapse of the reference group. For example, assume that in the past women used to compare themselves to other women only. As a result of a greater participation in the labor force, however, the reference group of women was extended to include men. Deprivation then increased without a change in gender inequality and moreover, it may increase even if inequality between men and women declines (Gurin, 1985). Clearly, mass-media, television, and globalization tend to widen the reference groups of individuals and therefore can increase deprivation even if inequality does not increase. Our conclusion is that if we accept Runciman's view that group identity is less important than the feelings of the individual with respect to his own reference group, then between-group component which is interpreted in Yitzhaki (2010) as representing polarization does not play a role. Luttmer (2005) presents the neighbors as a reference group, which supports Runciman's approach.

The alternative view that seems to be stressed in polarization is that group identity is the main determinant of deprivation. Therefore the appropriate element to concentrate on is between-group inequality. In this case the relevant measure for polarization is the BG component because it represents the deprivation between reference groups.

13.4 Summary

We have shown that the Gini can be used to form necessary conditions for dominance according to the social welfare approach, Yaari's approach, and the relative deprivation approach. We also showed the role of reference groups in RD theory. Sufficient and necessary conditions can be formed using the ALC. The next chapter deals with using those rules to get necessary and sufficient conditions for improvement according to those theories that can be useful for policy analysis.

Further research is needed in order to fully implement the relative notion in the relative deprivation approach (Layard, 1980, 2006). According to Runciman, the list of factors that lead to social deprivation includes status, power, and income.

[11] To see that note that G_b can be negative while the two terms involving G_{bp} in (13.24) cancel each other. G_b can be negative if one group includes poor people and a small number of extremely rich people so that the covariance between average rank and average income among groups is negative.

The factors that lead to social clashes and unrest are deprivation and power. We have analyzed the implications of between-group inequality, a concept that resembles polarization in a relative deprivation context and found that it does not lead to an increase in deprivation. However, we did not analyze differences in power, and it may well be that introducing power as a function of the group size increases the power in a nonlinear way. If this is the case, then polarization should be associated with power rather than deprivation or inequality in economic well-being. Esteban and Ray (1999) associate polarization with ethnic conflict. Montalvo and Reynal-Querol (2003, 2005) associate polarization with ethnic and religeous conflicts which is in line with the suggestion advocated in this chapter. Another topic that is missing from this chapter is the endogeneous formation of reference groups. In this chapter we have assumed that reference groups are given. Future research is needed to make the formation of reference groups an endogenous decision made by the individuals. Alesina and La Ferrara (2005) survey some of the approaches, Shayo (2007) presents an additional quantitative aspect, while Benabou (2000) seems to suggest the possibility of multiple equilibria.

An additional area that calls for additional research is the effect of economic growth on polarization and deprivation. A first step in this direction is offered in Wodon and Yitzhaki (2009) who argue that economic growth may lead to higher well-being but also to higher deprivation because it increases the spectrum of commodities that the individual feels she is deprived of.

Those areas are beyond pure economic theory that views the rational individual as a decision-making unit whose connection with the surrounding is mainly through the market. They are beyond the scope of this book.

Chapter 14
Policy Analysis

Introduction

The objective of this chapter is to introduce the use of the concentration curves and the Gini methodology in the areas of taxation and progressivity of public expenditures. Most of the literature in these areas considers the case of a representative individual, which means that issues of income distributions are ignored and the only issue that is considered is efficiency. For the Gini methodology and concentration curves to be useful we extend the model to include issues of income distributions as discussed by Diamond (1975) and Atkinson and Stiglitz (1980). We start with a crude characterization of optimal (mostly indirect) taxation which includes the issue of redistribution in addition to efficiency considerations. In a typical model the investigator assumes a social welfare function (SWF) and optimizes it subject to the behaviors of the individuals and to the instruments that are used by the government. Using those ingredients she gets the first-order conditions for optimization so that the relationships among the instruments in an optimal setting are determined (see, as a background, Atkinson &Stiglitz, 1980, pp. 386–393). Note that by assuming the existence of an SWF, the issues of horizontal equity and of comparisons of utilities of households with different structures are skipped because the mere existence of an SWF implies that one knows how to rank individuals according to economic well-being.[1]

[1] Mayshar and Yitzhaki (1996) extend the approach so that "economic well-being" can be affected by two parameters—ability and needs. Then, the investigator should be able to agree that given ability, the higher the needs the lower the well-being of the household, and given needs, the higher the ability the higher is its well-being. This kind of extension is useful for handling family size (needs increase with the number of children but one does not want to commit himself to a specific magnitude) or rural–urban distinctions (e.g., rural populations need less income to achieve a certain level of well-being than urban populations (Ravallion, 1993, p. 31)). There are some technical requirements for this extension to hold. This and other types of extensions are beyond the scope of this book.

S. Yitzhaki and E. Schechtman, *The Gini Methodology: A Primer on a Statistical Methodology*, Springer Series in Statistics 272, DOI 10.1007/978-1-4614-4720-7_14,

There are two major weaknesses in such an approach as a general approach for policy analysis: (1) the policy's recommendation depends on the assumed SWF, which depends on the social approach of the investigator. This means that two investigators who assume different SWFs may reach different recommendations with no easy way to compare the recommendations. (2) One has to know exactly the reaction of the individuals to the changes in the policy instruments that the government uses. Moreover, almost in all real-life situations some or all of the first-order conditions are violated. This complicates the problem even more because as pointed out by Corlett and Hague (1953), counting the number of violations is not meaningful because one violation can neutralize the other. As a result we do not have a good way to know how far we are from the optimum without assuming some debatable assumptions. The case of a known SWF is not covered in this book and the interested reader is referred to Lambert and Yitzhaki (1995).

In this chapter we start without imposing the social views on the problem to be solved in advance. Instead, we introduce the assumptions in a gradual way. If the deviation from an optimum is relatively large then we may be exempted from having to fully specify the SWF. Given the results from the data and our ambition we will add more assumptions on the shape and properties of the SWF. This gradual imposition of assumptions is useful because it allows one to evaluate the robustness and limitations of our conclusions. The larger the number and the scope of assumptions imposed, the less robust our conclusions are. We refer to this approach as "welfare dominance," which is actually an adaptation of the stochastic dominance literature into the area of taxation and public expenditure. (Welfare dominance was introduced in Chap. 13).

The main difference between the application of the welfare dominance approach in the area of public policy and the applications in other fields such as finance or econometrics is that in the area of public policy one has to take into account the reactions of the individuals to the changes in policy instruments, while in other areas this complication is not dealt with. The reason for this difference is that the nature of the problem changes from a direct application of the methodology into a Stackelberg game. In a democratic society, and to avoid the unrealistic assumption of lump-sum taxes, the government has to take into account the reactions of the taxpayers as utility maximizers. That is, the government can affect the prices by imposing taxes and subsidies, it can change the incomes of individuals by distributing allowances and it also has to take into account the adjustments that the individuals would make, given the changes in prices and incomes imposed by the government. The extensive literature on "excess burden" or "deadweight loss" is actually dealing with the effects of the changes in the reactions of the individuals to the changes in policy instruments by the government. Hence, in this field there is no point in applying the suggested methodology of welfare dominance without considering the reaction of the individuals. Therefore a significant part of the chapter will be devoted to issues that do not involve the Gini or the concentration curves but deal with excess burden.

There is an additional major restriction that is imposed on the analysis in this chapter. The analysis is restricted to be a marginal analysis. That is, instead of defining the conditions for optimum we only search for the direction that we should move with our instruments in order to improve the SWF. We do not define the magnitude of the change that will lead to the optimum. The limitation imposed by this restriction is that it does not allow us a full analysis of the optimal conditions and the steps needed to be taken in order to bring us to the optimum. On the other hand, it has several advantages: we do not assume that the government optimizes a specific target function, nor that the reaction of each taxpayer to the change in the government policy is known. Also, our analysis is relevant even when the economy is not in an optimum, and the most important thing—it does not require assumptions on the behaviors of the individuals. In essence, we are looking for a gradual reform that saves a lot of assumptions and simplifies the presentation. Obviously there is a price tag that goes with this approach: it gives less guidance to the optimal solution. We will show that this kind of approach enables us to derive more operational results with fewer assumptions.

The structure of this chapter is as follows: in Sect. 14.1 we introduce the concept of marginal analysis. Section 14.2 is devoted to the description of the economic model. In Sect. 14.3 we elaborate on distributional characteristics, namely the Gini income elasticity (GIE). An empirical illustration of the Dalton-improving (DI) reforms is given in Sect. 14.4. Section 14.5 concludes.

14.1 Marginal Analysis

Marginal analysis is defined as studying the effect of a small change in a policy instrument on the SWF or on several SWFs or other targets. By a small change we actually mean the derivative of the function of interest with respect to a change in the policy instrument. Hence, we can learn from it about the direction to take, but it does not specify the magnitudes of the steps needed in order to reach an optimum. Also, if the function is not a well-behaved function, i.e., may have more than one optimum, then we do not know whether the direction we selected is leading to the global optimum or to a local one. All we know (and care about) is that the chosen direction leads to an increase in the value of the SWF.

An alternative presentation of the marginal analysis is the restriction of the search to the direction of the desired tax reform (i.e., which taxes or subsidies to increase (decrease) and in what proportions) but not to the actual changes in the tax rates and government's expenditures. The assumption of a marginal tax reform is not required from a conceptual point of view; all concepts described in this chapter are applicable to non-marginal reforms as well. However, this assumption reduces the data required enormously. Unlike the analysis of non-marginal reforms that rely on fictitious or "cooked" data (because they require assumptions on unobserved data), the analysis of a marginal reform mostly relies on observed data.

The validity of a marginal analysis as an approximation to real-life reforms depends on the nature of the reform. From a technical point of view, a marginal analysis is applicable whenever first-order approximations are not leading to gross errors, but this is a technical (and not so useful) answer. The important point is that the ranking of households according to economic well-being is not seriously affected by the reform. A useful rule of thumb is that if the reform does not change incomes by more than 10% then a marginal analysis is reasonable. However if one intends to make the poor rich and the rich poor, then marginal analysis is not an appropriate method.

We now turn to the adjustments of the proposition concerning welfare dominance to the area of policy analysis.

14.1.1 Setting the Problem: Dalton and Gini Improvement Reforms

Similar to an optimal taxation problem, it is assumed that there is a SWF representing the decision-maker's preference, and it is based on individuals' indirect utility functions. That is, there exists an SWF with

$$W\left(v^1\left(y^1\right),\ldots,v^H\left(y^H\right),q_1,\ldots,q_m\right), \tag{14.1}$$

where v^h and y^h, $h = 1,\ldots, H$, are the indirect utility function and the exogenous income of the hth individual, respectively, and q_i , $i = 1,\ldots,m$, are the prices that the individuals face.

It is assumed that the social evaluation of the marginal utility of income is positive, that is,

$$\beta^h(y) = \frac{\partial W}{\partial y^h} = \frac{\partial W}{\partial v^h}\frac{\partial v^h}{\partial y^h} > 0 \tag{14.2}$$

for all h and that

$$\text{if } y^h = y^k \quad \text{then } \beta^h = \beta^k(h,k = 1,\ldots,H). \tag{14.3}$$

Equation (14.2) implies that an increase in the income of any member of the society increases the social welfare W, which is the Pareto principle, while (14.3) implies that two individuals with the same income have the same social evaluation of the marginal utility of income. This assumption allows us to ignore horizontal equity issues. Combining (14.2) with (14.3) means that we can omit the index h from β. Additional assumptions on the SWF will be imposed later.

We now describe the effect of a reform on the individual taxpayer. Consider an individual (or a household) with a well-behaved utility function $u^h(\ )$, unknown to us, and an observed allocation of his budget according to $y^h = \sum_i q_i x_i^h$, where q_i is the price of the ith commodity the individuals face, x_i^h is the quantity of commodity i consumed by individual h, and y^h is his given income. Assume that the vector of producers' prices, p, is given and that $t_i = q_i - p_i$ is the tax on commodity i, $i = 1, \ldots, m$. Then, the effect of a marginal tax reform on the individual, i.e., the marginal benefit (MB), is[2]

$$MB^h = -\sum_i x_i^h dq_i + dy^h, \tag{14.4}$$

which is the first-order approximation of the effect of the reform on household *h*, evaluated in monetary terms.

The effect of the reform on the SWF is the weighted sum of the effects on the individuals, weighted by the social evaluation of the marginal utility of income of the individuals

$$dW = \sum_{h=1}^{H} \beta^h(y) MB^h. \tag{14.5}$$

Having defined the marginal benefit to individual h ($h = 1, \ldots, H$) and to the society, we can classify the different criteria for evaluating the reform. The first criterion does not require the assumption of an SWF, while the others are based on it. The reforms are defined and listed in descending order of difficulty in finding them in practice.

Definition 14.1 Pareto improving (PI) reform: A reform will be called a Pareto improving reform if $MB^h \geq 0$ for all h, $h = 1, \ldots, H$.

Pareto improving reforms are mainly of a theoretical value because as far as we know they were not found in practice. Weaker concepts of reforms require the assumption on the SWF.

Definition 14.2 First-order welfare dominance (FWD) reform: A reform will be called an FWD reform if $dW > 0$ for all $\beta^h > 0$, $h = 1, \ldots, H$.

Definition 14.3 Dalton-Improving (DI) reform (second-degree stochastic dominance reform). A reform will be called a DI reform if $dW > 0$ for all $\beta^h > 0$, $h = 1, \ldots, H$, provided that if $y^j > y^k$ then $\beta^k(y) \geq \beta^j(y)$.

The assumption behind the definition of a DI reform is that the social evaluation of the marginal utility of income declines with an increase in income. Hence, if there are no additional constraints on the government, the optimal solution will be an egalitarian society. As far as we know, the DI reform is the only reform that may be relevant for practical purposes. It can be found in cases of extreme deviations

[2] One can derive this relationship under two alternative sets of assumptions: (a) the household is a utility maximizer and by Roy's identity $\partial v^h()/\partial t_i = -\lambda^h x_i^h$, where v is the indirect utility function and λ is the marginal utility of income. The marginal benefit is the income equivalent of the change caused by the reform, or (b) no optimization is carried out by the household and we are only interested in a Slutsky's compensation to the household.

from optimality, when the government totally ignores issues of income distributions or follows other considerations. In this section we will concentrate on this type of reform, while in Chap. 15 we will impose additional assumptions on the SWF.

14.1.2 Characterization of a Dalton-Improvement Reform

A DI reform can be viewed as consisting of the allocation of manna from heaven (due to a reduction in deadweight efficiency loss (i.e., excess burden) and transfers from rich households to poor ones). It is referred to as a DI because Hugh Dalton, a British economist, was the first to suggest that a transfer from a rich person to a poor one increases social welfare (Dalton, 1920; Mayshar & Yitzhaki, 1995). A DI reform is weaker than Pareto Improvement, which asserts that society benefits if the income of at least one member increases, provided that incomes of all other members do not decrease. It is also weaker than the first-order welfare dominance reform, which allows reducing the real income of an individual, provided that someone will get an income which is at least as high as that income. The DI reform allows for a decrease of the income of an individual, provided that the incomes of one or more poorer individuals increase by at least the same amount. A DI reform would be considered as an improvement of the existing tax system by everyone who accepts Dalton criterion. Bibi and Duclos (2007) and Duclos, Makdissi, and Wodon (2005, 2008) apply the DI reform to truncated distributions to deal with a reform concerned with poverty only.

A DI reform is actually an application of the second-degree stochastic dominance approach into the area of tax reforms. The Dalton criterion views any transfer from rich to poor as welfare increasing. To apply it we do the following steps: first we arrange the households from poorest to richest in ascending order. Then we consider the poorest (first) household. Under the Dalton criterion it must be that $MB^1 \geq 0$, otherwise it means that the poorest household did not get his share in the manna, nor a transfer from a richer individual. That is, if $MB^1 < 0$, then it is the poorest household that gives a transfer to others—in violation of the Dalton criterion.

Consider now the next to the poorest household. The restriction on the marginal benefit to the second household is $MB^1 + MB^2 \geq 0$. To see this note that if such a condition is violated then one can repeat the above argument with the poor being defined as the first and second households combined. So if $MB^1 + MB^2 < 0$, then the poor did not get their share in the manna and gave a transfer to the rich in the society. Note that MB^2 can be negative, in which case we can interpret the reform as a transfer from the second-to-the-poorest to the poorest household, in accordance with Dalton.[3]

[3] This distinguishes Pareto improving reform from DI reform. Under Pareto improving reform all MB^h, $h = 1,..,H$ must be nonnegative.

By the same reasoning, it can be shown that a DI reform requires that

$$\sum_{h=1}^{k} MB^h \geq 0, \quad \text{for all } k, k = 1, ..., H. \tag{14.6}$$

An additional interpretation of condition (14.6) is the following: consider a social planner who wants to decrease the poverty gap but does not know who is poor. To be on the safe side, it is best to follow the strategy to reduce poverty gaps for any possible poverty line. It is easy to see that condition (14.6) provides the necessary and sufficient conditions for such a strategy.

A simple change in condition (14.6) will allow us to adapt the DI reform to other social planners concerned about poverty. If the social planner is ready to commit herself to an upper bound to the poverty line, then all she has to do is restrict condition (14.6) to the first $P < H$ conditions, cover only households that are potentially poor, and aggregate the rest into one group.

To summarize: a tax reform which conforms with Dalton's principle has to fulfill the H conditions described in (14.6).

14.2 A Description of the Economic Model[4]

The economic model used in our empirical illustration relies on the following characteristics and assumptions:

(a) A fundamental requirement.
In order to search for a DI reform one has to be able to arrange the population in an increasing order of economic well-being. Without such an agreed-upon order it is simply impossible to determine who is "rich" and who is "poor," which is a necessary condition for determining whether a transfer increases (decreases) the social welfare. Our empirical illustration uses expenditure per capita as the indicator of economic well-being. Note, however, that expenditure per capita is only used to rank households. Hence, any monotonic transformation of expenditure per capita can be used without affecting the findings.

(b) The tax is shifted to the consumer.
Actually, this is an additional "data saving" assumption. There is nothing in the method that prevents the user from introducing tax shifting into the calculations. However, one needs a general equilibrium model to find the distributional effect of the portion of the tax absorbed by firms, households (the suppliers of factors of production), and the rest of the economy. Assuming that all production functions are homogeneous of degree one and that there is a perfect competition is equivalent to the assumption that taxes are borne by consumers. Most computational general equilibrium models (CGE models) utilize the above mentioned assumptions.

[4] This section is based on Yitzhaki and Lewis (1996).

(c) Revenue neutrality
It is easier to analyze revenue-neutral reforms because one can ignore the issue of the optimal size of government activity. However, this is another "assumption of convenience" and can be discarded provided that one has the appropriate data. By appropriate data we mean the information about the "willingness to pay" for the public goods for each household in the economy. An example of incorporating government expenditures into the reform is given by Slemrod and Yitzhaki (2001).

(d) Externalities are ignored
This is yet another simplifying assumption, intended to get rid of the need to explore the magnitudes and effects of externalities. The reason that we do not deal with it is that it is not related to the main topic of this book. It is related to the issue of "marginal efficiency cost of funds," discussed later (in Section 14.2.2). One way of including externalities can be found by Lundin (2001). As far as we can see it should not be difficult to incorporate such considerations into the calculations, provided that one can get hold of the appropriate data, or alternatively, that one is ready to make the necessary assumptions with regard to the effect of the externalities. If the externalities affect a public good (e.g., pollution and health hazards) then one should know the willingness to pay.

14.2.1 The Required Data and the Distributional Characteristics

Similar to any other economic problem, one has to define the target function and the constraints. The target function is defined in (14.6) while the constraint is that the reform is revenue neutral. The search for a DI reform requires two kinds of data for each tax parameter (referred to later as an instrument, because it may represent a tax rate, an exemption or any parameter of the tax function) involved in the reform. The first parameter reflects the effect on the target function. It will be referred to as the *distributional characteristic* of the instrument, a term coined by Feldstein (1972) (see below). The second parameter reflects the impact on the revenue constraint and will be referred to as the *marginal efficiency cost of public funds* (hereafter MECF). MECF reflects the cost to the society of revenue raised by changing the tax instrument.

Assume that individuals are ordered in a descending order of the social evaluation of the marginal utility of income. Inserting (14.4) into the left hand side of (14.6) we get

$$\sum\nolimits_{h=1}^{k} \mathrm{MB}^{h} = \sum\nolimits_{h=1}^{k} \left[-\sum\nolimits_{i} x_{i}^{h} dq_{i} + dy^{h} \right], \quad k = 1, \ldots, H. \tag{14.7}$$

By changing the order of the summation and dividing and multiplying by $X_i = \sum\nolimits_{h=1}^{H} x_i^h$, which is the "quantity demanded" of tax base (commodity in indirect taxes) *i*, we get

$$\sum_{h=1}^{k} MB^h = -\sum_i X_i dq_i \left\{ \sum_{h=1}^{k} (x_i^h / X_i) \right\} + \sum_{h=1}^{k} dy^h, \quad k = 1, \ldots, H. \tag{14.8}$$

The term in the curly brackets is the cumulative share of commodity *i* (tax base of instrument *i*) consumed (held) by the *k* poorest individuals. This term reflects the distributional characteristics of the tax instrument. It is portrayed by the relative concentration curve of the commodity. The data required for constructing concentration curves can be found in any survey of households' expenditures. The data set used in this chapter is the National Social Economic Survey in Indonesia (Central Bureau of Statistics of Indonesia, 1990). Because we will be dealing with indirect tax reforms, $dy^h = 0$ for all households.

14.2.2 Marginal Efficiency Cost of Funds

An important consideration in any tax reform is the expected change in tax revenue. This section is devoted to the derivation of revenue estimates. It will be shown that hidden in those estimates is the estimate of the marginal deadweight loss, which can be recovered.

The government tax revenue is

$$R(t,q,y) = \sum_i t_i X_i(p + t,y), \tag{14.9}$$

where y is a vector of incomes, the consumer prices are $q = p + t$, t is a vector of specific taxes, and $X_i(q, y) = X_i = \sum_{h=1}^{H} x_i^h$ is the demand for commodity *i*.

Revenue neutrality requires

$$dR = \sum_i MR_i dt_i = 0, \tag{14.10}$$

where $MR_i = \partial R / \partial t_i$ is the change in overall tax revenue as a result of a small change in the tax rate on commodity *i*. It turns out to be convenient to work with dollars of revenues rather than with tax parameters which may differ in their units. The change in tax revenue (denoted by δ_i) that results from a change in the tax rate on commodity *i*, dt_i, is then

$$\delta_i = MR_i dt_i. \tag{14.11}$$

The marginal tax reform, *dt*, could also be characterized by the vector of tax receipts, δ and the change in tax revenue would then be $dR = \sum_i \delta_i$.

Inserting (14.11) into (14.8) while taking into account that $dt_i = dq_i$ and $dy^h = 0$ for all *h* we get

$$\sum_{h=1}^{k} MB^h = -\sum_i [X_i/MR_i]\ \delta_i \left\{ \sum_{h=1}^{k} x_i^h \Big/ X_i \right\} \geq 0, \quad \text{for } k = 1, .., H \tag{14.12}$$

subject to $\sum_i \delta_i = 0$.

The term in the square brackets is the Marginal Efficiency Cost of public Funds. It answers the following question: what are the costs to the society of increasing the tax revenue by a dollar through a change in the ith instrument?[5]

To see that, let us concentrate on the last inequality in (14.12), the one with $k = H$. Then we get

$$\sum_{h=1}^{H} MB^h = -\sum_i [X_i/MR_i]\ \delta_i \geq 0 \text{ subject to } \sum_i \delta_i = 0. \tag{14.13}$$

A neutral tax reform, involving only two taxes will benefit the society if

$$\sum_{h=1}^{H} MB^h = -\{[X_1/MR_1]\ \delta_1 - [X_2/MR_2]\ \delta_1\} = \{MECF_2 - MECF_1\}\delta_1 \geq 0. \tag{14.14}$$

It can easily be seen that $MECF_2 > (<)\ MECF_1$ requires that $\delta_1 > (<) 0$ for the reform to have an efficiency gain.

To estimate the MECFs one has to have either a tax calculator that can evaluate, for each tax instrument, two parameters:

(a) The marginal change in revenue, MR_i.
(b) The tax base X_i, which is the expected change in tax revenue if no other change occurs.

In many practical estimations X_i is also used as an estimate of MR_i. The interpretation in those cases is that all MECFs are assumed to be equal, or that it is assumed that the tax is a lump-sum tax.

It should be emphasized that the MECF concept is different from the MCF concept used in many models of optimal taxation. The MCF is based on (14.5) which is the sum of the marginal benefits being weighted by the social evaluation of the marginal utility of income. The advantage of the MECF over the MCF is that it separates efficiency considerations from the distributional characteristics. Therefore it can be used for any target function that has no element of social welfare. For example, consider a government that will be interested in maximizing its probability of staying in power. Then the MECF continues to serve as an efficiency criterion while the social evaluation of the marginal utility of income will be substituted by the marginal propensity to vote for the government.

[5] An instrument is a parameter that the government can change.

For our empirical illustration we rely on a CGE model for Indonesia, written by Jeffrey Lewis (1993). This model is capable of estimating the MECFs needed for our illustration. However, as far as we can see, any CGE model with appropriate classifications is sufficient for providing estimates of MECF. Readers who are interested in this model are referred to Lewis (1993) or to Yitzhaki and Lewis (1996).

Having described the problem to be solved in the terminology used in public finance, let us describe it in the terminology of Lorenz and concentration curves. A search for DI reform implies a search for a second-degree stochastic dominance. That is, we search for a feasible distribution of real incomes which has an absolute Lorenz curve (ALC) that is not lower than the existing Lorenz curve. To do that we concentrate on the derivative of the Lorenz curve which describes the cumulative value of the MBs. This curve is actually represented by (14.6). However, the government can change the MBs curves through the taxes. Hence we have to look at the derivative of ALC with respect to the tax instrument. This derivative leads us to the absolute concentration curve (ACC) of each tax base with respect to income. This is reflected in the curly brackets in (14.8), or in other words, by the income elasticity of the tax base. The role of the MECF is the following: when the government reduces the tax on one tax base by a dollar and increases the tax on another tax base by a dollar (to keep the tax revenue intact) it may end up increasing or decreasing the incomes of the taxpayers because of the changes it creates in the deadweight loss. This is reflected by having shifted the appropriate ACCs. To sum up: we are searching for a weighted combination of shifted ACCs that will result in a nonnegative cumulative MBs curve. This is shown in Sect. 14.4.

14.2.3 The Characterization of the Solution

The problem to be solved is the one defined in (14.12): Find $\delta_i (i = 1,..., n)$ such that

$$\sum_{h=1}^{k} MB^h = -\sum_i [X_i/MR_i]\, \delta_i \left\{ \sum_{h=1}^{k} x_i^h/X_i \right\} \geq 0 \quad \text{for all k, k} = 1, ..., H$$

subject to $\sum_i \delta_i = 0$.

Because the trivial solution $\delta_i = 0$ for all *i* satisfies the constraints, one commodity should be chosen as a numeraire with its δ being equal to one or minus one. Also, it can be shown that any convex combination of two solutions of (14.12) is also a feasible solution. Therefore, if one finds two DI tax reforms, one with a positive change in the tax rate *j*, and the other with a negative change, then one can find a DI reform with no change in tax rate *j*. In this case we may conclude that instrument *j* is not essential for a DI reform.

In order to search for solutions, a numerical optimization algorithm should be used to solve the following problem

$$\text{Min}_{\delta} \sum_{k} \{\text{Max}[-\text{CMB}^{k}(\delta),0]\}^{2} \geq 0, \quad \text{s.t.} \sum_{i} \delta_{i} = 0; \delta_{1} \neq 0, \qquad (14.15)$$

where $\text{CMB}^{k} = \sum_{h=1}^{k} \text{MB}^{h}$, that is, the cumulative marginal benefits from the poorest k households. A feasible marginal tax reform, δ, will be considered a "solution" if the value of the objective function in (14.15) is zero. The algorithm used for the search is described by Yitzhaki (1982c), but any algorithm for numerical optimization can be used. Because the numerical algorithm cannot be interpreted easily, we have to present the necessary conditions for welfare dominance before moving to describe the results of the application of the model.

14.3 More on Distributional Characteristics: The Gini Income Elasticity

In many practical cases the search for a welfare dominance reform will fail to find a dominating reform. In those cases we may have to add more structure on the SWF. Another reason for imposing more assumptions is that in many cases one is only interested in the progressivity or regressivity of taxation, and not in calculating the MECFs of taxes which is much more complicated and requires more data than evaluating the distributional characteristics. The way forward that is also convenient to use in order to find an improvement in the SWF is to use necessary conditions for welfare dominance instead of using necessary and sufficient conditions as was done so far. Because we are dealing with the SWF and we are interested in the distributional characteristics, we will ignore the MECFs in this section. Note, however, that by doing this we ignore the efficiency considerations of the tax reform.

The distributional characteristic of a tax instrument is described by the relative concentration curve of the tax base with respect to economic well-being, which is represented by the inverse of the social evaluation of the marginal utility of income.

To be concrete, we will deal with an indirect tax imposed on commodity *j*. The concentration curve of commodity *j* depicts the cumulative share of aggregate expenditures on that commodity against the rank order of households, arranged in a decreasing order of the social evaluation of the marginal utility of income (i.e., in an increasing order of economic well-being). The area enclosed between the concentration curve of commodity j and the diagonal is equal to cov $(X_j, F(Y))/\mu_j$. The numerator is the (Gini) covariance (co-Gini) between the consumption of the commodity and income. Having connected the tax reform to the co-Gini we can move forward in the presentation in the following two alternative ways.

(a) Define a specific SWF of the type:

$$W = \mu\,(1 - G), \qquad (14.16)$$

where μ is the mean income and G is the Gini coefficient. Equation (14.16) can be used to derive both a necessary condition for DI reform and a legitimate SWF representing the theory of relative deprivation or Yaari's dual theory. In this case one can derive the first-order conditions of optimal taxation or marginal tax reforms.

(b) Realize that because we have ignored the efficiency consideration, a neutral tax reform will not affect the mean income so that we end up analyzing the effect of the tax reform on the inequality in real income. That is, we measure the progressivity of the tax system by the impact on the Gini coefficient of real income and evaluate the impact of the reform through its impact on the Gini coefficient. This approach is developed by Lerman and Yitzhaki (1985, 1994), Stark, Taylor, and Yitzhaki (1986), Wodon and Yitzhaki (2002b), and Yitzhaki (1994b). In this case one can derive the concepts needed directly from the Gini methodology. To avoid redundant replications, we will develop the appropriate parameters, relying on both approaches.

The difference in MECF between two taxes affects the mean income. Because we assumed that all MECFs are equal, or alternatively, that taxes are lump-sum taxes, we can ignore the mean income in (14.16) and restrict our attention to the effect of changing a tax rate (or any parameter in the tax function) on inequality.

Let $y = (y^1,\ldots,y^H)$ be the vector of incomes. Then the effect of a change in a tax on a commodity i on the Gini coefficient of real income is

$$\frac{dG_y}{dt_i} = \frac{\partial G_y}{\partial q_i}\frac{dq_i}{dt_i}. \tag{14.17}$$

Because it is assumed that for each individual all income is spent, then $y^h = \sum_{j=1}^{n} q_j x_j^h$, where x_j^h is the consumption of the jth commodity by the hth individual and n is the number of commodities. We get

$$G_y = \frac{2\mathrm{cov}(y,F(y))}{\mu_y} = \frac{2}{\mu_y}\sum_{h=1}^{H}\mathrm{cov}(y^h, F(y)) = \frac{2}{\mu_y}\sum_{h=1}^{H}\mathrm{cov}\left(\sum_{j=1}^{n} q_j x_j^h,\ F(y)\right)$$
$$-\frac{2}{\mu}\sum_{j=1}^{n} q_j\mathrm{cov}\left(\sum_{h=1}^{H} x_j^h,\ F(y)\right) = \frac{2}{\mu_y}\sum_{j=1}^{n} q_j\mathrm{cov}(X_j, F(y)), \tag{14.18}$$

where X_j is the consumption of the jth commodity. An equivalent presentation is to use X_j as representing the expenditures on the jth commodity. For simplicity, let us assume that the income distribution is discrete, the tax is an advalorem tax (i.e., $q_i = p_i(1 + t_i)$), and using $\mu_y = \sum_{j=1}^{n} q_j\,\mu_j$, where μ_y denotes the average income while μ_j is the average value of commodity j, we get

$$\frac{dG_y}{dt_i} = s_i G_y\left[\frac{\mathrm{cov}(x_i, F(y))}{\mathrm{cov}(y, F(y))}\frac{\mu_y}{\mu_i} - 1\right], \tag{14.19}$$

where $s_i = \mu_i/\mu_y$ is the share of expenditure on commodity i in the budget.[6]

The square brackets in (14.19) can be interpreted in the following way.

Definition 14.4 The "Gini income elasticity" (GIE) of a tax base (commodity) is

$$\eta_i = \frac{\mathrm{cov}(x_i, F(y))\ \mu_y}{\mathrm{cov}(y, F(y))\ \mu_i} = \beta^N_{X_i,y} \frac{\mu_y}{\mu_i}, \tag{14.20}$$

where $\beta^N_{X_i,y}$ is the semi-parametric Gini regression coefficient of X_i on Y.

The properties of the GIE are the following.

(a) It is composed of the Gini regression coefficient of the consumption of commodity i on income, divided by the share of the commodity in the overall consumption. Interpreting the regression coefficient as the marginal propensity to spend on commodity i then the right hand side of (14.20) can be interpreted as the marginal propensity to spend with respect to income divided by the average propensity to spend. Hence, by definition it represents the income elasticity of the Engel curve of consumption as a function of income. The title Gini is added because the overall marginal propensity to consume is a weighted average of the marginal propensities to consume between adjacent observations of income, weighted by the weighting scheme of the Gini regression (see Chap. 7). The explanation for this approach can be found in optimal tax theory (see Diamond 1975). There are major differences between Diamond's general approach and the application presented here: (a) While Diamond's marginal utility of income is a general one, here it is derived implicitly from the regression methodology used in estimating the Engel curve. Using Gini regression implies the use of the marginal utility implied by the Gini. (b) Diamond (1975) derives the optimal tax rate while we are concerned with the effect of a small change in the tax rate. For application of this theory one only needs the covariance between the consumption of the commodity and the social evaluation of the marginal utility of income. To apply the economic model one does not need to specify the curvature of the Engel curve of the commodity. All that is needed for the economic model is a weighted average of slopes of the Engel curve (the marginal propensity to spend) between adjacent observations, weighted by the social evaluation of the marginal utility of income. We have used the double role of the GMD, both as a measure of variability and as a component in the SWF to get a nonparametric estimate of the marginal propensity to spend. Equation (14.20) represents an income elasticity because we divided the marginal propensity to spend by the average propensity to spend. Hence, the term GIE.

(b) It also represents the elasticity of the Gini coefficient in real income for an equal amount of tax revenue collected through the ith instrument. To see this note that the revenue change, measured in terms of average income and ignoring second-order effects is

[6] For the derivation of (14.19) see Lerman and Yitzhaki (1985) or Stark, Taylor, and Yitzhaki (1986).

$$dR = \frac{1}{\mu_y} \frac{\partial \sum_{j=1}^{n} t_j \mu_j}{\partial t_i} dt_i = \frac{\mu_i}{\mu_y} dt_i = s_i dt_i. \tag{14.21}$$

By using (14.19) we get

$$\frac{dG_y}{G_y dt_i} \bigg/ \frac{dR}{dt_i} = (\eta_i - 1). \tag{14.22}$$

The left hand side represents the percentage change in the Gini coefficient in real income, divided by a percentage change in tax revenue through a change in the ith instrument, which is the elasticity of the Gini with respect to a change in tax revenue. The right hand side is the GIE minus one.

Equation (14.22) enables us to introduce the following definitions of progressivity/regressivity of a tax system.

(a) If $\eta_i > 1$, that is, if the GIE of a tax instrument is greater than one then increasing the tax will be progressive while reducing it will be regressive.
(b) If $\eta_i = 1$ then the tax instrument will be neutral.
(c) If $0 < \eta_i < 1$ then increasing the tax will be regressive while decreasing it will be progressive.
(d) If $\eta_i < 0$ then the tax base is an inferior good.

In all the above definitions progressivity or regressivity are determined according to whether the Gini coefficient in real income decreases or increases as a result of an increase (decrease) in the tax. The properties of (14.22) quantify the simple logic of tax progressivity. The greater the income elasticity of the tax base, the greater the progressivity of the tax. There are two qualifications to this simple relationship: (a) It is defined on the margin. (b) The way progressivity is measured should be identical to the estimates of income elasticities along the Engel curve of the tax base that are aggregated into one coefficient.

To sum-up: the GIE determines whether a change in the tax on the commodity increases or decreases the Gini coefficient of income inequality. (Lerman and Yitzhaki 1994). By comparing the (Gini) income elasticities among themselves and to one, the impact of a tax on the Gini coefficient of economic well-being can be evaluated.

The above discussion holds for the extended Gini as well. That is, using the same methodology, one can define the extended GIE and replicate the section with respect to the EG. As for elasticity, it obeys all the rules that are relevant to income elasticities. Note, however, that the use of an OLS linear regression to estimate the income elasticity of a tax base for the purpose of measuring progressivity of taxation may contradict the assumption of a declining social evaluation of the marginal utility of income. To avoid such a contradiction one has to use regression methods based on variability measures that can form necessary conditions for welfare dominance.

The EG income elasticity is also useful when studying the curvature of the Engel curve of the tax base. That is,

Let $\eta_i(\nu)$ be the EG income elasticity with an extended Gini parameter ν. Then

(a) If $\eta_i(\nu)$ is an increasing function of ν then the Engel curve of the tax base is a concave function. (To see this, note that the larger ν the higher the weight attached to the slope of the curve at lower incomes).
(b) If $\eta_i(\nu)$ is a decreasing function of ν then the Engel curve of the tax base is a convex function.
(c) If $\eta_i(\nu)$ is a constant regardless of the value of ν then the Engel curve of the tax base is linear.

Hence, one can use $\eta_i(\nu)$ to conclude about the curvature of the Engel curve of the tax base.

To find a DI reform without having to estimate efficiency gains we can look at the GIEs of the tax bases for several EGs (that is, for different choices of ν) to find necessary conditions for DI reforms.

14.4 An Empirical Illustration of DI Reforms

The empirical illustration of how to find a DI reform is based on Yitzhaki and Lewis (1996) who search for a DI reform in Indonesia's energy sector. The reason for choosing this reform is that it results in a reform that contradicts the conventional wisdom which relied only on efficiency considerations and was advocated by the World Bank.

14.4.1 Distributional Characteristics of Commodities in Indonesia

The data used came from the Indonesian sample of Family Expenditure 1990. Expenditure per capita is used as a proxy for economic well-being of the household and the individual is considered as the relevant unit, by assigning a weight according to the number of individuals in the household. In order to simplify the calculations, the population is divided into 94 cells, each with 500 households (except for the last cell which consisted of 79 households).

Figure 14.1 presents the concentration curves of these commodities together with the Lorenz curve for expenditure per capita. The lowest curve is the concentration curve of gasoline, indicating that the (Gini) income elasticity of gasoline is the highest and it is larger than one. Above it is the concentration curve of electricity which is also below the Lorenz curve, which means that the income elasticity of electricity is, on average, larger than one and lower than the income elasticity of gasoline. The concentration curve of kerosene is below the diagonal

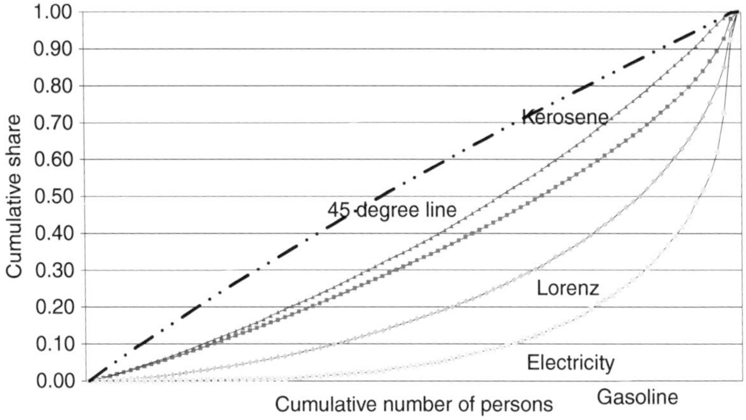

Fig. 14.1 Concentration curves of energy producing items. Source: Yitzhaki and Lewis (1996), Fig. 2, p. 551. Reprinted with permission by Oxford Journals

and above the Lorenz curve, which means that the income elasticity is bounded between zero and one.

Table 14.1 presents the (Gini) income elasticities, which are also the ratios of the areas between the 45° line and the concentration curves to the area between the 45° line and the Lorenz curve. As can be seen the income elasticity of kerosene is 0.59 indicating that kerosene is a necessity while electricity has an income elasticity of 1.61 and for gasoline it is 2.30. As shown in Chaps. 9 and 10 these estimates are consistent estimates of the appropriate population parameters and their asymptotic distributions converge to the normal under regularity conditions. As one can gather from the standard errors, the income elasticities of these commodities differ significantly from one another.

The last two columns of Table 14.1 report the income elasticity of regular rolled cigarettes. They are included in order to demonstrate some of the properties of the analysis. The fourth column shows the income elasticity of the quantity sold of regular rolled cigarettes while the fifth column reports the income elasticity of expenditure on this commodity. The difference between the two (0.56 0.26) = 0.3 indicates that the income elasticity of the price paid is 0.3.[7]

14.4.2 The Marginal Efficiency Costs of Funds

As explained in Sect. 14.2.2 the MECF requires the use of a tax calculator. The tax calculator should be able to provide estimates of the changes in overall tax revenue

[7] To see this note that E = PQ, where P is price and Q is quantity. Let Y be income then $\Delta E \approx P\Delta Q + Q\Delta P$. Hence $\frac{\Delta E E}{\Delta Y Y} \approx \frac{\Delta Q}{\Delta Y}\frac{Y}{Q} + \frac{\Delta P}{\Delta Y}\frac{Y}{P}$.

Table 14.1 Gini income elasticities of commodities[a]

Commodity	Kerosene	Electricity	Gasoline	Q-cig.[b]	V-cig.[c]
Income elasticity	0.59	1.61	2.30	0.26	0.56
Standard error	(0.07)	(0.05)	(0.14)	(0.09)	(0.07)

Source: Yitzhaki and Lewis (1996), Table 1, p. 551

[a]"Income elasticity" is calculated with respect to expenditure per capita. The population includes all Indonesian households grouped into 94 groups, each one with 500 observations except for the last one. Households are ordered according to expenditure per capita. Jackknife standard errors are reported in parentheses

[b]Q-cig. is the income elasticity of quantity of regular rolled cigarettes

[c]V-cig. is the income elasticity of expenditure on regular rolled cigarettes

under two alternative assumptions: (a) under the assumption of behavioral response to the change in the tax parameter and (b) under the assumption of no behavioral response. The MECF is calculated as the ratio in tax revenues collected under the alternative assumptions. The MECFs were calculated using a CGE model of the Indonesian economy. A major problem that arises when using two separate sources of data is mismatching of classifications and levels of aggregation. On the one hand the survey of family expenditure does not allow the distinction between imported and locally produced goods, while the CGE model was primarily designed to deal with trade issues. Also, the CGE model includes only one food sector while the level of disaggregation in the survey distinguishes between different types of cigarettes. The search for DI reforms was confined to three commodities with equal levels of aggregation in both data sources: electricity, kerosene, and gasoline.

14.4.3 Simple Dalton-Improving Reforms

Let us start the search for DI reforms by a riddle which demonstrates an important property: is it possible to have a revenue-neutral DI reform which is based on the change in the taxation of one commodity?

As the reader may have gathered the question would not be raised if the answer was negative. Consider the following case: a specific subsidy (that is a subsidy to the quantity consumed) and an ad-valorem tax are simultaneously imposed on the same commodity. The rates are defined in such a way that the reform is neutral ($\delta_{\text{Quant.}} = -\,\delta_{\text{Expend.}}$). Because one can expect the marginal costs of funds of those two taxes to be equal, the way to find such a tax reform is by plotting the difference between the concentration curve of the quantity consumed and the concentration curve of the expenditure on that commodity. The vertical difference between the concentration curves is the cumulative marginal benefit from the reform. If the curve of the differences is nonnegative then a revenue-neutral, one-commodity tax reform is found. Figure 14.2 presents the difference between the concentration curve (cumulative marginal benefit) of quantity and the one of expenditure of regular rolled cigarettes,

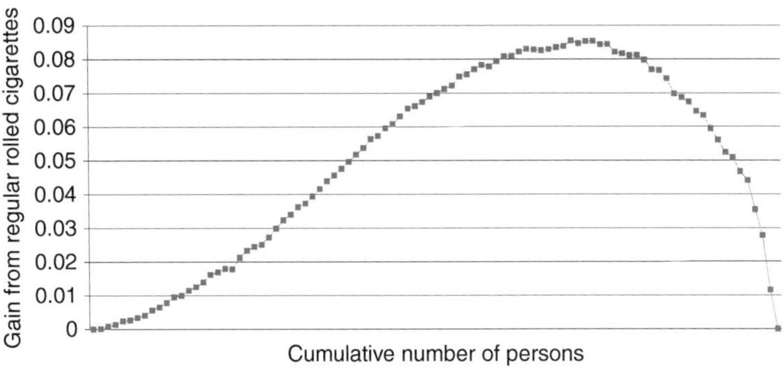

Fig. 14.2 The cumulative gain from a combination of a subsidy and a tax. Source: Yitzhaki and Lewis (1996), Fig. 3, p. 553. Reprinted with permission by Oxford Journals

showing that a DI reform of this kind exists. The difference-in-concentration curve (DCC) reports the cumulative marginal benefits to the population. As can be seen from Fig. 14.2 the cumulative gain is positive, which means that a DI reform exists. The bottom 70% of the population may gain up to eight cents for each dollar of tax paid by the top 30% of the population. The intention in presenting this example is to demonstrate the importance of separating the tax instrument from the commodity. The same commodity may have different tax bases depending on the nature of the change in the tax function. A good example is motor vehicles—several countries impose taxes on different properties such as quantity (a constant amount per vehicle), weight, engine's size and, of course, the value of the motor vehicle. Those taxes may have significant differences in their distributional characteristics, a property that can be exploited to increase the variety of tax instruments.

14.4.4 Dalton-Improving Reforms

In order to search for DI reforms which involve more than one commodity, the MECFs of the commodities are required. The CGE model was not constructed to search for DI reforms. Therefore only three commodities (electricity, gasoline, and kerosene) could be easily matched in the CGE model and in the survey of family expenditure.

Table 14.2 reports the MECFs and other relevant parameters. The first column reports the tax base, the second—the marginal revenue of a change of a 5% in the tax rate (e.g., the tax on kerosene was changed from −0.48 to −0.43; electricity from 0.0 to 0.05). The third column, which is derived as a ratio of the first two parameters, is the marginal cost of raising a dollar of revenue. The fourth column reports the effective tax rate on the commodity.

A comparison between Tables 14.1 and 14.2 reveals that the ranking of MECFs is equal to the ranking of the (Gini) income elasticities. This means that the

Table 14.2 Parameters for MECF calculations[a]

	X^b	MR	MECF	Tax rate
Kerosene	62.2	57.4	1.08	−0.48
Electricity	83.5	73.4	1.13	0.0
Gasoline	104.7	38.7	2.70	0.3

Source: Yitzhaki and Lewis (1996), Table 2, p. 553

[a]From a CGE model for Indonesia. See Yitzhaki and Lewis (1996)

[b]X, the tax base, is measured in billions of 1985 Indonesian Rupiah

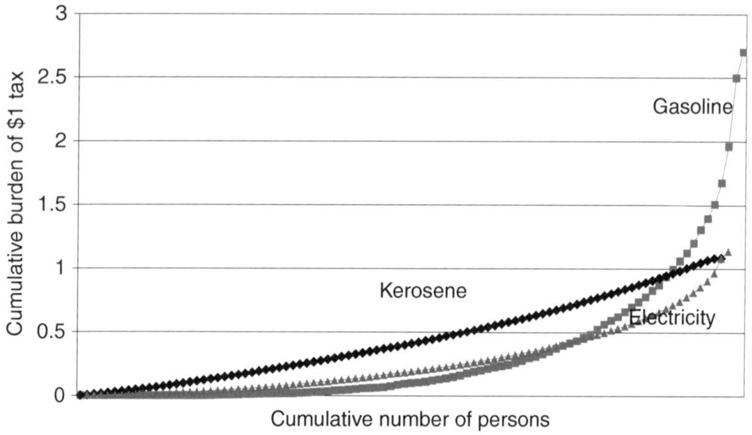

Fig. 14.3 Distributional burden of energy taxes. Source: Yitzhaki and Lewis (1996), Fig. 4, p. 554. Reprinted with permission by Oxford Journals

structure of taxation of those commodities is reasonable in the sense that the higher the share of the tax paid by the poor the lower is the marginal excess burden. Hence, one has to reach the conclusion that the Indonesian tax system does take into account distributional considerations. We are allowed to reach this conclusion because would the government care only about efficiency, then the optimal policy would be to change the tax rates so that all MECFs are equal.

Figure 14.3 presents the distributional burden of a marginal dollar of taxes raised from those commodities. The distributional burden is the concentration curve of the commodity with respect to economic well-being multiplied by the MECF of the commodity. Let us consider gasoline. The cost to society, at the margin, of raising a dollar of revenue from the gasoline tax is 2.7 dollars (the curve reaches 2.7). The burden on the poorest 50% of the population is twelve cents and the rest is borne by the upper half of the population. On the other hand, a dollar of revenue raised through kerosene costs the society only 1.08 dollars but the lower 50% of the population pay 40 cents. An efficiency-guided economist will

Table 14.3 Dalton-improving reforms

	Kerosene	Electricity	Gasoline
Reform D	−1.0	1.035	−0.035
Reform E	−1.0	3.26	−2.26

Source: Yitzhaki and Lewis (1996), Table 3, p. 555
Reprinted with permission by Oxford Journals

recommend reduction of the subsidy to kerosene and a decrease of the tax on gasoline, which would save the society more than a dollar and a half for each dollar of reduction in subsidy to kerosene. On the other hand, the economist who cares about distribution will point out that the burden of those taxes is not shared in a fair way. Most of the burden of kerosene is borne by the poor, so that a transfer of a dollar of taxation from gasoline to kerosene can be viewed as manna from heaven to the rich accompanied by a transfer from the poor to the rich. This is another way of saying that the Indonesian taxes seem reasonable. Although efficiency considerations call for reducing the gap in tax rates, distributional concerns point in the other direction. Note that all curves intersect, which means that it is impossible to find a DI neutral tax reform that involves only two commodities. On the other hand, intersection of the concentration curves point to the possibility of finding two SWFs, both conforming with Dalton criterion, so that one will justify raising the tax on one commodity and decreasing the tax on the other, while the second shows the opposite.

Economists are not equipped nor entitled to handle issues of fairness. In order to reach specific conclusions the social planner has to be more specific with regard to her social preference. All we can say is that the Indonesian taxation of those commodities takes into account distributional concerns.

Having failed to find a DI reform that is based on two commodities, we have to search for three-commodity reforms. In this case, one has only one free tax to determine. (One change in a tax on a commodity is used as a numeraire and its value is either 1 or −1. Another change in a tax rate is determined by the budget constraint). Because only one variable is free, and the DI efficient set of reforms is a cone (Mayshar & Yitzhaki, 1995, 1996) one can deduce that if a given set of DI reforms is not empty, it forms a closed section on an interval.

The results of applying the optimization algorithm are reported in Table 14.3. It is found that DI revenue-neutral reforms have the following structure: an increase of the tax on electricity and a decrease of the taxes (subsidies) on kerosene and gasoline. Table 14.3 presents the two extreme reforms, referred to as reforms D (distribution) and E (efficiency). Reform D can be described as follows: for each dollar of reduction in tax on kerosene it reduces the tax on gasoline by 3.5 cents and increases the tax on electricity by 1.035 dollars. The other extreme reform is reform E, which raises the tax on electricity by 3.26 dollars and reduces the tax on gasoline by 2.26 dollars. Any convex combination of those reforms is a DI revenue-neutral reform as well.

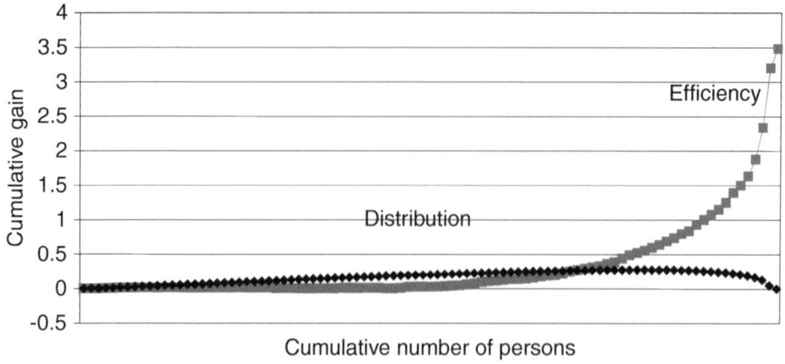

Fig. 14.4 Cumulative gain of two extreme DI reforms. Source: Yitzhaki and Lewis (1996), Fig. 5, p. 556. Reprinted with permission by Oxford Journals

Figure 14.4 presents the cumulative gain resulting from those reforms. As can be seen, reform D does not result in efficiency gain to the society,[8] but it results in a gain to low income groups. On the other hand, reform E results in an efficiency gain of 3.5 dollars for every dollar reduction of tax on kerosene, but the gain is limited to high income groups.

Having found a set of DI reforms allows us to consider the appropriate reform by adding other considerations. The nature and type of considerations can vary from one country to another. Let us state a few possible considerations.

(a) One may argue that DI reforms should not discriminate between regions or ethnic groups. The way to incorporate such a consideration is to impose a restriction that reforms will be DI for each region.
(b) The set of DI reforms was derived under the assumption that expenditure per capita represents the ranking of economic well-being. Most economists will probably agree that family size can affect economic well-being, but there may be a disagreement about its quantitative effect. The same argument can be raised with respect to a rural/urban distinction. Mayshar and Yitzhaki (1996) show how to incorporate these kinds of considerations.
(c) An additional consideration is popular support. To satisfy both the economist and the politician one can search for DI reforms that maximize the number of beneficiaries from the reform. The major problem in incorporating such a constraint is that unlike the search for DI reform that can be done with tabulated data, searching for majority requires the whole sample to be examined. Needless to say that it is a negligible constraint for modern computers.
(d) Another issue that is being ignored is how to incorporate externalities. It can be shown that externalities affect only the MECF. They do not affect the concentration curve. As such it is not covered in this book. The interested reader is referred to Lundin (2001).

[8] To see the amount of efficiency gain, add up the MECFs multiplied by the revenue changes: $1.0 \times 1.083 - 1.035 \times 1.083 + 0.035 \times 2.70 = 0.008$.

14.4.5 Sensitivity Analysis

Most of the analyses in this chapter were carried out without any parameterization. No assumptions were imposed on the curvature of the Engel curves of the commodities, and the assumptions concerning the SWF are pretty mild. The only parameters that are results of a heavy modeling effort are the MECFs. However, revenue estimates are routinely performed by almost all treasury departments and there is no way to avoid them when dealing with tax reforms. Because revenue estimates are based on intuition, complex modeling, and art, there is no point in performing statistical testing, especially if the number of parameters is large.

One way to investigate the robustness of our conclusions is to arbitrarily change the MECFs and check what happens to the set of DI reforms. Because the results are only sensitive to ratios of MECFs, a multiplicative bias in all estimates should not affect the conclusions. Also, it is clear that if the MECF of kerosene is reduced and those of electricity and gasoline are increased there will be no qualitative change in the conclusions.

The most suspicious MECF is that of gasoline (2.7). That is, a dollar of tax collected costs the society 2.7 dollars. Hence, we can arbitrarily reduce it to see if there is a qualitative change in our conclusions. Such calculations showed that small changes in this MECF did not change the results qualitatively and even a reduction of 40% to 1.62 produced a DI reform: Kerosene -1.0, gasoline -1.67, electricity $+2.66$ and the overall efficiency gain was 0.75 dollar.

The next step was to reduce and increase MECF of electricity by 10%. No qualitative changes occurred. An explanation for the robustness of the results can be found in Figs. 14.1 and 14.2. Figure 14.1 shows that if the MECFs are equal then there can be three types of DI reforms: (a) reduce the tax on kerosene and increase the tax on electricity, (b) reduce the tax on kerosene and increase it on gasoline, or (c) reduce the tax on electricity and increase it on gasoline. Hence, if the MECF of electricity is lower than the MECF of kerosene, the DI reform will consist of subsidizing kerosene and taxing electricity. Therefore the conclusion that one should subsidize kerosene will be unaffected even if the MECFs of electricity and gasoline are lowered to the level of the MECF of kerosene. On the other hand, it is worth reducing the tax on gasoline because of its high MECF. This result will not be affected if the MECF of gasoline continues to be higher than the MECF of electricity.

14.4.6 Non-neutral Reforms

In order to analyze non-neutral reforms one has to know the willingness to pay for the public goods produced by the government. Because some of the DI reforms resulted in efficiency gain, the question arises as to whether it is possible to split the gain between the public and the government, subject to the extreme and

Table 14.4 Non-neutral reforms

Reform number	Revenue	Kerosene	Electricity	Gasoline
D.5	0.50	−1.0	1.91	−0.41
E.5	0.50	−1.0	2.88	−1.38
R.70S	0.70	−1.0	2.38	−0.68
R.75F	0.75	−1.0	2.56	−0.81

Source: Yitzhaki and Lewis (1996), Table 4, p. 558
Reprinted with permission by Oxford Journals

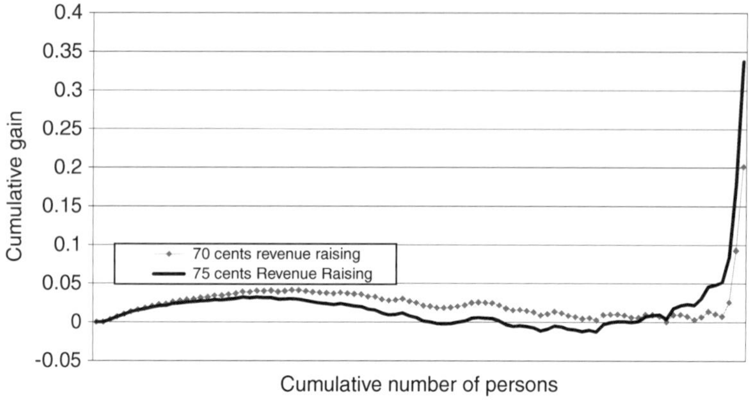

Fig. 14.5 Cumulative gain of revenue raising reforms. Source: Yitzhaki and Lewis (1996), Fig. 6, p. 559. Reprinted with permission by Oxford Journals

(un)realistic assumption that the government wastes its share. To search for such a reform one has to change the revenue constraint (14.10) so that $dR \geq c$, where c is a constant larger than zero.

Table 14.4 reports the results concerning four reforms. The first and second reforms D.5 and E.5 are the extreme DI reforms, subject to the constraint that the reform has to raise 50 cents on every dollar of subsidy to kerosene. Similar to the reforms reported in Table 14.3, reform D.5 increases the income of the poor, while reform E.5 is mainly concerned with efficiency gain. Comparison of Table 14.3 with Table 14.4 reveals that the set of DI reforms that raise revenue by 50 cents is encompassed by the set of DI reforms that are revenue neutral. The range of changes in electricity taxation has declined from [1.035, 3.26] to [1.91, 2.88], which may indicate that the distributional oriented reforms are more sensitive to revenue neutrality than efficiency concerned reforms.

Being able to raise revenue and at the same time to satisfy the Dalton criterion, one is tempted to ask how much more revenue can be extracted. The last two lines of Table 14.4 report the results of such an investigation. When the revenue required was raised to 70 cents, one could still find DI reforms. Reform R.70S reports the changes in taxes that are required. On the other hand, an attempt to raise 75 cents failed to find a DI reform. Reform R.75F is the best found (although not a DI reform). Figure 14.5 presents the cumulative gains to the public (ignoring additional revenue) of reforms R.70S and R.75F. As can be seen, reform R.70S continues to generate a cumulative gain of 35 cents, mainly for the top three deciles and some gain of 5 cents for the bottom two deciles and the same magnitude of loss to the middle class. Reform R.75F generates 75 cents but it fails to bring forth a DI reform. The second to the sixth deciles are hurt (the cumulative gain curve is declining), while the rich are still making 30 cents gain.

14.5 Summary

The DI reform is a tax reform that is based on the Dalton principle which says that a transfer from a rich person to a poor one is desirable, without having to state how much the government is ready to pay for the implementation of the principle. To place the DI reform in the context of welfare dominance, it is the application of second-degree welfare dominance in the field of tax reforms and optimal taxation. When applied to Indonesia, it is found that the structure of energy taxes in Indonesia is reasonable, but that the country could benefit from a further subsidy to kerosene accompanied by a tax on electricity and a reduction of the tax on gasoline. These conclusions are robust to changes in the parameters representing the Indonesian economy. By taking into account distributional concerns, it can be shown that the rules for optimal taxation that take into account only efficiency considerations may lead to wrong conclusions.

The application of the methodology is a bit cumbersome and requires data on the MECF of each tax. This may present a difficulty in implementation of the methodology. However, if one is concerned with distributional issues only, then one can use the GIE in order to find out the progressivity/regressivity of different taxes. The extended Gini can be used to check for the sensitivity of the estimates to different assumptions concerning the effect of the social evaluation of the marginal utility of income on the progressivity of taxes and transfers.

An important conclusion is that for the purpose of analyzing tax reforms there is no need to assume a functional form of the Engel curve because all that is needed is a weighted average of the marginal propensities to spend, weighted by the social evaluation of the marginal utility of income. To prevent contradictions between economic theory and statistical estimation it is worth to avoid using variability measures that are not compatible with welfare dominance.

It is worth pointing out that the literature offers some extensions: Duclos and Makdissi (2004) and Duclos, Makdissi, and Wodon (2005) apply the methodology to poverty-dominant reforms. Makdissi and Mussard (2008) apply the methodology to Yaari's-type SWFs.

Finally, identifying models that ignore distributional considerations is relatively easy. Any model that includes a representative individual is a model that ignores distributional concerns.

Chapter 15
Policy Analysis Using the Decomposition of the Gini by Non-marginal Analysis

Introduction

The objective of this chapter is to demonstrate the usefulness of several decompositions of the Gini (and the EG) in order to analyze the strengths and the weaknesses of various policies. We concentrate on distributional issues. The other component of the problem of tax reform—the estimation of the marginal cost of taxation—is identical to the description given in Chap. 14 hence it will not be repeated here.

There is one major advantage to using decompositions in the area of income distribution. Many of the variables that are used in this area are highly correlated with income. Therefore there is always the danger of double or even triple counting, which implies an exaggeration of the effect of a specific variable. For example, low income tends to be correlated with low level of education, low level of health, high level of unemployment, low quality of housing, etc. The advantage of using the decomposition of the variability of the income is the elimination of the possibility of double counting of the effects of the variables which compose it. The reason is because each part of the variability that occurs in the dependent variable (income) is accounted for and attributed to one and only one variable (or to the interaction between them, via the correlations). By increasing the effect of one variable, the effect of another variable has to decline. It is true that one can still attribute the effects of variables that do not participate in the decomposition to those that participate, but this deficiency is relevant to all methods of analysis. In this chapter we distinguish between three types of decompositions: (a) according to income sources, (b) according to population subgroups, and (c) decomposition over time. Decomposition according to sources can be divided into two groups: marginal and non-marginal. Technically the marginal decomposition is based on a decomposition according to the variables (one-stage decomposition), while the non-marginal decomposition is based on decompositions according to the variables and their ranks in the population (two-stage decomposition). We refer to the one-stage decomposition as a marginal one because it is useful for analyzing marginal

S. Yitzhaki and E. Schechtman, *The Gini Methodology: A Primer on a Statistical Methodology*, Springer Series in Statistics 272, DOI 10.1007/978-1-4614-4720-7_15,

changes only. If one wants to analyze structural changes then one has to refer to the non-marginal (two stage) decomposition, as will be done in this chapter.

We start this chapter with an application that requires non-marginal decomposition according to income sources. It analyzes the following problem: in most countries the government affects the income distribution directly by using both direct taxation and direct transfers (e.g., child benefits, allowances for the elderly, etc.). Usually the taxes and the transfers are run by different agencies. It is obvious that some individuals are getting transfers from one branch of the government while paying taxes to the other. We argue that if the government is giving money to one pocket of the individual while taking (some of) it from another pocket then this is a non-coordinated policy. We refer to such a policy as non-coordinated in order not to confuse it with an inefficient policy, a term that usually refers to excess burden. It is clear that a non-coordinated policy is also an inefficient one. It is also clear that some degree of non-coordination is reasonable whenever there are different criteria governing the applications of the policy. Because we do not have an objective way to define the "appropriate" level of non-coordination we compare the performances of two countries: Ireland and Israel. This way we avoid the need to define the optimal degree of non-coordination. Next we use the same data set for the decomposition by subpopulations, and lastly we illustrate the decomposition over time.

The structure of the chapter is as follows. Sect. 15.1 deals with the performances of Ireland and Israel via the decomposition by sources. Section 15.2 uses the same data set in order to perform decomposition according to population subgroups. Section 15.3 uses the same methodology as in Sect. 15.1, except that in this section the decomposition is done over time. The demonstrations do not exhaust all possibilities of analysis, but they are intended to demonstrate different types of possible uses of the decompositions to analyze and evaluate policies. Section 15.4 concludes.

15.1 Decomposition by Sources: Analyzing the Coordination Between Direct Benefits and Taxation[1]

Our interest in this section is to illustrate the application of the non-marginal decomposition of the Gini in order to evaluate the degree of coordination between the direct taxation and transfers of the government. Because there is no "natural" value that represents a good coordination, we compare the performances of two administrations: Ireland and Israel.

Disposable (net) income can be presented as the sum of random variables (non-governmental income, government transfers, and direct taxes). Therefore we can decompose the Gini of net income in a way that resembles the decomposition of the coefficient of variation, plus some additional terms which reflect the deviations of the

[1] This section is based on Carty, Roshal, and Yitzhaki (2009).

underlying distributions from being "exchangeable up to a linear transformation."[2] This enables us to estimate the Gini correlations between different types of income sources and to learn how those correlations affect transition from one income definition to another. In dealing with the transition from "before-tax" to "after-tax" income, most of the literature perform "before" and "after" comparisons. For example, Burkhauser, Frick, and Schwarze (1997) compare economic well-being and inequality between the USA and Germany using Theil and Gini indices of inequality. However they perform the comparison for before- and after-tax incomes without decomposing the inequality measure into the contribution of each of the sources. Wolff (1996) compares wealth inequality over time between eight industrialized countries. Wolff and Zacharias (2007a) analyze the changes in the inequality before and after the addition of fiscal components such as taxes and transfers. However they perform a different decomposition and concentrate on wealth rather than on income inequality. Aaberge et al. (2002) compare income inequality and income mobility between Scandinavian countries and the USA. Slemrod (1992) performs the decomposition in order to analyze the change over time in the redistributive effect of tax policy in the USA. Lazaridis (2000) analyzes households expenditures in Greece, Keeney (2000) performs it for farm incomes in Ireland, while Berri et al. (2010) apply the decomposition to the transportation sector in three European countries.

Several recent papers focus on empirical evidence on factor or subpopulation decompositions of the coefficient of variation for explaining trends over time in income inequality in one or several countries. Jenkins (1995) used the decomposition of the mean logarithmic deviation and the coefficient of variation across factors' components and population subgroups (household type, age of household head, etc.) in the UK. Jäntti (1997) conducted a cross-country study of factor decomposition of the coefficient of variation using the Luxembourg Income Studies (LIS) data for five countries—Canada, the Netherlands, Sweden, UK, and USA. Breen et al. (2008) extended the work of Jäntti to eight countries, using a double decomposition of the squared coefficient of variation, where the decomposition by age groups is nested within the source decomposition (referred to as factor decomposition). Nolan and Maitre (2000) compare the inequality in Ireland between 1987 and 1997 using income deciles and decompose it by income sources. Before we proceed to the decomposition itself, a brief description of the data is called for.

15.1.1 The Basic Data of Ireland and Israel

Both countries have exhibited a high rate of economic growth for about two decades, until the financial crisis in 2008. In Ireland this was due to low corporate taxation and to EU membership. The EU membership enlarged the external market and brought

[2] In a less technical language, exchangeability up to a linear transformation means that the joint distributions of the two variables are symmetric with respect to each other.

in EU aid which increased investments in education, increased participation in the labor market and introduced a policy of restraint in government spending. Israel on the other hand has benefited from large waves of Jewish immigration, many of whom are highly educated, from the introduction of market-oriented structural reforms and from the increase in foreign investment due to the peace process which started in the 1990s.

The total GDP (PPP, by the World Bank) for both countries is very close—171,862 billion dollars for Ireland (ranked 51) and 169,847 for Israel (ranked 52). However, Ireland has one of the highest GDP per capita (nearly $43,000 PPP) and is ranked 9 by the IMF (2006), while Israel is ranked 33 in the GDP per capita list (IMF, 2006), with a GDP per capita that is almost half that of Ireland's (nearly $26,000 PPP). Ireland has experienced unprecedented rapid economic growth (of nearly 142% in 16 years[3]), which started in the 1990s, during which it was transformed from a mainly agrarian and manufacturing-based country to one of the wealthiest in Europe, on the basis of hi-tech and international trade.

The Israeli GDP has grown by 100% from 1990 to 2006. Like Ireland's, Israel's economy has moved from being mainly agrarian and traditional manufacturing-based into being based mostly on hi-tech, pharmaceutical, and chemical industries, with a high percentage of the GDP being spent on research and development in these areas.

The demographic structure of the two countries is barely comparable. Table 15.1 presents the basic comparison of household composition between Ireland and Israel. The average household size in Israel (3.3) is larger than in Ireland (2.8), indicating that there are many more persons living in large households in Israel: nearly 45% of the persons in Israel live in households consisting of five persons and more, compared to only 30% in Ireland. Large households (of six persons or more) comprise only 5% of the households in Ireland but 12% in Israel.[4]

On the other hand, nearly half of the households in Ireland comprise one or two persons (50% of households and 27% of persons), whereas in Israel such households make up 42% of the total (20% of persons).

Although in 2006 the percentage of elderly (aged 65+) in both countries was similar (11% in Ireland and 9.6% in Israel), the percentage of children aged less than 15 was much lower in Ireland—20.4% in comparison to 28.3% in Israel.[5]

This basic comparison suggests that one of the major differences in the demographic structure of the two countries is that the Israeli household includes more children, which implies lower percentage of working-age persons (Table 15.2). Households with children comprise 40.2% of Ireland's households (58.2% persons),

---

[3] 117% from 1995 to 2006, adding it up to 25% from 1990 to 1995, according to Tables 4 at http://www.cso.ie/releasespublications/documents/economy/2006/nie2006tables1995-2006excel.xls and http://www.cso.ie/releasespublications/documents/economy/HistoricalNIETables1970-1995excludingFISIM1.xls.

[4] Source: QNHS (Ireland) and HES (Israel).

[5] Source: Statistical Yearbook of each country.

Table 15.1 Distribution of households and persons living in households by household size[a]

	Ireland		Israel	
Sample size (households)	5,836		6,259	
Sample size (persons)	14,634		20,743	
Households in population (thousands)	1,494		2,027	
Persons in population (thousands)	4,253		6,733	
Children (under 15)	20.4%		28.3%	
Children (under 18)	26.6%		33.0%	
Elderly (65 and above)	11.0%		9.6%	
Average household size	2.8		3.3	
Average size of large households (6+)	6.5		7.0	
Maximum household size	12		25	
Population distribution by household size	% of households	% of persons	% of households	% of persons
1	21.9%	7.7%	18.4%	5.5%
2	27.9%	19.6%	23.4%	14.1%
3	17.2%	18.1%	16.2%	14.6%
4	17.7%	24.8%	16.9%	20.4%
5	10.3%	18.1%	12.8%	19.3%
6+	5.1%	11.8%	12.3%	26.1%
	100.0%	100.0%	100.0%	100.0%

Source: Carty, Roshal, and Yitzhaki (2009)

[a]Figures in the table are calculated using the data from the QNHS (Ireland) and HES (Israel) and may differ slightly from the demographic figures published in the Statistical Yearbooks

Table 15.2 Distribution of households with and without children (aged <18)

	Ireland		Israel	
	% of households	% of persons	% of households	% of persons
1 adult, no children	21.9%	7.7%	18.4%	5.5%
2 adults, no children	25.0%	17.5%	22.0%	13.2%
3+ adults, no children	12.9%	16.6%	13.4%	14.5%
Adults only	59.8%	41.8%	53.8%	33.3%
1 adult with children	6.2%	6.2%	3.0%	2.6%
2 adults with 1–3 children	21.9%	29.3%	24.4%	29.1%
2 adults with 4+ children	1.5%	3.4%	6.1%	12.5%
Others with children	10.5%	19.3%	12.8%	22.5%
Households with children	40.2%	58.2%	46.2%	66.7%
Total	100.0%	100.0%	100.0%	100.0%

Source: Carty, Roshal, and Yitzhaki (2009)

and 46% of Israel's (66.7% persons). As a consequence, we should expect that inequality comparisons between the two countries will be sensitive to the scale of adult-equivalent income used (Coulter, Cowell, & Jenkins, 1992; Cowell, 1984).

The decomposition of the Gini coefficient according to the contributions of each household's size to the overall inequality enables us to assess the inequality in each group and to differentiate it from the contribution of different distributions of household's size.

15.1.2 The Impact of Equivalence Scales

The determinants of economic well-being include income and household size. Returns to scale in consumption and different needs of the household members suggest the use of an adult-equivalent scale that enables the comparison of the economic well-being of households with different needs and structures.

In order to calculate equalized income, one divides the household income by the number of equivalent adults (De Vos & Zaidi, 1997; Friedman, 1951; OECD, 2006). The households are then weighted by the number of equivalent adults.[6] Thus, if we compare two households with the same income, the larger one has the *lower* equalized income, but *higher* weight in the population, because it represents a larger number of persons. Higher equivalence scale *lowers* large households' income per equivalent adult while attributing a larger weight in the population to large households.

Ireland's national equivalence scale attributes a weight of one to the household head (first adult), 0.66 to each subsequent adult (aged 14+), and 0.33 to each child aged less than 14. The equalized household size is the sum of these weights.

Israel's national equivalence scale does not make a distinction between adults or children, and considers only the total number of persons in the family. Thus, in Israel, given the same household size, the equalized size of households with or without children is the same, whereas in Ireland the higher weight is attached to the household containing a higher proportion of adults. Hence, the Israel's scale gives much more weight than the Ireland's one to families with children. Accordingly, considering that adults are likely to be those who bring money to the family, higher-income households are more likely to be attributed a higher weight by the Irish scale.

Table 15.3 compares the equivalence scales of Ireland and Israel and a scale calculated as the square root of the household size (used by Nolan & Smeeding, 2005). In order to compare the Ireland's and Israel's scales, in Table 15.4 we normalize Israel's scale to 1 by dividing it by 1.25. In most cases, especially for large households, this scale is the lowest of the three. The Ireland's scale is the highest, especially when considering households without children.

---

[6] Some researchers tend to use equalized income but to weight unequalized persons. This causes internal contradiction leading to ambiguous results (see Ebert 1997, 1999). Throughout this chapter we weight households by their number of equalized persons only.

Table 15.3 Equivalence scales of Ireland and Israel

Household composition	Persons in household	# of adults	# of children	Normalized Israeli scale[a]	CSO—Ireland	Square root of household size
1 Adult	1	1	0	1.00	1.00	1.00
1 Adult + 1 child	2	1	1	1.60	1.33	1.41
2 Adults	2	2	0	1.60	1.66	1.41
1 Adult + 2 children	3	1	2	2.12	1.66	1.73
2 Adults + 1 child	3	2	1	2.12	1.99	1.73
3 Adults	3	3	0	2.12	2.32	1.73
1 Adult + 3 children	4	1	3	2.56	1.99	2.00
2 Adults + 2 children	4	2	2	2.56	2.32	2.00
4 Adults	4	4	0	2.56	2.98	2.00
2 Adults + 3 children	5	2	3	3.00	2.65	2.24
3 Adults + 2 children	5	3	2	3.00	2.98	2.24
5 Adults	5	5	0	3.00	3.64	2.24
2 Adults + 4 children	6	2	4	3.40	2.98	2.45
6 Adults	6	6	0	3.40	4.30	2.45
2 Adults + 5 children	7	2	5	3.80	3.31	2.65
2 Adults + 6 children	8	2	6	4.16	3.64	2.83

Source: Carty, Roshal, and Yitzhaki (2009)
[a]Original Israeli scale divided by 1.25

Table 15.4 Gini coefficient, using different equivalence scales

Income	Ireland	Israel	Difference
Economic, nongovernmental			
Total household income	0.5643	0.5324	−5.65%
Equalized by Ireland's scale	0.5092	0.5117	0.49%
Equalized by Israel's scale	0.5166	0.5145	−0.41%
Equalized by square root	0.5218	0.5105	−2.18%
Net disposable income			
Total household income	0.3962	0.3934	−0.71%
Equalized by Ireland's scale	0.3256[a]	0.3752	15.23%
Equalized by Israel's scale	0.3339	0.3796[b]	13.69%
Equalized by square root	0.3391	0.3704	9.21%
Difference between own-scale Gini for equalized net income			16.58%

Source: Carty, Roshal, and Yitzhaki (2009)
[a]Officially published inequality index for Ireland is 0.324. This difference is due to a slightly different calculation procedure
[b]Here should stand the officially published inequality index for Israel, which is 0.390. However, the official Gini is calculated using the different (Income) survey instead of HES. We use HES throughout the chapter, as it provides more household characteristic indicators

The question we would like to address now is—what is the estimated impact of the use of different equivalence scales on the Gini coefficient? Table 15.4 presents the various Gini coefficients calculated by the three definitions of equivalence scales for Ireland and Israel. When the whole household's income is taken into account, there is higher inequality in economic incomes in Ireland than in Israel. Inequalities between households in the two countries are similar for net disposable incomes. However, the move to equalized income yields a striking change. The differences between the two countries in inequality for net incomes range from 9.2% to 15.2%. The impact of the equivalence scales on the two countries is completely different; whereas in Ireland the Ireland's scale produces the lowest inequality and the square root produces the highest, in Israel the order is reversed.

For the purpose of comparability, in the rest of this section only Ireland's national scale is used.

15.1.3 Decomposition of the Gini According to Income Sources

Almost all countries use two types of instruments to improve the income distribution: the income tax handles the distribution at the upper end of the distribution, while demo-grants and other allowances are intended to deal with horizontal equity and the lower end of the distribution. In most countries those two branches of the government are under different governmental ministries. A key question in determining the effectiveness of the government in handling the redistribution in a coordinated way is whether those two branches of the government are coordinated or not. By coordination it is meant that if one branch of the government is on the giving side to the family, the other branch of the government should not be on the taking side. Such coordination is not needed in the level of abstractions of Mirrlees (1971) and in almost all optimal taxation models because in those models there is no cost attached to transferring resources between the individual and the government. However in a more realistic model we should expect transaction costs and also that the Marginal Efficiency of the Cost of Funds (MECF) of government transfers are higher than the MECF of tax reductions because tax reductions are based on giving up on collecting funds from the individual, while transfers (sometimes referred to as benefits) use funds that were already collected (Slemrod & Yitzhaki, 2002). Reducing such back and forth transfers saves administrative cost and reduces the excess burden of taxation.

Usually, the rules for providing benefits by the government can be different than the rules of taxation. The reason may be that having one rule that governs all government branches is too complicated to administer. Also, it is not clear from a normative point of view whether those rules should be identical or not. These two issues are beyond the scope of this book. The important point for us is to agree that coordination between allowances and taxation should be maximized, but it is also reasonable to expect some degree of non-coordination. Both countries publish the Gini coefficient of net (i.e., after tax and benefits) income as the measure of inequality. For our purpose, which is presenting the methodology, we ignore other technical differences that affect the measurement of inequality such as differences in the accounting period. In addition, the simple comparison does not provide details regarding the factors that affect the final result, namely—the choices of the income definition and weighting schemes. Technical issues such as the period over which the income is measured are also important to be taken into account.[7]

To compare the governmental policies, the Gini coefficient of after-tax income is decomposed into the contributions of three different sources:

$$\text{Net income} = \text{economic income} - \text{taxes} + \text{allowances}. \tag{15.1}$$

The data source for Ireland is the EU-SILC, the voluntary Survey on Income and Living Conditions of different types of households. All persons aged 16 and above are required to participate. Information is collected continuously throughout the year, with up to 130 households surveyed each week to give a total sample of 5,000–6,000 households in each year. In 2006 the achieved sample size was 5,836 households and 14,634 individuals. The income reference period for EU-SILC is the 12 months prior to date of interview.

The data source for Israel is the Household Expenditure Survey (HES). The survey aims to obtain data on the components of households budgets, as well as additional data that characterize various aspects of the living standards of the households. Information is obtained throughout the year on the 3-month income of the household members aged 15 and above. In 2006 the sample size of the HES was 6,259 households and 20,743 individuals.

15.1.4 *Estimates of the Gini Coefficient*

In 2006 the Gini coefficients that were officially published by national institutions (based on the total equalized disposable income) were 0.324[8] for Ireland and 0.390[9] for Israel, nearly 20% higher. These coefficients were calculated using equalized disposable income and based upon national equivalence scales. As mentioned above different equivalence scales result in different measures of inequality and poverty. Disposable income includes nongovernmental income (from work, business, etc.) plus government transfers minus taxes (shares of each component are indicated in Table 15.5). Therefore the tax and benefits schemes of each country might affect the inequality even further. Table 15.5 demonstrates the difference

[7] The Irish accounting period is a year, while the Israeli one is 3 months. Finkel, Artsev, and Yitzhaki (2006) find that the Gini coefficient calculated from a 3-month accounting period was by nearly 4% higher than the index based on a 12-month period. Other estimates are provided by Creedy (1979, 1991), Burkhauser and Poupore (1997), and Gibson, Huang, and Rozelle (2001).

[8] Ireland, Central Statistics Office, EU Survey on Income and Living Conditions (EU-SILC 2006), Table 1. (http://www.cso.ie/releasespublications/documents/eu_silc/current/eusilc.pdf).

[9] Israel, Central Bureau of Statistics, Press Release 13 August 2007. (http://www.cbs.gov.il/hodaot2007n/15_07_150e.pdf). As mentioned above, this figure is based on the Income Survey, which we do not use here. The same calculation based on the Household Expenditure Survey would yield the Gini coefficient = 0.380.

Table 15.5 Gini coefficient of inequality equalized by Ireland's scale[a]

	Ireland			Israel		
	Share (out of net income)	Gini	Change	Share (out of net income)	Gini	Change
Nongovernmental (economic) income	107%	0.5092		107%	0.5117	
Economic income − taxes	−26%	0.4827	−5%	−23%	0.4760	−7%
Gross (economic + government transfers)	20%	0.3782	−26%	16%	0.4224	−17%
Net (economic + transfers − taxes)	100%	0.3256	−36%	100%	0.3752	−27%

Source: Carty, Roshal, and Yitzhaki (2009)
[a]In Ireland the accounting period is 12 months, in Israel—3 months. As noted earlier, this adds nearly 4% to the Gini in Israel

between Ireland and Israel in the Gini coefficients, calculated by using the same (Ireland's) equivalence scale, by economic (or nongovernmental) income, economic income excluding transfers and excluding taxes; gross income defined as economic income plus government transfers; and finally, the disposable (net) income.

The table shows that when measured by nongovernmental, before-tax income only and equalized by the same scale, both countries have similar levels of inequality. The Irish tax and benefit system collects higher taxes and distributes more benefits than the Israeli tax and benefit system, while the net collections to the governments' coffers are equal. Although it collects less than the Irish one, the tax system in Israel reduces the inequality slightly more than in Ireland, but government benefit programs are much more effective in reducing inequality in Ireland.

Overall, the difference between the two countries in the Gini coefficient when calculated using the same equivalence scale is less dramatic than the original: 13% compared to 16.5% using national scales.

Figure 15.1 shows the distribution of households by equalized nongovernmental incomes for both countries. Mean income of each country is set to 1 and all other incomes are in reference to the mean. The medians for both countries are around 70% of the mean. However, there is a significant proportion of households with zero income, very high—nearly 23% in Ireland and 16% in Israel. In Israel, on the other hand, there are more households with lower than average incomes than in Ireland, and also more households with very high incomes, twice the country's average.

Figure 15.2, presenting the disposable income distributions, shows how governmental benefits and taxes manage to "correct" the initial distributions. The median of each distribution has moved to be located slightly over 80% of the mean, and in Ireland most of the households are located around it. In Israel there is a large portion of households with an income lower than 40% of the mean income. The gaps in the two distributions lie in the low income regions.

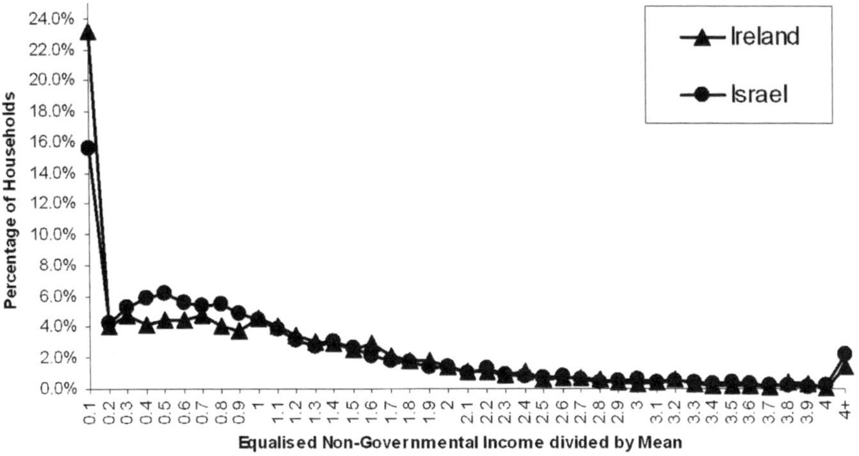

Fig. 15.1 The distribution of nongovernmental (equalized) income between households, Ireland and Israel, 2006. (Mean income of each country = 1. Ireland's equivalence scale is used for both countries)

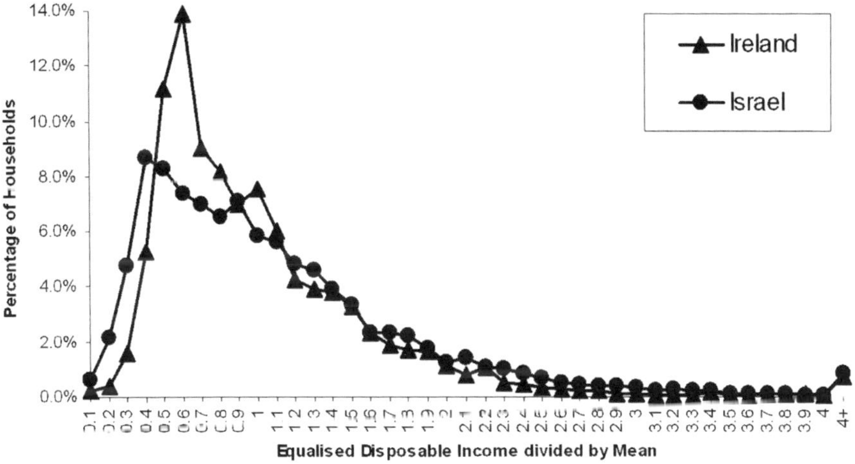

Fig. 15.2 The distribution of disposable (equalized) income between households, Ireland and Israel, 2006. (Mean income of each country = 1. Ireland's equivalence scale is used for both countries). Source: Carty, Roshal, and Yitzhaki (2009)

Comparing the tax-welfare regimes of the two countries, it can be noted that in Ireland the extent of the income tax is higher than in Israel (20% of the net income in Ireland as opposed to 16% in Israel), but the amount of welfare transfers is higher as well (26% of the net income in Ireland and 23% in Israel). In relation to the Gross Domestic Product, social welfare expenditure sums up to 7.8% of the GDP in Ireland and 7% in Israel.

15.1.5 A Full Decomposition of Gini by Income Sources: Empirical Findings

Tables 15.6 and 15.7 present the components of the decomposition for Ireland and Israel, as described in Sect. 15.1.4. Net disposable income is presented as the sum of three components: economic nongovernmental income, government transfers, and taxes (mostly negative values).

The first line presents the Gini for each one of the variables. The Gini calculated on the net disposable income is 0.3256 for Ireland and 0.3752 for Israel, the lowest of the components. The second line presents the jackknife standard errors.

The third line presents the share of each of the components in the disposable income. The share of economic income is larger than 1, as it is usually higher than the disposable income. The share of taxes, however, is negative, balancing the difference. Government transfers comprise nearly 20% of the net income in Ireland and only 16% in Israel. The fourth line is the proportion of the population which receives any (even negative) part of the given component. For example, in Israel, almost 100% of the population pay taxes (or receive transfers), compared to only 78% of the population in Ireland. Government transfers reach almost the same proportion of the population with Ireland's figure being slightly higher, and non-governmental income is received by a higher proportion of the population in Israel (90%) than in Ireland (87%).

Table 15.6 The components of Gini of equalized disposable (net) income: economic income, government transfers, and taxes—Ireland

	Gini indices of inequality (Gi) and income shares (ai)				
	Economic	Transfers	Taxes	Sum = net income	
Gini coefficient	0.5092	0.5412	0.6404	0.3256	
St. error	0.0085	0.0065	0.0083	0.0096	
Income share	1.066	0.196	−0.262	1.000	
Proportion of nonzero	0.872	0.853	0.782	1.000	
Gini of nonzero	0.4372	0.4625	0.5388	0.3255	
	Gini correlations matrix (Γij)				
i/j	Economic	Transfers	Taxes	Sum = net income	
Economic		−0.6368	−0.9070	0.9489	
Transfers	−0.7161		0.6998	−0.4221	
Taxes	−0.9427	0.6385		−0.8632	
Sum = net income	0.9484	−0.4402	−0.7814		
	Jackknife standard errors for the difference between Γij and Γji				
i/j	Economic	Transfers	Taxes	Sum = net income	
Economic		0.0164	0.0054	0.0021	
Transfers			0.0216	0.0212	
Taxes				0.0148	
	G_O	G_O^2	$G_O \Sigma a_i D_{iO} G_i$	$a_i^2 G_i^2$	$\Sigma\Sigma a_i a_j G_i G_j \Gamma_{ij}$
Net	0.3256	0.1060	−0.0037	0.3340	−0.2224

Source: Carty, Roshal, and Yitzhaki (2009)
Ireland, households weighted by household weight × equivalent adult: 2006

Table 15.7 The components of Gini of equalized disposable (net) income: economic income, government transfers, taxes—Israel

	Gini indices of inequality (G_i) and income shares (a_i)				
	Economic	Transfers	Taxes	Sum = net income	
Gini index	0.5117	0.6214	0.6760	0.3752	
St. error	0.0057	0.0053	0.0053	0.0052	
Income share	1.068	0.158	−0.226	1.000	
Proportion of nonzero	0.898	0.836	1.000	1.000	
Gini of nonzero	0.4560	0.5472	0.6759	0.3752	
	Gini correlations matrix (Γ_{ij})				
i/j	Economic	Transfers	Taxes	Sum = net income	
Economic		−0.4410	−0.9323	0.9528	
Transfers	−0.4569		0.4209	−0.0655	
Taxes	−0.9651	0.4821		−0.9095	
Sum = net income	0.9435	−0.2181	−0.8404		
	Jackknife standard errors for the difference between Γij and Γji				
i/j	Economic	Transfers	Taxes	Sum = net income	
Economic		0.0146	0.0044	0.0020	
Transfers			0.0205	0.0166	
Taxes				0.0075	
	G_O	G_O^2	$G_O \Sigma a_i D_{iO} G_i$	$a_i^2 G_i^2$	$\Sigma\Sigma a_i a_j G_i G_j \Gamma_{ij}$
Net income	0.3752	0.1408	0.0036	0.3316	−0.1930

Source: Carty, Roshal, and Yitzhaki (2009)
Israel, households weighted by household weight × equivalent adult: 2006

The fifth line is the Gini coefficient calculated over those who have the given component in their income.

The second part of each table presents the Gini correlation coefficients between each pair of components, followed by the statistical tests for the differences between each pair of coefficients. The coefficients have the expected signs—for example, the higher the economic income is, the higher are the governmental transfers, but here the progressivity of Ireland's governmental transfers is noticeable—in Ireland the correlations are much stronger (in absolute values) than in Israel, indicating that the transfers are given to a really poorer population, and taxes are paid by the richer ones.

Another result derived from this part is that there are differences between each two correlation coefficients, meaning that the joint distributions are not exchangeable. This implies that the marginal distributions of the net income components change in a nonspecific way, which is, again, not surprising given the progressive type of taxation.

However, in Ireland the differences between the correlation coefficients of the net income with economic income and government transfers are not significant. In Israel the insignificant difference is only between economic income and government transfers.

The third part of the tables presents the decomposition of the Gini according to (4.10). The term that distinguishes the structure of the decomposition of Gini from the one in the decomposition of the coefficient of variation, namely $G_O \sum_{t=1}^{T} a_t D_{t0} G_t$,

Table 15.8 Pearson and Spearman correlation coefficients, net income, Ireland (Pearson in the lower left triangle, Spearman in the upper right triangle)

i/j	Economic	Transfers	Taxes	Sum = net income
Economic	1	−0.7180	−0.8834	0.8906
Transfers	−0.3890	1	0.6403	−0.4301
Taxes	−0.7960	0.3395	1	−0.7590
Sum = net income	0.9445	−0.1698	−0.5947	1

Table 15.9 Pearson and Spearman correlation coefficients, net income, Israel (Pearson in the lower left triangle, Spearman in the upper right triangle)

i/j	Economic	Transfers	Taxes	Sum = net income
Economic	1	−0.5077	−0.9165	0.9296
Transfers	−0.2356	1	0.4527	−0.2423
Taxes	−0.8777	0.1883	1	−0.8374
Sum = net income	0.9477	−0.0072	−0.7265	1

Source: Carty, Roshal, and Yitzhaki (2009)

subtracts 0.0037 of the (square of the) Gini in Ireland and adds 0.0036 to the (square of the) Gini in Israel. However the magnitude of this element does not seem to be important. For any reasonable purpose of comparison, we can simply ignore this component.

Schechtman and Yitzhaki (1999) have pointed out that the properties of the Gini correlation are a mixture of the properties of Spearman's and Pearson's correlation coefficients. Tables 15.8 and 15.9 present the Spearman and Pearson correlation coefficients. As can be seen, the Gini correlations are almost identical in magnitude to the appropriate Spearman correlation coefficients, but may sharply differ from the Pearson correlation coefficients. For example, while in Ireland the Spearman correlation coefficient between net income and government transfers is around −0.4, Pearson coefficient is −0.17. The difference between Spearman and Pearson coefficients reaches 0.33 in Ireland and 0.26 in Israel (between nongovernmental income and government transfers, in both countries). The reason is that while Pearson's correlation coefficient measures a linear relationship, Spearman's coefficient measures a general monotonic trend, not necessarily linear.

In conclusion, there are two major differences between Ireland and Israel: Ireland collects more tax revenue and also redistributes more than Israel, and the Gini correlation between (minus) taxes and transfers is higher in Ireland than in Israel. Comparisons of the Gini correlations reveal that on average the correlation in Ireland is higher by roughly 17%. Comparison of the contribution of the correlation term in reducing the square of the Gini as reported in Tables 15.6 and 15.7 reveals a 0.03 difference between the two countries (−0.22 vs. −0.19), which translates into a higher percentage points in the reduction of the Gini inequality coefficient. Hence, we can conclude that the Gini correlation between taxes and welfare is the major contributor to the better effectiveness of the Irish system.

We view this correlation as a quantitative measure of the *tax-welfare churn* (Saunders, 2005; Whiteford, 2006, and the references therein). To see this note the following: In a Mirrlees's optimal income tax framework (1971), taxpayers both pay taxes and receive transfers from the government. In a well-coordinated system we should expect the tax to increase and the transfers to decline with income. In this ideal tax system the correlation between the (minus) taxes and transfers should be equal to one. The Gini correlation describes the effect on the target function (reducing inequality) and is well-suited to measure correlation when the relationship is nonlinear. Hence, by decomposing the Gini we are also able to offer an empirical measure of the effectiveness of the tax-welfare churn.[10]

15.2 Decomposition by Population Subgroups

The decomposition of the inequality measure with respect to the contributions of different subgroups leads to slightly different questions—the effects of the demographic structures on governmental policies.

In this section we compare the demographic structures of inequality and the income compositions between two countries: Ireland and Israel. The choice of these two countries is not coincidental: our aim was to compare countries which are as close in their demographic and economic parameters as we can find. Of all the European countries, Ireland is the most similar to Israel by the combination of the following factors: size, economic development in the recent two decades, and demographics that is highly influenced by religion. In comparison to the OECD countries, Ireland has a rather high overall level of inequality, though below Israel (and the USA). There is also high percentage of large families, which is the distinguishing factor of the Israeli society.

15.2.1 Background

By choosing an inequality measure one is actually choosing a social welfare function to represent the society.[11] The official statistics of both countries include reporting the Gini coefficient of after-tax income. Because countries could have chosen to publish alternative measures of inequality and they did not do so, we assume that the revealed preference and consequently the data available for public debates concerning policy evaluations should be measured by the effect of the policy on the Gini coefficient. By decomposing the Gini coefficient to the contributions of taxes and welfare we are able to differentiate between the effect of taxes and the effect of allowances. The correlation between those effects enables

[10] We are indebted to Joel Slemrod for pointing out this issue.

[11] See Chap. 13.

us to find out the level of coordination between separate organs of the government. This issue is referred to in public debates as the *tax-welfare churn* (Saunders, 2005; Whiteford, 2006, and the references therein). In our investigation, it turned out that the major factor affecting the performance of governmental policy is the tax-welfare churn. We note that there may be other factors that we have ignored. However, although we do not deal with mobility here, the technique presented can be formulated in terms of mobility as well (see, e.g., Yitzhaki & Wodon, 2004; Beenstock, 2004; Beenstock & Felsenstein, 2007).

In addition, our aim is to evaluate the effect of the demographic structure on inequality. The demographic structure is given and even identical policies can result in different performances if the demographic structures are different. Therefore in evaluating policy performance one has to take into account the demographic structure. The methodology we use is the decomposition of the Gini coefficient by population subgroups. Beblo and Knaus (2001) apply the decomposition of the Theil index of inequality to evaluate the country's contribution to the overall inequality in the 11 countries of the European Monetary Union. The decomposition of the Gini coefficient (based on demographic subgroups) allows to evaluate the contribution of each subgroup to the inequality in the entire population and thus to determine the extent to which the demographic structure of each country affects its Gini coefficient. Achdut (1996) performed similar analysis for Israel for the years 1979 and 1993. However the method she used does not allow one to find the coordination between tax and welfare systems, or the effect of the demographic structure, which are the major issues in this chapter.

Differences in the demographic structure are important factors in any analysis because it is well-known that income levels correlate with certain demographic characteristics such as age, education level, marital status, family size, the fraction of rural population, the number of immigrants and their levels of education, and so on. The composition of demographic groups differs in every country due to different historical, cultural, and religious characteristics. Can these demographic differences then explain differences in the inequality as measured by the Gini coefficient?

It may happen, for example, that in a certain demographic subgroup the inequality is consistently higher than in the rest of the population. In the group of single people, for example, one finds students at the very beginning of their professional life, people who are at the top of their career, as well as divorcees and widowers. If a country exhibits a higher ratio of such a group, could this contribute to a higher inequality measure? Or consider a situation where there might be a high inequality *between* groups. For example, if there are special government benefits for large families, who therefore enjoy higher incomes, these benefits can increase the income gap between these families and the others.

15.2.2 Empirical Findings

Following the methodology presented in Chap. 4, we now move to the decomposition of the inequality in the two countries by population subgroups.

First we divided the population into subgroups based on household size, where households of six persons or more are grouped together. Table 15.10 summarizes the results of the decomposition by household size for Ireland and Israel, using nongovernmental equalized income.

Column (II) presents the number of observations in each group, column (III) is the proportion of the specific category in the whole country's population of equivalent adults, column (IV) is the mean income of the group (in national currency), column (V) is the same mean, but calculated in relation to the mean income of a two-persons household. This is the mean income that can be compared between the two countries.

Column (VI) requires more explanation. It presents the mean rank of the incomes of the group members in the overall income. Each individual is given a rank in the income distribution, from 0 to 1, so that the average rank of the entire population within itself is always 0.5. Theoretically, the average rank of the group of relatively poor people would be lower than 0.5 because their incomes lie at the left end of the range of the entire income distribution. Therefore, if for any group the mean rank of its members is lower than 0.5 we conclude that they are relatively poor.

Column (VII) is the group's income share in the total income, the total income being set to 1. The group-specific Gini is listed in column (VIII), with the overall Gini indicated in the last line, and the overlapping index with the whole population is presented in column (IX) (see Chap. 4 for the definition of the overlapping coefficient).

The contribution of each group to the overall Gini is the product of columns (VII), (VIII), and (IX) and it is listed in column (X).

Finally, the within- and between-groups Ginis are calculated and presented below the table, as well as the overall Gini. The within-group Gini is the sum of column (X), the between-groups Gini, Pyatt's between-groups Gini, and the overall Gini are calculated separately.

As expected when using only direct, nongovernmental incomes the inequality is very high (nearly 0.51), and this is true for the two countries. There are, however, subpopulations that are more equal than others.

Looking at column (VIII) in both countries, the lowest inequality is among medium-sized (3–4) households. Their mean incomes are the highest. Singles is the least equal group (i.e., they have the highest Gini coefficient). They also possess the lowest (or second-to-the-lowest) income in terms of mean (column (IV)) and distributional rank (column (VI)). When comparing between the countries, in Ireland larger households are more equal than in Israel. Also, in Israel large families are, on average, much poorer than in Ireland: their mean income (per equalized person) is half of the income in two-persons households and their rank is also the lowest. This is not surprising, taking into account that these large households are actually much larger in Israel, implying that there are fewer earners per household. In Ireland the poorest group is the group of singles.

Age differences also play a significant role in explaining inequality. Households of singles, for example, may exhibit higher inequality if there is a high percentage of younger persons (students or people at the beginning of their careers) and pensioners, i.e., concentration at the ends of the age distribution. Table 15.11 shows the distribution of the households of singles by age. It can be seen that in Israel nearly 14% of those living alone are under the age of 30, compared to only

Table 15.10 Decomposition of Gini by household size, nongovernmental income

HH size (I)	Obs (II)	Pi (III)	Mean income (IV)	Relative mean (V)	F0i: mean rank (VI)	Si: income share (VII)	Gi: Gini of the group (VIII)	Oi: overlap with the whole population (IX)	S*G*O (X)
(A) Ireland—nongovernmental income									
1 Person	1,816	0.11	16,415	0.71	0.35	0.08	0.717	1.27	0.070
2 Persons	1,695	0.23	23,259	1.00	0.46	0.23	0.615	1.12	0.157
3 Persons	867	0.18	25,299	1.09	0.55	0.20	0.453	0.95	0.085
4 Persons	786	0.23	26,079	1.12	0.58	0.26	0.397	0.86	0.090
5 Persons	450	0.16	22,640	0.97	0.54	0.16	0.391	0.85	0.053
6+	222	0.10	18,278	0.79	0.43	0.08	0.507	0.92	0.036
Within-group Gini							0.491	(96%)	
Between-groups Gini							0.018	(4%)	
(G_B Pyatt)							(0.073)		
Total							0.509		

(B) Israel—nongovernmental income									
1 Person	1,112	0.09	4,537	0.75	0.42	0.08	0.632	1.17	0.056
2 Persons	1,545	0.17	6,047	1.00	0.51	0.20	0.564	1.05	0.120
3 Persons	1,012	0.16	6,244	1.03	0.58	0.19	0.447	0.90	0.076
4 Persons	1,045	0.19	6,049	1.00	0.57	0.23	0.437	0.90	0.089
5 Persons	796	0.17	5,523	0.91	0.55	0.18	0.436	0.90	0.072
6+	749	0.21	3,037	0.50	0.36	0.12	0.526	0.95	0.062
Within-group Gini							0.474	(93%)	
Between-groups Gini							0.038	(7%)	
(G_B Pyatt)							(0.118)		
Total							0.512		

Source: Carty, Roshal, and Yitzhaki (2009)

Table 15.11 Distribution of the households of singles by age (2006)[a]

Age range	Ireland (%)	Israel (%)
24 or younger	2.5	5.5
25–29	5.7	8.6
30–34	7.4	7.8
35–39	5.1	5.9
40–44	5.6	4.3
45–49	6.2	3.3
50–54	9.4	4.7
55–59	8.6	8.0
60–64	9.5	7.2
65+	40.0	44.7
Total	326,500	372,532

Source: Carty, Roshal, and Yitzhaki (2009)
[a]Calculations based on EU-SILC (Ireland) and HES (Israel)

8% in Ireland. Also, there are more pensioners (above the age of 65) in Israel living in single households than in Ireland (44.7% and 40.0%, respectively).

The overlapping index in column (IX) tells us whether one or more of the groups form distinctive strata. If this index equals the proportion of the group (column (III), then this group forms a perfect stratum. The lower the index, the less the group's distribution overlaps with that of the whole population. There are no distinctive groups in either one of the countries. However, medium-sized households have their distinctive place in the overall distribution. The distributions of the 2- and 3-person households mimic the distribution function of the whole country's population.

Singles in Ireland, being the least equal group, have the highest overlapping index, meaning they are divided into two separate strata and are present in both the highest and the lowest ends of the income distribution of the entire population. This result is less noticeable for Israel.

The between-groups inequality is very low and contributes very little to the overall inequality—4% in Ireland and 7% in Israel. The relatively high Pyatt's between-groups component shows that the overlapping has reduced the between-groups component. Pyatt (1976) calculates inequality between groups as if all members in a group had the same income.

According to the last column, the highest contribution (23% in Israel and 30% in Ireland) to the overall inequality in both countries is that of the 2-person households. This is due to the high income share of this group, the high inequality and the high overlapping component altogether.

Table 15.12 presents the average rank of members of one group in terms of the other (that is, had they been ranked within the ranking of the other group). The diagonal presents each group in its own ranking, which is 0.5 by definition. The ranking, unlike mean income, is not sensitive to extreme values. An average ranking above 0.5 means that on average, households in a given group have higher ranks in the other group's distribution than in their own, implying they are richer. For example, 4-person household, which is relatively poor in its own group, will be considered rich in terms of singles. Again, one can see that large households are the poorest in terms of other household groups in both countries (although they are richer than singles in Ireland), but they are much poorer in Israel.

Table 15.12 Average rank of one household size in terms of another, nongovernmental income

Household size	1	2	3	4	5	6+
Ireland						
1	0.5	0.389	0.314	0.286	0.309	0.368
2	0.611	0.5	0.414	0.384	0.416	0.502
3	0.686	0.586	0.5	0.474	0.517	0.628
4	0.714	0.616	0.526	0.5	0.547	0.669
5	0.691	0.584	0.483	0.453	0.5	0.629
6+	0.632	0.498	0.372	0.331	0.371	0.5
Israel						
1	0.5	0.411	0.353	0.356	0.375	0.521
2	0.589	0.5	0.442	0.445	0.469	0.636
3	0.647	0.558	0.5	0.503	0.532	0.723
4	0.644	0.555	0.497	0.5	0.529	0.719
5	0.625	0.531	0.468	0.471	0.5	0.700
6+	0.479	0.364	0.277	0.281	0.300	0.5

Source: Carty, Roshal, and Yitzhaki (2009)

Table 15.13 Overlapping between household sizes, nongovernmental income

Household size	1	2	3	4	5	6+
Ireland						
1	1.0	1.167	1.291	1.348	1.380	1.387
2	0.806	1.0	1.170	1.239	1.248	1.180
3	0.636	0.815	1.0	1.071	1.080	0.991
4	0.538	0.718	0.925	1.0	1.005	0.886
5	0.535	0.709	0.902	0.976	1.0	0.928
6+	0.601	0.800	0.965	1.034	1.050	1.0
Israel						
1	1.0	1.105	1.209	1.214	1.237	1.168
2	0.868	1.0	1.126	1.131	1.136	0.951
3	0.723	0.860	1.0	1.005	1.014	0.756
4	0.721	0.855	0.996	1.0	1.014	0.778
5	0.718	0.846	0.977	0.981	1.0	0.795
6+	0.768	0.865	0.970	0.968	1.004	1.0

Source: Carty, Roshal, and Yitzhaki (2009)

Table 15.13 presents the overlapping matrix between household groups. The rows represent the groups whose distributions are used as the base distributions. For example, when 4-person households are used as the base, singles in both countries form a stratum (0.538 and 0.721 for Ireland and Israel, respectively). The interpretation of the two overlapping indices is that there are relatively more large households in the distribution of singles (i.e. poor) than there are singles in the distribution of large households (1.387 vs. 0.601 in Ireland and 1.168 vs. 0.768 in Israel).

We now turn to the analysis of the net disposable income (Table 15.14). Comparing the two countries, we see that in each subgroup the inequality is lower in Ireland than

Table 15.14 Decomposition of Gini by household size, disposable income

HH size (I)	Obs (II)	Pi (III)	Mean income (IV)	Relative mean (V)	F0i: mean rank (VI)	Si: income share (VII)	Gi: Gini of the group (VIII)	Oi: overlap with the whole population (IX)	S*G*O (X)
(A) Ireland-disposable income per equivalent adult									
1 Person	1,816	0.11	19,225	0.82	0.41	0.10	0.359	1.15	0.040
2 Persons	1,695	0.23	23,460	1.00	0.51	0.25	0.372	1.02	0.093
3 Persons	867	0.18	22,881	0.98	0.54	0.19	0.308	0.95	0.056
4 Persons	786	0.23	22,281	0.95	0.55	0.24	0.285	0.95	0.065
5 Persons	450	0.16	19,905	0.85	0.50	0.15	0.258	0.91	0.035
6+	222	0.10	18,040	0.77	0.38	0.08	0.340	1.06	0.029
Within-group Gini							0.318	(98%)	
Between-groups Gini							0.008	(2%)	
(G_B Pyatt)							0.047		
Total							0.326		

(B) Israel-disposable income per equivalent adult									
1 Person	1,112	0.09	5,124	0.85	0.51	0.09	0.395	1.02	0.037
2 Persons	1,545	0.17	6,059	1.00	0.57	0.22	0.393	0.95	0.080
3 Persons	1,012	0.16	5,624	0.93	0.59	0.18	0.322	0.88	0.051
4 Persons	1,045	0.19	5,264	0.87	0.56	0.21	0.326	0.92	0.063
5 Persons	796	0.17	4,763	0.79	0.51	0.17	0.333	0.94	0.053
6+	749	0.21	3,026	0.50	0.31	0.13	0.373	0.97	0.048
Within-group Gini							0.332	(89%)	
Between-groups Gini							0.043	(11%)	
(G_B Pyatt)							(0.114)		
Total							0.375		

Source: Carty, Roshal, and Yitzhaki (2009)

Table 15.15 Average rank of one household size in terms of another, disposable income

Household size	1	2	3	4	5	6+
Ireland						
1	0.5	0.398	0.374	0.375	0.396	0.517
2	0.602	0.5	0.463	0.461	0.499	0.625
3	0.626	0.537	0.5	0.495	0.547	0.671
4	0.625	0.539	0.505	0.5	0.552	0.675
5	0.604	0.501	0.453	0.448	0.5	0.640
6+	0.483	0.375	0.329	0.325	0.360	0.5
Israel						
1	0.5	0.433	0.421	0.449	0.494	0.694
2	0.567	0.5	0.493	0.520	0.567	0.755
3	0.579	0.507	0.5	0.528	0.579	0.776
4	0.551	0.480	0.472	0.5	0.550	0.748
5	0.506	0.433	0.421	0.450	0.5	0.710
6+	0.306	0.245	0.224	0.252	0.290	0.5

in Israel. Interestingly the between-groups component in Israel contributes more than in Ireland.

Most of the results based on the nongovernmental income still hold: the singles have the highest inequality, although in Israel they improved their situation significantly. Two-person households are now the richest group in terms of mean income. It is interesting to note their position relative to the 4-persons group: in terms of nongovernmental income, the 4-person households were the richest group in Ireland by mean income and by mean rank, and the second richest group in Israel.

In Ireland there is also a change in the relative positions of singles versus large households: while the relative position of singles has improved, it has worsened for the large households. No meaningful improvement has occurred in the relative position of large households in Israel.

In Israel 3-person households have the highest mean rank, the lowest Gini—and the lowest overlapping index, making them the most distinctive stratum. In Ireland, these are the 4-person households which stand out of the whole distribution.

Table 15.15 allows us to see the improvement in the position (mean rank) of households when passing from nongovernmental to net disposable income. This is not surprising, although one would not expect the change in the relative rank between several groups of households. Those who had the higher rank are expected to preserve their relative position. However, the position of singles in Ireland has improved much more significantly than expected. In relation to large (6+) households, by nongovernmental income they had a mean rank of 0.368. Now they are even richer than the large households, with the mean rank of 0.517. Two-person households have also improved their position "on the account" of larger families, with mean rank of 0.625 (compared to 0.502) by nongovernmental income. Lerman and Yitzhaki (1995) showed how the impacts of policies are not only in narrowing gaps between people at different income levels. In fact, these policies also affect the positions of people, moving some who are initially at low incomes to pass those at higher incomes.

Table 15.16 Overlapping between household sizes, disposable income

Household size	1	2	3	4	5	6+
Ireland						
1	1.0	1.137	1.181	1.179	1.235	1.079
2	0.833	1.0	1.081	1.084	1.094	0.892
3	0.794	0.921	1.0	1.008	1.018	0.854
4	0.805	0.921	0.994	1.0	1.027	0.879
5	0.759	0.873	0.948	0.948	1.0	0.886
6+	0.948	1.050	1.080	1.078	1.112	1.0
Israel						
1	1.0	1.047	1.129	1.099	1.084	0.798
2	0.930	1.0	1.076	1.041	1.003	0.677
3	0.850	0.913	1.0	0.968	0.941	0.634
4	0.896	0.939	1.027	1.0	0.989	0.722
5	0.909	0.938	1.021	0.998	1.0	0.772
6+	0.973	0.930	0.964	0.965	1.012	1.0

Source: Carty, Roshal, and Yitzhaki (2009)

In Israel the change in relative ranking has occurred when comparing households of 2 with households of 4 and 5 persons. In comparison to these households, couples are now richer than 4- and 5-person households, while in terms of nongovernmental income they were relatively poorer.

Analysis of the overlapping indices (Table 15.16) shows how closer household's distribution functions are now to the whole population—all the numbers are closer to 1 than in Table 15.12.

In conclusion: by decomposing the Gini coefficient of inequality by household size we find that (1) for market income the level of inequality is similar for both countries. However small households (1–3 persons) have higher inequality in Ireland than in Israel and larger households have lower inequality in Israel; (2) For net disposable income, the inequality in Israel is higher than in Ireland by 15%, and in each household group the inequality is also higher in Israel than in Ireland, which means that governmental tax and benefit programs are more progressive in Ireland; (3) Correcting for the household composition differences, we find that for net disposable income, the inequality in Ireland increases by 0.5%, but in Israel it is reduced by 2%. This means that our conclusions with respect to progressivity cannot be attributed to the differences in demographic structure.

15.3 Decomposition Over Time: Non-marginal Analysis: Mobility, Inequality, and Horizontal Equity[12]

When discussing issues related to social welfare, sociologists have concentrated their attention on mobility as a factor determining inequality of opportunity. In contrast, economists have focused on inequality in income or consumption, often (although not always) without specific reference to mobility. Formally, if we

[12] This section is based on Yitzhaki and Wodon (2004).

consider a bivariate distribution representing initial and final distributions, an inequality index is a summary statistic defined over each marginal distribution, i.e., the initial and the final distributions. In contrast, a mobility index describes the transition process between these two distributions. Some economists (e.g., Shorrocks, 1978a; Atkinson, 1983; Dardanoni, 1993; see also Atkinson & Bourguignon, 1992, for a review of empirical studies of earnings mobility) have approached mobility indices as complements to the tools used by statisticians and sociologists (e.g., Prais, 1955; Bartholomew, 1982; see also Bibby, 1975, for a review of the sociological literature). But in most cases these mobility indices have been developed using properties of transition matrices, independently of the concepts of inequality and equity.

In this section we define and discuss the properties of the Gini index of mobility—which together with the Gini coefficient of inequality provides an overall consistent framework for the analysis of mobility, inequality, and horizontal equity. In so doing we follow up on a few papers devoted to the relationships between the three concepts. For example, Shorrocks (1978b) shows how income mobility reduces inequality over time. King (1983) develops an index of inequality which can be decomposed into two components: one is related to mobility and the other is related to horizontal equity.

To motivate and illustrate the relationship between mobility and inequality, consider the system of job rotation in the early days of the Kibbutz. Members of the Kibbutz rotate jobs. Hence, although at each given period of time there is no equality among members, inequality vanishes over time. Inequality is observed only because snap-shots are used to describe an otherwise lengthy process. Another example of the impact of the period of measurement on inequality is the distribution of income over the life cycle. If one is interested in life-time inequality, then yearly inequality is inappropriate. Although individuals may have exactly the same pattern of income flow over the life cycle, one will observe inequality simply because the time period used for measurement is too short. A third type of transition over time is associated with uncertainty. If the distribution of income is affected by random shocks, the resulting process can be analyzed formally in the same way as job rotation, except that in the case of job rotation the transition is known in advance to the individuals, while in the case of uncertainty it is not.[13] In all these cases, a mobility index can help in predicting the appropriate level of inequality over a period of time from a series of snap-shots at any given point in time.

Mobility is also related to horizontal equity. Usually transition processes take time, but we may also think of instantaneous transition processes with no time dimension attached to them. This applies to changes in incomes due to a reform in taxation. Traditionally the changes in individual rankings before and after taxes

[13] This type of problems may also occur when incomes are registered on a cash flow base rather than on an accrual basis. Different sources of income such as capital gains, farm income, and other types of capital income, which are registered according to realization, may have different accumulation and distribution patterns over time. Relying on snap-shots of the distribution may exaggerate the impact of those incomes on inequality in the long run.

have been analyzed through the concept of horizontal equity. The main principle of horizontal equity in tax reforms is defined by Feldstein (1976, p. 95) as "*if two individuals would have the same utility level if the tax remained unchanged, they should also have the same utility level if the tax is changed.*" The implication of Feldstein's definition of horizontal equality (and to the best of our knowledge of all measures of horizontal equity) is that rank switching, i.e., the change in the rankings of individuals between the initial (before the reform) and final (after the reform) distributions is an undesired property. The violation of this norm is the target of horizontal inequity measurement. Measures of horizontal equity are discussed among others by Atkinson (1979) and Plotnick (1981). But clearly the same rank switching is also the target of mobility measurement, except that mobility is viewed as a desired property to have. The information needed for calculating an index of horizontal equity is identical to that needed for calculating an index of mobility. As will be shown, the Gini mobility index is equivalent to the Atkinson–Plotnick measure of horizontal inequity.

The Gini mobility and inequality indices enable the identification of three separate factors at work when there is a change between two income distributions: growth, inequality, and mobility. Growth is interpreted as a constant percentage increase in all incomes. A change in inequality occurs if one distribution deviates from the other by more than a multiplication by a constant. Mobility occurs when individuals change positions along the distribution.[14] Note that growth may occur without affecting inequality and mobility, a change in inequality may occur without affecting growth and mobility, and mobility may occur without affecting growth and inequality. Clearly, we could use alternative measures to capture inequality or mobility. The advantage of using an overall consistent framework is to insure that concepts are not mixed up in the process of measurement.

Another advantage of using the overall framework of the Gini is that mobility is not defined as an independent concept. Therefore there is no need to derive a separate axiomatic justification for it. Any set of axioms that supports the Gini such as those proposed by Ebert and Moyes (2000) can also be used to support the mobility index. We will return to this argument in Sect. 15.3.2.

The Gini index of mobility has one additional property which is useful for discussing inequality. Consider the case of having two criteria for ranking the population, such as income and wealth. Each criterion enables the evaluation of a marginal distribution by a measure of inequality. Changing the criterion changes the observed level of inequality. What can be said if one is interested in a composite measure of inequality taking both criteria into account? A possible solution is to create a scale which will weigh the two criteria and then compute the measure of inequality as a weighted average of the two criteria. But in many circumstances there is no a priori agreed upon weighting scheme and one has to search for an appropriate weighting scheme. To evaluate whether this is a serious problem, we

[14] For an alternative and interesting view, see Fields and Ok (1996, 1999) who present an axiomatic characterization to absolute changes in incomes.

state the conditions that enable us to predict the level of inequality when using a weighted average of two criteria without assuming specific weights. That is, given the range of the weighting scheme, the Gini mobility index can also be useful in describing the change in ranking when one moves from using one criterion to the other. The upshot from the above discussion is that many issues that involve transitions from one distribution to another can be represented by the same index. This is illustrated using panel data from rural Mexico.

The structure of the discussion is as follow: In the first section (Sect. 15.3.1) we introduce the concept of the Gini index of mobility in both its symmetric and asymmetric forms. We discuss the relationship with transition matrices and illustrate it on a panel data on income, land owned, land cultivated, and cash transfers to rural farmers from a survey conducted by the World Bank in collaboration with the Secretaria de Reforma Agraria of Mexico. Next, in Sect. 15.3.2 we discuss the issue of predicting inequality for a linear combination of variables. Then we illustrate the use of the Gini mobility index for predicting composite measures of inequality when the analyst is not able or not willing to specify the weights to be attributed to each factor in the composite measure of welfare. This is done by using information on rural Mexican households for income and wealth as measured by land ownership.

The empirical analysis in this section is devoted to the inequality of a combination of two variables: one depending on income and wealth, and the other depending on income at two points in time. Section 15.3.3 shows the equivalence between the Gini index of mobility and the Atkinson–Plotnick measure of horizontal equity, with an application to the impact on inequality and horizontal equity of a Mexican cash transfer program for farmers. More specifically, we show the impact of Procampo, the program of cash transfers to farmers, on income inequality in rural Mexico, and on horizontal equity.

15.3.1 The Gini Index of Mobility

15.3.1.1 Definitions and Properties

The most convenient way to define the Gini index of mobility is by using continuous distributions. However because we want to analyze the relationship between the index and the literature on mobility, which relies on transition matrices, we will move back and forth from continuous to discrete variables to allow for matrix notation.

Let (Z_1, Z_2) denote a bivariate income distribution in states 1 (initial) and 2 (final), respectively. It is assumed that first and second moments exist. Define $Y_j = Z_j/\mu_{Zj}$ as the income in terms of mean income. Then (Y_1,Y_2) have a bivariate distribution with $\mu_1 = \mu_2 = 1$. A mobility index should describe the association between observations in distributions 1 and 2. We distinguish between symmetric and asymmetric indices of mobility. An index S_{12} defined over distributions 1 and

2 is symmetric if it satisfies $S_{12} = S_{21}$ for any two distributions. The advantage of this property is that the index does not suffer from the index number problem that is typical to directional movement from one state to the other. The disadvantage of a symmetric index is that it requires more information than an asymmetric one because one needs to have all the components of the two distributions in order to estimate it. The Gini symmetric index of mobility is defined as

$$S_{12} = \frac{COV[(Y_1 - Y_2), (F_1(Y) - F_2(Y))]}{COV(Y_1, F_1(Y)) + COV(Y_2, F_2(Y))}, \tag{15.2}$$

where $F_j(Y)$ is the (marginal) cumulative distribution of Y_j.

By collecting terms, (15.2) can be written in a more convenient way as

$$S_{12} = \frac{G_1(1 - \Gamma_{12}) + G_2(1 - \Gamma_{21})}{G_1 + G_2}, \tag{15.3}$$

where Γ_{ij} is the Gini correlation coefficient between Y_i and Y_j. (Recall that Y_1 and Y_2 are normalized so that $\mu_1 = \mu_2 = 1$).

The intuition behind (15.2) and (15.3) is straight forward: the Gini index of mobility is based on the association between the change in the incomes between two periods and the change in the rankings of the same incomes over the same periods. The higher the association, i.e., the more the change in income is also associated with a change in the position of the household in the population, the higher the mobility.[15]

One can also define the (directional) asymmetric mobility index $M_{js} = (1 - \Gamma_{js})$, where j is the initial state and s is the final state. Then the symmetric index of (15.3) is a weighted average of the two asymmetric indices

$$S_{12} = w_1 M_{12} + w_2 M_{21}, \tag{15.4}$$

where $w_i = G_i/(G_1 + G_2)$, i = 1,2 is the share of the inequality of distribution (or period) i in the sum of the inequalities in the two distributions (or periods).

The properties of the Gini mobility index (15.3) (and (15.4)) can be derived from the properties of the Gini coefficient and the Gini correlations.

(a) $S_{12} \geq 0$. This property is based on the bounds on the Gini correlations: $1 \geq \Gamma_{js} \geq -1$.[16]

---

[15] Fields and Ok (1999) refer to this type of mobility measures as *correlation-based* mobility index. An example of such an index is Hart's index discussed by Shorrocks (1993).

[16] An important property of the Gini correlation is that the bounds are identical for all marginal distributions. This property does not hold for the Pearson correlation coefficient (Schechtman & Yitzhaki, 1999). This means that one minus Pearson's correlation coefficient cannot serve as an index of mobility because a change in the shape of one of the marginal distributions, that does not affect the transition process of the ranks, may change the value of the correlation.

(b) The range of the Gini mobility index is between 0 and 2.

(b.1) Minimum Mobility: $S_{12} = 0$ if both Gini correlations are equal to one. This is the extreme case of no mobility. Although the marginal distributions may change, the rankings of the individuals do not. If the transition process has not changed the ranking of the units, then the mobility index equals zero. This corresponds to the immobility axiom by Shorrocks (1978a, 1978b). Note, however, that inequality can change between the initial and the final distributions. Examples of such cases are abundant: the application of a pure income tax so that the ranking of before-tax income is identical to the ranking of after-tax income[17] or, alternatively, the effect of a tax reform that does not change the ranking of after-tax income. Another example is economic growth that affects all units by a monotonic increase of their incomes, as can be the case when the returns to schooling are changing (that is, distances between adjacent incomes increase or decrease), but the order (the education endowments) is not reversed. Still another example is a macroeconomic shock that affects all individuals without causing changes in ranks. Note that although inequality can change even if there is no change in the rankings, inequality cannot change between the two distributions if there is no change in incomes.

(b.2) Midpoint: If Y_1 and Y_2 are statistically independent then $S_{12} = S_{21} = M_{12} = M_{21} = 1$. Because in most cases of mobility the correlation between the initial and final marginal distributions tends to be positive, some investigators (e.g., Prais (1955) and his followers) defined independence as the extreme case of mobility. Shorrocks (1978a, 1978b) on the other hand prefers to define property (b.3) as the extreme case. This distinction is not relevant for our purposes.

(b.3) Maximum mobility: The maximal value is 2. It occurs when $\Gamma_{12} = \Gamma_{21} = -1$, implying that $S_{12} = S_{21} = M_{12} = M_{21} = 2$. Maximum mobility occurs if there is a total reversal in the ranks. That is, the richest in distribution 1 is the poorest in distribution 2, the second richest in distribution 1 becomes the second poorest in distribution 2, etc. In this case the final distribution is derived from the initial distribution by a declining monotonic transformation. Note that in this case mobility is independent of whether the overall inequality increases or decreases between the initial and final distributions.[18]

(c) Higher mobility: this property corresponds to an increase in the Gini mobility indices. That is, the lower the Gini correlations between the initial and the final distributions, the higher the mobility.

---

[17] See Feldstein (1976). See also the measures of progression in the income tax (Lambert 2001, Chap. 6).

[18] One could divide the symmetric and asymmetric Gini indices of mobility by two in order to keep the indices between zero and one.

In this section we have derived the mobility index as based on the Gini correlations. Because some of the mobility indices are based on transition matrices, we will also base the Gini mobility index on transition matrices.

15.3.1.2 The Relationship with Transition Matrices

The traditional way to study mobility is by analyzing transition and turnover matrices.[19] The main interest in this literature is occupational mobility, while the interest in the present section is in the impact of mobility on inequality. This difference in interest calls for slightly different approaches.[20] For convenience and without loss of generality, we divide the initial and the final populations into equiproportional groups, so that the difference between a transition and a turnover matrix is a multiplication by a constant. In this section we show that provided that one is interested in the impact on the Gini coefficient of inequality, the components of the Gini indices of mobility Γ_{js} and Γ_{sj} are sufficient statistics for the information contained in turnover and transition matrices. This means that transition matrices do not add information over the informational content of the mobility indices. To show the relationships between the Gini indices of mobility and transition matrices, it is convenient to rely on discrete distributions.[21] Let Y_{jk} ($j = 1,2$; $k = 1,..,K$) be the normalized income (so that the mean income equals to one) and let F_{jk} be its normalized rank (the value of the empirical distribution, a number between zero and 1) of observation k in state j. In addition let Y_j and F_j be $K \times 1$ vectors of the normalized incomes and their ranks in state j, respectively. Without loss of generality we assume that the observations F_{jk} are arranged in an increasing order of the ranking of the first period. That is, F_1 is the only vector whose elements must be arranged in a nondecreasing order. Because we are dealing with normalized incomes with unit means, the Gini coefficient of distribution j can be written as

$$G_j = 2\, Y_j' F_j - 1, \tag{15.5}$$

---

[19] A turnover matrix is a matrix the elements of which add up to one. A transition matrix is a matrix of which the elements in the rows add up to one. Usually transition matrices represent the conditional probabilities, while the elements of a turnover matrix represent the joint probability distribution of the two variables.

[20] The mobility index that is the closest to the one suggested in this chapter is Bartholomew's (1982) index of mobility which is based on the expected value of the absolute difference in the values attached to categories in the initial and final distributions. However, Bartholomew's index is sensitive to the initial and final marginal distributions, and therefore may give a misleading picture of the transition process. For example, assume that everyone in the society is promoted by one category. Bartholomew's index would indicate transition although there is no change in the ranking of the members. On the other hand, the Gini mobility index is not affected by linear transformations of the marginal distributions. See Boudon (1973, pp. 51–54) for a discussion of the properties of Bartholomew's index.

[21] Note, however, the important contribution by Geweke, Marshall, and Zarkin (1986) who analyze mobility indices in a continuous time framework.

where Y'_j is the transpose of the vector Y_j. Using the same procedure, the Gini correlation Γ_{js} is

$$\Gamma_{js} = \left(Y_j'F_s - 0.5\right) / \left(Y_j'F_j - 0.5\right). \tag{15.6}$$

Let T_{js} represent the transition matrix of size $K \times K$ from period j to period s. Because we are interested in an inequality index, aggregation of observations into groups may cause the loss of intra-group inequality. Therefore the size of the matrix has to be the size of the sample (K). In the sample, the transition matrix will be a permutation of the identity matrix and it can take any shape that transition matrices are allowed to have.[22] Let $t_{n,m}$ be an element in the transition matrix T_{js}. Then $t_{n,m} = 1$ if the observation with rank n in state j moved to rank m in state s. Otherwise $t_{n,m} = 0$. It is easy to see that for the vector of ranks we have $F'_s = F'_j\, T_{js}$, and $F'_j = F'_s\, T'_{sj}$ where j and s represent the initial and final distributions, respectively, and T'_{sj} is the transpose of T_{js} (because T_{js} is a permutation of the identity matrix, its inverse is identical to its transpose). The Gini correlation coefficient Γ_{js} is defined as a function of the transition matrix as follows

$$\Gamma_{js} = \left(Y_j'F_s - 0.5\right) / \left(Y_j'F_j - 0.5\right) = \left(Y_j'F_jT_{js} - 0.5\right) / \left(Y_j'F_j - 0.5\right). \tag{15.7}$$

The Gini correlation Γ_{sj} is obtained in a similar way, and the Gini symmetric mobility index which includes both Γ_{js} and Γ_{sj} relies both on the transition matrix and its transpose. Furthermore, assume that a population goes through two consecutive transitional processes, described by the transition matrices T^1 and T^2. Then, the accumulated transition process over the two periods is $A = T^1T^2$. In order to compute the Gini indices of mobility over the two periods one can proceed as before, using the matrix A as representing the overall transition process. An extension to more than two periods can be done in a similar way. This implies that one can study convergence and ergodic properties by using a series of Gini indices of mobility instead of the more complicated series of underlying transition matrices. The convergence of transition matrices to a given matrix will lead to the convergence of the Gini mobility index to a given constant. Although we will not

[22] The transition matrix is a special case of a doubly stochastic matrix, where each column and each row add up to one, as discussed by Marshall and Olkin (1979, Chap. 2), although each element should be multiplied by a constant. A similar situation arises when the variable is a binary variable: although the probability is a continuous variable, the realization of the variable in the sample is either one or zero. Traditionally transition matrices have been applied to discrete distributions, due to grouping. The fact that we are not dealing with groups is not due to an inability to handle groups. Rather, we define the transition matrix without grouping in order to avoid the loss of intra-group differences in ranks and thereby inequality, which may be relevant for calculating the inequality index. Note that since the mobility index is a sufficient statistic for the informational content of the transition matrix for our purpose, there is no need to construct the transition matrix and therefore its size is irrelevant in practice.

work with transition matrices in what follows, it is worth to briefly describe the special cases of the Gini mobility indices in terms of the transition matrices.

(a) Minimum mobility ($S_{12} = 0$) occurs if the transition matrix is the identity matrix. Hereafter the I matrix.
(b) Maximum mobility ($S_{12} = 2$) occurs if the transition matrix is composed of ones in the diagonal which is opposite to the main diagonal and zeros elsewhere. Hereafter the M matrix.
(c) Midpoint ($S_{12} = 1$) occurs if the transition matrix is composed of identical rows and columns and each entry equals 1/K. (Note that this case can only be described for the population, with probabilities as the elements of the matrix. In the sample, the entries are either zero or one.)
(d) Higher mobility: the lower the Gini correlation, i.e., the lower the value of (15.7), the higher the value of the mobility index of the transition process. Clearly the value of (15.7) is determined by the entries in the transition matrix, so that (15.7) provides a ranking of transition matrices according to mobility. Higher mobility implies lower absolute values for the Gini correlations and higher absolute values for the Gini indices of mobility.

If one has information on the Gini coefficient of one marginal distribution, then the Gini correlations represent the only informational content of the mobility matrix that is relevant for predicting the Gini coefficient of the other distribution. This does not imply, however, that having information on incomes and ranks in the first period, as well as information on the changes in ranks from the transition matrix as summarized by Gini indices of mobility will be sufficient to predict inequality in the second period. To predict inequality in the second period, it would be necessary to know the incomes in the second period. But the Gini correlations remain sufficient statistics of the transition matrix with respect to the information that is available in the transition matrix for analyzing the Gini coefficient of income inequality in the second period.

15.3.1.3 An Empirical Illustration

The illustration of using the Gini mobility indices (asymmetric and symmetric) is based on a panel data on income, land owned, land cultivated, and cash transfers to rural farmers from a survey conducted by the World Bank in collaboration with the Secretaria de Reforma Agraria of Mexico.[23] The survey was carried out in 1994 and 1997 in rural areas, in the so-called ejido sector. Until recently Mexico's ejido sector was functioning under a system of communal property whereby land could not be alienated, rented, or mortgaged, and usufructuary rights were contingent on

[23] Using the framework proposed in this section, Beenstock (2002a, 2002b) analyzes intergenerational mobility in Israel; Fisher and Johnson (2006) apply the methodology using consumption data from the USA, Wodon (2001) applies the methodology to mobility and risk during the business cycle in Argentina and Mexico, and Wodon and Yitzhaki (2003a) look at wage inequality over time in Mexico.

Table 15.17 Summary statistics for net income, Procampo transfer (1997 only) land owned and land cultivated for years 1994 and 1997

Variable (PC = per capita)	Mean	Std. dev.	Min	Max
Year 1994				
PC net income, pesos	1,537.32	4,089.88	−731.08	71,661.95
PC land owned, hectares	1.97	3.25	0.00	65.00
PC land cultivated, hectares	1.85	3.19	0.03	65.00
Year 1997				
PC net income, pesos	1,770.30	4,202.08	−1,257.07	51,207.98
PC Procampo transfer, pesos	332.51	522.93	0.00	7,878.00
PC land owned, hectares	2.20	5.44	0.00	188.75
PC land cultivated, hectares	2.02	5.74	0.00	201.00

Source: Yitzhaki and Wodon (2004), p. 188. The sample is restricted to households for which all variables are available (1027 observations)

occupation and cultivation of the land. A land titling reform was initiated in 1992 as part of the broader liberalization of Mexico's rural economy, enabling ejidatarios (those living in the ejidos) to own their land on an individual basis. Moreover, in line with the North American Free Trade Agreement requirements, government support programs for agricultural inputs (subsidies) and outputs (guaranteed prices) were terminated. To enable farmers to adjust, the government created as of 1994 a temporary cash transfer program named Procampo, whereby eligible farmers receive a fixed sum of money per hectare cultivated for up to 15 years (see Cord and Wodon (2001) for details).

The subset of the survey data that will be used in order to illustrate the properties of the Gini indices of mobility consists of information on per capita incomes (1994 and 1997), per capita land owned (1994 and 1997), per capita land cultivated (1994 and 1997), and per capita transfers from Procampo (1997 only; in 1994 the households did not yet receive the transfers). Summary statistics for all the variables of interest are given in Table 15.17. The mean quarterly per capita income is slightly higher in 1997 than in 1994 (in constant terms). The lack of growth in income between the 2 years is in large part due to Mexico's devaluation in December 1994 and subsequent economic downturn in 1995. There are a few households for which per capita incomes are negative due to the possibility of losses in any given quarter for some farmers (the cost of farm inputs may be larger than the revenues from the sales of outputs). These negative values do not represent any problem for the analysis, provided it is recalled that the Gini coefficient of inequality can then be greater than one when the variable of interest has negative values (one such case will appear in the empirical analysis). On average, households own and cultivate two hectares of land per person. The standard deviation of the distribution of land is larger in 1997 than in 1994, as well as the maximum value of the land owned or cultivated. Finally, in 1997 Procampo payments amount on a per capita basis to 332.5 pesos per person on average, which is about 18.9% of average per capita income.

Table 15.18 Gini indices of inequality and mobility (symmetric and asymmetric)

	Income	Land owned	Land cultivated
Gini coefficient of inequality, 1994	0.818	0.567	0.576
Gini coefficient of inequality, 1997	0.830	0.603	0.628
Asymmetric index of mobility, 1994–1997	0.261	0.297	0.335
Asymmetric index of mobility, 1997–1994	0.267	0.312	0.330
Symmetric index of mobility	0.264	0.305	0.333

Source: Yitzhaki and Wodon (2004), p. 189. The sample is restricted to households for which all variables are available (1,027 observations)

Table 15.18 provides the Gini coefficients of inequality for per capita income, land owned, and land cultivated in both years, as well as the various Gini indices of mobility. Because there are some negative values, the Gini coefficients of inequality for per capita income are fairly high, at 0.818 in 1994 and 0.830 in 1997. The corresponding measures for land owned and cultivated are somewhat lower but high as well. They range from 0.567 for land owned in 1994 to 0.628 for land cultivated in 1997. Inequality is higher in 1997 than in 1994 for both income and land (whether owned or cultivated).

There is also substantive mobility between the 2 years both in terms of income and in terms of land. The highest level of mobility is observed in land cultivated, perhaps in part because of the impact of land reform. Now that farmers can own their land, it is easier for them to give it for cultivation to others without loosing their property. There is also a relatively high level of mobility in land owned, indicating that there are sales going on, also in part as a result of the land titling reform. Mobility is somewhat lower for per capita income, but nevertheless substantial given that only 3 years separate the two periods. The mobility in per capita income may be due in part to the fact that households having a bad quarter may have negative values in one year but not in the other. The two asymmetric indices of mobility are fairly close to each other in all cases, which is an indication that there is likely to be exchangeability between the distributions. The relatively high level of mobility hints that yearly observations suffer from high volatility and that extending the time span of measurement can reduce the measured inequality significantly.

15.3.2 Predicting Inequality of a Linear Combination of Variables

15.3.2.1 Definitions and Properties

One useful property of the Gini indices of mobility is that the indices help in estimating composite measures of inequality whereby the analyst is interested in the inequality of a weighted sum of attributes. Let $Y_{(\alpha)} = \alpha Y_1 + (1 - \alpha)Y_2$ with

$0 \leq \alpha \leq 1$. If α is known, then the Gini coefficient (or GMD) for $Y_{(\alpha)}$ can be directly calculated. However, if Y_1 and Y_2 represent two different attributes such as land ownership and income, or if they represent incomes at two different points in time, then one might prefer not to be forced to assume a particular value for α, but to evaluate the sensitivity of the inequality of the linear combination to hypothetical reasonable ranges of α. In this case, the Gini of $Y_{(\alpha)}$ is bounded as follows

$$\mathrm{Max}[0,\ \alpha G_1\Gamma_{12} + (1-\ \alpha)G_2\Gamma_{21}] \leq G_{Y(\alpha)} \leq\ \alpha G_1 + (1-\ \alpha)G_2. \tag{15.8}$$

In a typical case, (15.8) would provide a meaningful range for predicting composite inequality. The upper bound of $Y_{(\alpha)}$ is achieved under perfect Gini correlations between Y_1 and Y_2 whereby the ranks in the distributions are the same. That is, Y_1 and Y_2 are monotonically related. The lower bound takes into account the Gini correlations between the two variables. The larger the Gini correlations between Y_1 and Y_2 (assuming they are positive, as will be the case in the empirical illustration), the higher the lower bound, and the shorter the interval of possible values for $G_{Y(\alpha)}$.

It should be emphasized that the asymmetric mobility index is measuring the change in the ranks between the base period and the final period with income in the base period as the weighting scheme. Because the non-weighted average change in ranks is zero by definition, the deviation of the index from zero is caused by the correlation between the changes in ranks and the levels of the incomes in the base period. For any given household, an increase (decrease) in rank typically takes place together with an increase (decrease) in income. The asymmetric mobility index does not take this simultaneous change in incomes into account because it takes only the baseline incomes into account. Hence, the index (slightly) underestimates mobility for observations whose ranks increase, and it (slightly) overestimates the impact for observations whose ranks decrease. This is again the index number problem, and the lower bound in (15.8) is constructed by taking this property into account (see Appendix 15.1 for the proof). This property is also responsible for the fact that the two asymmetric mobility indices of the Gini need not have the same sign (i.e., the signs of the asymmetric mobility indices may depend on the choice of the base period).

Equation (15.8) enables the evaluation of the inequality of a weighted average of variables when one is unwilling to quantify the relative importance of the two variables in the overall distribution. However, if one is ready to state the role of each variable exactly, then one can fully determine the impact of each variable on the overall inequality by decomposing the Gini of a linear combination of variables.

Equation (4.6) can be used to clarify the argument that having defined mobility in a Gini framework implies that one can use the axiomatic justification of the Gini to serve as the axiomatic characterization of mobility. To see this, assume that one adopts Ebert and Moyes' (2000) set of axioms to justify the use of the

Table 15.19 Intervals for composite inequality measures

	Income and land ownership in 1997		Income in 1994 and 1997	
	Gini correlation for income and land	Gini correlation for land and income	Gini correlation for income 1994 and 1997	Gini correlation for income 1997 and 1994
Γ_{sj}	0.491	0.381	0.739	0.733
α	Lower bound estimate	Upper bound estimate	Lower bound estimate	Upper bound estimate
0.0	0.408	0.830	0.608	0.830
0.1	0.390	0.807	0.608	0.829
0.2	0.372	0.785	0.607	0.828
0.3	0.355	0.762	0.607	0.826
0.4	0.337	0.739	0.607	0.825
0.5	0.319	0.716	0.606	0.824
0.6	0.301	0.694	0.606	0.823
0.7	0.283	0.671	0.606	0.822
0.8	0.266	0.648	0.605	0.820
0.9	0.248	0.625	0.605	0.819
1.0	0.230	0.603	0.605	0.818

Source: Yitzhaki and Wodon (2004) p. 192

Gini index to measure inequality in each period and to measure inequality in lifetime income, which is represented by Y_0. Mobility is then an index that connects yearly inequality to lifetime inequality. Therefore, the only axiomatic adjustment needed is that one accepts the present value of lifetime income as an indicator of welfare.

15.3.2.2 An Empirical Illustration

Assume that both income and wealth determine the well-being of an individual in the society, but the relative weights of the two factors can only be approximated. The joint distribution of income and wealth is known. Then one may want to evaluate the inequality of a combined index of these two variables without having to exactly specify the weight attached to each factor. In the case of rural Mexico, we can take the amount of land owned per capita as a proxy for wealth. Using the values in Table 15.18, and considering different values for α, one gets the results in the first part of Table 15.19. The lower and upper bounds were computed using (15.8). It can be seen that in the case of income and land ownership, the lower and upper bounds provide a relatively wide interval because the Gini correlation between per capita income and per capita land ownership is not very high. An index of well-being which would take into account both income and land would thus result in a substantial reduction in the measured level of inequality.

Table 15.20 Predictive power of the exchangeability assumption for composite income Gini

α	Midpoint estimate from the interval in (15.8) [1]	Variance-like[a] estimate from a full decomposition equation [2]	Actual value of the composite Gini index [3]	Difference between actual and midpoint [3]–[1]	Difference between actual and variance-like [3]–[2]
0.0	0.719	0.830	0.830	0.111	0.000
0.1	0.718	0.809	0.816	0.098	0.007
0.2	0.717	0.792	0.805	0.088	0.013
0.3	0.717	0.780	0.796	0.079	0.016
0.4	0.716	0.771	0.790	0.074	0.019
0.5	0.715	0.768	0.786	0.071	0.018
0.6	0.714	0.769	0.783	0.069	0.014
0.7	0.714	0.774	0.784	0.070	0.010
0.8	0.713	0.785	0.789	0.076	0.004
0.9	0.712	0.799	0.799	0.087	0.000
1.0	0.711	0.818	0.818	0.107	0.000

Reprinted with permission by Emerald

[a]By variance-like from a full decomposition we mean constructing $Y_{(\alpha)}$ under the exchangeability assumption.

Source: Yitzhaki and Wodon (2004), p. 192

Another application relates to inequality over time. If one wants to take into account two or more periods for computing the Gini coefficient of inequality, one can do so without specifying the weights (in this case the discount rate) for the two periods provided that one has computed the Gini asymmetric indices of mobility. The results of the calculations obtained for various values of α with the per capita incomes of the two periods are given in the second part of Table 15.19. The predicted interval is smaller due to the relatively large Gini correlations of income and land for the 2 years. Still, given the results in Table 15.19, it can be seen that income inequality could decrease by a maximum of 25% if two time periods were taken into account for estimating the indicator of economic well-being instead of one time period only.

Table 15.20 and Fig. 15.3 provide the midpoint interval estimates for the composite Gini index taking into account incomes in both 1994 and 1997 as well as the predicted values under the assumption of exchangeability. The estimates under the exchangeability assumption perform very well and better than the midpoints for all values of α, so that the assumption can probably be used in a number of empirical studies.

15.3.3 Mobility and Horizontal Equity

15.3.3.1 Definitions and Properties

As pointed out by King (1983) measures of mobility can be applied to horizontal inequity as well. In our case, it turns out that the Atkinson–Plotnick index of

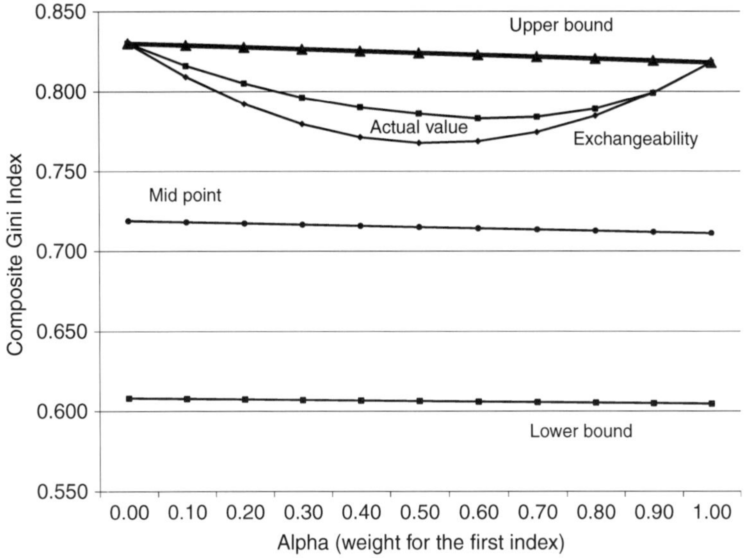

Fig. 15.3 Composite Gini: income in 1994 and 1997. Source: Yitzhaki and Wodon (2004), p. 193. Reprinted with permission by Emerald

horizontal inequity is a special case of the asymmetric Gini mobility index. That is, it can be shown that

$$AP = (1/2)(1 - \Gamma_{ba}), \tag{15.9}$$

where AP indicates the Atkinson–Plotnick (Atkinson (1979) and Plotnick (1981)) index of horizontal inequity, while b and a represent "before" and "after" reform distributions.[24] As shown by Lerman and Yitzhaki (1995), the other Gini correlation coefficient (i.e., Γ_{ab}) is also a key parameter in another index of horizontal inequity proposed by Kakwani (1984). It was mentioned earlier that the asymmetric mobility index may underestimate mobility when ranks increase and overestimate mobility when ranks decrease because of the index number problem. Given the similarity between the concepts of mobility and inequity, the index number problem also appears in indices of horizontal inequality. If one wants to impose symmetry on an index of inequity, it will be appropriate to use the symmetric version of the Gini mobility index.

[24] Silber (1995) developed an index of the intensity of change in ranking, which is equal to twice the Plotnick index. Using (15.9) it means that it is also equal to the Gini mobility index.

Table 15.21 Impact of Procampo cash transfers on income inequality and horizontal equity

Inequality	
Per capita income Gini index with Procampo	0.830
Per capita income Gini index without Procampo	1.002
Mobility	
Asymmetric index between PC income with and without Procampo	0.023
Asymmetric index between PC income without and with Procampo	0.035
Symmetric index of mobility	0.030
Horizontal equity	
Atkinson–Plotnick measure of horizontal inequity	0.011

Source: Yitzhaki and Wodon (2004), p. 195

15.3.3.2 An Empirical Application

Table 15.21 provides the results of the impact of Procampo, the program of cash transfers to farmers, on income inequality in rural Mexico, and on horizontal equity. Without Procampo, the Gini coefficient of inequality is 1.02 in 1997 (recall that with negative income values the Gini coefficient can be greater than one). With Procampo, the Gini coefficient of inequality is reduced to 0.830. The Gini correlations between the incomes with and without Procampo are very high, so that the mobility indices are small, at 0.023 and 0.035 depending on which distribution is taken as the base. The index of horizontal equity of Atkinson–-Plotnick, which is half of the asymmetric mobility index when using the incomes without Procampo as the base is small at 0.011, implying that Procampo results in fairly limited reranking in the population, in part because so many farmers benefit from the program in proportion to the land they cultivate (which is itself positively correlated with per capita income).

15.4 Summary

In this chapter we have presented several different interpretations and indices that exist in the literature which can be traced to parameters that determine the decomposition of the Gini coefficient of a linear combination of variables. Among those indices are indices of mobility and horizontal inequity. When dealing with mobility, we have shown that the Gini correlation can be presented as a summary statistic for transition matrices, provided that one is interested in changes in the Gini coefficient. Implicitly we have argued that mobility and horizontal inequity can be viewed as representing the same formal process, except that mobility implies a positive attitude, while horizontal inequity implies a negative one. It is in the eyes of the beholder (or in the subject matter) to determine whether the attitude toward an increase in the index should be positive or negative. Another (well-known) conclusion from the discussion in this chapter is that marginal distributions (snapshots or distributions that are based on one attribute) tend to exaggerate overall inequality. Hence, if the interest is in some kind of convex combination of the marginal distributions, we should expect the inequality of the combination to be lower than the inequality observed from the marginal distributions. In other words, following on the work of Shorrocks (1978a, 1978b), Atkinson (1983), King (1983), and Atkinson and Bourguignon (1992), it turns out that analyzing mobility can be interpreted as adding a dynamic and/or additional dimension to inequality analysis. In the framework of this chapter, the links between mobility, inequality, and horizontal equity have been made explicit for the special case of the widely used Gini coefficient.

The empirical applications, based on data from Ireland, Israel, and Mexico, have shown the wide applicability of the index. In the first application we compared Israel and Ireland using three types of decompositions: (a) according to income sources, (b) according to population subgroups, and (c) decomposition over time. In the second application we have measured the extent of inequality and income mobility in the ejido sector of rural Mexico between the 2 years 1994 and 1997; the impact of cash transfers programs on inequality and mobility, with a discussion of horizontal inequality; and how the tools presented can be applied to generate bounds for composite indices of inequality when the weights of the various components of the measure of welfare (such as income and land ownership) are not known.

Appendix 15.1

Proof of (15.8)

The proof consists of finding upper and lower bound for $G_{Y(\alpha)}$. The upper bound is

$$\begin{aligned} G_{Y(\alpha)} &= 2COV[\alpha Y_1 + (1-\alpha)Y_2, F(Y(\alpha))] \\ &= 2\alpha COV[Y_1, F(Y(\alpha))] + 2(1-\alpha)COV[Y_2, F(Y(\alpha))] \\ &\leq 2\alpha COV[Y_1, F(Y_1)] + 2(1-\alpha)COV[Y_2, F(Y_2)] = \alpha G_1 + (1-\alpha)G_2. \end{aligned}$$

Recall that Y_1 and Y_2 are normalized with $\mu_1 = \mu_2 = 1$. The derivation of the upper bound is based on Cauchy-Shwartz inequality, which can be utilized to show that for all Y_j and Y_k, $COV[Y_j, F(Y_k)] \leq COV[Y_j, F(Y_j)]$.

The lower bound is obtained from

$$\begin{aligned} G_{Y(\alpha)} &= 2COV[\alpha Y_1 + (1-\alpha)Y_2, F(Y(\alpha))] \\ &= 2\alpha COV[Y_1, F(Y(\alpha))] + 2(1-\alpha)COV[Y_2, F(Y(\alpha))] \\ &\geq Max[0, 2\alpha COV[(Y_1, F(Y_2)] + 2(1-\alpha)COV[Y_2, F(Y_1)] \\ &= Max[0, \alpha G_1\Gamma_{12} + (1-\alpha)G_2\Gamma_{21}]. \end{aligned}$$

Chapter 16
Incorporating Poverty in Policy Analysis: The Marginal Analysis Case

Introduction

The main purpose of this chapter is to expose the reader to additional tools that can be helpful in analyzing the distributional impact of a governmental policy. Assuming that one accepts the Gini coefficient of after-tax income as representing the social attitude toward the income distribution then one can summarize the effects of actions taken by the government by the Gini income elasticity (GIE). Decomposing the GIE by the contributions of the different sections of the income distribution enables one to both use the Gini as representing the social attitude and at the same time target the policy to sections of the distribution. The decomposition of the GIE presented is actually identical to the decomposition of the Gini regression coefficient applied to the Gini coefficient. The main message is that analyzing the effect of public policy by concentrating only on the poor population is not an appropriate approach because it violates the Pareto principle of efficiency and therefore leads governments and researchers to adopt and recommend policies that contradict the verbal declarations of the targets of the policies. On the other hand, by using a decomposition approach of the Gini coefficient or of the EG coefficient, the policy is consistent with the Pareto principle of efficiency and is based on additional useful information that is thrown away when dealing with traditional poverty analysis. An additional type of decomposition is needed whenever one is interested in targeting. We call a policy a targeted one whenever the policy instrument affects only a portion of the population. In this case we will want to decompose the effect of the policy to the contributions of two instruments: the choice of the subpopulation affected (i.e., targeting) and the effect on the subpopulation affected. The issue of targeting is not covered in this book. We refer the interested reader to Wodon and Yitzhaki (2002a, 2002b).

The structure of the chapter is as follows: Sect. 16.1 deals with analyzing the distributional impacts of programs intended to reduce poverty. Next, in Sect. 16.2 we present the arguments for constructing the poverty line. Section 16.3 shows the decomposition of the Gini coefficient, one component of which is Sen's poverty index.

S. Yitzhaki and E. Schechtman, *The Gini Methodology: A Primer on a Statistical Methodology*, Springer Series in Statistics 272, DOI 10.1007/978-1-4614-4720-7_16,

Section 16.4 decomposes other instruments based on the Gini (e.g., the GIE). Section 16.5 presents an empirical illustration, while Sect. 16.6 illustrates the evaluations of policies with the methodology. Section 16.7 concludes.

16.1 Analyzing the Distributional Impact of Programs Intended to Reduce Poverty[1]

The past four decades have witnessed a drastic change in the field of income distribution. There were shifts in interest and research effort from income distribution issues to poverty issues. One result of this change is the use of poverty indices rather than inequality indices, although the new poverty indices are rather reminiscent of the familiar inequality indices.[2] The main argument in this section is that there is little to be gained from investigating the properties of poverty measures. Because by definition the cumulative distributions of the incomes of the poor and the rich do not overlap, one can decompose any income inequality index into:

(a) Inequality within the group of the poor
(b) Inequality within the nonpoor (the "rich"), and
(c) Between-groups inequality.

The between-groups inequality can be further decomposed into three components: (i) a poverty-gap, (ii) an affluence gap[3], and (iii) a poverty-affluence-lines gap.

These decompositions give all the information supplied by the poverty measure and provide additional information that is useful in the analysis of poverty. The secondary decomposition of the between-groups component enables one to pinpoint the share of the poor in the population, the poverty gap and the inequality among the poor from which all the components of a poverty measure can be identified. Alternatively, one can use the secondary decomposition in order to decompose the inequality index into a poverty index [components (a) and (c.i)], an affluence index which is a mirror image of the poverty index [components (b) and (c.ii)] and between poverty line–affluence line inequality [component (c.iii)]. To save space, only the former decomposition will be developed in this section.

---

[1] This section is based on Yitzhaki (2002).

[2] Lest the above be construed as our original observation, note that Amartya Sen's 1976 path-breaking paper states that his poverty index "is essentially a translation of the Gini coefficient from the measurement of inequality to that of poverty" (Sen, 1976a, p. 226). See also Sen (1986).

[3] The affluence gap is a mirror index of the poverty gap; it reflects the difference between the mean income of the rich and the affluence line. The affluence line is defined as the residual with which the rich would be left if all the poor had been on the poverty line.

Truncating the distribution at the poverty line is often liable to limit one's ability to analyze the implications of a poverty-alleviating policy. Whenever it is impossible to use means-testing (i.e., testing the ability to pay), one must rely on indirect instruments—taxes and subsidies, or on direct governmental expenditures, where the targeting of the policy to the poor may be limited. In those cases the effectiveness of the policy instrument may depend on the differential (between poor and rich) incidence of the policy. Hence, comparison of the incidence among the poor with the incidence among the rich is essential for policy evaluation.

The approach to poverty measurement in this section adopts the viewpoint of a national policy-oriented economist. It is assumed that the interest in poverty measurement originates from the need for a yardstick with which to evaluate alternative poverty alleviation programs. Clearly, if the interest in poverty measurement originates from a different point of view (say, an interest in international comparisons of poverty) some of the arguments presented in this section may not apply.

The aim of this section is twofold. The first is to illustrate the decompositions of the Gini coefficient and Sen's poverty index (Sen, 1976a). The second aim is to apply the decomposition to statistics that characterize the distributional impact of a change in policy on income inequality. Instead of using simulations to evaluate the incidence of a policy change, one may evaluate the impact of the policy change on the inequality (and poverty) indices directly. It turns out that the most important summary statistic in the decomposition of the Gini according to income sources is the (Gini) income elasticity (Garner, 1993; Karoly, 1994; Lerman, 1999; Millimet & Slottje, 2002; Yitzhaki, 1994b). It enables the evaluation of the effect of a small change in a tax or subsidy or a small change in governmental expenditures (e.g., on education, on the elderly) on the Gini coefficient of inequality of income. The decomposition of the Gini enables the investigator to also evaluate the impact of a tax change on the Gini coefficient among the poor, on the Gini coefficient among the rich, and on the between-groups Gini. The advantage of summarizing the performance of the policy by parameters rather than by running simulations is that the former can offer statistical tests on the parameters that can help in evaluating the significance of the results.

One seemingly weakness of the between-groups elasticity is that it is based on only two observations—the share of the "commodity"[4] in poor people's incomes versus its share in the income of the rich. Because it is based on two observations, it is insensitive to the index of inequality used. That is, this elasticity determines the effect of a change of a tax or subsidy on between-groups inequality regardless of the inequality index used for the analysis. As we will show, the direction of the general incidence or that of the incidence among the poor can sometimes be opposite to the direction of the between-groups incidence, which means that a policy can be progressive for the poor but regressive for the overall population. The decomposition may enrich the analysis

[4] The appropriate interpretation of the term "commodity" in this chapter should be a tax base. The use of the term "commodity" enables us to borrow the terminology of Engel curves and consumption functions.

of the distributional impact of tax and benefit reforms. Moreover, application of the decomposition to data from Romania indicates that "caring more about the poor" should be modeled as attaching higher weight to the between-groups inequality.

16.2 The Usefulness of a Poverty Line

A well-defined poverty line is the cornerstone of poverty measurement. A classic definition of the poverty line is "… the *cutoff living standard level* below which a person is classified as poor" (*Poverty Reduction Handbook*, World Bank, 1991, p. 13; italics in the original). Anyone on or below the poverty line is defined as poor and is covered by the poverty index. The rest of the population is ignored.

In assessing the need for an official poverty line[5] it is useful to distinguish between political, administrative and social-welfare-measurement points of view.[6] Declaring an official poverty line and preparing a yearly report on the status of the poor may have political implications (Haveman, 1987, 1993; Sen, 1979, 1983). However, these implications are not relevant to the subject of this chapter because the decomposition of an inequality measure can be carried out with or without an official poverty line. Obviously, any administration needs a poverty line (or a cut-off point)—it lets welfare departments define their constituencies, it reduces arbitrary decisions, and it enables welfare departments to evaluate the success or failure of their programs. It is not at all clear, however, whether all welfare programs should necessarily employ the *same* cut-off point. Having a single cut-off point (i.e., an official poverty line) might be useful if cash transfers were deemed a satisfactory solution to the poverty problem and if the poor were easily identifiable. But if society is differentially sensitive to different dimensions of inequality, and/or if markets are incomplete, and/or if not all people are rational (or even share the same tastes), then a case could be made in favor of different cut-off levels for different programs or different groups.

In fact, both theory and practice suggest precisely such a differentiation. Tobin (1970) argues that society may be more sensitive to inequality in some areas (health and children) than in others, which can be handled by cash transfers. Society will probably be less tolerant to inequality in the opportunities open to "naive" individuals (e.g., children's education) than to "self-inflicted" poverty (e.g., as a result of gambling or drinking). This means that the poverty line, which can be interpreted as the income below which an individual is entitled to get help, may differ depending on the issue at hand. In practice, the US National School Lunch

[5] Atkinson (1993) discusses the merits of having an official poverty line. Fisher (1992) describes the development of the official poverty line in the USA. Haveman (1987) evaluates the impact of social science research on the development of the poverty line.

[6] Ravallion (1994a, 1994b) deals with the co-movement of inequality and poverty indices. He finds that in many cases inequality and poverty rankings are similar.

Program and the School Breakfast Program provide free meals to families with incomes under 130% of the official poverty line (according to Atkinson (1993) who provides additional examples). Another example is the Earned Income Tax Credit (EITC) scheme, designed to increase the incentive to work among poor families: beneficiaries of this program are entitled to have an income of up to 170% of the poverty line.

Even if a poverty line is a necessary administrative device and an effective political tool, it does not necessarily follow that it serves any useful purpose from a social welfare function point of view.[7] There are several arguments against classifying persons as either poor or rich.

(a) The theoretical justification of the poverty line can be found in the *Focus* axiom (Sen 1976a) that a poverty index should meet. Poverty measures that obey the *Focus* axiom concentrate on incidence among the poor and ignore the rich. Concentrating on the poor is a good strategy for welfare departments, because they are required to evaluate the relative neediness of the individuals they deal with. In practice, however, since the adoption of poverty alleviation as the official target of international institutes, poverty measures have increasingly served to evaluate policies that affect the population as a whole, rich and poor. Focusing on poverty alleviation need not necessarily imply ignoring the rich; rather, it implies that one is unconcerned with redistribution among the rich. But this argument should be modeled as implying a positive constant social evaluation of the marginal utility of income among the rich, not that the rich should be ignored. Ignoring the rich violates Pareto's principle of economic efficiency and hence may lead to Pareto-inefficient policies. To see this, consider two alternative poverty alleviation policies with identical effects on the poor but different effects on the rich. Concentrating exclusively on the poor may lead the policy maker to the conclusion that the programs are equivalent, leading to the choice of an inefficient program.

(b) Because there is no substantive difference between someone who is just an epsilon above the poverty line and someone who is just an epsilon below it, it would appear that a continuous function would describe the impact of poverty alleviation better than a discontinuous one. To see the kind of problems that may arise as a result of discontinuity at the poverty line, consider an economist who advises a government on how to reduce the number of poor people, subject to a revenue constraint. The economist—by design or otherwise—will naturally be inclined to recommend helping those who are close to the poverty line (from below) and ignoring (or possibly taxing) those who are even worse off, because such an "optimal" policy yields the largest decrease in the target function subject to given resources. Note that anyone who evaluates policies according

[7] Atkinson distinguishes between a poverty line that serves a criterion for benefit eligibility, and a definition for purely statistical purposes. If the purpose is purely statistical, "then a certain amount of rough justice may be acceptable" (1993, p. 24).

to their effectiveness may fall into this trap of applying an inappropriate (implicit) social welfare function.[8]

(c) Because it is the rich who pay for alleviating poverty, one cannot close the model without taking them into account. Closing the system requires a model that includes the whole economy.[9]

(d) Most countries do not rely exclusively on means-tested programs (instruments directed solely at the poor) for poverty alleviation. Instead, they use fiscal instruments directed at the entire population. The interest of the analyst in analyzing the effect of a general fiscal instrument should also be focused on differential incidence, which requires comparison of the incidence between the poor and rich. Truncating the distribution at the poverty line inhibits such analysis.

There are two viable substitutes for poverty indices.[10] One is to use inequality indices that can stress the lower portion of the income distribution; these include Atkinson's index of inequality (Atkinson, 1970) and the extended Gini coefficient (Chap. 6). The main property of these indices is that by changing a parameter one can increase the sensitivity of the index to transfers at the lower end of the income distribution. Such inequality indices can be useful measures of poverty without having to cope with the drawbacks entailed in truncating the income distribution. However, this alternative is not useful for someone who wishes to single out the poor as a distinct group.

The other substitute for poverty indices is to decompose an index of income inequality by poor, nonpoor (rich), and poor-versus-rich inequality (which we will refer to as "between-group" inequality). If inequality among the rich is not a consideration, one can simply ignore that component. This approach has several advantages.

(a) It provides more information than using poverty indices alone because it enables the investigator to also look at differential incidence. No piece of information is lost in the process of decomposition.

[8] This type of problems arises whenever there is a discontinuity in the implied social evaluation of the marginal utility of income at the poverty line. Some indices of poverty, e.g., Foster, Greer, and Thorbecke (1984) are normalized to have a zero implied marginal utility of income and hence do not suffer from this shortcoming. See Pyatt (1987) for a discussion and solution of this problem.

[9] See Chand and Parthasarathi (1995) for an example of incorporating a poverty measure in a national framework. Because average national income is included in the target function, the implied social welfare function possesses constant marginal utility above the poverty line.

[10] A third possibility is to use a Welfare Dominance approach with a large range of possible poverty lines as in Atkinson (1987) and Foster and Shorrocks (1988). Unlike the approach based on using a poverty index, this approach does not offer a complete ordering of possibilities and hence can sometimes be of a limited practical value (see Lambert (2001)). Also, it does not overcome the problem caused by ignoring the population above the poverty line. In any case, the same methodology suggested in this section can be applied to the stochastic dominance approach as well (Mayshar and Yitzhaki, 1995).

(b) Poverty lines are arbitrary in the sense that one can rarely determine poverty with precision, and one has to allow for inadvertent changes over time. When the poverty line is not adjusted correctly, having information about the rich enables the investigator to determine whether a change in the incidence of poverty was caused by an inadvertent shift in the poverty line or by a change in the income distribution. This issue will be elaborated upon later.

Finally, it is worth mentioning that the role played by the poverty line is different under the two approaches. Under the poverty measurement approach, whether an observation is above or below the poverty line is crucial. Under the decomposition approach the only thing that is determined by the poverty line is the classification of the observation into poor or nonpoor. An error in misclassification does not affect the overall inequality. Therefore the analysis is less sensitive to the poverty line.

The next section illustrates the decomposition approach and illustrates the drawbacks in omitting the rich from the analysis.

16.3 The Decompositions of the Gini Coefficient and Sen's Poverty Index

This section shows how the Gini coefficient can be decomposed into three components: Sen's poverty index, an affluence index (a mirror image of the poverty index), and an index of the between-groups (poverty line–affluence line gap) inequality. The following notation is used: μ_i, i = {p, r, o}, is the mean income of group i, where {p, r, o} are the poor, the rich, and the overall population, respectively, Z is the poverty line, and P_i,i = {p, r} is the proportion of group i in the population. The decomposition is first carried out geometrically, and then algebraically. Because the geometrical proof is based on Lorenz curves, which may represent hypothetical distributions, the geometrical proof can be applied to any index of inequality.

Figure 16.1 portrays a typical Lorenz curve, OHGIB. The percentage of the poor in the society is determined by finding the point on the Lorenz curve where the slope is equal to Z/μ_o. In Fig. 16.1 the percentage of the poor in the population is denoted by OE. That is, OE = Pp. The slope of the (dashed) line OJG is μ_p/μ_o; the slope of OKF is Z/μ_o, and the slope of ODB is 1. The area of the triangle OFG is the poverty gap and the area enclosed by OHGJ is the inequality within the poor. Ignoring the normalization of Sen's index,[11] the area between the Lorenz curve and OKF is Sen's index of poverty.

One can also define an index of inequality within the rich that is a mirror image of Sen's poverty index (call it a Sen-like affluence measure). Define the line FMB as the

[11] One can always divide and multiply by the normalization factor to get the exact index weighted by the normalization factor.

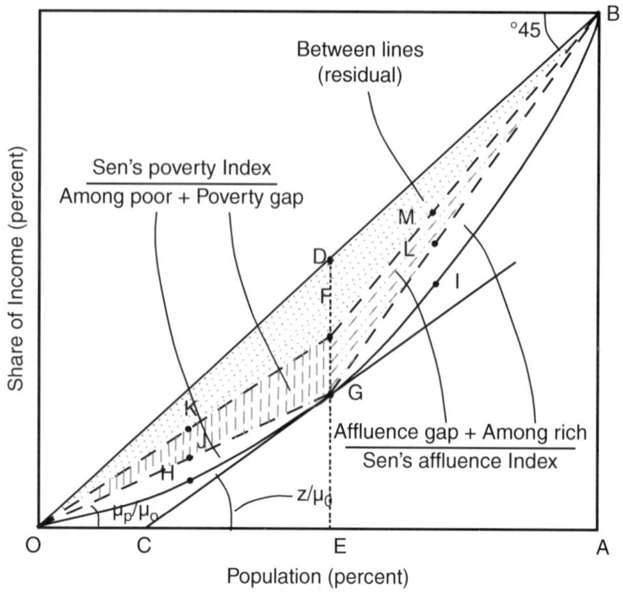

Fig. 16.1 Gini coefficient and Sen's index. Source: Yitzhaki, 2002, p. 67. Reprinted with permission by Elsevier

"affluence line." Similar to the poverty index and ignoring the normalization, the affluence index is the area FMBIG. The affluence index is composed of the area FMBLG, which is the "affluence gap" and the area GIBL which is the inequality within the rich. The remaining area (the triangle OFB) is between poverty–affluence lines inequality, where all the poor are assumed to have the same income—exactly the poverty-line level—and likewise all the rich have the same income ["affluent income," defined by the poverty line as: $Z_r = (\mu_o - P_pZ)/P_r$, where Z_r is the affluence line]. Thus, the affluent income line is actually derived from the poverty line.[12]

This same procedure can be used to decompose the Gini coefficient into the sum of Sen's poverty index (the area between the OKF line and the Lorenz curve), Sen's affluence index (the area between FMB and the Lorenz curve) and the between poverty line–affluence line inequality (the triangle OFB).

An alternative decomposition of the Gini coefficient stresses the difference between the traditional intra- and inter-group inequalities. That is, the area OHGJ is the inequality within the poor, GIBL is the inequality within the rich, while the triangle OGB is the between-groups component.

The difference between the above two decompositions is the way that the poverty gap (the triangle OFG) is treated. Using Sen's index the poverty gap is

[12] It should be pointed out that if the poverty line is not equal to the mean income, then not all the rich will be above the affluence line.

used as a component of the poverty index while under the group decomposition it is viewed as a component of between-group inequality. In this chapter we will rely on the intra- and inter-groups decomposition.

It is important to stress that the exact same decomposition can be carried out with any index of inequality. The intuitive explanation to this argument is that the division of the society to poor and rich implies no overlapping, and therefore the Gini and the variance decompose in an identical way. To see this, note that each curve (or line) in Fig. 16.1 represents an income distribution, so that one can repeat the same decomposition using the same grouping. Specifically, note that exactly the same figure would serve in order to decompose Atkinson's (1970) index into its components: the Foster, Greer, and Thorbecke (1984) index, the FGT "affluence index," and the between poverty–affluence lines inequality.

We contend that there is a lot to be gained from presenting the whole decomposition. For example, assume that we observe an increase in the proportion of poor in the society. The increase in poverty however can be caused by a deterioration in the status of the poor, or alternatively, by an inappropriate adjustment of the poverty line in response to relative price changes. Assume that the poverty line has inadvertently increased. The percentage of poor people will rise and inequality within them will increase and the poverty gap and between-groups inequality will increase.[13] The natural conclusion would be that there is an increase in poverty. But looking at the rest of the distribution can change the conclusion. Assume that one also observes an increase in Sen's affluence index. Because an increase in the poverty line cannot cause Sen's affluence index to increase, one can safely conclude that there has been a deterioration in the overall income distribution and in the status of the poor. If, on the other hand, one observes a decrease in Sen's affluence index, a decrease which can result from an upward movement of the poverty line, then one should suspect that an inadvertent increase in the poverty line is the cause of the deterioration of the poverty index. In other words, if both Sen's poverty index and Sen's affluence index increase it is clear that inequality has increased; but if Sen's affluence index declines, the between-groups inequality declines, while the Sen's poverty *increases*, it may indicate an undue upward slide of the poverty line. The obvious lesson is that truncating the distribution may be a bad policy even if one is interested only in poverty. This is an important observation because in many cases the poverty line is close to the mode of the distribution, making the poor and rich populations sensitive to the exact position of it.

The next section repeats the decomposition using the covariance formula of the Gini coefficient. This decomposition allows the derivation of policy implications regarding the components of the Gini coefficient.

[13] To see this, note that the slope of the Lorenz curve at the poverty line is less than one; hence, any movement of the poverty line to the right increases the triangle representing between-groups inequality. The intuitive explanation of why raising the poverty line increases the gap between the rich and the poor is that transforming the poorest among the rich into the richest among the poor increases the average income of the rich by more than it increases the average income of the poor.

16.4 Decompositions

16.4.1 Decomposition of the Gini Coefficient

The decomposition of Gini to poor/rich is a special case of using ANOGI, where there is no overlapping between the groups (i.e., perfect stratification). This decomposition is elaborated upon in Chap. 4. Therefore we do not replicate the proof but we do change the notation.

Assume that the society is divided into two administrative groups: (1) the poor, whose income is $y \le Z$ (Z is the poverty line), and (2) the rich, whose income is $y > Z$. Then, applying the decomposition by population subgroups ((4.27), with $O_i = 1$ for all i, and $G_B = G_{BP}$) we get

$$G_o = P_p S_p G_p + P_r S_r G_r + G_B, \tag{16.1}$$

where G_i denotes the Gini coefficient of y_i, the income of subgroup i, for $i = \{p, r, o\}$ for the poor, the rich, and the overall population, respectively, P_i is the share of subgroup i in the population, $S_i = P_i\, \mu_i / \mu_o$ is subgroup i's share in total income, and μ_i denotes the mean income of subgroup i. (When needed, S_{yi} and S_{xi} will denote the shares of the income (y) and the commodity (x), respectively).

G_B is the between-groups inequality. Because the distributions of the poor and the rich do not overlap, the Pyatt's (1976) and the Yitzhaki and Lerman's (1991) between-groups Ginis are equal. Note that $S_r = 1 - S_p$ and $P_r = 1 - P_p$. Also, it is easy to show that[14]

$$G_B = P_p - S_p, \tag{16.2}$$

which means that the between-groups inequality is equal to the share of the poor in the population *minus* their share in income.[15] G_B is an increasing (decreasing) function of the poverty line, depending on whether $Z < (>)\ \mu_o$. Hence, for all practical purposes G_B is an increasing function of the poverty line. This result should be treated with caution because an increase of inequality within the poor and an

[14] The between-groups component is equal to twice the difference between the area of the triangle OBA and the area enclosed by OGBA. By presenting the areas in terms of S_p and P_p one gets (16.2).

[15] To get the components of Sen's poverty index, the between-groups inequality triangle (multiplied by 2) should be further decomposed into the following triangles

(a) Poverty gap = area OFG = $P_p \times P_p (Z - \mu_p)/\mu_o$;
(b) Affluence gap = area FBG = $P_r \times P_p (Z - \mu_p)/\mu_o$;
(c) Between-lines inequality = area OFB = $P_p - P_p Z/\mu_o$.

This will complete the decomposition into all components in Sen's poverty index.

increase in between-groups inequality may simply be the result of (unintentionally) raising the poverty line.

Sen's poverty index includes three components: the proportion of the poor, P_p, the poverty gap $Z - \mu_p$, and the inequality among the poor, G_p.

In the notation used in this chapter, Sen's index is

$$SEN_p = P_p \left[\frac{Z-\mu_P}{Z} + \frac{\mu_P}{Z} G_p\right] = \frac{\mu_o}{P_p\ Z}\left[P_p\ P_p \frac{Z-\mu_P}{\mu_o} + P_p\ S_p\ G_p\right], \tag{16.3}$$

where the first component on the right is the normalization factor, and the factors in the square brackets are the poverty gap (the area of a triangle) and the within-poor Gini. By using (16.1), (16.3), and footnote 14 it is easy to see the decomposition (16.1). Note, however, that because changing the poverty line affects several components simultaneously, one has to exercise caution when deriving the impact of such a change on the components of the inequality.

Finally, as will be seen later, the impact of a policy measure on inequality is a function of its effect on each component, weighted by the component's share in income inequality. We define the weight of each component in the inequality as

$$w_P = \frac{P_P\, S_P\, G_P}{G_O}; \quad w_r = \frac{P_r\, S_r\, G_r}{G_O}; \quad w_B = \frac{G_B}{G_O}, \tag{16.4}$$

where w_i is the weight of the component for group i in the Gini coefficient ($i = p,r$) and w_B is the weight of the between-groups component.

From (16.4) it clearly follows that

$$1 = w_p + w_r + w_B. \tag{16.5}$$

A typical example of the weights is as follows: let $P_p = 0.3$ and $S_p = 0.05$, then $G_B = 0.25$ and an overall Gini of less than 0.5 implies that $w_B > 0.5$. If $G_p < G_o$ we get $w_p < 0.015$. Clearly, the lion's share of the weight is given to the between-groups inequality in this case. Therefore one may be interested in the between-groups component to shed more light on the distribution among the poor.

16.4.2 *The Decomposition of the (Gini) Income Elasticity*

Descriptive inequality indices are not sufficient for policy analysis. For an evaluation of a policy one needs a way to evaluate the effect of the change in a policy parameter on inequality and poverty. One way to do that is to simulate the effects of different policies. Another approach would be to evaluate the derivative of the inequality index with respect to the policy instrument so that rough-and-ready evaluation can be performed. This approach was suggested by Lerman and Yitzhaki (1985, 1994) and

Yitzhaki (1994b), who developed a technique for answering the following question: let dt_x be the change (in percentage terms) in a tax on "commodity" x. That is, the consumer price of x changes from P_x to P_x $(1 + dt_x)$, where P_x is the original price, which may or may not include a tax component. How will the change affect the Gini coefficient of income inequality?

In this section we are interested in decomposing the GIE (14.21) into the effects on the poor and on the rich in the society (see Chap. 14 for several alternative interpretations of the GIE). Rewriting (14.22), the effect of a change in the price of x on the Gini coefficient in the overall population is

$$\frac{\partial G_o / \partial t_x}{G_o} = S_{xyo} \left(\eta_{xyo} - 1 \right), \tag{16.6}$$

where ∂t_x is a small change in the tax on x, $S_{xyo} = \mu_{xo}/\mu_{yo}$ is the average propensity to spend on x in the overall population, and η_{xyo} is the (Gini) income elasticity of commodity x.

As shown in Chap. 14, the direction of the effect of a change in the tax on x depends on one parameter—the (Gini) income elasticity of x. If the GIE equals (is greater than or is lower than) one, then the Gini coefficient will not be affected (decrease, increase).

The components of the (Gini) income elasticity are

$$\eta_{xy} = \frac{\text{cov}\ [x, F(y)]}{\text{cov}\ [y, F(y)]} \bullet \frac{\mu_y}{\mu_x} = \frac{b_{xy}}{S_{xy}}, \tag{16.7}$$

where b_{xy} is the (Gini) regression coefficient of the Engel curve, x is the dependent variable, and y is the explanatory variable. For consumption expenditures, the term b_{xy} is "the marginal propensity to spend on x," but it is also the (Gini) simple regression coefficient of consumption of commodity X on income.

It is convenient to decompose the components of the GIE in two steps. First we decompose the numerator—the Gini regression coefficient (the marginal propensity to spend)—and then we apply this decomposition to get the decomposition of the GIE. As shown in Chap. 7, when there are two nonoverlapping groups the Gini regression coefficient decomposes neatly into three components—the between-groups regression coefficient and two intra-group regression coefficients. Formally

$$b_{xyo} = w_p b_{xyp} + w_r b_{xyr} + w_b\, b_{xyb}, \tag{16.8}$$

where the w_i , i = {p,r,b} are the shares of the components in the Gini inequality in income, while b_{xyb}, b_{xyp}, and b_{xyr} are between-groups and intra-group regression coefficients of x on y, respectively. The immediate implication of (16.8) is that the overall (Gini) marginal propensity to spend is a weighted average of intra- and inter-groups propensities to spend. Note that b_{xyp}, b_{xyr}, and b_{xyb} do not necessarily have the same sign and hence the sign of b_{xyo} can be opposite to the signs of b_{xyr} and b_{xyp}.

Having decomposed the numerator, the decomposition of the overall income elasticity is straightforward. Using the definition of income elasticity in (16.7), we get

$$\eta_{xyo} = \frac{w_p\ S_{xyp}}{S_{xyo}}\ \eta_{xyp} + \frac{w_r\ S_{xyr}}{S_{xyo}}\ \eta_{xyr} + w_b\ \ \eta_{xyb}, \tag{16.9}$$

where η_{xyp}, η_{xyr}, and η_{xyb} are intra-group and between-groups elasticities and S_{xyp} is the share of x held by the poor. Equation (16.9) presents the overall GIE as a weighted sum of intra- and inter-groups elasticities. Note that each income elasticity has the same implication on the appropriate Gini inequality component as the overall elasticity: for example, if $\eta_{xyp} > 1$, then an increase in the tax on x decreases income inequality within the poor.

Although the decomposition is restricted to elasticities, one may argue that other components of a poverty measure are also of interest. It is argued that the entire effect on all components can be evaluated. The empirical illustration, however, is restricted to decomposing the Gini income elasticities.

Finally, it is worth mentioning that when the interest is in economic welfare then one may also be interested in evaluating the impact of the change in the tax on x on absolute changes in the level of well-being. In this case it is recommended to decompose $\mu_y\ (1 - G_y)$, where μ_y is mean income and G_y is the Gini's inequality measure. Because the decomposition of the mean is trivial it is easy to apply the decomposition of this chapter to this alternative setting.

16.5 An Empirical Illustration

16.5.1 The Data and the Main Findings

The data source used to illustrate the decomposition is the Family Expenditure Survey of Romania for 1993 (which includes 8,999 observations). The survey suffers from several limitations such as (a) the unemployed are underrepresented, (b) the inflation rate during the sample period was around 300%, rendering the nominal data meaningless. Therefore consumption had to be constructed from quantities and real prices (see Rashid (1995) for a detailed analysis of the construction of the data). Such limitations are common in many developing countries. The estimators of the Gini income elasticities are robust because they rely on ranks rather than on variate values, which gives them a relative advantage in handling contaminated data.

Economic well-being is represented by expenditures (i.e., consumption) per capita. This variable was chosen as a proxy for permanent income, which is an indicator of economic well-being. Household size is taken into account by using consumption per capita and weighting each household by its size.

Table 16.1 Components of the Gini coefficient of expenditure per capita: Romania, 1993

	All	Poor	Rich	Between
P-proportion of the population	28,766	0.20	0.80	
Mean expenditure per capita (Lei per month)	30,189	16,031	33,726	
Gini coefficient	0.22507	0.10637	0.18048	0.09373
Weight in overall Gini	1.0	0.010	0.573	0.416

Source: Table 1, Yitzhaki, 2002, p. 73
Reprinted with permission by Elsevier

Table 16.1 presents the decomposition of the Gini coefficient of expenditure per capita. The poverty line used is 20,087 lei per capita per month. The first line presents the proportion of individuals who are poor (20%) and nonpoor. The second line presents the mean expenditure per capita in each group, with the average expenditure of the rich being 210% higher than the average expenditure of the poor. The third line presents the Gini coefficients. The overall Gini is 0.225—typical to East European countries. Inequality among the poor is only 0.11, which is lower than inequality among the rich. Between-groups inequality is 0.09, which is 42% of the overall Gini.[16] The last line reports the w_i, the weight of each component in the Gini. The weight attached to inequality within the poor is less than 1% of the total weight, which means that the contribution of the inequality within the poor to the overall inequality is meager. However, this does not mean that the poor have no impact on the overall inequality. Their impact is expressed in the weight attached to the between-groups inequality, which is 42% of the total.

Having described the weight of each component we move on to analyze the effect of government policy on each component of inequality. The parameter that reflects this effect is the (Gini) income elasticity, which enables us to answer the following question: assume that the price of a "commodity" is increased by a small percentage point, what will the impact on the Gini index of inequality (i.e., on economic well-being) be? If the (Gini) income elasticity equals (is greater than, less than) one, the tax will not affect (decrease, increase) inequality. In general, the higher the elasticity, the more progressive the tax will be. (The usual rules that apply to the elasticity of a sum of commodities apply to the GIE as well). The decomposition of the income elasticity of "commodities" into poor, rich, and between-groups components enables us to see the effect of the tax on inequality among the poor, the rich, and between the groups.

Table 16.2 presents the decompositions of the Gini income elasticities for selected commodities in Romania, where income is defined as expenditures per capita. For completeness, Table 16.3 presents the average propensity to spend in each group, that is the S_{xyi} ($i = p,r,o$). The first line in Table 16.2 presents the (Gini)

[16] The poverty gap and the secondary decomposition can be computed using the data reported in the table. For example, the poverty gap in terms of average income is $0.2(20{,}087 - 16{,}031) \div 30{,}189 = 0.0269$.

Table 16.2 Gini income elasticities: Romania, 1993

	All	Poor	Rich	Between
Family size	−0.52	−0.48	−0.48	−0.67
	(0.01)	(0.05)	(0.01)	
Wage income	1.05	1.89	0.91	1.21
	(0.02)	(0.12)	(0.03)	
Agriculture income	1.08	0.45	1.16	0.99
	(0.03)	(0.16)	(0.05)	
Pension income	1.19	1.61	1.05	1.34
	(0.07)	(0.43)	(0.10)	
Child allowance	−0.70	0.34	−0.92	−0.64
	(0.03)	(0.15)	(0.04)	
Unemployment benefits	−0.67	0.42	−0.80	−0.72
	(0.09)	(0.39)	(0.12)	
Social assistance	0.60	0.67	0.61	0.62
	(0.14)	(0.76)	(0.19)	
Tobacco	1.01	1.56	0.99	1.02
	(0.03)	(0.16)	(0.05)	
Petrol	1.84	2.78	1.76	1.78
	(0.05)	(0.55)	(0.07)	
Wood + coal + oil	1.31	0.52	1.49	1.06
	(0.07)	(0.37)	(0.09)	
Electricity	0.70	1.72	0.46	1.01
	(0.04)	(0.28)	(0.05)	
Gas	0.68	1.90	0.43	0.98
	(0.03)	(0.23)	(0.05)	
Transportation	1.01	1.58	1.00	1.18
	(0.03)	(0.16)	(0.03)	
Wage tax	1.13	1.90	0.86	1.26
	(0.02)	(0.12)	(0.04)	
Farmers-incidence	0.60	0.46	0.58	0.62
	(0.01)	(0.06)	(0.02)	
Farmers per capita-incidence	0.03	−0.01	0.02	0.04
	(0.006)	(0.01)	(0.008)	

Jackknife standard errors appear in parentheses
Source: Yitzhaki, 2002, Table 2, p. 74

income elasticity of family size. The overall income elasticity is −0.52, which means that had we given a lump-sum subsidy to each person in the household, then the overall inequality would have declined. Assuming that the subsidy accounts for 1% of the expenditures, the overall Gini of the expenditures would have declined by 0.52%. Income elasticity among both the poor and the rich equals −0.48, while the between-groups elasticity is −0.67. Hence, a per capita subsidy will have the greatest effect on between-groups inequality.

We first apply the methodology to taxation of income sources. The second line in Table 16.2 presents the income elasticity of wage income. An across-the-board increase in wage income will mildly increase overall inequality (1.05), but will

Table 16.3 Income shares: Romania, 1993[a]

	All	Poor	Rich
Wage income	0.73	0.59	0.74
Agricultural income	0.55	0.56	0.55
Pension income	0.04	0.03	0.04
Child allowance	0.018	0.044	0.015
Unemployment compensation	0.01	0.03	0.01
Social assistance	0.01	0.01	0.01
Tobacco	0.023	0.022	0.023
Petrol	0.013	0.004	0.014
Wood + coal + oil	0.01	0.01	0.01
Electricity	0.01	0.01	0.01
Gas	0.02	0.02	0.02
Transportation	0.02	0.02	0.02
Wage tax	0.16	0.12	0.16
Income	1.54	1.44	1.55

[a]Entries present the shares of the item in the expenditures of the relevant group. For example, income is 154% of expenditures, while wage tax is 16%

Source: Yitzhaki (2002), Table 3, p. 75

have a large regressive impact on inequality among the poor (1.89), will decrease inequality among the rich (0.91), and will increase the between-groups inequality (1.21).

Agricultural income presents almost the opposite example. An increase in agricultural income mildly increases overall inequality (1.05), decreases inequality among the poor (0.45), increases inequality among the rich (1.16) and does not affect between-groups inequality (0.99). An increase in pension income increases all components of inequality. However, the large standard error among the poor indicates that this result is not robust.

The next group of items is intended to evaluate the effect of changes in public expenditures: an across-the-board increase in child allowances decreases all components of inequality with the smallest effect on inequality among the poor. Unemployment benefits display a similar pattern: although an increase in social assistance reduces inequality, it is clearly an ineffective program.

The next group of commodities represents classical commodities for indirect taxation. An increase in the price of tobacco will not affect overall inequality (1.01), between-groups inequality (1.02), and inequality among the rich (0.99), and will reduce inequality among the poor (1.56), making it an incidence-neutral ideal target for taxation. An increase in the price of petrol decreases all components of inequality, while an increase in taxes on electricity and gas is regressive for the population as a whole and progressive among the poor. Taxing wood-coal and oil will be progressive for the overall population but regressive among the poor. However, the standard errors indicate that this result is not robust. Transportation is progressive in all components.

The next item to check is the effect of a change in the wage tax. This turns to be a bit surprising. An across-the-board increase in wage tax will decrease overall (1.13), among the poor (1.9) and between-groups (1.26) inequalities, but will increase inequality among the rich (0.86). This result is consistent with the earlier finding that an increase in wages will decrease inequality among the rich.

The last two lines illustrate how to incorporate poverty incidence into inequality analysis. Farmers (anyone with income from agriculture) are assigned a dummy variable of one while the rest of the population gets a zero. A head tax on farmers is imposed and the (Gini) income elasticity is calculated. The results are reported in the line "Farmers-incidence." As can be seen, subsidizing farmers is more effective than subsidizing agricultural income as far as reducing between-groups inequality and overall inequality are concerned, but it is not more effective in reducing inequality among the poor.

The last line in Table 16.2 reports the effect of subsidizing each member of a household of farmers. As expected, this policy is more effective than handing an equal amount to each farmer. Surprisingly it is much less effective, in all components, than an allowance to family size which is given to all households (see the first line). The conclusion is that the incidence of poverty among farmers is lower than the incidence in the population as a whole.

The different applications of the (Gini) income elasticities in Table 16.2 to direct, indirect, and hypothetical taxes enable the user to compare the effectiveness of various policy measures in a quantitative and unified manner.

16.5.2 Sensitivity Analysis

The main theme of this chapter is that the (Gini) income elasticity contains all the needed information for analyzing the distributional impact of taxes, subsidies, and other government programs on different groups in the population. But this elasticity depends on an implicit and specific welfare function, which is embodied in the Gini coefficient. The question we seek to answer is: how sensitive are the policy conclusions to the reliance on the Gini coefficient?

As shown in Chap. 6, the Gini coefficient is one member of a family of extended Gini coefficients. This family of inequality measures can be written as

$$G(v) = v(v-1)\int_{0}^{\infty}(1-F)^{(v-2)}\,[F - LC(F)]dF, \tag{16.10}$$

where v is a parameter that is determined by the investigator and LC(F) is the Lorenz curve. If v equals 2, we get the regular Gini coefficient. The higher v gets,

Table 16.4 Weights attached to the between-groups component

	Extended Gini parameter						
	2	4	6	8	10	12	14
Share of between-groups	0.42	0.62	0.70	0.76	0.73	0.68	0.62
Between-groups Gini	0.094	0.249	0.327	0.358	0.363	0.353	0.334
Overall Gini	0.225	0.367	0.431	0.470	0.497	0.518	0.535

Source: Table 4, Yitzhaki (2002), p. 78

the greater the emphasis on the lower portion of the Lorenz curve. In the extreme case, when ν approaches infinity, the index represents someone who only cares for the very poorest of the poor.

The extended Gini can also be decomposed into three components: inequality within the poor, inequality within the rich, and between-groups inequality. To see this, let us define $LC_B(F)$ as the between-groups Lorenz curve, which is composed of linear segments where all the poor receive the poverty-line income and all the rich receive the affluent-line income (lines OG and GB in Fig. 16.1).

By adding and subtracting $LC_B(F)$ we get

$$G(v) = v(v-1)\int_0^\infty (1-F)^{v-2}\,[[F - LC_B(F)] + [LC_B(F) - LC(F)]]dF, \quad (16.11)$$

where the first term is the contribution of the between-groups component, while the second term can be decomposed into the contribution of the inequality among the poor and inequality among the rich.

One can repeat the question regarding the impact of a change in the price of a commodity on inequality among the poor, inequality among the rich, and between-groups inequality, and then derive the extended Gini income elasticities. Note, however, that the between-groups (Gini) elasticity is based on two observations and is therefore not affected by the index of inequality that is used in the analysis. Hence, an important factor that determines the sensitivity of the (Gini) income elasticities is the weight attached to the between-groups Gini component.

Table 16.4 presents the weight attached to the between-groups component for various values of the parameter of the extended Gini coefficient. It is based on the between-groups component of the extended Gini, which is equal to

$$G(v) = \frac{P_P - S_P}{P_P}[1 - (1 - P_P)^{v-1}]. \quad (16.12a)$$

The second line in the table shows that the weight attached to the between-groups component increases until ν reaches 8 and then declines. When ν equals 14, the weight is equal to the weight when ν = 4. Thus, for all practical purposes one can clearly view the between-groups elasticities as the relevant elasticities for a poverty-oriented planner. We are unable to supply such rules of thumb for inequality among the poor and among the rich; one has to calculate them.

An empirical conclusion that can be drawn from Table 16.4 is that for reasonable values of ν the term "caring more about the poor" is translated into attributing the lion's share of the weight to between-groups elasticity. To see this, note that ignoring the Gini coefficient which is analyzed in the previous section, in all cases reported the share of between-groups inequality is above 60%. This means that for a poverty-alleviating national policy the most important question to ask is how much is transferred to the poor and this issue should be given more weight than the question of allocation among the poor. It is not clear how general this result is. Also, because the weight attached to inequality among the poor increases, one has to check the income elasticity among the poor too: if it differs substantially from other elasticities, it might affect the overall elasticity as well.

The sensitivity of the (Gini) income elasticity to the level of the poverty line depends on the curvature of the Engel curve. If the Engel curve is linear, it is easy to show that the between-groups elasticity is equal to overall elasticity and that elasticity among the poor is greater than elasticity among the rich. One can view the difference between the overall and between-groups elasticities as an indicator of deviation from linearity. If the Engel curve is linear, an increase in the poverty line decreases the income elasticities among the poor and the rich without affecting between-groups elasticity.

16.6 Policy Analysis

From a policy-oriented economist's viewpoint, the major purpose of calculating a poverty or inequality index is to enable the evaluation of distributional implications of reforms in government policies. Because reforms are usually composed of changes in several parameters it is important to be able to evaluate the implications of a comprehensive policy change. The GIEs obey the basic rules governing elasticities and they can therefore be used to evaluate the impact of several changes in parameters. This section illustrates the decomposition of the (Gini) income elasticity to evaluate the effect of a reform on inequality among the poor, inequality among the rich, and between-groups inequality.

Before proceeding with the illustration, it is only fair to point out that if one is interested in simplifying policy analysis, income elasticities, and indices of inequality are not a good starting point. All inequality measures in use are relative, in the sense that incomes are normalized by mean income. Policy analysis can be conducted more conveniently with a social welfare function that is absolute in nature. If one is interested in analyzing tax policy using a Gini coefficient, it is convenient to define a social welfare function of the type $\mu_y\,(1 - G_y)$, where μ_y is the mean real income and G_y is the Gini coefficient.[17]

To illustrate, consider first the following policy reform: (a) The price of petrol is raised by 10% (of the existing consumer price), while the price of tobacco is increased by 5%. Then a first-order approximation to the decline in real income (in terms of average expenditures) is $(0.05s_T + 0.1s_P)$, where T and P represent Tobacco and Petrol, respectively. The combined effect on the Gini coefficient is (see Yitzhaki (1994a, 1994b))

$$\frac{dG_y}{dR_1} = G_y\,\frac{S_T(\eta_T - 1)0.05 + S_P(\eta_P - 1)0.1}{S_T\,0.05 + S_P\,0.1}, \tag{16.12b}$$

where R_1 is a one dollar tax revenue collected by the proposed reform. Equation (16.12b) is a weighted average of the derivatives of the Gini coefficient with respect to the taxes imposed, multiplied by the changes in taxes. The denominator normalizes the effect to give the effect of "one dollar of combined taxes." Since G_y is predetermined, it is ignored. The impact of this reform on the different components of inequality can be found by using the appropriate shares and elasticities from Tables 16.2 and 16.3.

The first line in Table 16.5 reports the changes in real income (expressed as percentages of consumption expenditure for each group); the second line presents the income elasticities of (a dollar of) taxes of this reform *minus* one (to make it easy to see whether the reform is progressive or not). As expected, all income groups are expected to suffer from such a reform, but since all elasticities are greater than one, all Ginis of consumption are expected to decline, with the biggest change being in inequality among the poor.

Consider now an alternative reform [reform (b)] in which in addition to the changes suggested in reform (a), the government also intends to decrease child benefits across the board by 4%. While reform (a) is clearly progressive, reform (b) includes a regressive element (reducing child benefits), and the final result is not clear, *a priori*. The third line in Table 16.5 presents the first-order approximation to changes in real income (the Slutsky compensation) in terms of the average income of each group. As can be seen from the fourth line, reform (b) is mildly regressive in the sense that overall inequality, between-groups inequality, and inequality among the rich slightly increase while inequality among the poor slightly declines.

---

[17] This form was first suggested by Sen (1973). Yitzhaki (1982a) shows that it may be interpreted as representing the theory of Relative Deprivation (Runciman, 1966). Yitzhaki (1982b) shows that it forms necessary conditions for second-degree stochastic dominance, which means that for two income distributions z and y, if the mean social welfare for z is greater than the mean social welfare for y for any concave social welfare function, then $\mu_z\,(1 - G_z) > \mu_y\,(1 - G_y)$ holds. Kakwani (1995) shows that one can decompose the "income elasticity" derived from this function into an income effect and an inequality effect and that the inequality effect can be interpreted as a progressivity index.

Table 16.5 The effect of tax reforms on real income and inequality

	Effect on			
	All	Poor	Between	Rich
Reform (a)				
Change in real income[a]	−0.245	−0.15		−0.255
(Gini) income elasticity *minus* 1	0.45	0.88	0.42	0.41
Reform (b)				
Change in real income[a]	−0.317	−0.326		−0.315
(Gini) income elasticity *minus* 1	−0.038	0.051	−0.045	−0.032

[a]Percent of mean income
Source: Table 5, Yitzhaki (2002), p. 79
Reprinted with permission by Elsevier

Incorporating other changes can be done in a similar fashion. For example, Millimet, Slottje, Yitzhaki, and Zandvakili (2003) analyze the distributional impact of investment in schooling. This kind of extension is beyond the scope of this book.

16.7 Summary

This chapter deals with tools that can be helpful in analyzing the impact of public policies. The tools are based on decompositions of inequality indices. The two main arguments of this chapter are as follows.

(a) Poverty indices are redundant in the sense that their contents can be derived by a decomposition according to subgroups of the appropriate index of inequality; and
(b) The main interest of policy-oriented economists when calculating indices of inequality is to find out how changes in economic policy affect their values.

By claiming that poverty indices are redundant we do not imply that the poverty line is redundant too. Having a poverty line and an institute which reports every year the status of the poor may have political implications. The analysis of the political implications is beyond the scope of this book.[18] The poverty line can be used in the decomposition of the inequality measure. The sensitivity of the inequality measure

[18] The following anecdote shows how difficult would it be to evaluate the political impact. Every year, in November, the National Insurance Institute (NII) of Israel publishes the official number of poor people in Israel. Because of population growth, this number tends to increase every year. This event marks the beginning of about a 3-week ritual with television shows interviewing poor people, parliamentary debates, and newspaper articles criticizing the government for its failure to eliminate poverty. In 1999, the NII was late in publishing its report. However, a resourceful politician published a number, and the well-oiled political machinery started to roll without the official report. News about the hardship of the poor were reported on television, the prime minister's office blamed his predecessor, etc. It took about a week to realize that the real number has not been published yet.

to the choice of the poverty line tends to be lower than the sensitivity of the appropriate poverty measure, and this may reduce the white noise. The arguments are illustrated by decomposing the Gini coefficient into Sen's poverty index, a Sen-like affluence index, and between-groups inequality.

To be effective, one should not concentrate only on measurement issues, but on pointing out which policy instruments can be effective for poverty alleviation. The discussion in this chapter points out the usefulness of the GIE, which is a summary statistic describing the distribution of potential tax and subsidy bases among income groups. The (Gini) income elasticity contains all the needed information for describing the impact of economic policies on poverty and inequality, provided that one uses the Gini coefficient as the measure of inequality.

An interesting empirical finding for Romania is that for reasonable values of inequality aversion "caring more about the poor" should be modeled as attaching the lion's share of the weight to between-groups inequality. This means that the dominant consideration in any poverty alleviation program should be devoted to how much is transferred to the poor. The way by which the subsidies are allocated among the poor plays a secondary role.

Chapter 17
Introduction to Applications of the GMD and the Lorenz Curve in Finance

Introduction

The purpose of this part of the book is to expose the reader to applications of the Gini methodology in financial theory. Those applications are relevant whenever one is interested in decision making under risk or in reducing the incompatibility between financial theory and econometric applications. Risky situations are characterized by having to make decisions without knowing what the exact outcome is going to be. This definition covers almost every decision a person makes.

When dealing with decisions under risk it is important to distinguish between two possible structures.

(a) Trade can take place after the realization of the random variables.
(b) No trade is possible after the realization.

The major difference between the two cases is that if (a) holds then all decisions under risky situation can be treated as dealing with one dimension, namely wealth or income as a proxy for wealth. That is, although there may be many dimensions to be taken into account, as is the case in portfolio theory, the risk aversion one has to deal with is relevant to one variable only. In finance the usual assumption is that trade is possible—a fact that simplifies the analysis. If, on the other hand no trade is possible after the realization then a multivariate (in the sense of multidimensional) analysis is required.

Most of the early arguments in economic theory were developed under the assumptions of full information and certainty. For example, the basic concept of efficiency is defined under the condition of certainty. The typical extension of economic theory to cover risky situations is to assume a probability distribution on the possible outcomes, which either reflects objective or subjective probabilities, and to aggregate the results in one way or another as is the case under expected utility theory.

S. Yitzhaki and E. Schechtman, *The Gini Methodology: A Primer on a Statistical Methodology*, Springer Series in Statistics 272, DOI 10.1007/978-1-4614-4720-7_17,

This aggregation brings to some formal similarities between models representing the main considerations in welfare economics and finance. The first similar factor is the aggregation. In both cases there is some kind of aggregation: in welfare economics it is an aggregation over individuals, while in finance it is an aggregation over the states of nature. The second similarity is that in both areas the aggregations are performed by using increasing concave functions. In welfare economics it is the social welfare function, while in finance it is the utility function. These two points of similarity explain why we should expect that models that are used in finance will share some of the characteristics of the models that were developed in the areas of social welfare and income distribution. An additional similarity which is the most distinguished characteristic from our point of view is the asymmetric approach to the data. In the area of income distribution this asymmetry is caused by the assumption of the declining marginal utility of income. In the area of finance it is caused by the assumption of risk aversion that can also be traced, in the expected utility model, to the declining marginal utility of income. These similarities between the fields enable us to apply almost the same formal models in the two fields. To attest that, note Atkinson's (1970) path-breaking paper in which he borrowed the concept of stochastic dominance that was developed in finance (Hanoch & Levy, 1969; Rothschild & Stiglitz, 1970) into the area of income distribution. Another example is Yaari (1987, 1988) who applied the same formal approach to the two fields using, of course, different arguments. Atkinson's contributions can be traced to two issues that concern this book.

(a) Atkinson has shown that the second-degree stochastic dominance (SSD) rules can be presented by using Lorenz curves. This observation led to the development of the Marginal Conditional Stochastic Dominance (MCSD) rules which can be viewed as the extensions of the rules from the univariate into the multivariate dimensions, or alternatively, from the variability measures to the covariability measures. This, in turn, leads the user to search for variability measures that can be used to construct necessary conditions for stochastic dominance. The GMD and the EG are measures with this property, as was shown in Chaps. 2–6.
(b) Atkinson has shown that if the Lorenz curves of two distributions (with equal means) intersect then it is possible to find two legitimate social welfare functions that will rank them in reversed orders. This implies that stochastic dominance rules cannot offer a complete ordering of outcomes. Incomplete order of outcomes is problematic and it is restricted in its applications because it leaves the user without a recommendation in many real-life cases. This property suggests the use of the Gini because it can complement the order into a complete ordering without violating the stochastic dominance rules, as was shown in Chap. 5.

In addition to the similarities there is one major difference that makes financial theory more complicated than welfare theory. Most models in finance include elements of additive and multiplicative relationships. To see this, consider the return on a portfolio. The transfer of funds from one asset to the other is additive, while the returns of different periods of time are multiplicative. This complication creates difficulties in the estimation of the expected return on a portfolio. This problem is exacerbated when we have to deal with the correlations between asset's returns over time because the Pearson's correlation coefficient measures linear relationships, and applying multiplicative models can turn a strong correlation into a weak one or even into no correlation whenever the product in based on a large number of periods (Embrechts, Lindskog, & McNeil, 2003; Levy, Guttman, & Tkatch, 2001; Levy & Schwarz, 1997).

Many of the models in finance are best described as time-series models and require tools for time-series analyses. The role of the GMD in time-series analyses has not been fully developed yet (see Chap. 23), but the first steps towards this direction can be found by Serfling (2010) and Shelef and Schechtman (2011).

Several properties of the Gini suggest that it may play an important role in analyzing time series and financial data. The relevant properties are (1) The GMD of a linear combination can be decomposed into the contributions of the individual components and the (Gini) correlations among them (the decomposition of the variance is a special case of this decomposition). (2) The GMD is based on L1 metric therefore it shares all the asymptotic properties that the variance has (because the variance is based on L2 metric). There are additional reasons for using the GMD in finance. Two of them are relevant in the area of welfare economics as well: (a) the GMD is compatible with the expected utility theory and with Yaari's dual theory and (b) it reveals more. An additional and crucial reason is because it helps overcome some of the econometric problems of using additive and multiplicative models in finance.

The reasons mentioned above are of a totally different nature. The first reason—compatibility with the expected utility theory and with Yaari's dual theory—is the following: Atkinson's proof that if two Lorenz curves intersect then it is impossible to determine whether inequality (or risk) has increased or declined created a wedge between the economic and statistical theories. The traditional way in econometrics is that the economist writes a model in a general form, identifying the major variables and the relationships among them and then comes the econometrician, who can be the same person wearing a different hat, and specifies the exact relationships among the variables by imposing assumptions that enable her to estimate the parameters and to test the validity and robustness of the conclusions. But Atkinson's findings imply that one's social views concerning the income distribution (or risk aversion) may determine the direction of the findings. Hence it is natural to impose the social views on the statistical analysis, which is actually the idea behind the extended Gini family of variability measures. In finance it implies that the risk aversion of the decision maker should be taken into account when estimating the model. This point needs a clarification. We do not argue that the classical way of analysis described above suffers from an internal inconsistency. However, we do argue that if some of the assumptions imposed on the data by the econometrician do not hold in the data then they may contradict the assumptions or the logic of the economist. To see this, note that expected utility theory defines the beta of an asset as a weighted average of the slopes of the security characteristic curve, weighted by the marginal utility of income. The higher the risk aversion the greater the decline of the marginal utility of income is. This implies that the higher the risk aversion the more weight should the estimation procedure attach to states of nature that result in low return on the portfolio. As long as the relationship between the return on an asset and the return on the portfolio (income in our example) is linear, the method of estimation of the beta need not take into consideration the risk aversion of the decision maker, because the slope of the regression curve is a constant. A problem arises when the linearity assumption is violated by the data. As shown in Chap. 7 the OLS attaches large weights to extreme observations, and it is not sensitive as to whether the extreme observations are at the lower end of the curve or at the upper end. As a result a wedge is created between the issues stressed by economic theory and the estimation procedure. While the risk-averse investor cares about the lower section of the curve, the estimation procedure is symmetric around the mean. If the linearity assumption is violated by the data, then it is possible that the econometrician (when using OLS) may estimate the regression curve by imposing large weights on the segments of the curves that the economist cares the least about. As a result the estimation procedure may supply the wrong ranking of assets. To solve this discrepancy, economic theory should be selected first and then imposed on the estimation procedure. Using the GMD or the EG regressions enables introducing the risk aversion into the estimation process.

The second reason for preferring the use of the Gini is that it reveals more. It enables the user to see whether the imposition of a symmetric correlation between variables holds in the data or not. Also, as we showed in Chap. 4 the decomposition of the Gini of a linear combination of random variables enables the user to see whether the combination belongs to the same family of distributions as the distributions of the individual variables and whether it converges to the normal distribution (this will be discussed in Chap. 23).

The additional reason for preferring the use of the Gini is that portfolio theory may involve combinations of multiplicative and additive models, a case in which the Pearson's correlation coefficient fails.

The rest of this chapter deals with the demonstration of the above-mentioned reasons, while Chap. 18 demonstrates the imitation of portfolio theory using mean-Gini, deals with the construction of Mean-extended Gini portfolios and introduces investors with different risk aversions and their effects on the betas.

The structure of the chapter is the following: Sect. 17.1 deals with the role of variability in calculating the rate of return. In Sect. 17.2 we restrict the discussion to additive processes and apply stochastic dominance to the portfolio selection problem. Section 17.3 develops the necessary and sufficient conditions for stochastic dominance in a portfolio context, while Sect. 17.4 shows how the concepts based on the Gini are related to the classical beta that is used in finance and Sect. 17.5 concludes.

It should be stressed that there is an additional well-developed application of the GMD in finance which is related to hedging. See among others Cheung, Sherman, Kwan, and YIP (1990) and Lien and Tse (2002) and the literature surveyed there. Applications of the concept of MCSD, intended to improve indices used in Finance can be found by Clark, Jokung, and Kassimatis (2011) and Clark and Kassimatis (2012). We do not cover this area of research because we restrict our discussion to basic issues. In principle, we could analyze almost every area in finance that uses the variance by replacing the variance with the square of the GMD.

17.1 The Role of Variability in Calculating the Rate of Return

It is well accepted that variability is associated with risk. The question we intend to ask in this section is whether it is also associated with return. This problem is important because if variability is also associated with an increase in the return then it may be that risk-averse investors do seek variability.

Let us start with several examples. In the first example consider an investor who is faced with a portfolio which is divided equally between two assets, one delivers a constant return of 10% for a period and the other delivers (−10) %. The investor has to hold the portfolio for two periods. The question is what is the rate of return on the portfolio? This is an example of a combination of additive and multiplicative models: additive because the portfolio consists of the sum of two assets and multiplicative because after two periods the rates are raised to the second power.

The answer is that the rate of return after two periods is given by $0.5 \times (1.1)^2 + 0.5 \times (0.9)^2 - 1 \approx 0.01 > 0$. Would we multiply the periodic rates of return by 2, that is the periodic rates of return would be 20% and (−20) % for the first and second assets, respectively, the rate of return after two periods will be higher: $0.5 \times (1.2)^2 + 0.5 \times (0.8)^2 - 1 = 0.04$. That is, the more diverse the rates of return are, the higher the portfolio's two period rate of return will be.

In the second example assume a continuous time framework. Given an investment for one period, assume that the instantaneous rate of return is normally distributed with parameters (μ, σ^2), and it is uncorrelated over time. It can be shown that for this case the value of investment is lognormally distributed hence the expected value of a dollar invested in this project at the end of the period is equal to $e^{\mu+0.5\sigma^2}$, which means an expected growth rate of $\mu + 0.5\sigma^2$. This implies that the higher the variance is, the higher the expected return will be. In other words, the variability plays a positive role in the expected value of return.

The explanation to both examples is the following: when dealing with compounded interest, the base for calculating the return in the second (and above) period changes and therefore the mean of the periodical rate of return is not the only contributor to the overall rate of return on the investment; the variance plays a role as well. To grasp the magnitude of the influence of the variability on the rate of return note Yitzhaki (1987) who found that the rate of return of the rich on investments in the stock markets is twice the rate of return of the poor. Almost all of the difference comes from the component of $0.5\sigma^2$ which was small among the poor but it is roughly equal to μ among the rich. Hence running regressions on rates of return may yield biased results if the variances of the rates of returns (i.e., of the different groups of investors) are different.

Table 17.1 Pearson correlation coefficient between X and X^a for a choice of values of a

a	0.1	0.5	0.7	0.9	1.1	2	5	10
Pearson	0.9035	0.980	0.994	0.999	0.999	0.968	0.821	0.661

One way to estimate a rate of return on a portfolio is to evaluate its value at the end of the period, and only then to estimate the rate of return by comparing the value at the end with the value at the beginning of the period. However, this method is not free of problems. Because the Gini cannot contribute to the discussion we do not elaborate on the subject and refer the interested reader to Yitzhaki (1987).

The contribution of the Gini method is related to another aspect of the combination of multiplicative and additive models, namely the effects on the correlation coefficient. Several authors have pointed out the inadequacy of the Pearson correlation coefficient in representing the association in such cases.

Let us start with the example of a normally distributed rate of return. The value of the asset is distributed according to the lognormal distribution. Consider two portfolios of equal values for which the distributions of the instantaneous rates of return are identical and jointly follow the bivariate standard normal distribution with $\rho = -1$. In theory, the two portfolios create a safe asset with constant return. However, because the distributions of the values of the portfolios are lognormal then according to De Veaux (1976) the correlation between two lognormal random variables cannot be lower than (−0.368). Therefore, there is a contradiction between the two approaches.

In order to further illustrate the problem with Pearson correlation in case of multiplicative models, consider the following example: Let X be a uniformly distributed variable on (0,1). Obviously corr(X,X) = 1. Apply a transformation $Z = X^a$ to one of the variables. Then the Spearman and the two Gini correlations between X and Z are all equal to one. The Pearson correlation coefficient depends on *a*.

$$\rho(Z,X) = \frac{0.5a}{(a+1)(a+2)\,[(1/12)\,\mathrm{var}(Z)]^{0.5}}$$

where

$$\mathrm{Var}(Z) = \frac{1}{(2a+1)} - \frac{1}{(1+a)^2} = \frac{(1+a)^2 - (2a+1)}{(2a+1)(a+1)^2} = \frac{a^2}{(2a+1)(a+1)^2}.$$

Plugging Var(Z) in the presentation of $\rho(Z,X)$ we get

$$\rho(Z,X) = \frac{(2a+1)^{0.5}}{(a+2)\,[(1/3)\,]^{0.5}} = \frac{[3(2a+1)]^{0.5}}{(a+2)}.$$

The Pearson correlation coefficient reaches its maximal value of 1 at a = 1. Table 17.1 presents the rate of decline as *a* gets further from 1.

Would we apply different transformations to the two variables, so that $Z_1 = X^a$ and $Z_2 = X^b$ then the rate of convergence of the Pearson correlation coefficient to zero will behave according to the following equation

$$\rho(Z_1,Z_2) = \frac{(a+1)(b+1)-(a+b+1)}{(a+b+1)(a+1)(b+1)[\mathrm{var}(Z_1)\mathrm{var}(Z_2)]^{0.5}} = \frac{[(2a+1)(2b+1)]^{0.5}}{(a+b+1)}$$

which converges faster to zero as a goes to infinity as long as $b > 1$.

Another example of the difficulties that arise when using Pearson correlation coefficient in a multiplicative model is given by Levy and Schwarz (1997). They refer to time series data and claim that when the sequences of observations are based on partial sums, the correlation coefficient will be independent of the differencing interval (that is, 1-day, 1-week, etc.) but in the multiplicative model, the more periods one takes into account the lower the Pearson correlation coefficient will be, except for the case where the periodical correlation is one. More precisely, let $(S_1,T_1),(S_2,T_2),\ldots$ be a sequence of independent, identically distributed pairs of variables and let $X_k = S_1 \times S_2 \times \ldots \times S_k$ and $Y_k = T_1 \times T_2 \times \ldots \times T_k$. Levy and Schwarz show that the Pearson correlation between X_n and Y_n tends to zero as n approaches infinity except when $Y = kX$ for some positive k (in which case the correlation is 1 for all n).

The proof of Levy and Schwarz (1997) is complicated and will not be given here. We note that preliminary work shows that the Spearman and Gini correlations are less vulnerable to this kind of criticism. However, further research is needed in order to be able to resolve this issue. We comment on this issue in Chap. 23.

An additional criticism of using the Pearson correlation coefficient in finance comes from advocates of modeling dependence between random variables with Copulas (Embrechts, Lindskog, & McNeil, 2003; Embrechts, McNeil, & Straumann, 2002) who prefer Kendall's tau or Spearman correlation coefficient. They present an example with two lognormally distributed variables with parameters (0, 1) and $(0, \sigma^2)$, respectively, and argue that if σ tends to infinity, then Pearson's correlation coefficient between them, ρ, converges to zero. The conclusion they reach is "Hence, the linear correlation coefficient can be almost zero, even if X and Y are comonotonic or countermonotonic" (2001, example 3.2, p. 15).

Additional support for a copula dependence in finance is offered by Dennenberg and Leufer (2008). In the abstract they argue that "it is argued that the dual volatility and dependence parameters are better suited than the classical parameters for applications in finance and insurance." They even define Gini covariance and "Gini copula correlation" but theirs are different than the Gini correlations used in this book. As far as we know it is not clear whether the Gini covariance defined by Dennenberg and Leufer (2008) can be used to decompose the Gini of a linear combination of random variables, which is an important property to have.

To sum up: we mentioned several problems in applying the Pearson correlation and even the expected rate of return in cases where multiplicative and additive operations are involved. It is clear to us that we only touched the top of the iceberg.

17.2 Stochastic Dominance, Lorenz Curves, and Gini for the Additive Model[1]

In this section and in the rest of our discussion about the role of the GMD in finance we will return to the main stream in finance which is based on an additive model. That is, we assume an investor who is an expected utility maximizer. The investor is interested in maximizing the expected one-period utility or the discounted value, using a risk-free rate of discount of several periods. In short, we ignore the multiplicative nature of compounded return.

Our interest is to apply the tools offered by the Gini methodology to portfolio theory.

The essence of portfolio optimization is to find a combination of safe and risky assets that maximizes the expected utility (or another function) of the investor. Because the utility function is not known, this target is substituted by maximizing the expected return while keeping the risk at a bearable minimum. This is the rationale behind the mean-risk models and in particular the mean-variance (MV) model which was originally derived as a special case of expected utility (EU) maximization. Although the conditions for which MV is analytically consistent with EU seldom hold in practice,[2] MV is widely accepted as the theory that makes sense from a practitioner's point of view because it captures two attributes: maximizing expected returns and minimizing risk.

One trade-off for its intuitive attractiveness is the dependence of mean variance on a specific measure of risk. A more general approach that relies on expected utility theory without fully specifying a utility function is stochastic dominance that is expressed in terms of probability distributions rather than the usual parameters of risk and return which are used in MV. Second degree stochastic dominance (SSD) rules apply a general form of expected utility theory assuming risk-averse expected utility maximizers. The outcomes thus apply to a wider group of investors.

Unfortunately generalizing the theory complicates the rules to the point that they seem intractable to most practitioners (see, for example, Thistle, 1993). Moreover, when the rules are applied to a portfolio of assets, which is the most relevant case for an investor, they cannot be reasonably explained and one must rely on faith in them and on the algorithm producing the optimal portfolios.

The aim of this section is to express SSD rules in terms of the traditional concepts used in portfolio theory. In other words, we will interpret SSD rules in terms of expected return and systematic risk (beta) so that portfolio managers can better grasp the rules. We do this by using absolute Lorenz curves in place of the typical cumulative distribution functions. This lets us express SSD conditions in terms of return and risk, and reconcile them with the capital asset pricing model (CAPM).

---

[1] This section is based on Shalit and Yitzhaki (2010).

[2] For example, multivariate normal probability distributions of returns or quadratic utility functions.

Using the EG will enable us to extend the CAPM to include investors with different attitudes toward risk.

Besides adjusting SSD rules to fit the problems of interest to portfolio managers, we extend SSD to MCSD rules. These rules state the conditions under which all risk-averse investors holding a specific portfolio will prefer to increase the share of one asset over the share of another. MCSD is a less demanding concept than SSD because it considers only marginal changes of holding risky assets in a given portfolio.[3]

It is worth noting that as far as we can see, stochastic dominance cannot bring, without additional assumptions, further results beyond the restriction to marginal analysis. To see this, note that an interior optimal portfolio and non-marginal dominance cannot coexist because if they both exist, then one asset dominates the portfolio and therefore we will end up with a corner solution.

17.2.1 Expected Utility, Stochastic Dominance, and Mean-Gini Rules

To achieve portfolio efficiency under expected utility maximization we must use utility functions and know the probability distributions of returns of all assets. To alleviate the need for specific utility functions in constructing optimal portfolios we propose using the rules of stochastic dominance, which are expressed in terms of cumulative probability distributions. If we confine the discussion to the class of all risk-averse expected utility maximizers, an appropriate mechanism would be SSD theory that states the necessary and sufficient conditions under which a portfolio is preferred to another by all risk-averse expected utility maximizers.

SSD conditions were developed independently by Hanoch and Levy (1969), Hadar and Russell (1969), and Rothschild and Stiglitz (1970). SSD rules are typically obtained by comparing the areas under the cumulative distributions of portfolio returns and are defined as follows (see Levy, 1992, 2006): consider two risky portfolios *A* and *B* with cumulative distributions F_A and G_B, respectively.

Definition For all risk-averse investors with nondecreasing concave utility functions U with $U' \geq 0$ and $U'' \leq 0$, SSD states that A dominates B if $E_F U(A) \geq E_G U(B)$, where E_F and E_G are the expectations using F_A and G_B, respectively.

Proposition *A necessary and sufficient condition. SSD rules state that A dominates B if and only if* $\int_{-\infty}^{z} [G_B(x) - F_A(x)]dx \geq 0$ *for all z which belong to the range of returns on A and B.*

---

[3] Yitzhaki and Mayshar (2002) showed that the assumption of continuity in the portfolio space implies that if there is no portfolio that dominates a given portfolio under MCSD, then there will be no other portfolio (among all of portfolios, not just marginal ones) that dominates the given portfolio.

Proof Hanoch and Levy (1969). See Chap. 5.

Definition The SSD efficient set is defined as the set of all the portfolios which are not dominated by other portfolios according to SSD rules.

The above-mentioned necessary and sufficient conditions are based on the areas under the respective cumulative probability distributions. The rules say that for all returns the area under the cumulative distribution for the preferred portfolio is always smaller than the area under the cumulative distribution for the dominated portfolio.

SSD rules are not easy to interpret especially by capital market practitioners who are used to evaluate risk and return (see Best, Hodges, & Yoder, 2006). This is further enhanced when constructing portfolios of risky and safe assets, because it is difficult to evaluate cumulative distributions of portfolios whose compositions are changing constantly. Therefore linear programming and numerical optimization methods are commonly used to build efficient SSD portfolios, most of them relying on discrete distributions (see, e.g., Chow, Riley, & Formby, 1992; Dentcheva & Ruszczyński, 2006; Ogryczak & Ruszczyński, 2002; Post, 2003; Ruszczyński & Vanderbei, 2003). As these techniques are based on numerical optimization methods, it is virtually impossible to check and interpret the results intuitively in terms that are used by the practitioners.

Financial economists and practitioners are used to visualize the analysis as done in the classroom. Hence, we suggest an easier way: presenting SSD conditions by means of absolute Lorenz curves, following the formulations given by Shorrocks (1983) and Yitzhaki and Olkin (1991).[4] These curves enable us to see the contribution of every asset to the expected return and the risk of the portfolio.

The Lorenz curve expresses the cumulative expected return on the portfolio as a function of the cumulative probability distribution of the return on the portfolio. Given a portfolio with a cumulative distribution F(x), the absolute Lorenz curve (the *Lorenz*) is defined as

$$\mathrm{ALC(p)} = \int_{-\infty}^{x_p} x\, f(x)dx \quad \text{for} -\infty \le x_p < \infty \quad \text{where } x_p \text{ is defined by } p = \int_{-\infty}^{x_p} f(x)dx \tag{17.1}$$

and f() is the density function of the return on the portfolio. The properties of the absolute Lorenz curves and the relationship with the GMD are listed and proved in Chap. 5.

We can now use the *Lorenz* to compare portfolios. According to SSD rules, portfolio *A* dominates portfolio *B* if and only if

$$\mathrm{ALC_A(p)} \ge \mathrm{ALC_B(p)} \text{ for all } 0 \le p \le 1. \tag{17.2}$$

[4] Shorrocks (1983) calls these curves *generalized Lorenz curves*.

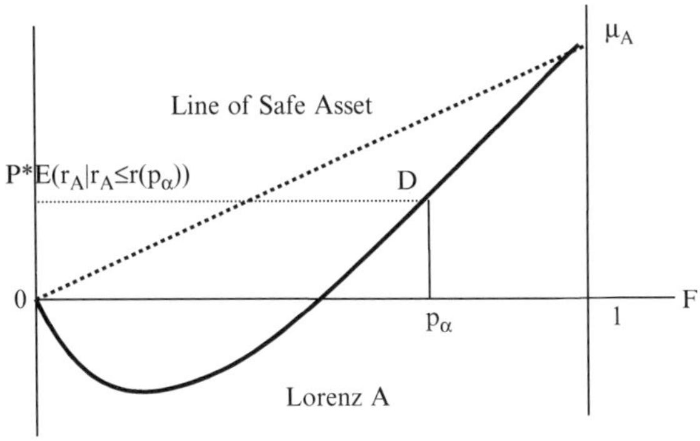

Fig. 17.1 SSD and absolute Lorenz curves. Source: Shalit and Yitzhaki (2010), Fig. 1, p. 434

The rationale for using absolute Lorenz curves to describe the properties of risky portfolios can be seen in Fig. 17.1. The *Lorenz* of a portfolio enables us to represent the expected return and the risk of the portfolio geometrically. As returns of a risky portfolio are ranked in increasing order, the shape of the *Lorenz* is convex with the lowest returns being at the left. In addition, the returns represent the slopes of the *Lorenz*. The curve starts at (0, 0) and ends at (1, μ), where μ is the expected return on the portfolio.

A safe asset with the same expected return μ will have a linear *Lorenz* that starts at (0, 0) and ends at (1, μ). In Fig. 17.1 the *Lorenz* of this asset is drawn as the straight dotted line which we label "the line of safe asset" (LSA). It is represented by the expected return multiplied by the probability *p*, where p is defined in (17.1). That is, LSA(p) = μp. We note that LSA coincides with the line of independence (LOI) as detailed in Chap. 5.

We can express the risk of a portfolio as a function of the vertical difference between LSA (that yields the same expected return) and its *Lorenz*. Indeed, for every probability *p*, investing in the portfolio provides a cumulative expected return expressed by the *Lorenz,* while investing in the riskless asset yields the same cumulative mean as given by the LSA.

Therefore, the farther the LSA from the *Lorenz* is, the greater the risk assumed by the portfolio. One possible measure of risk is the Gini's mean difference (GMD) of the portfolio which is obtained from the distances between the LSA and the *Lorenz*. Equation (17.3) shows that the area between the LSA and the *Lorenz* is equal to one-fourth of GMD

$$\int_0^1 [\mu\ p - \mathrm{ALC(p)}]dp = \mathrm{cov(r,F(r))} = \frac{1}{4}\Delta_r = \frac{1}{2}\ \Delta_r^*, \tag{17.3}$$

where μ p is the LSA(p), ALC(p) is the *Absolute Lorenz curve* (hereafter, the *Lorenz*) and Δ_r^* is half the GMD of the portfolio. Other measures of risk such as the extended Gini and even the variance can be obtained as functional of the vertical difference between the LSA and the ALC. (See Chap. 5).

We can gain additional insight from Fig. 17.1. The horizontal axis depicts the probabilities ranked from those generating the lowest portfolio returns and yielding the highest marginal utility to those generating the highest returns with the lowest marginal utility. Thus, the (equal) probabilities on the horizontal axis are ranked according to declining marginal utility. Because utility is defined over wealth, ranking probabilities with respect to portfolio returns yields the same result as if the rankings were according to declining marginal utility for each investor. All investors concur with this ranking because it is based only on portfolio returns, and their portfolio is assumed to be their only risky wealth.

While investors who hold the same portfolio and the same set of probabilities with respect to states of nature may not exhibit the same marginal utility from portfolio returns, they all agree upon the *ranking* of the marginal utilities of these returns. Hence, ranking with respect to portfolio returns is the only information we need in order to rank portfolios with respect to marginal utilities. The vertical axis in Fig. 17.1 shows the cumulative portfolio returns up to a specific state of nature, where states of nature are ordered according to the returns associated with their occurrences. The vertical difference between the LSA and the *Lorenz* of the portfolio represents the returns that, multiplied by the marginal utility, make up the expected utility. In other words, the loss in expected utility due to riskiness is the sum (integral) of the marginal utility multiplied by the distance between the LSA and the *Lorenz*. Different investors have different marginal utilities, so the loss due to riskiness differs among investors.

The connection between SSD and the nonintersection of *Lorenz* curves can be explained as follows. If one chooses to use a linear utility function, a necessary condition for the portfolio to be preferred by all expected utility maximizers is that it is preferred by the risk-neutral investor whose marginal utility is a constant. In this case one needs to look only at the last point on the *Lorenz*, which equals the portfolio's expected return.

Another necessary condition is that the area below the *Lorenz* of the dominating portfolio will be greater than the area below the *Lorenz* of the dominated portfolio. This area is one-half the expected returns minus one-fourth of the GMD. This is the logic behind the mean-Gini (MG) necessary conditions for SSD (Chap. 5) which are expressed as

$$\begin{aligned} \mu_A &\geq \mu_B \\ \mu_A - \Delta_A^* &\geq \mu_B - \Delta_B^*. \end{aligned} \tag{17.4}$$

These conditions state that if portfolio *A* is SSD preferred to portfolio *B*, then the mean and the risk-adjusted mean return of *A* cannot be less than the mean and the risk-adjusted mean return of *B* when risk is measured by the Gini of the portfolio.[5]

[5] Yitzhaki (1982a) also shows that the mean-Gini conditions for SSD are sufficient whenever the cumulative probability distributions intersect at most once.

Conditions (17.4) are stated in terms of the GMD. They hold for all EG variability measures relevant for the risk-averse investor as well. Note, however, that one can define an infinite number of necessary conditions by requesting that the dominating portfolio will have a higher *Lorenz* than the dominated one at any given p.

17.2.2 Absolute Concentration Curves and Marginal Conditional Stochastic Dominance

So far we described the necessary conditions for SSD in terms of risk-adjusted mean returns, treating each portfolio with a given composition of assets. The next step is to apply SSD to the appropriate composition of assets in the portfolio. The core of portfolio theory is the idea that diversification of asset holdings reduces an investor's exposure to risk. SSD in a portfolio must be applied in an environment where investors can change the choice of assets. For this purpose we rely on absolute concentration curves (ACCs). Because SSD rules are much more complex in a portfolio context than in applications to individual assets, one must recognize their limitations as noted by Shalit and Yitzhaki (1994) and formulate a simpler question.

Rather than defining rules for dominance, one might ask whether a given portfolio *A* belongs to the SSD efficient set. The inquiry to belong to an SSD efficient set proceeds in several consecutive conditions:

(a) Is it possible to find an alternative portfolio *B* in the neighborhood of *A* that differs from *A* by changing the shares of only two assets and SSD dominates portfolio *A*?
(b) If it is impossible to find such a portfolio, is it possible to find an alternative portfolio *B* in the neighborhood of *A* that differs from *A* by more than two assets and SSD dominates *A*?
(c) Finally, provided that we have failed to find portfolios that dominate *A* according to (a) and (b), is it possible to find an alternative portfolio *B* that SSD dominates *A*?

A portfolio that is not dominated by another portfolio according to these conditions belongs to the SSD efficient set. We address each question (i.e., condition) separately.

The first question is answered using the concept of MCSD as defined by Yitzhaki and Olkin (1991) and Shalit and Yitzhaki (1994). MCSD states the conditions under which all risk-averse investors holding a given portfolio *A* will prefer to increase the share of one asset over another. MCSD is more confining than SSD because it considers only marginal changes in holding risky assets in a given portfolio and restricts the change to involve two assets only.

To make MCSD operational, we develop the concept of ACC as follow.

Consider a portfolio of *n* risky assets $\left\{\alpha\middle|\sum_{i=1}^{n}\alpha_i = 1\right\}$ whose return r_α is defined by $r_\alpha = \sum_{i=1}^{n}\alpha_i\, r_i$, where r_i is the return on asset i, i = 1,. . .,n, and let f_α be the density function of the portfolio. Let $\mu_i(t) = E(r_i|r_\alpha = t)$ be the conditional expected return on asset *i*, given the portfolio return *t*. The ACC of asset *i* with respect to portfolio $\{\boldsymbol{\alpha}\}$ is defined as the cumulative conditional expected returns on asset *i* as a function of the portfolio's cumulative distribution $\mathrm{p} = \mathrm{F}_{\boldsymbol{\alpha}}(\mathrm{r}_{\alpha,\mathrm{p}})$

$$\mathrm{ACC_i(p)} = \int_{-\infty}^{\mathrm{r}_{\alpha,\mathrm{p}}} \mu_\mathrm{i}(\mathrm{t})\mathrm{f}_\alpha(\mathrm{t})\mathrm{dt} \quad \text{for} - \infty \leq \mathrm{r}_{\alpha,\mathrm{p}} \leq \infty \tag{17.5}$$

where $\mathrm{r}_{\propto,\mathrm{p}}$ is the pth quantile of the return distribution $\mathrm{F}_\propto$ defined as

$$\mathrm{p} = \int_{-\infty}^{\mathrm{r}_{\alpha,\mathrm{p}}} \mathrm{f}_\alpha(\mathrm{t})\mathrm{dt} = \mathrm{F}_\alpha(\mathrm{r}_{\alpha,\mathrm{p}}).$$

When not confusing, the subscript p will be omitted.

Similarly, from (17.1), the *Lorenz* of portfolio $\{\alpha\}$ is

$$\mathrm{ALC}_\alpha(\mathrm{p}) = \int_{-\infty}^{\mathrm{r}_{\alpha,\mathrm{p}}} \mathrm{tf}_\alpha(\mathrm{t})\mathrm{dt} \quad \text{for} - \infty \leq \mathrm{r}_{\alpha,\mathrm{p}} \leq \infty. \tag{17.6}$$

Following the definition of the portfolio, its *Lorenz* can be written as the weighted sum of the assets' ACCs held in the portfolio and can be expressed as

$$\mathrm{ALC}_\alpha(\mathrm{p}) = \sum_{\mathrm{i}=1}^{\mathrm{n}} \alpha_\mathrm{i}\mathrm{ACC_i(p)} \quad \text{for } 0 \leq \mathrm{p} \leq 1. \tag{17.7}$$

Figure 17.2 depicts the ACC of asset *i*. The horizontal axis represents the cumulative distribution of the portfolio's return and the vertical axis measures the cumulative expected return of asset i. The ACC of asset *i*, which is an asset that does not need to be included in the portfolio, relates the cumulative expected return on that asset to the cumulative probability distribution of the portfolio. The ACC of asset *i* is the solid curve. The dashed straight line is the LSA that connects the origin $(0, 0)$ with the point $(1, \mu_\mathrm{i})$, where μ_i is the unconditional expected return of asset *i*. The LSA represents an asset whose returns are independent of the performance of the portfolio and that has the same unconditional expected return as asset *i*.[6]

We now state the main theorem to determine MCSD using ACCs.

MCSD Theorem *(Shalit & Yitzhaki, 1994): Given portfolio {α}, asset k dominates asset j for all concave utility functions if and only if*

$$ACC_k^\alpha(p) \geq ACC_j^\alpha(p) \quad \textit{for all } 0 \leq p \leq 1, \tag{17.8}$$

with at least one strong inequality.

[6] LSA coincides with the Yitzhaki and Olkin (1991) line of independence (LOI). Samuelson (1967) shows that independent assets that are not included in the portfolio would be added to it if they have the same expected returns.

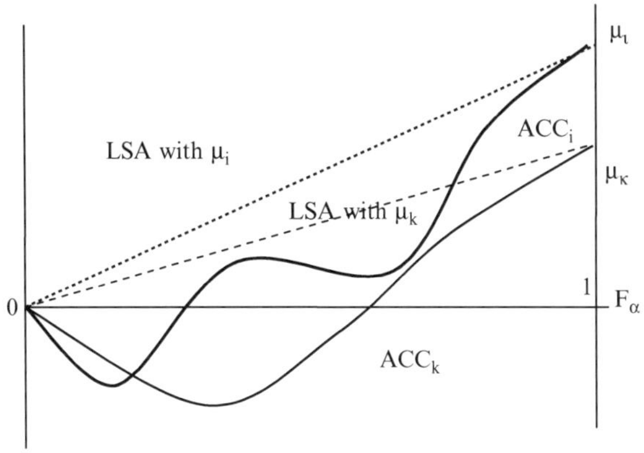

Fig. 17.2 Absolute concentration curves. Source: Shalit and Yitzhaki (2010), Fig. 2, p. 437

A proof is given in Chap. 5. Here an intuitive proof is given.

Intuitive proof Equation (17.7) provides a very simple proof for the theorem. Given the share of each asset in the portfolio, the ACC of asset j is the derivative of the *Lorenz* of the portfolio with respect to asset j. To increase the share of one asset on behalf of another in order for the new portfolio to SSD-dominate the given portfolio, the derivative of the *Lorenz* of the portfolio with respect to the dominating asset has to be greater than the derivative of the dominated asset everywhere.

To derive the necessary conditions for MSCD and relate them to the fundamental ideas in finance, we describe the basic properties of the ACC.

(1) The ACC of asset *i* passes through the points (0, 0) and $(1, \mu_i)$.
(2) The derivative of the ACC of asset *i* with respect to *p* is $\mu_i(t) = E_i(r_i|r_{\alpha,p} = t)$. Consequently, the ACC increases if and only if $\mu_i(t) > 0$.
(3) The ACC is convex, straight, or concave if and only if $\partial\mu_i(t)/\partial t \begin{Bmatrix} > \\ = \\ < \end{Bmatrix} 0$.
(4) When the returns r_α and r_i are independent the ACC$_i$ $[p_\alpha]$ coincides with the LSA (where p_α is the cumulative distribution of the portfolio).
(5) The area between the LSA and the ACC of asset i is equal to $\text{cov}[r_l, F_\alpha(r_{\alpha,p})]$, the covariance of the return on asset *i* and the cumulative probability distribution of portfolio $\{\alpha\}$. The area below the ACC is the area of the lower triangle minus that area between the LSA and the ACC, namely,

$$\int_0^1 ACC_i(p)dF_\alpha = \frac{1}{2}(\mu_i - 2\text{cov}(r_i, F_\alpha(r_{\alpha,p})) = \frac{1}{2}(\mu_i - \beta_i^G \Delta_\alpha^*),$$

where $\beta_i^G = \frac{2\text{cov}(r_i, F_\alpha(r_{\alpha,p}))}{\Delta_\alpha^*}$ is the Gini regression coefficient of asset *i* on the portfolio $\{\alpha\}$ and $\Delta_\alpha^* = 2\text{cov}(r_\alpha, F_\alpha(r_{\alpha,p}))$ is one-half of the GMD of the portfolio. Note that the beta is the well-known beta that is used in the CAPM except that it is based on the GMD rather than on the variance.

These properties allow one to state the necessary conditions for MCSD, namely, that if asset *j* dominates asset *k* conditional of holding portfolio $\{\alpha\}$ then

$$\begin{aligned}&\mu_j \geq \mu_k\\&\mu_j - \beta_j^G\, \Delta_\alpha^* \geq \mu_k - \beta_k^G\, \Delta_\alpha^*.\end{aligned} \tag{17.9}$$

The first condition implies that the dominating asset has a higher expected return than the dominated asset, regardless of the risk involved. The second necessary condition is more meaningful, as it states that a preferred asset has a higher risk-adjusted expected return than the risk-adjusted expected return of the less favored asset. Indeed β_j^G expresses the systematic risk in the mean-Gini model (MG-CAPM)[7]. Hence $\mu_j - \beta_j^G\, \Delta_\alpha^*$ is the risk-adjusted expected return, which is defined as the mean minus the beta calculated in Gini terms while the multiplication by Δ_α^*quantifies the risk.

From the second necessary condition one can obtain

$$\frac{\mu_j - \mu_k}{\Delta_\alpha^*} \geq \beta_j^G - \beta_k^G, \tag{17.10}$$

that is, when a security dominates another by MCSD, the difference between the two securities' expected returns per unit of portfolio risk must be greater than the difference in their systematic risks defined in terms of MG–CAPM.

Using the mean and the risk-adjusted mean return, this result allows for a complete ordering of investment alternatives. MCSD criteria using ACCs establish only a partial ordering. A complete ordering is an advantage when no dominance can be assessed by using ACCs, but a decision maker nevertheless wants to rank investment alternatives. In that case, the mean-Gini necessary conditions for MCSD provide an investment ranking that does not necessarily satisfy the sufficient conditions.

The above discussion has shown the criteria for finding a portfolio that will dominate the given portfolio by changing two assets at a time. It was based on the fact that the MCSD theorem is restricted to substituting one asset by another.

We now extend the method to several assets. The extension can be done in a relatively simple manner: instead of looking at two assets, we look at two linear combinations of assets.

The question asked is:

Is it possible to find an alternative portfolio B in the neighborhood of A, that SSD dominates A and differs from it in more than two assets?

According to (17.7), a combination of ACCs of several assets defines a new ACC that is a linear combination of the individual ACCs. Hence, to address MCSD involving more than two assets one needs to search for a linear combination of

---

[7] Whenever the CAPM is mentioned, it is interpreted as the reference portfolio held by the investor and not necessarily the market portfolio.

assets whose ACC is not below the ACC of the linear combination of other assets. This can be solved numerically as in Shalit and Yitzhaki (2003), and then the optimal ACC can be delineated.

Till now the search was restricted for a portfolio that SSD dominates a given portfolio in the neighborhood of the given portfolio. The next question to be asked is:

Is it possible to find an alternative portfolio B that SSD dominates A?

Yitzhaki and Mayshar (2002) have shown that if a portfolio is not MCSD-dominated by another portfolio it is also not SSD-dominated by any other portfolio. To understand the intuition that leads to this conclusion, let us consider two portfolios A and B, where B SSD-dominates A. In that case, for all risk-averse utility functions:

$$E[U(B)] \geq E[U(A)]. \tag{17.11}$$

To prove the argument it must be shown that if (17.11) holds, there is also a portfolio in the neighborhood of A that SSD dominates A. First note that:

$$\lambda E[U(B)] + (1 - \lambda)E[U(A)] \geq E[U(A)] \quad for\, 1 \geq \lambda \geq 0. \tag{17.12}$$

Because U is concave, we know that:

$$E\{U[(1 - \lambda)A + \lambda B]\} \geq \lambda E[U(B)] + (1 - \lambda)E[U(A)]. \tag{17.13}$$

Combining (17.12) and (17.13) we get:

$$E\{U[(1 - \lambda)A + \lambda B]\} \geq E[U(A)] \quad for\, 1 \geq \lambda > 0. \tag{17.14}$$

We now apply (17.14) for $\lambda \to 0$ and $\lambda > 0$, by which we find a portfolio in the neighborhood of A that SSD-dominates A. Therefore, it is impossible to have a portfolio that SSD dominates A without also having a portfolio in the neighborhood of A that SSD dominates A. Thus, we may conclude that if A is not MCSD-dominated then A is not SSD-dominated. Zhang (2009) presents an example in which it is shown that in a portfolio with more than two risky assets, even if a portfolio passes all pairwise MCSD dominances for assets, it still may be inefficient because it can be dominated by a portfolio that differs from the original portfolio by more than two assets. This means that numerical optimization must be involved when dealing with efficiency in portfolios.

17.3 Risk Aversion, Extended Gini, and MCSD

The extended Gini enables one to introduce the degree of risk aversion by adding one parameter into the calculation of the measure of dispersion. Indeed, with the parameter ν, which represents risk aversion, the extended Gini variability measure characterizes risk-averse investors ranging from risk-neutral ($\nu = 1$) to highly risk-averse max–min individuals ($\nu = \infty$). Other necessary conditions for MCSD that are specific to risk-averse agents can then be derived using the mean and systematic risk (now measured using the extended Gini). The MCSD-dominating asset has to have a higher risk-adjusted expected return than the dominated asset for every risk-averse investor. The risk-adjusted expected return is based on using the mean-extended Gini CAPM. For each asset and risk aversion coefficient, the extended Gini beta is calculated and used to adjust the expected return for risk.

The extended Gini specifies increasing risk aversion by stressing the lower segments of the distribution of portfolio returns. Similar to the GMD, which is defined as the vertical difference between the LSA and the *Lorenz* of the portfolio, the extended Gini is the *weighted* vertical difference between the LSA and the *Lorenz*. Using the parameter ν to adjust the area definition, we define the extended Gini for asset X as:

$$\Delta(\nu, X) = \nu(\nu - 1)\int_0^1 (1 - p)^{\nu-2}(p\mu_X - ALC_X(p))dp \tag{17.15}$$

where $ALC_X(p) = \int_{-\infty}^{X_p} xf_x(x)dx$ is the *Lorenz*, X_p is indirectly determined by $p = \int_{-\infty}^{X_p} f_x(x)dx$, $\nu(\nu - 1)(1 - p)^{\nu-2}$ is the weight associated with each portion of the area, and $p\mu_X$ is LSA(p). The parameter ν (>0) is being established by the researchers.[8]

There are some special cases of interest for the extended Gini:

For $\nu = 2$ (17.15) becomes one-half of GMD.

For $\nu \to \infty$ the extended Gini reflects the attitude of a max–min decision maker who wants to express risk in terms of only the worst outcome.

For $\nu \to 1$ (17.15) converges to zero, representing a risk-neutral investor who does not use any measure of dispersion to evaluate risk.

For $0 < \nu < 1$ the extended Gini is negative and models a risk-loving investor. For ease of presentation and because we are dealing with risk-averse investors, we assume that $\nu > 1$, although many of the results reported can be applied without modification to risk-loving investors. While in stochastic dominance the relevant definition of extended Gini is through the concentration curve, in financial analysis the covariance formula for the extended Gini is more convenient, because it enables imitating the variance based models:

$$\Delta(\nu, X) = -\nu cov(X, [1 - F(X)]^{\nu-1}). \tag{17.16}$$

The equivalence of the two definitions is derived in Chap. 6.

The weighted area under the Lorenz curve is equal to:

$$\mu - \nu cov(r,[1 - F(r)]^{\nu-1}). \tag{17.17}$$

---

[8] See Aaberge (2000) and Kleiber and Kotz (2002) on additional connections between the Lorenz curve and the extended Gini. Graves and Ringuest (2009) offer a tutorial for stochastic dominance.

We refer to (17.17) as RAR(ν)—the risk-adjusted expected return of an asset using the extended Gini Δ(ν,X). As shown in Chap. 6 (6.6), (17.17) is also a special case of Yaari's (1987) dual utility function.

One can introduce risk aversion differentiation into the SSD and MCSD necessary conditions and make them specific to various investors. A necessary condition for SSD is that the RAR(ν) of the dominating portfolio will be at least as high as the RAR(ν) of the dominated portfolio. Hence, the conditions for the portfolios shown in (17.4) become:

$$\begin{aligned} &\mu_A \geq \mu_B \\ &\mu_A - \Delta_A(\nu,X) \geq \mu_B - \Delta_B(\nu,X). \end{aligned} \tag{17.18}$$

The necessary conditions for MCSD developed in (17.9) can be replicated with the extended Gini as follows:

If asset *j* MCSD dominates asset *k* conditional on holding portfolio {α} then

$$\begin{aligned} &\mu_j \geq \mu_k \\ &\text{and } \mu_j - \beta_j(\nu)\ \Delta_\alpha(\nu, X) \geq \mu_k - \beta_k(\nu)\ \Delta_\alpha(\nu, X)\,. \end{aligned} \tag{17.19}$$

Note that this time β_j (ν) is defined in terms of the extended Gini as follows:

$$\beta_j(\nu) = \frac{\mathrm{cov}(r_j, [1 - F_\alpha(r_\alpha]^{\nu-1})}{\mathrm{cov}(r_\alpha, [1 - F_\alpha(r_\alpha)]^{\nu-1})}, \tag{17.20}$$

and Δ_α(ν,X) is the extended Gini as shown in (17.16).

Interpretation of (17.19) remains the same as of (17.9), except that the necessary conditions depend on the investor's specific coefficient of risk aversion (that is, on the choice of ν). This is the main point: if asset *j* dominates asset *k* according to MCSD, then it must be that the risk-adjusted expected return of *j* is higher than the risk-adjusted expected return of *k*, where risk is measured by extended Gini betas, for any choice of risk aversion parameter ν. In other words, if asset *j* MCSD dominates asset *k* for a given portfolio α, there is no extended Gini beta for *k* such that for any possible ν the RAR(ν) of *k* will be greater than the RAR(ν) of *j*. These conditions, however, are merely necessary and not sufficient, because the family of extended-Gini utility functions does not cover all possible risk-averse utility functions. For example, they do not include a change in the parameter of risk aversion ν on a given point along the distribution of returns.

The above discussion introduces the extended Gini to express the necessary conditions for SSD and MCSD. One issue is left unresolved: how do we choose the risk aversion parameter ν? Hence, the question to be asked is really how one can choose a utility function that represents a specific investor. By gathering information on the investor's decision making under risk, presumably one can estimate the parameter ν specifically for a particular investor, but this is a question for further research. See for example, Shalit and Yitzhaki (1989) where one estimates the risk aversion of the market using the market portfolio over time.

17.4 Beta and Capital Market Equilibrium

Stochastic dominance was developed in order to construct portfolios for specific investors' classes. As such, it ignores the notion of capital market equilibrium.[9] On the other hand, the concept of beta emerged as the equilibrium price of non-diversifiable risk carried by an asset in a competitive financial market situation. The beta is the regression coefficient of the return on the asset on the returns of the portfolio (the market if everyone holds the market portfolio). In this section we comment briefly on the connection between the beta and MCSD.

Consider first a financial market of risky assets where all returns follow a multivariate normal distribution. The agents in this market may have different levels of risk aversion. In this case investors will hold an identical portfolio of risky assets, i.e., the "market portfolio," and the only difference between them will come about in the allocations of their wealths between the risky portfolio and the risk-free asset. In this theoretical textbook case, the settings for deriving SD and beta will be identical.

As we have shown in (17.9) a necessary condition for MCSD of asset j over asset k is that if asset *j* dominates asset *k* conditional on holding portfolio $\{\alpha\}$ then

$$\mu_j \geq \mu_k$$
$$\mu_j - \beta_j^G \Delta_\alpha^* \geq \mu_k - \beta_k^G \Delta_\alpha^*.$$

Note that the first condition is not affected by the risk aversion of the investor. The second condition says that the risk adjusted expected return on asset j is greater than the risk adjusted expected return on k. But this condition has to hold for all EG and actually for all concave utility functions. Therefore we may conclude that stochastic dominance rules do not invalidate the concept of beta. They only demand that we define the concept for each utility function. We will analyze the implications of assuming investors with different risk aversions in Chap. 18.

17.5 Summary

The objective of this chapter is to introduce the terminology and the theoretical results that are needed for the applications of the Gini methodology in financial theory. Those applications are relevant whenever one is interested in decision making under risk, or in reducing the incompatibility between financial theory and econometric applications. There are several reasons for using the Gini in finance: (a) it is compatible with the expected utility theory and with Yaari's dual theory (b) it reveals more and (c) it helps overcome some of the econometric problems in finance.

---

[9] This topic is dealt in Chap. 18.

The main tools are based on the stochastic dominance rules. It is shown how to use these rules in classifying efficient portfolios. Stochastic dominance rules can be derived by curves or numerical optimization. For an economist and a practitioner who are used to think in terms of risk and return, a major weakness of the models based on numerical optimization is their inability to express the results intuitively. The remedy is to characterize the rules geometrically by using absolute Lorenz curves (ALC) for SSD and absolute concentration curves (ACC) for MCSD. One can then interpret the rules in terms of risk-adjusted mean returns depending on different measures of risk aversion. The curves enable the user to see the contribution of every asset to the expected return and the risk of the portfolio. For example if we denote the ALC of a safe asset (having expected return μ) by LSA (that is, LSA(p) = μ p) then one can express the risk of a portfolio as a function of the vertical difference between LSA (that yields the same expected return) and its ALC: the farther the LSA from the ALC is, the greater the risk assumed by the portfolio. One possible measure of risk is the GMD of the portfolio which is obtained from the distances between the LSA and the ALC: the area between the LSA and ALC is equal to one—fourth of GMD. Chapter 18 will be devoted to the applications of the concepts derived in this chapter to construct optimal portfolios.

How does systematic risk explain stochastic dominance efficiency? Beta, which is used by practitioners in finance, measures systematic risk as the covariance between asset return and market return.[10] The concept is rooted in mean-variance theory as it prices security risk in capital market equilibrium. The measure depends mainly on the validity of MV and its compatibility with maximizing expected utility when returns are multivariate normally distributed or when the investor's utility function is quadratic. The presence of fat tails and skewness in financial data precludes normality of returns, and quadraticity of preferences leads to unwarranted results. Hence, alternative measures of systematic risk are called for. For example one can look at the covariance between asset return and marginal utility to express undiversifiable risk correctly. Hence systematic risk depends upon the choice of the risk measure chosen by the investors. In the case of Gini's mean difference and the extended Gini, the resulting betas are the mean-extended Gini betas used in the necessary conditions for stochastic dominance. Gregory-Allen and Shalit (1999) have shown that the mean extended Gini (MEG) betas, which depend upon the investor's degree of risk aversion, subside to the standard MV betas only when returns are normally distributed. As it is seldom the case that normality holds, we advocate MEG betas to be used for stochastic dominance.

[10] In general the term market's return should be interpreted as the portfolio's return. See Shalit and Yitzhaki (2009) concerning CAPM with heterogeneous risk-averse investors.

Chapter 18
The Mean-Gini Portfolio and the Pricing of Capital Assets

Introduction

Since its development by Markowitz (1952, 1970), the mean-variance (MV) model for portfolio selection has become the standard tool by which risky financial assets are allocated. MV has gained a prominent place in finance because of its conceptual simplicity and ease of computation. Many authors, however, have challenged the model's assumptions, primarily the normality of the probability distributions of the assets' returns or the quadraticity of the preferences. MV validity has been reasserted by Levy and Markowitz (1979) and by Kroll, Levy, and Markowitz (1984), who showed that MV faithfully approximates expected utility.

The challenge to the validity of MV has led researchers to seek alternative solutions to efficient portfolio selection, resulting in approaches such as the three-moments, lower partial moments, semi-variance, value-at-risk, stochastic dominance and mean-Gini (MG) models, to name only a few. Still, no other model has managed to attain the popularity of MV by practitioners so far, owing to the lack of intuitive reasoning and to the complex computations required by the alternative models.

The MG analysis provides a consistent alternative to MV modeling whenever investment returns are not normally distributed or when the investor's utility is not quadratic (and therefore MV is not applicable). The MG approach in finance is superior to the MV approach because by supplying necessary conditions for stochastic dominance MG efficient set is a subset of an SSD efficient set. That is, every portfolio in the efficient set could be derived by a maximization of an expected utility function.

The mean-extended Gini (MEG) offers a simple way to include risk aversion in the construction of an efficient portfolio by providing a family of variability measures that depend on one parameter. (See Chap. 6 for details). By varying the EG parameter the investigator modifies the risk aversion and offers an efficient

This chapter is based on Shalit and Yitzhaki (2005, 2010).

S. Yitzhaki and E. Schechtman, *The Gini Methodology: A Primer on a Statistical Methodology*, Springer Series in Statistics 272, DOI 10.1007/978-1-4614-4720-7_18,

frontier that suits the risk preference of the investor. The efficient frontier can then supply the price attached to a unit of risk, given a level of risk aversion.

The objective of the first part of this chapter is twofold. First, we present the MG and MEG portfolio models as workable alternatives to MV. We compute the MG and the MEG efficient frontiers numerically and compare the results to the MV frontier. Second, we derive the MG efficient frontier analytically in a way that is similar to the derivation of the MV efficient frontier so it can be used in practice as easily as MV (under some assumptions to be detailed later). Generally, deriving MG portfolios is complex mainly because of the additional information which the Gini infers on properties of the distributions. If one is ready to forgo this additional information, finding and interpreting MG portfolios can be as simple as constructing and analyzing MV portfolios. In the second part of the chapter we deal with financial market equilibrium, where investors differ in their risk aversion. In particular, we prove that the result of the MV-Capital Asset Pricing Model (CAPM), namely that all investors hold the same market portfolio is caused by the assumption that all investors have the same risk aversion. Once this assumption is removed then each type of risk-averse investors holds the same risky portfolio, but it may be that no one holds the market portfolio. In some sense it turns out that characterizing the equilibrium in the capital market is not different than characterizing the equilibrium in the markets for commodities.

The uses of the MV and the MEG models follow two similar steps but they differ by the order of actions. In the MEG procedure one first chooses a utility function, (i.e., a parameter representing risk aversion) and then the utility function determines the measure of variability that represents the risk and the appropriate correlation coefficients. In the MV procedure one first chooses the variance as the measure of risk. Hence, one is actually selecting a specific utility function (i.e., quadratic) which imposes a specific type of risk aversion. This difference composes the base of our criticism of the MV model. The advantage of the MEG approach is that it enables the estimation of both the CAPM and the risk aversion of the market, while the MV approach imposes the risk aversion and given that, estimates the CAPM.

The structure of the chapter is the following: the first part (Sects. 18.1 and 18.2) deals with the construction of the MG and the MEG portfolios, while the second part (Sects. 18.3 and 18.4) deals with financial market equilibrium, where investors differ in their risk aversion. Section 18.5 concludes.

18.1 The Mean and Mean-Extended Gini Efficient Frontiers

In the MG model investors use the portfolio's Gini Mean Difference (GMD) as the measure of risk to be minimized, subject to a given mean return. The most convenient presentation of the GMD to be used is the covariance presentation:

$$\Delta^* = 0.5\Delta = 2\text{cov}[\text{r},\text{F}(\text{r})] \qquad (18.1)$$

where r is the return, Δ is the GMD, and F(r) is the cumulative distribution function (cdf) of the return.(See Chap. 2 for the various presentations of the GMD).

The advantage of the Gini over the variance as a measure of risk is rooted in the necessary and sufficient conditions for second-degree stochastic dominance (SSD) in the following way: consider two portfolios (1 and 2) yielding returns r_1 and r_2, with means μ_1 and μ_2, and Ginis Δ_1 and Δ_2. Then $\mu_1 \geq \mu_2$ and $\mu_1 - \Delta^*_1 \geq \mu_2 - \Delta^*_2$ are necessary conditions whereby no risk-averse expected utility maximizer will prefer portfolio 2 to portfolio 1. If one restricts the distributions of the portfolios to the family of cumulative distributions that intersect at most once, these conditions are also sufficient. Because MV is not compatible with expected utility theory, MG ranks risky alternatives consistently even whenever MV might fail (Shalit & Yitzhaki, 1984).

The implication of this result is that the efficient set of MG is included in the efficient set of risk-averse investors, so that every efficient MG portfolio maximizes the expected value of a utility function. This result does not hold for the MV efficient set. To see this, consider the choice between two portfolios. The first portfolio offers a return between zero and one dollar, and the second offers returns between one million and two million dollars. Both portfolios are included in the efficient MV set because the first offers a lower variance and the second offers a higher expected return. Thus, if one relies only on the mean and the variance, one may end up choosing the portfolio that every expected utility risk-averse investor would reject. The necessary conditions for stochastic dominance prevent MG users from making this mistake. We now move to constructing MG portfolios.

Consider a portfolio p whose returns r_p are obtained by $\mathrm{r_p} = \sum_{\mathrm{i}=1}^{\mathrm{n}} \alpha_\mathrm{i} \mathrm{r_i}$, where α_i and $\mathrm{r_i}$ are the share and the return to asset i, respectively. Then, the (half of the) GMD of the portfolio is:

$$\Delta_p^* = 2\mathrm{cov}(\mathrm{r_p}, \mathrm{F_p}) = 2\sum \alpha_\mathrm{i}\mathrm{cov}(\mathrm{r_i}, \mathrm{F_p}). \tag{18.2}$$

We can now obtain the MG-efficient frontier by solving the following optimization problem:

$$\begin{aligned} &Min\ \Delta_\mathrm{p}^* \\ \text{s.t.}\ \ &\mu_\mathrm{p} = \sum \alpha_\mathrm{i}\mu_\mathrm{i} \\ &1 = \sum \alpha_\mathrm{i} \\ &\alpha_\mathrm{i} \geq 0\ \ \mathrm{i} = 1,\ldots,\mathrm{n} \end{aligned} \tag{18.3}$$

where the last set of inequalities is optional, and is applicable when short sales are not allowed. Problem (18.3), although similar in structure to the MV optimization problem, is much more complicated than the MV problem because the cumulative distribution of the portfolio is not a simple function of the distribution functions of the returns of the individual assets.

MG analysis can be extended to include the investor's preference toward risk. This is done by introducing the extended Gini as a measure of risk. The extended Gini attaches higher weights to the lower portions of the probability distribution of the return of the portfolio. This implies that higher risk aversion attributes more weight to the lower payoff realizations than will lower risk aversion.

The extended Gini is defined much like the definition in (18.1):

$$\Delta(v,r) = -v\ \mathrm{cov}\{r, [1 - F(r)]^{v-1}\}, \tag{18.4}$$

where ν is a parameter determining the relative weight attributed to various portions of the probability distribution. The parameter ν ranges from 1 to infinity, with $\nu \to 1$ implying variability as viewed by a risk-neutral investor. For $\nu = 2$ we obtain the standard Gini risk aversion and for $\nu \to \infty$ we allow for the max–min investor who wants to avoid the worst possible outcome.

The utility function that is implied by using the extended Gini can be viewed as a special case of the utility functions suggested by Yaari's (1987) dual theory of risk aversion that distinguishes the notion of declining marginal utility of income from behavior under risk. In the case where ν is a positive integer, the link of ν to risk aversion can be shown as follows: the extended Gini equals the mean return minus the expected least outcome from ν independent random draws from the return distribution:

$$\Delta(v\ ,\mathrm{r}) = \mu - \mathrm{E}[\mathrm{Min}(\mathrm{r}_1,\ldots,\mathrm{r}_v)]. \tag{18.5}$$

(See Chap. 6 for this and other ways to express the extended Gini). Equation (18.5) can then be used to develop additional necessary conditions for stochastic dominance. In particular, comparing risk adjusted return $\mu - \Delta(\nu,r)$ of a risky portfolio with the return on a safe portfolio enables one to view $\mu - \Delta(\nu,r)$ as the certainty equivalent of the portfolio's return. With a higher ν one assigns higher odds to obtaining bad outcomes. Hence, the certainty equivalent of the portfolio is lowered. This, in turn, raises the risk premium required by the investor.

This interpretation of certainty equivalence relates risk aversion to the discounting of probabilities of good outcomes, which does not originate from assuming declining marginal utility of income as is the case in expected utility theory. Rather, the higher the risk aversion, the more the investor tends to amplify the probability of bad events and to discount the probability of good events.

In applications the empirical cumulative distribution function is used as an estimator of the cumulative distribution. It is obtained by ranking the returns of the portfolio in increasing order and dividing the rank of each observation by the number of observations. Because a ranking procedure is invoked each time the portfolio's GMD (or extended Gini) is calculated, nonlinear programming techniques should be used with caution.[1]

[1] When applied to empirical data, the problem is one of a piece-wise linear optimization. See Okunev (1991) and Okunev and Dillon (1988) for a linear programming solution.

There are several numerical methods to solve the MG portfolio selection. As far as we know the simplest and best way is offered by Cheung, Kwan, and Miu (2007). They use Excel Solver Macro to choose the optimal portfolio. This program can be easily adjusted to solve the MEG portfolio optimization, with and without short sales.

18.2 Analytic Derivation of the Mean-Gini Frontier

As mentioned above, under certain conditions the MG-efficient portfolios can be derived analytically in a manner similar to the derivation of MV portfolios. In this section it is shown that if one is ready to impose restrictions on the underlying asset distributions then one can derive the MG portfolios using the same technique that solves the MV constrained minimization problem. To see the parallel between the MG and the MV derivations, only the differences between the two dispersion measures have to be explored because all the other components of the optimization problems are identical.

The Gini and the variance derive their properties from the covariance. The variance is calculated as the covariance of the return with itself, while the GMD is the covariance of the return with its cdf. We note that GMD's reliance on the return and its cumulative distribution complicates its use, but this relationship does enable the GMD to extract more information about the underlying distribution.

The main concern in portfolio analysis is that the Gini is associated with two Gini correlation coefficients between each pair of returns, while the variance is related to one correlation coefficient (the Pearson's correlation coefficient). The two Gini correlation coefficients between r_i and r_j, are given by

$$\Gamma_{ij} = \frac{\mathrm{cov}[r_i, F_j(r_j)]}{\mathrm{cov}[r_i, F_i(r_i)]} \qquad \Gamma_{ji} = \frac{\mathrm{cov}[r_j, F_i(r_i)]}{\mathrm{cov}[r_j, F_j(r_j)]}. \tag{18.6}$$

Both correlation coefficients are needed in order to decompose the portfolio's GMD into the contributions of individual assets. (See proposition 18.1 below for the decomposition and Chaps. 3 and 4 for the properties of the Gini correlations). For our purpose it is important to note that the two correlation coefficients are not necessarily equal. They are equal if the distributions of r_i and r_j are exchangeable up to a linear transformation. Intuitively, exchangeability up to a linear transformation requires as a necessary but not a sufficient condition that the shapes of the marginal distributions of assets are equal up to a linear transformation. A disparity in the two correlations means different shapes of the two marginal distributions (of asset returns).

The correlation coefficients allow us to decompose the portfolio's Gini as follows:

Proposition 18.1 *Let* $r_p = \sum_{i=1}^{n} \alpha_i r_i$. *Then*

$$\Delta_p^2 - \Delta_p \sum_{i=1}^{n} \alpha_i D_{ip} \Delta_i = \sum_{i=1}^{n} \sum_{j=1}^{n} \alpha_i \alpha_j \Delta_i \Delta_j (\Gamma_{ij} + \Gamma_{ji}) \tag{18.7}$$

where $D_{ip} = \Gamma_{ip} - \Gamma_{pi}$ $(i = 1, \ldots, n)$ *is the difference between the two Gini correlations defined by the return of the portfolio and the return of asset i.*

Proof See Chap. 4.

Assuming exchangeability up to a linear transformation between the distribution of each asset and the portfolio implies that D_{ip} equals 0 for all i. Hence, exchangeability among the portfolio and the assets leads to:

$$\Delta_p^2 = \sum_{i=1}^{n} \alpha_i^2 \Delta_i^2 + \sum_{i=1}^{n} \sum_{j \neq i}^{n} \alpha_i \, \alpha_j \Delta_i \Delta_j \, \Gamma_{ij}. \tag{18.8}$$

Note that the decomposition of the variance of a linear combination of random variables is a special case of the above decomposition where each (square of) Gini in (18.8) is replaced by the variance and each Gini correlation is replaced by the Pearson correlation coefficient. Because the rest of the optimization problem (18.3) is identical to the MV optimization problem, one can adapt the textbook derivation of MV and apply it to MG (see, for example, Huang and Litzenberger (1988, p. 63) and Merton (1972)).[2] By substituting the GMD by the EG we can replicate the models an infinite number of times. That is, for each EG risk aversion parameter we produce the appropriate optimal portfolio.

Ignoring sampling variability, the MG, MEG, and MV solutions will be identical if the underlying distributions are multivariate normal. However, even if the distribution of only one asset diverges from normality, the solutions of the MG, MEG, and MV will differ.

The numerical optimization algorithm offered by Cheung, Kwan, and Miu (2007) which uses Excel Solver Macro to choose the optimal MG portfolio solves the portfolio optimization problem quickly, with and without short sales and with and without the exchangeability assumption. Also, this program can be easily adjusted to solve the MEG portfolio optimization, with and without short sales.

18.3 Capital Market Equilibrium with Two Types of Investors[3]

Having established that one can replicate MV portfolio with MEG portfolios, we turn now to see the implication of this development on the classical results in finance, namely the Capital Asset Pricing Model (CAPM) and the beta.

---

[2] To be precise, for each covariance in the MV framework we substitute a Gini correlation multiplied by the appropriate Gini.

[3] This section is based on Shalit and Yitzhaki (2009).

The CAPM that has dominated finance since the 1960s is a two-parameter model that is convenient and appealing to most investors, practitioners, and financial theoreticians, as it is simple and can present the choice between return and risk in a transparent way. While the contingent markets approach of Arrow and Debreu (1954) provides a theoretical alternative to capital market equilibrium with heterogeneous investors, most financial practitioners prefer to characterize the distribution of risky assets by two summary statistics: one for the mean return, and one for risk. The most popular measure for the latter is the variance.

In the standard two-parameter approach (such as when using an MV utility function) heterogeneity among investors devolves with risk aversion as in the trade-off between risk and mean return and not through the individual's perception of the distribution of asset returns. In fact, heterogeneous MV investors view risky assets homogeneously, as the probability distributions are the same for all investors and the correlations are identical as well.

Capital market equilibrium is reached under the CAPM mutual fund separation theorem which asserts that investors hold a selection of risky assets known as the market portfolio which is composed of all risky assets and identical for all investors. As the price of risk increases, investors hold a greater proportion of the risk-free assets and reduce their holdings in the mutual fund of risky assets whose proportions remain unchanged.

Review of actual investors' positions reveals considerable challenge to the market portfolio single equilibrium. Canner, Mankiw, and Weil (1997) are a notable example. They note that popular advice on asset allocation among cash, bonds, and stocks contradicts CAPM and MV financial theory.

Some of the results obtained by the CAPM are due to the assumption of identical investors with respect to the way risk is defined. Once we allow different attitudes toward risk then the main result that all investors hold the market portfolio is not supported by the model. For equilibrium we have to assume that investors hold the same expectation concerning the future distribution of returns.[4] Otherwise, they will simply trade and therefore we are not in equilibrium. Our aim in this section is to show that there is capital market competitive equilibrium in a two-parameter model with a market portfolio but that heterogeneous investors who differ in risk aversion will not have to hold it. Only when investors define risk exactly in the same way can they hold the market portfolio of risky assets.

Using the MEG we challenge the existence of the mutual fund separation theorem which claims that in equilibrium all investors should hold the same market portfolio, even under heterogeneity. In fact, if investors are heterogeneous in the sense that they perceive the risk of uncertain returns differently, it is shown in Shalit and Yitzhaki (2010) that at equilibrium no one should necessarily hold the market

[4] Actually, it is possible to have an equilibrium under different expectations. As will be seen later, the adjustment needed is that given the different expectations, the marginal rates of substitution between two assets are the same between all assets and all investors.

portfolio of risky assets. The market will clear, in the sense that one set of prices (as expressed by mean returns) will be revealed, but the proportions of risky assets held by investors will be quite different.

Following the first section of this chapter, we assume that the heterogeneous investors are MV or MEG investors. The prime question is how we model capital market equilibrium with such heterogeneous participants. We set up the problem in MEG terms and provide a solution using a simple Edgeworth box. Although the discussion is characterized in geometric terms, the results are compelling. Capital market equilibrium with heterogeneous investors reveals that each will hold a different efficient portfolio of risky assets but no investor has to hold the market portfolio. MV implies homogeneity as investors perceive risk similarly. Hence, the only possible equilibrium solution is that each participant holds the same portfolio as the market portfolio.[5]

In a world of identical expectations of the distributions of asset returns, the MEG approach enables us to differentiate between two separate problems:

1. How is risk perceived and measured?
2. How much is one ready to pay to reduce exposure to risk?

The first question is answered by the type of variability measure that risk-averse investors use to capture risk. This measure quantifies and qualifies risk.

The second question as to what price investors are ready to pay to reduce risk exposure is answered by setting one risk price in the market so that the marginal rate of substitution between risk and expected return will be the same for all assets. For homogeneous investors the marginal rates of substitution are equal when investors hold the same portfolio. Heterogeneous investors, on the other hand, perceive and measure risk differently, even though the return distribution is expected to stay the same. In this case the marginal rates of substitution for different investors and different assets can be equal to the price that is set by in the market only if the investors hold different portfolios. Our argument is that adding the assumption of heterogeneous risk-averse investors leads to a presentation of the CAPM as similar to the presentation of the equilibrium in commodities market: at equilibrium all marginal rates of substitutions are equal.

The structure of the arguments is as follows: first we present the investor's problem using expected utility maximization, and discuss stochastic dominance and the two-parameter MG approach. We then elaborate on the MEG ordering functions. Using an Edgeworth box, we solve the capital market equilibrium-first for homogeneous investors and then for heterogeneous investors-and explain the main results.

[5] Under a MV framework, both Harris (1980) and Nielsen (1990) use the Edgeworth box to model capital market equilibrium, the first by analyzing the trade-off between risk and return, and the second by characterizing allocation risk.

18.3.1 The Two-Parameter Investment Model

We set the basis for establishing the ranking function in a standard two-period portfolio choice model. Facing n risky assets with random returns r_i for $i = 1, \ldots, n$ and initial wealth w_0 the investor chooses a portfolio X, that is: the shares $\{\alpha_i\}$ of the assets $\{r_i\}$, $i = 1, \ldots, n$, respectively, such that $\sum_{i=1}^{n} \alpha_i = 1$ and the choice maximizes the expected utility of the final wealth:

$$\begin{aligned} &\text{Max} \quad E[U(w)] \\ &\text{subject to} \quad w = w_0\left(1 + \sum_{i=1}^{n} \alpha_i r_i\right) \quad \text{and} \quad \sum_{i=1}^{n} \alpha_i = 1. \end{aligned} \tag{18.9}$$

We assume initially that optimal choice of assets generates a distribution of feasible portfolios which solve problem (18.9) (and also that short sales are allowed). Once feasible portfolios are created, one can compare them by considering increasing and concave utility functions that are known only to the investors. For two portfolios X and Y whose cumulative distributions are given by F and G, the notion of maximum expected utility states that X dominates Y (second degree stochastic dominance) if and only if:

$$E_F U(X) \geq E_G U(Y). \tag{18.10}$$

Because we do not know the utility function, we apply the laws of second-degree stochastic dominance (SSD) in order to determine the set of efficient portfolios. As Hadar and Russell (1969), Hanoch and Levy (1969), and Rothschild and Stiglitz (1970) propose, SSD expresses the conditions under which all risk-averse investors prefer one portfolio over another. SSD states that X dominates Y if and only if:

$$\int_{-\infty}^{z} [G(t) - F(t)]dt \geq 0 \quad \text{for all } z \in (-\infty, \infty). \tag{18.11}$$

(See Chap. 5). Various methods have been used to apply the conditions expressed by (18.11). One way to use SSD is to compare the areas under the cumulative distributions of portfolio returns. Alternatively, one can compare the absolute Lorenz curves, which are the cumulative expected returns on the portfolio, following Shorrocks (1983) and Shalit and Yitzhaki (1994). In essence, for all risk-averse investors to prefer one portfolio of assets over another, the Lorenz curve of the dominating portfolio must lie not lower than the Lorenz curve of the dominated one.

We note that neither approach provides practical results in large portfolios because both involve infinite numbers of pair-wise comparisons of portfolios. SSD also provides researchers with a partial ordering, forcing the imposition of additional restrictions on the investor's preferences.

Another way to resolve differences between the Expected utility-stochastic dominance approach and a two-parameter approach is to restrict the distribution of returns to two-parameter probability distributions, usually the mean and the variance. Meyer (1987), for example, restricts the distribution to a family that differs only by location and scale parameters. Levy (1989) extends Meyer's results to show the distributional restrictions that guarantee the equivalence of SSD and MV efficient sets. See also Wong and Au (2004).

The essence of our approach is to select the parameters from a set of parameters that form the necessary conditions for SSD rules. We thus ensure that the complete ordering of portfolios produced by the two-parameter approach does not contradict the partial ordering produced by SSD rules. In other words, the efficient set generated by the two-parameter approach is guaranteed not to include SSD dominated portfolios.

Unfortunately, MV cannot be considered as a potential model. Indeed, MV is compatible with EU and SSD in limited instances and so the MV-efficient set includes SSD-dominated portfolios.[6] However, our two-parameter analysis can include the MV as a way of incorporating risk.

We construct the ranking function as follows. Let μ_X (μ_Y) and Δ_X (Δ_Y) be the mean and the GMD of portfolio X (portfolio Y) then:

$$\mu_X \geq \mu_Y \quad \text{and} \quad \mu_X - \Delta_X^* \geq \mu_Y - \Delta_Y^* \tag{18.12}$$

are necessary conditions for portfolio X to SSD-dominate portfolio Y. The first inequality in (18.12) compares the mean returns of the two portfolios. The second inequality in (18.12) compares the risk-adjusted mean returns of the two portfolios, where the portfolio's GMD represents the risk for the investor.

The advantage of using the set (μ, $\mu - \Delta^*$) over the set (μ, Δ^*) as parameters in the ranking function is that these two parameters are defined as "good" while (μ, Δ^*) is a combination of "good" and "bad." This allows us to borrow without adjustment many microeconomic theory results. To generate results that are compatible with the financial models, however, we use (μ , Δ^*) as well. Both presentations include the same parameters, and we will use them interchangeably.

18.3.2 The Mean-Extended Gini Ordering Function

We define the ranking function V (μ, $\mu - \Delta^*$), where μ is the mean and Δ^* is one-half of the Gini's mean difference of the risky prospects. We extend the function to

---

[6] To illustrate this issue, assume that *X* is uniformly distributed between [0, 1] while Y is uniformly distributed between [1,000, 2,000]. Clearly, all investors prefer Y over *X*, but both of them are included in the efficient MV set. Consequently, relying on MV to analyze portfolios may produce efficient portfolios that are inconsistent with expected utility theory.

depend on the EG so that the ranking function is V(μ, μ – Δ (ν)), where Δ (ν) is the EG defined as,

$$\Delta(v) = \mu - \mathrm{a} - \int_{\mathrm{a}}^{b} [1 - \mathrm{F(w)}]^{v} \mathrm{dw} \tag{18.13}$$

and $v \in (1,\infty)$ reflects the investor's aversion toward risk. (See proposition 6.1 for details).

For a risk-neutral investor $v = 1$ and the extended Gini is zero. For $v = 2$, (half of) the standard GMD is obtained.

For $\nu \rightarrow \infty$, the Gini represents risk as viewed by a max–min investor (see Chap. 6 on discussions about the extended Gini).

We relate SSD to the MG model by using the function $\delta(v)$ defined as $\delta(v) = \mu - \Delta(v) = \mathrm{a} + \int_{\mathrm{a}}^{\mathrm{b}} [1 - \mathrm{F(w)}]^{v} \mathrm{dw}$, that is: the mean minus the extended Gini. This value can be interpreted as the certainty equivalent of the distribution valued by the type v investor, which can also be interpreted as the risk-adjusted mean return. The construction of an ordering function that ranks distributions with respect to SSD is based on Proposition 18.2.

Proposition 18.2 *Conditions $\delta_X(1) \geq \delta_Y(1)$ and $\delta_X(v) \geq \delta_Y(v)$ for all $v \in (1, \infty)$ are necessary for X to dominate Y according to SSD.*

Proof See Chap. 6 and Yitzhaki (1982a, 1983).

Some properties of δ(ν) (and of the extended Gini) that are needed to pursue our arguments are:

(i) $\delta(\nu) = \mu - \Delta(\nu)$ may be interpreted as the risk-adjusted mean return (or the certainty equivalent).

(ii) $\frac{\partial\, \delta(\nu)}{\partial\, \nu} \leq 0$. That is, δ(ν) is a non increasing function of v. This property implies that the higher the risk aversion the lower the certainty equivalent of the portfolio.

(iii) The values of $\delta(v)$ for specific choices of ν are:

$\delta(0) = \mathrm{b}$

$\delta(1) = \mu$ because $\Delta(1) = 0$.

$\delta(2) = \mu - \Delta^*$, where Δ^* is (one-half of) the Gini's mean difference.

$\lim_{v \to \infty} \delta(v) = \mathrm{a}$. This property implies that in the extreme risk aversion case, the lower bound of the distribution is the certainty equivalent of the portfolio.

(iv) If $w = c$, where c is a constant (i.e., the risk-free asset), then:
$\delta(\nu) = \mathrm{c}$ for all $v > 0$ because $\Delta(v) = 0$.

(v) If $\mathrm{w_i} = \mathrm{c_0\, w_j} + \mathrm{c_1}$, where $\mathrm{c_0} > 0$ and $\mathrm{c_1}$ are given constants, then:
$\delta_{\mathrm{i}}(\nu) = \mathrm{c_0}\, \delta_{\mathrm{j}}(\nu) + \mathrm{c_1}$ because $\Delta_{\mathrm{i}}(\nu) = c_0\, \Delta_{\mathrm{j}}(v)$ and $\mu_{\mathrm{i}} = \mathrm{c_0}\mu_{\mathrm{j}} + \mathrm{c_1}$.

(vi) If $\mathrm{w_3} = \mathrm{c_0\, w_1} + \mathrm{c_1\, w_2}$, where $\mathrm{c_0} > 0$ and $\mathrm{c_1} > 0$ are given constants and if the correlation coefficient between $\mathrm{w_1}$ and $\mathrm{w_2}$ is $-1 \leq \rho_{12} < 1$, then:
$\Delta_3(\nu) < \mathrm{c_0}\, \Delta_1(\nu) + \mathrm{c_1}\, \Delta_2(\nu)$.
Properties (iv)–(vi) are similar to the properties of the standard deviation.

(vii) For a portfolio $w = \sum_{i=1}^{n} \alpha_i r_i$, where the α_i are given constants,

$$\Delta_w(\nu) = -\nu \sum_{i=1}^{n} \alpha_i \mathrm{cov}\{r_i, [1 - F(w)]^{\nu-1}\}$$

where F(w) is the cumulative distribution of w. For the case of the Gini, $\nu = 2$, we get:

$$\Delta_w(2) = \Delta_w = 2\sum_{i=1}^{n} \alpha_i \mathrm{cov}[r_i, F(w)].$$

(viii) Assume that ν's are integers such as $\nu = 1, 2, 3, \ldots$, then $\delta_w(\nu) = E[\min(w_1, \ldots, w_\nu)]$. That is, $\delta_w(\nu)$ is the expected minimum of ν independent draws from the distribution F(w). This property is useful when estimating the extended Gini, as it relates the Gini to the rank-order statistics.

(ix) With property (viii) and assuming a normal distribution, $\delta_w(\nu) = \mu - C(\nu)\sigma$ where C(ν) is a constant that depends on ν and σ is the standard deviation. (For $\nu = 2$, $C(\nu) = 1/\sqrt{\pi}$).

(x) The extended Gini of a sum of random variables can be decomposed similarly to the way the variance is decomposed. (See Schechtman & Yitzhaki, 2003).

To sum up, one can view δ(ν) as the certainty equivalent of a distribution with mean μ where Δ(ν) represents the risk premium. When $\nu \to \infty$, it means that using δ(ν) to evaluate the portfolio is identical to the evaluation of the max–min investor. When $\nu = 1$ investors evaluate assets as if they were risk-neutral. In the extreme case of risk lovers (defined by $\nu < 1$), $\nu \to 0$ investors are interested only in the maximum value of a distribution as defined by max–max investors. Given the properties of δ(ν) one can construct the two-parameter ranking function V.

Proposition 18.3 *The function $V[\delta(\nu_1), \delta(\nu_2)]$ with $\nu_1 \geq 1$, $\nu_2 > 1$ and with $\partial V/\partial\,\delta(\nu_1) > 0$, $\partial V/\partial\,\delta(\nu_2) > 0$ ranks risky alternatives with respect to SSD criteria.*

Proof Assume that F(w) stochastically dominates G(w) according to SSD. Thus, following Proposition 18.2, $\delta_F(\nu_1) \geq \delta_G(\nu_1)$ and $\delta_F(\nu_2) > \delta_G(\nu_2)$; hence, $V[\delta_F(\nu_1), \delta_F(\nu_2)] > V[\delta_G(\nu_1), \delta_G(\nu_2)]$.

The term δ (ν) is a special case of Yaari's (1987) dual utility function, so the function $V[\delta(\nu_1), \delta(\nu_2)]$ also ranks portfolios with respect to Yaari's utility function. To use V(,) following the MV model, we restrict the discussion to $\nu_1 = 1$ and $\nu_2 > 1$, so that V can be written as

$$H[\mu, \mu - \Delta(\nu)] = V[\delta(1), \delta(\nu)] \qquad \text{for} \quad \nu > 1. \tag{18.14}$$

The function H enables one to use μ and Δ to represent mean return and risk, respectively. H ranks distributions as follows: if two distributions have the same certainty equivalent, the one with the higher mean return is preferred. If the two distributions have the same mean return, the one with the higher certainty equivalent is preferred.

We move the investor's problem represented by (18.9) into the space (μ, Δ) and now solve the problem with the function H. Instead of using a utility function, investors minimize the portfolio's GMD subject to a given mean return. From property (vii), the Gini Δ_w of the portfolio is:

$$\Delta_w = 2\sum_{i=1}^{n} \alpha_i \text{cov}\,[r_i\,, F(w)]. \tag{18.15}$$

In addition to the n risky securities, investors are allowed to borrow or save a risk-free asset whose rate is r_f. Hence, investors choose the portfolio $\{\alpha_i\}$ that minimizes Δ_w subject to a mean return:

$$\mu_w = r_f + \sum_{i=1}^{n} \alpha_i(\mu_i - r_f). \tag{18.16}$$

Alternatively, investors can choose a portfolio that maximizes H [μ_w, μ_w – Δ_w (ν)].

The necessary conditions for a maximum are given by:

$$(H_1 + H_2)\,(\mu_i - r_f) - H_2\, d\Delta_w/d\,\alpha_i = 0 \qquad i = 1, \ldots, n. \tag{18.17}$$

where H_k is the partial derivative of H with respect to its *k-th* argument, k = 1,2.

Because the Gini (and EG) is homogeneous of degree one in α, Euler theorem states that:

$$\Delta_w = \sum_{i=1}^{n} \alpha_i\, \partial\Delta_w/\partial\,\alpha_i. \tag{18.18}$$

Hence, adding the necessary conditions (18.17) after they are multiplied by their respective α_i leads simply to.

$$\frac{(H_1 + H_2)}{H_2}(\mu_w - r_f) = \Delta_w$$

or

$$\frac{(\mu_w - r_f)}{\Delta_w} = \frac{H_2}{H_1 + H_2}. \tag{18.19}$$

The solution shows the optimal portfolio as the one whose slope equals the slope of the function H [μ_w, μ_w – Δ_w (ν)] in the space [μ , Δ(ν)].

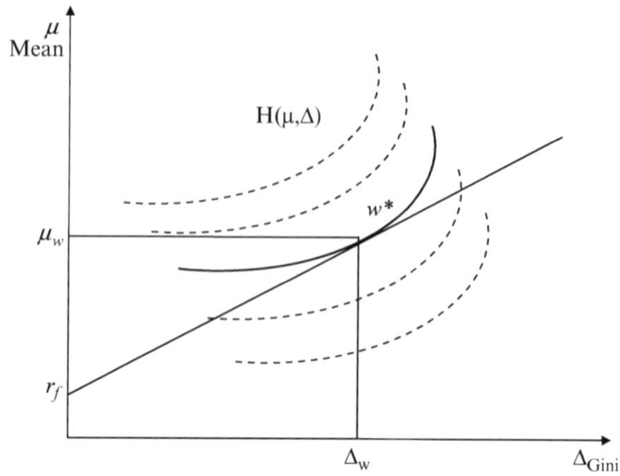

Fig. 18.1 Optimal Portfolio in (μ, Δ) Space. Source: Shalit and Yitzhaki (2009), p. 762, Fig. 1

Figure 18.1 shows that the solution on point w* is unique because of the convexity and non-satiation (i.e., no maximum without constraints. No nirvana). The slope of the indifference curves is given by:

$$\frac{d\,\mu}{d\,\Delta}\Big|_{H=\text{constant}} = \frac{H_2}{H_1 + H_2} > 0. \tag{18.20}$$

By the maximization of H the second order conditions guarantee that:

$$-2\,H_{12}\,H_1\,H_2 + H_{22}\,H_1^2 + H_{11}\,H_2^2 < 0. \tag{18.21}$$

where H_{kj} are the second derivatives of H with respect to the k, j arguments, k, j = 1,. . .,n.

Hence, convexity is obtained by:

$$\frac{d^2\,\mu}{d\,\Delta^2}\Big|_H = \frac{1}{(H_1 + H_2)^3}\left(2H_{12}\,H_1\,H_2 - H_{22}\,H_1^2 - H_{11}\,H_2^2\right) > 0. \tag{18.22}$$

From the properties of δ (ν), H_1 is the marginal utility produced by increasing the portfolio's mean return, while the certainty equivalent is held constant. Similarly, H_2 is the marginal utility of increasing the certainty equivalent, given a constant mean return. In other words, H_2 expresses the marginal utility of reducing risk along the same mean return. Hence, $H_1 + H_2$ is the marginal utility of increasing the portfolio without incurring risk, because adding a constant to the portfolio increases μ and δ (ν) by the same amount, as seen from property (v) above.

18.4 Equilibrium

To demonstrate the existence of a competitive equilibrium in a capital market with heterogeneous investors, the basic Edgeworth box is used. This concept allows characterizing the equilibrium for an exchange economy of heterogeneous agents who have different amounts of risky assets and different preferences toward risk.

The geometric representation of the Edgeworth box requires three components: (1) two types of agents, each with a utility function characterized by convex indifference curves; (2) an initial distribution of assets to be traded; (3) a willingness to trade in order to improve one's utility by bilateral bargaining that leads to efficient allocation and eventually to market equilibrium. For the Edgeworth box to represent a competitive market, it is assumed that there are numerous investors for each type of investors.

We adapt the standard Edgeworth box model and consider three assets, two risky ones and one safe. The box consists only of the risky assets that form the box axes. The risk-free rate r_f is treated as a residual investment which determines the budget constraint. Instead of a utility function we use the special case of Yaari's, (i.e., Gini) function obtained when investors minimize the extended Gini of a portfolio, $\Delta_w(v)$, subject to the given mean return $\mu_w = r_f + \sum_{i=1}^{n} \alpha_i(\mu_i - r_f)$. The resulting iso-risk indifference curves are functions of only the risky assets that appear in the box. Investors choose $\{\alpha_i\}$ such that $\sum_{i=1}^{n} \alpha_i = 1$ (all non negative unless short selling is allowed) to maximize $-\Delta_w(v) = -\sum_{i=1}^{n} \alpha_i \mathrm{cov}\{r_i, -v[1 - F(w)]^{v-1}\}$ subject to $\mu_w = r_f + \sum_{i=1}^{n} \alpha_i(\mu_i - r_f)$.

The first-order conditions of that optimization are:

$$\frac{\partial \Delta_j(v)}{\partial \alpha_i} = \lambda_j(\mu_i - r_f) \quad \text{for all } i = 1, \ldots, n \tag{18.23}$$

where λ_j is the Lagrangean multiplier associated with investor j's mean return constraint and $\Delta_j(v)$ is his extended Gini . As the Gini is homogeneous of degree one, by Euler theorem:

$$\Delta_j(v) = \lambda_j \sum_{i=1}^{n} \alpha_i(\mu_i - r_f). \tag{18.24}$$

The term $1/\lambda_j$ as the price of investor j is:

$$\frac{\mu_w^j - r_f}{\Delta_w^j(v)} = \frac{1}{\lambda_j}, \tag{18.25}$$

which is also the slope of the tangent in Fig. 18.1. Hence, the first-order conditions for investor j become:

$$\mu_i - r_f = \frac{\mu_w^j - r_f}{\Delta_w^j(v)} \frac{\partial \Delta_w^j(v)}{\partial \alpha_i} \quad \text{for all } i = 1, \ldots, n. \tag{18.26}$$

Under homogeneity, investors have the same attitudes toward risk and the same price of risk, which can be expressed using the market portfolio M as $\frac{\mu_M - r_f}{\Delta_M(v)} = \frac{1}{\lambda}$. Recall also that:

$$\frac{\partial \Delta_w^j(v)}{\partial \alpha_i} = -v \operatorname{cov}\{r_i, [1 - F_w^j(r_w)]^{v-1}\}.$$

Therefore

$$\mu_i = r_f - (\mu_M - r_f)\, v \operatorname{cov}\{r_i, [1 - F_M(r_M)]^{v-1}\}/\Delta_M(v).$$

Or equivalently

$$\mu_i = r_f + (\mu_M - r_f)\, \beta_{iM}(v) \tag{18.27}$$

where $\beta_{iM}(v)$ is the extended Gini regression coefficient of the rate of risky asset i on the market portfolio M. This is the standard MEG CAPM when all investors have the same type of risk aversion characterized by v. The equation prices the mean return into the systematic risk using an identical measure of risk for all risky assets and all investors.

Under heterogeneity, groups of investors have different v and therefore each type has a different definition of risk. However, relative prices are equal among investors. For each investor j, the first order conditions for risky asset i with respect to risky asset k are obtained as:

$$\frac{\frac{\partial \Delta_w^j(v)}{\partial \alpha_i}}{\frac{\partial \Delta_w^j(v)}{\partial \alpha_k}} = \frac{\mu_i - r_f}{\mu_k - r_f} \quad \text{for all } i, k = 1, \ldots, n. \tag{18.28}$$

Between investors, the equilibrium conditions amount to:

$$\mu_i - r_f = \frac{1}{\lambda_j} \frac{\partial \Delta_w^j(v_j)}{\partial \alpha_i} = \frac{1}{\lambda_l} \frac{\partial \Delta_w^l(v_l)}{\partial \alpha_i}, \tag{18.29}$$

for all j, l investors and all i assets.

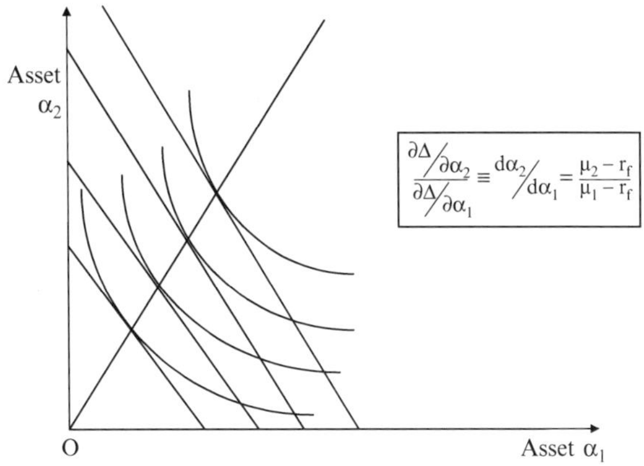

Fig. 18.2 Optimal Gini Indifference Curves in Asset Space. Source: Shalit and Yitzhaki (2009), Fig. 2, p. 763

Assuming differentiability, the last conditions (18.29) can be expressed as:

$$\mu_i - r_f = -\frac{1}{\lambda_j}\, \nu_j \mathrm{cov}\{r_i, [1 - F_w^j(r_w)]^{\nu_j - 1}\} = -\frac{1}{\lambda_l}\, \nu_l \mathrm{cov}\{r_i, [1 - F_w^l(r_w)]^{\nu_l - 1}\}\,.$$

The equilibrium conditions can now be written as:

$$\frac{\mathrm{cov}\{r_i, [1 - F_w^j(r_w)]^{\nu_j - 1}\}}{\mathrm{cov}\{r_k, [1 - F_w^j(r_w)]^{\nu_j - 1}\}} = \frac{\mathrm{cov}\{r_i, [1 - F_w^l(r_w)]^{\nu_l - 1}\}}{\mathrm{cov}\{r_k, [1 - F_w^l(r_w)]^{\nu_l - 1}\}} = \frac{\mu_i - r_f}{\mu_k - r_f} \tag{18.30}$$

for all j, l investors and all i, k risky assets. That is, the marginal rates of substitutions between two assets are equal to the relative prices (returns). Note that this condition is not different from the equilibrium in the market for tomatoes and cucumbers.

Second-order conditions are guaranteed by the quasi-convexity of the Gini function (this follows from the properties that are listed below proposition 18.2). In the space defined by the risky assets $\{\alpha_i\}$, conditions (18.28) indicate that the slopes of the indifference curves of the Gini function are equal to the respective ratios of excess asset mean returns. This is a standard solution that occurs when investors choose a portfolio that maximizes H $[\mu_w, \mu_w - \Delta_w(\nu)]$.The results are expressed in the indifference curves drawn in the space (μ, Δ) shown in Fig. 18.1.

As the Gini is quasi-convex and homogeneous of degree one with respect to portfolio weights of the risky assets$\{\alpha_i\}$, the indifference curves are equally spaced convex isoquants as shown in Fig. 18.2 for two risky assets and one risk-free asset.[7] Because of the homogeneity of a given Gini function, the slopes of the isoquants are constant along rays through the origin.

[7] Only the shares allocated to risky assets are shown on the axes. The share of the risk-free asset determines the location of the "budget constraint"; the farther it is from the origin, the lower the share of risk-free asset in total wealth is.

Because we can define an explicit Gini function for a specific ν, conditions (18.28) construct a distinct linear expansion path that is the locus of Gini-minimization portfolios. As wealth allocated to risky assets increases, the portfolio's mean return increases together with the Gini of the portfolio which moves to a new isoquant. This is obtained by increasing the shares of the two risky assets and reducing the share of the risk-free asset. Indeed, the wealth shown in the Edgeworth box is only the wealth allocated to the risky assets after deducting the share of wealth held in the risk-free asset. As long as asset mean returns are constant, the expansion path is a straight line through the origin. The slope of the expansion path defines the ratio of risky assets held by the investor. As the slope depends upon the Gini function, the ratio of risky assets varies with the perception of risk as expressed by ν.

It is worth mentioning that Fig. 18.2 applies also to MV investors who use the variance as a measure of risk. In this case the isoquants represent the variance of the portfolio of risky assets. Hence, we can include MV investors as a special group in the capital market.

We first examine homogeneous investors which have identical perceptions of risk.[8] We claim that homogeneity of risk perception leads investors to hold identical portfolios of risky assets.[9] If, furthermore, portfolios are duplicated under the assumption of constant returns to scale, investors will exhibit identical ratios of risky assets. In classical financial market theory, the "market portfolio" represents the shares held by all investors. This is basically the ratio of all risky securities. Thus, all investors hold the identical market portfolio.

The basic CAPM result is demonstrated in Fig. 18.3 using the Edgeworth box. Here we consider a market with only two risky assets that total $\overline{\alpha}_1$ and $\overline{\alpha}_2$. We have two types of investors (A and B) who are homogeneous in the sense that they have identical perceptions of risk, but differ in their initial endowments of risky assets such that $\alpha_1^A + \alpha_1^B = \overline{\alpha}_1$ and $\alpha_2^A + \alpha_2^B = \overline{\alpha}_2$. This initial endowment is shown by point I. As with the standard Edgeworth box geometry, the origin of preferences of type A investors is O_A and the one of type B investors is O_B.

The equal Gini indifference curves, which are identical for A and B, show that the two types of investors would benefit by trading among those in the same categories until they reach the Pareto efficient allocation E. At the initial endowment I, the initial mean return ratios are different for the type A and type B investors, and do not allow for trading. Thus, mean returns change until the same price ratio is tangent to the two Gini indifference curves as shown at point E. This point is the competitive equilibrium located on the diagonal as the expansion paths of the two types of investors are identical.

The market portfolio is the slope of the diagonal $\overline{\alpha}_1/\overline{\alpha}_2$ which represents the same ratio of risky assets held by each type of investors. At the equilibrium E,

[8] Although investors are homogeneous in the way they perceive risk, they can be heterogeneous in the way they price risk, as reflected by the risk-free-to-market portfolio ratios.

[9] Homogeneity of risk perception implies that all MG investors have the same ν, or that all investors are MV investors. See property ix of δ.

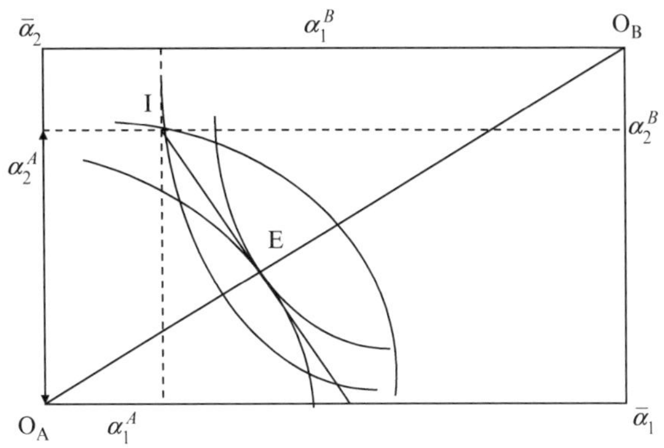

Fig. 18.3 Capital Market Equilibrium with Homogeneous Investors. Source: Shalit and Yitzhaki (2009), Fig. 3, p. 764

the price ratio as expressed by the unique slope of the tangent is the ratio of mean returns:

$$\frac{\frac{\partial \Delta}{\partial \alpha_2}}{\frac{\partial \Delta}{\partial \alpha_1}} = \frac{d\alpha_2}{d\alpha_1} = \frac{\mu_2 - r_f}{\mu_1 - r_f}.$$

For MV investors the indifference curves (like those for the Gini in Fig. 18.2) are derived from the variance of a portfolio of risky assets whose covariance matrix is unique and identical for all. Hence, the MV isoquants are the same for type A and type B MV investors; their shape depends on the covariance between α_1 and α_2. Hence, for MV investors, the only equilibrium solution is located on the diagonal of the box, implying that they hold the same market portfolio of risky assets.

The Edgeworth box in Fig. 18.3 reflects the capital market equilibrium in the case of MV or for homogeneous MG investors. Our first result summarizes this equilibrium.

Result 1 At equilibrium, homogeneous investors, either extended Gini or MV homogeneous investors, hold the market portfolio of risky assets as expressed by the slope of the diagonal of the Edgeworth box.

The problem of equilibrium is different with heterogeneous investors.[10] Heterogeneity results from the way distributions are taken into account, implying a different quantification of the risk measure. Hence, we consider two types of investors: Type A and type B investors, with different values of ν: The two types, each with a different endowment are shown in the Edgeworth box in Fig. 18.4.

[10] If returns are multivariate normal, heterogeneity is reduced to homogeneity and the standard MV result is obtained.

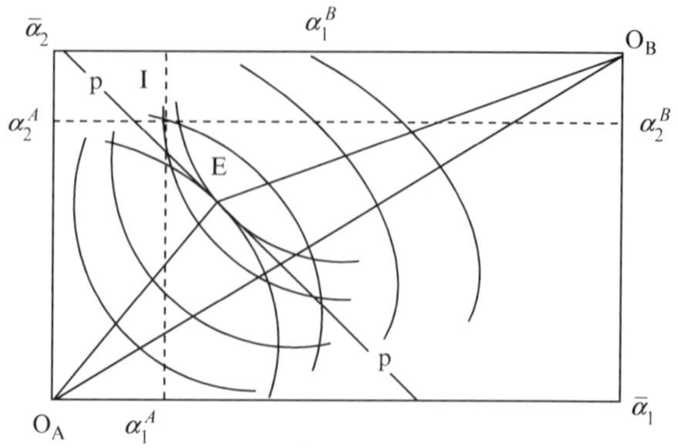

Fig. 18.4 Capital Market Equilibrium with Heterogeneous Investors. Source: Shalit and Yitzhaki (2009), Fig. 4, p. 764

Because the types have different aversions toward risk, their indifference extended Gini curves are not the same and they produce different expansion paths. From the initial endowment allocation at point I investors trade to improve their positions and move to new indifference extended Gini curves. The higher the equal extended Gini curve, the higher the portfolio's mean return. Hence, investors choose to trade, resulting in changes of the price ratio of mean returns. Extended Gini measures are minimized until investors reach the Pareto-efficient competitive equilibrium at point E where equal Gini curves are tangent to each other with slopes equal to the mean return ratio as shown by line p − p. To state this formally:

$$\left.\frac{d\,\alpha_2}{d\,\alpha_1}\right|_A = \left.\frac{d\,\alpha_2}{d\,\alpha_1}\right|_B = \frac{\mu_2 - r_f}{\mu_1 - r_f}.$$

This price ratio defines a unique equilibrium. Because the equal extended Gini curves are not identical, the optimal expansion paths for type A and type B investors are different. Therefore, the ratio of risky asset optimal portfolio held by each type of investors is different, and no investor will hold the "market portfolio" that is represented by the slope of the diagonal of the Edgeworth box. In other words, the equilibrium is expressed as:

$$\left.\frac{\alpha_1}{\alpha_2}\right|_A \neq \left.\frac{\alpha_1}{\alpha_2}\right|_B \neq \frac{\bar{\alpha}_1}{\bar{\alpha}_2}$$

This leads us to the second result:

Result 2 Unless risky asset returns are all multivariate-normal, at equilibrium heterogeneous extended Gini investors hold different portfolios of risky assets and no one has to hold the market portfolio as expressed by the slope of the diagonal of the Edgeworth box.

The contract curve is the locus of all undominated equilibria following various initial endowments. Income distribution comes about in the relative size of the investors' initial endowments. From welfare economics analysis we draw the next two results:

Result 3 As the extended Gini is homogeneous of degree one in asset shares, the contract curve is either identical to the diagonal of the Edgeworth box or lies on one side of the diagonal.[11]

This result implies that once a type of investor tends to invest relatively more in one asset, he will continue to do so under all market circumstances. (This fact applies to all types of investors). Thus it is possible to identify and relate types of assets with classes of investors.

Result 4 Expected returns on assets depend directly upon the income distribution across types of investors.

In some sense this result moves us back to traditional microeconomic theory that asserts the significance of income distribution when consumers have different tastes. Yet, this result clearly contradicts the CAPM, which claims that asset returns are determined solely as a function of the demand of a representative investor.

Result 5 Heterogeneous investors who have the same v will hold an identical portfolio of risky assets.

The MEG model has been shown to be richer than the mean variance in that it enables the researcher to construct an infinite number of "capital asset pricing models" for v homogeneous markets. It is shown in Shalit and Yitzhaki (1984, 1989) that if investors have the same degree of risk aversion, one can estimate capital asset pricing model betas for every v and then, using the holding of the market portfolio, find the v that fits the data best. The heterogeneous model with many v differs considerably from these results as conditions (18.28) establish specific equilibrium relations between asset returns and risk as viewed by all investors in the market.

18.5 Summary

The MEG approach is used to characterize the equilibrium in a capital market with heterogeneous risk-averse investors as a two-parameter model. As it is compatible with maximizing expected utility, MEG provides necessary and sometimes sufficient conditions for stochastic dominance theory. Standard capital market equilibrium assumes homogeneous investors with identical perceptions of risky assets. In these models, heterogeneity comes about with the different trade-offs between the risk-free asset and a portfolio of risky assets.

In the MEG model it is demonstrated how homogeneity of risk preferences leads to the mutual fund-portfolio separation results that all investors hold the same market portfolio ratio of risky assets. This is the standard MV result. When there

[11] This result is derived from the homogeneity property of the isoquants. The contract curve cannot cross the diagonal, as it can only be the diagonal itself or lie on one side of it.

are different perceptions about risk, a more general capital market equilibrium emerges.

Heterogeneous investors do not hold the same portfolio of risky assets. Furthermore, no investor must hold the "market portfolio" in order for capital markets to be in equilibrium. Asset prices are characterized by their mean returns and the various perceptions of risk. Each group of investors with its unique attitude toward risk defines its positions according to the specific extended Gini (i.e., their specific ν).

Although the model is simple it can be constructed only if one recognizes that by using a variability measure, both the risk aversion of the investor and her utility function are determined. Economists have used the Edgeworth box for some time to depict competitive interactions in competitive markets, welfare economics, and international trade, and to show Walras general equilibrium. The box is so well established in microeconomics that it is quite surprising it has not been used before to solve the basic issues of capital market equilibrium.

It seems that there is only one explanation. Financial economics has been captivated by the MV paradigm that is very simple and very intuitive to use. But the MV is an appropriate model only if we restrict ourselves to multivariate normal distributions or quadratic utility functions. If those requirements are violated then the model is not compatible with expected utility theory, nor with Yaari's dual approach. The MG and the MEG models share the simplicity of the MV and are compatible with the leading theories of behavior under risk. Using the Edgeworth box it is possible to show that the assumption of identical risk-averse investors allows only the diagonal as the contract curve, leading to the identical "market portfolio" solution.

Chapter 19
Applications of Gini Methodology in Regression Analysis

Introduction

Ordinary least squares (OLS) regression is based on the fact that the variance of a linear combination of random variables can be decomposed into the contributions of the individual variables and to the contributions of the correlations among them. The fact that one can imitate this decomposition (under certain conditions) when decomposing the GMD of a linear combination of random variables enables one to take any OLS-based econometric textbook and replicate each chapter using the GMD instead of the variance. Practically, this means doubling the number of models because every OLS econometric model can be replicated by the GMD, resulting in different estimates of the parameters Moreover, we present via examples (Chap. 21) that the estimates can differ in sign. This means that two investigators who use the same variables, the same model, and the same data may come up with contradicting results concerning the effect of one variable on the other. The only difference between the two researchers lies in the measure of variability they use—the GMD or the variance. Needless to say that in many cases of policy decisions the debate is on the magnitude of a parameter, which is much more vulnerable than the sign and not on the sign itself. And to make life even more complicated any regression model that is estimated by the GMD can be replicated with the EG. This means moving from doubling the number of possible estimates to an infinite number of estimates.

The simple OLS and Gini regression coefficients can be interpreted as weighted averages of slopes between adjacent observations of the explanatory variable. The difference between the two methods lies in the variability measure used, which is reflected in the weights. Both the OLS and the Gini regression coefficients can be viewed as nonparametric estimates of the "weighted average" slope of the curve. These slopes can serve as the slopes of the linear approximation to an unknown regression curve. If the curve is not monotonic or even only nonlinear then one should expect non-robustness of the estimates.

S. Yitzhaki and E. Schechtman, *The Gini Methodology: A Primer on a Statistical Methodology*, Springer Series in Statistics 272, DOI 10.1007/978-1-4614-4720-7_19,

Generally the Gini methodology offers additional parameters on top of the ones offered by the variance-based methodology. For example, in decomposing the Gini of a linear combination of random variables there are two correlation coefficients between each pair of variables and these coefficients are not necessarily equal. These can be used to check for robustness of the econometric "evidence." In some sense, this means using the additional parameters in order to reveal the implicit and hidden assumptions behind the OLS and other methodologies in order to find out whether those assumptions are driving the results, as will be shown in the rest of this chapter.

The complete implementation of the Gini methodology will take years to develop and requires many research teams. At this point it is simply impossible to cover all the fields that will benefit from adopting it. Hence, our empirical application is restricted to just a few topics. We list other areas of applications that can benefit from using this old-new technique in Chap. 23.

We concentrate on the ability of the Gini regression to uncover some of the implicit and whimsical assumptions of the regression (Leamer, 1983). We believe that at this point this is the best use of the existing theory.

Among the assumptions that lie behind the OLS regression we concentrate on two assumptions: (1) the model is linear and (2) the correlation coefficient between explanatory variables is symmetric. We note in passing that the second assumption is somewhat disturbing because the regression method itself is based on a structure which is asymmetric. To attest that, note the asymmetry between the dependent and the explanatory variables. The extreme violation of the linearity assumption is non-monotonic relationship between the variables. Therefore, we illustrate the methodology of finding non-monotonic relationship.

The existence of two correlation coefficients between two variables enables us to test for the linearity of the model. Usually each optimization results in an orthogonality condition which in the OLS is referred to as the normal equation. The orthogonality conditions in OLS and Gini regression imply that the appropriate covariance between the residuals and the explanatory variable is set to zero. Testing whether the other covariance is also equal to zero provides a specification test for the model. (Note that this is only possible under the Gini regime where there are two covariances between each pair of variables.) This can be done for each explanatory variable so we may be able to conclude that the model is well specified for certain variables but not for the others.

Another useful tool is the mixed OLS and Gini regression. By "mixed" we mean that some variables are treated by one method, while others are treated by another method. The reason that it is useful is because as far as we know this is the only regression method that enables us to move from one regression methodology to the other in a stepwise way. It enables the investigator to move from one regression to the other gradually so that the variable or variables that are causing the non-robustness of the sign of the regression coefficient will be detected.

It is worth remembering that those issues do not cover all properties of the Gini methodology. Chapter 23 is devoted to possible extensions.

The richness of possibilities of extensions and replications forces us to restrict our empirical illustration to a few topics. At the end of this part we will list several immediate applications, without going into them.

The structure of this part of the book is as follows: Chap. 19 presents and illustrates the properties of the Gini methodology in the simple regression case, Chap. 20 presents the multiple Gini regression, while Chap. 21 presents the mixed OLS–EG regression.

The structure of Chap. 19 is the following: Sect. 19.1 presents simulated results to illustrate the ability of the simple EG regression to trace the curvature of a curve, Sect. 19.2 illustrates the use of the LMA curve to trace the curvature of the regression curve. In Sect. 19.3 we illustrate the decomposition of the regression coefficient into the contributions of different sections, while Sect. 19.4 illustrates all the above issues using an example from the labor market. In Sect. 19.5 we point out the impacts of common data manipulations on the estimates, while Sect. 19.6 concludes.

19.1 Tracing the Curvature by Simple EG Regression: Simulated Results[1]

In this section we illustrate the fact that the EG regression estimates a weighted average of slopes. The larger the value of ν is, the more weight is attached to the lower levels of the explanatory variable.

In order to clarify the point we use a simple illustration of a nonlinear regression curve with a single explanatory variable, where we can obtain the analytical expression for the curvature and thus we can illustrate the one-to-one relation between the curvature and the EG regression coefficient as ν varies.

The model we use is the following: let X be uniformly distributed on [0,1] and let $Y = X - X^2$. The regression curve is bell-shaped, symmetric around 0.5. The regression coefficient is the "average" slope of the curve. The derivative of the curve at x is a function of x (i.e., it is the slope at x), and it is given by $Y'_x = 1 - 2x$. Its range is [−1, 1] and it is uniformly distributed on [−1,1].

The formal presentation of the extended Gini regression coefficient is[2]

---

[1] This section is based on Ben Hur et al. (2010).

[2] The reader who is interested only in the application presented in this section may question the wisdom of using $(1 - F(x))^{\nu}$ instead of $F^{\nu}(x)$. The reason for this cumbersome definition is that it enables us to define monotonic increasing concave functions that are similar to utility functions required in finance or inequality measurement (see Shalit and Yitzhaki, 2002; Wodon and Yitzhaki, 2002b; Yaari, 1987; Yitzhaki, 1983). However, as pointed out in Chap. 6 one can think of other forms for the extended Gini.

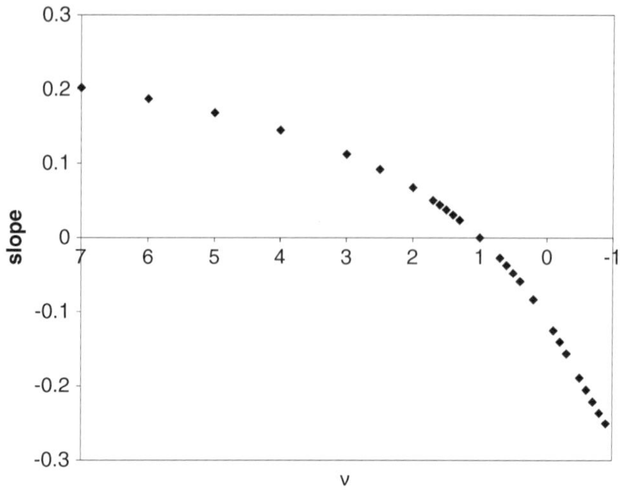

Fig. 19.1 The extended Gini regression coefficient of the regression curve Y = X − X² as a function of v. Source: Ben Hur et al. (2010), Fig. 1, p. 37. Reprinted with permission by MILI Publications

$$\beta(v) = \frac{\mathrm{cov}(\mathrm{Y}, -[1 - \mathrm{F(X)}]^v)}{\mathrm{cov}(\mathrm{X}, -[1 - \mathrm{F(X)}]^v)}.$$

We now calculate β(v) for the example above.

$$\mathrm{Cov}\left(\mathrm{X}^2, -[1 - \mathrm{F(X)}]^v\right) = \frac{(v+2)(v+3) - 6}{3(v+1)(v+2)(v+3)}.$$

Similarly,

$$\mathrm{Cov}(\mathrm{X}, -[1 - \mathrm{F(X)}]^v) = \frac{v}{2(v+1)(v+2)}.$$

Combining the two results, we get that the regression coefficient is

$$\beta(v) = \frac{\mathrm{cov}(\mathrm{Y}, -[1 - \mathrm{F(X)}]^v)}{\mathrm{cov}(\mathrm{X}, -[1 - \mathrm{F(X)}]^v)} = 1 - \frac{2[(v+2)(v+3) - 6]}{3v(v+3)}. \tag{19.1}$$

Figure 19.1 shows the relationship between the EG regression coefficient, which is the average slope (the average of the marginal propensities) on the Y-axis, and v on the X-axis.

The term "marginal" means the derivative. For example, marginal propensity to consume (derivative of consumption with respect to income), marginal product of labor, marginal utility (derivative of utility with respect to income), etc.

It can be seen that for $\nu = 1$, (i.e., for the GMD) the average slope is zero. It is monotonically decreasing with ν and as expected it is positive for $\nu > 1$ and negative for $\nu < 1$. Note, however, that because it is a weighted average of all slopes, even if one pursues ν to its extreme values it will not reach the extreme values of the slopes (minus and plus 1).[3] We note that the X-axis in the plot seems to be reversed (from large values of ν to small values). The reason for this is because large values of ν correspond to emphasizing small values of X. This enables to see (from the figure) the change in the slope of the curve with respect to a change in X, the explanatory variable.

To sum up: we have shown the ability of the simple EG regression coefficient to follow the curvature of a regression curve by changing the weights given to different percentiles of the explanatory variable (via $(1 - F(X))^{\nu}$). Ben Hur, Frantskevich, Schechtman, and Yitzhaki (2010) argue that this property is unique to the EG, because other methods such as quantile regression change the weighting scheme along the distribution of the residuals which is the conditional distribution of the dependent variable. Note, however, that would we minimize the EG of the residuals, which is the other regression that is based on the EG (which is not investigated in this book), then we would have an EG version of regression that imitates quantile regression.

19.2 Tracing the Curvature of a Simple Regression Curve by the LMA Curve

The use of the EG to trace the curvature of a regression curve relies on aggregating slopes by a predetermined set of weights which are functions of the cumulative distribution of the explanatory variable. As such, we always end up with one parameter. Therefore it is possible that we might miss ranges in which there is a minor change in the curvature of the curve.

For this purpose we suggest to use the LMA curve which is derived from the connection between the co-Gini and the ACC (to be defined below). The derivation of the curve and its properties are detailed in Chaps. 5 and 8. Here we only sketch the properties needed for the applications. The LMA curve is useful whenever one is interested in a gross evaluation of the effects of the following actions.

1. Omitting a group of observations.
2. Omitting irrelevant observations and using only a subgroup of the population—e.g., imposing bounds on observations that participate in the regression.

[3] See Yitzhaki (1983, Table 1, p. 623) for an investigation of the properties of the weighting scheme.

3. Substituting a continuous variable by a discrete one (given in intervals) with the data entries taken to be either midpoints or averages. Referred to as binning (Wainer, Gessaroli, & Verdi, 2006).
4. Applying a monotonic nondecreasing transformation to the explanatory variable.

While omitting observations sounds suspicious, using a transformation seems natural and is being used in practice quite often. Binning can be considered as a classical method since it was used by Ernst Engel to find "Engel's law" which states that the poorer a family is, the larger the budget share on necessities (Chai & Moneta, 2010). Binning is also used in nonparametric statistics in order to overcome the need to specify a functional form for the regression curve (Chai & Moneta, 2010; DiNardo & Tobias, 2001).

Note that (1) and (2) are formally identical. The difference lies in the reasoning for the omission. Also, (3) can be viewed as a special case of (4).

Our main focus is on the effects of the above-mentioned commonly used practices on the sign of the regression coefficient. The reason for stressing sign change is that it may reverse the conclusion reached. Instead of positive (negative) effect it may turn the effect into a negative (positive) one.

19.2.1 Definitions and Notation

In the regression context there are two potential uses for concentration curves. The first use is to learn about the curvature of the simple regression curve and the second use is to learn about the weighting scheme. (see Heckman, Urzua, and Vytlacil (2006a, b) and Heckman (2010) for the derivations of weighting schemes for many different econometric models).

Let $g(x) = E\{Y|X = x\}$ be the conditional expectation of Y given X. We will refer to it as the regression curve. We start with three definitions.

Definition of ACC: The *absolute concentration curve* (ACC) of Y with respect to X denoted by $A_{Y \circ X}(p)$ is

$A_{Y \circ X}(p) = \int_{-\infty}^{x_p} g(t)\, dF_X(t)$, where x_p is implicitly defined by $p = \int_{-\infty}^{x_p} dF_X(t)$.

For simplicity of exposition, we write ACC instead of $A_{Y \circ X}(p)$ for the absolute concentration curve.

Definition of LOI: Connect the points (0,0) and $(1,\mu_Y)$ by a straight line $LOI = \mu_Y p$. Yitzhaki and Olkin (1991) call this line *the line of independence* (LOI).

The LOI can be interpreted as an absolute concentration curve plotted between two independent variables. That is, if Y and X are independent, then the ACC curve coincides with the LOI.

Because we are interested in "deviations from independence" we will be interested in a curve which is the LOI minus the ACC.

Definition of LMA: LMA is defined as the *LOI minus the absolute concentration curve of Y with respect to X*. Formally, $LMA_{YoX}(p) = LOI - ACC = \mu_Y p - A_{YoX}(p)$.

The properties of ACC and LMA, relevant for this section, are as follows.

(a) The ACC passes through the points (0, 0) and $(1, \mu_Y)$. Property (a) enables us to define a variation of the ACC (the LMA) that will make the analysis of the regression curve easier.
(b) The derivative of the LMA with respect to p (at x_p) is $\mu_Y - E_Y(Y|X = x_p)$. This follows directly from the definitions of the LOI and ACC. As a consequence the $LMA_{YoX}(p)$ is increasing (decreasing, constant) if and only if $\mu_Y - g(x_p) > (<, =)\ 0$.
(c) The LMA is concave at p (convex, straight line) if and only if $dg(x_p)/dp = [dg(x_p)/dx_p][dx_p/dp] > (<, =)\ 0$ (which is equivalent to $dg(x_p)/dx_p > (<,=)\ 0$ because $dx_p/dp > 0$).
(d) If X and Y are independent then ACC is a straight line which coincides with the LOI, and the LMA curve coincides with the horizontal axis. Properties (c) and (d) enable the user to identify sections with constant, increasing, and decreasing slopes of the regression curve: linearity of LMA implies a horizontal regression curve, concavity of LMA means an increasing regression curve, while convexity means a decreasing regression curve.
(e) The area between the LMA and the horizontal axis is equal to $cov(Y, F_X(X))$ (Chap. 5). Note that if the curve intersects the horizontal axis then the sign of $cov(Y, F_X(X))$ depends on the magnitudes of the areas above and below the horizontal axis.
(f) The LMA is above the horizontal axis for all *F* if and only if $cov(Y, T(X)) > 0$ for all continuous differentiable monotonically increasing functions T(X).

The advantage of using the LMA (instead of the ACC) is that it is easy to detect what will happen to cov(Y,F(X)) (and hence to the sign of the Gini regression coefficient) if sections of observations of X are omitted from the regression, as will be illustrated later.

For the purpose of analyzing the effect on the OLS regression coefficient one needs a modified LMA curve for which the area beneath it will be equal to cov(Y,X). It is shown in Yitzhaki (1998) that a simple transformation can make the curve applicable to OLS: if one substitutes the horizontal axis to be X instead of F_X, then the area between the new curve and the horizontal axis will be equal to cov(Y,X). However, the nature of the curve changes and further research is needed to study its properties. For our purposes it is sufficient that property (f) holds in the transformed curve, hence one can change the sign of a regression coefficient in an OLS regression if and only if the LMA curve intersects the horizontal axis. Note, however, that because $F(X) = F(T(X))$, it is impossible to change the sign of a Gini regression by a monotonic transformation of X.

19.2.2 The Simple Gini Regression Coefficient and the Concentration Curve

The Gini regression coefficient is a ratio of two covariances (Chap. 7):

$$\beta^{G}_{Y.X} = \frac{\mathrm{cov}(Y, F(X))}{\mathrm{cov}(X, F(X))}.$$

It resembles Durbin's (1954) estimator. The denominator is always positive hence the sign of the regression coefficient is determined by the numerator. By property (d) of the LMA curve one can see whether there are sections with different signs along the regression curve. To ease the analysis of the contribution of different sections of the curve to the regression coefficient, we normalize the LMA curve by dividing it by cov(X, F(X)). We call the curve NLMA (normalized LOI minus ACC). The additional property of this curve is that the area between the curve and the horizontal axis is equal to the Gini regression coefficient.

Because the analysis is relatively simple, we will list the steps for practical use and then illustrate by two examples from the labor market. The steps are the following.

(a) Plot the data using NLMA.
(b) Is the relationship monotonic over the entire range? Look for convex and concave parts (properties (c) and (d) above). If the curve is not entirely convex or concave, then one may be able to change the sign of the regression coefficient by omitting "redundant" observations.
(c) Does the curve intersect the horizontal axis? If not—there is no monotonic transformation that can change the sign of the OLS regression coefficient. If yes, one can find a transformation that will change the sign of the OLS regression coefficient. (Property (e) above).

19.3 The Decomposition Approach

The discussion below is based on the decomposition of regression coefficients. The regression coefficients we are dealing with are the OLS and Gini regression coefficients. Let (Y,X) be a bivariate random variable.

The least squares regression coefficient is given by

$$\beta^{OLS}_{Y.X} = \frac{\mathrm{cov}(Y, X)}{\mathrm{cov}(X, X)}$$

and the Gini regression coefficient is given by

$$\beta^{G}_{Y.X} = \frac{\mathrm{cov}(Y,\ F(X))}{\mathrm{cov}(X,\ F(X))},$$

where F(X) is the cumulative distribution function of X.

Assume that the observations are partitioned into M disjoint groups, denoted by $m = 1,\ldots, M$ and let $p_m = n_m/n$ be the relative size of group m.

As shown in Chap. 7 the OLS regression coefficient can be decomposed as follows

$$\beta^{OLS} = \sum\nolimits_{m=1}^{M} w_m\ \beta_m + w_B\ \beta_B, \tag{19.2a}$$

where

$$w_m = p_m \frac{\mathrm{cov}_m(X,X)}{\mathrm{cov}(X,\ X)},\ \beta_m = \frac{\mathrm{cov}_m(Y,\ X)}{\mathrm{cov}_m(X,\ X)},\ w_B = \frac{\mathrm{cov}_B(\overline{X}_m,\overline{X}_m)}{\mathrm{cov}(X,X)},\ \beta_B = \frac{\mathrm{cov}(\overline{Y}_m,\overline{X}_m)}{\mathrm{cov}(\overline{Y}_m,\overline{X}_m)}$$

and $\overline{X}_m$ and $\overline{Y}_m$ denote the vectors of group means. If the groups do not overlap, i.e., the ranges of the groups' explanatory variables do not overlap, then (19.2a) is applicable to the Gini regression as well, after adjusting the weights and the regression coefficients to the Gini counterparts. The adjustment in cases of overlapping is described next. We start with the decomposition of the Gini covariance, which is given by

$$\mathrm{cov}(Y,F_o(X)) = \sum_{m=1}^{M} p_m \mathrm{cov}_m(Y,F_o(X)) + \mathrm{cov}_B(\overline{Y}_m,\overline{F}_o(X)),$$

where $F_o(X)$ is the overall cumulative distribution of the explanatory variable, $\overline{Y}_m$ is the vector of group means, and $\overline{F}_0(X)$is the vector of the means of the ranks of groups. The next step is dividing and multiplying by the same factors. That is,

$$\begin{aligned}\mathrm{cov}(Y,F_o(X)) &= \sum\nolimits_{m=1}^{M} p_m \frac{\mathrm{cov}_m(Y,F_o(X))}{\mathrm{cov}_m(X,F_o(X))}\frac{\mathrm{cov}_m(X,F_o(X))}{\mathrm{cov}_m(X,F_m(X))}\mathrm{cov}_m(X,F_m(X)) + \mathrm{cov}_B(\overline{Y}_m,\overline{F}_o(X))\\ &= \sum\nolimits_{m=1}^{M} p_m \frac{\mathrm{cov}_m(Y,F_o(X))}{\mathrm{cov}_m(X,F_o(X))} O_m G_m + \mathrm{cov}_B(\overline{X}_m,\overline{F}_o(X))\frac{\mathrm{cov}_B(\overline{Y}_m,\overline{F}_o(X))}{\mathrm{cov}_B(\overline{X}_m,\overline{F}_o(X))},\end{aligned}$$

where $O_m = \dfrac{\mathrm{cov}_m(X,\ F_o(X))}{\mathrm{cov}_m(X,\ F_m(X))}$ is the overlapping between group m and the overall population, $G_m = \mathrm{cov}_m(X,F_m(X))$ is (one-fourth of) the Gini mean difference of X_m, while $\beta_m = \dfrac{\mathrm{cov}_m(Y,\ F_0(X))}{\mathrm{cov}_m(X,\ F_0(X))}$ is a kind of Gini regression coefficient.

Using the above notation we get

$$\beta^{G} = \sum\nolimits_{m=1}^{M} w_m\ \beta_m + w_B \frac{\mathrm{cov}_B(\overline{Y}_m,\overline{F}_o(X))}{\mathrm{cov}_B(\overline{X}_m,\overline{F}_o(X))}, \tag{19.2b}$$

where $w_m = \frac{p_m O_m G_m}{G}$ is the contribution of group m to the overall variability, $G = \mathrm{cov}(X,F(X))$ is (one-fourth of) the Gini's mean difference of X and $w_B = \frac{\mathrm{cov}_B(\overline{X}_m, \overline{F}_o(X))}{\mathrm{cov}(X,F(X))}$.

19.4 An Illustration: Labor Force Participation by Gender and Age[4]

Figure 19.2 presents the normalized line (of independence) minus the absolute concentration curve (NLMA) which is the LMA divided by cov(X, F(X)). As discussed above, the total area enclosed by the curve and the horizontal axis is equal to the Gini regression coefficient. The horizontal axis depicts the cumulative distribution according to age, while the vertical axis depicts the difference between the cumulative value of participation in the labor force would participation be independent of age (i.e., LOI) and the actual cumulative value of participation, divided by the Gini of the age distribution. The data are taken from Income Survey, 2005, conducted by the Israeli Central Bureau of Statistics. Similar results were obtained for the years 2003 and 2004.

Figure 19.2 is based on 12,685 observations for women and 11,213 for men. The Y-axis is the NLMA, while the X-axis is the cumulative distribution of age (which can be translated into the corresponding age. For example $F(x) = 0.3$ for males

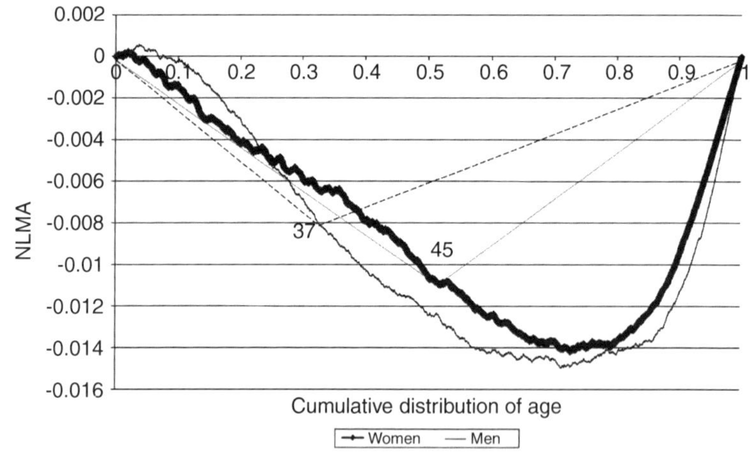

Fig. 19.2 NLMA curve for participation in the labor market versus age according to gender. (For individuals working 30 h and more. Cumulative distribution of age is depicted on the horizontal axis.) Source: Yitzhaki and Golan (2010)

[4] The section is based on Yitzhaki and Golan (2010) and discussed in Chap. 5.

Table 19.1 Regression coefficients of participation on age in the different sections by gender

	GINI				OLS			
	Men		Women		Men		Women	
Section	Weight	Beta	Weight	Beta	Weight	Beta	Weight	Beta
I	0.032	0.015	0.178	0.001	0.026	0.015	0.158	0.001
II	0.316	−0.016	0.090	−0.023	0.349	−0.017	0.112	−0.024
Between	0.651	−0.006	0.732	−0.008	0.624	−0.006	0.729	−0.008
Overall	1.000	−0.009	1.000	−0.008	1.000	−0.010	1.000	−0.009

Source: Yitzhaki and Golan (2010)

corresponds to age = 37). The curve enables us to detect regions with positive or negative slopes according to whether the curve is concave or convex, and according to whether the regression coefficient of each section is contributing positively or negatively to the overall regression coefficient. If the curve is above (below) the horizontal axis then this range has a positive (negative) contribution to the regression coefficient. In addition a concave (convex) section implies that would we take this section alone we will find a positive (negative) regression coefficient. Table 19.1 presents the regression coefficients in each section.

As can be seen from Fig. 19.2 and Table 19.1 both overall regression coefficients (Gini and OLS) are negative. The curves are composed of two parts —concave and convex. It is concave for males for the lower 30% of the observations (up to age 37) and for females for the 55% of the observations with smallest ages (up to age 45). As a result in these sections the regression coefficients of both OLS and Gini regressions are positive. On the other hand, in the second section the curves are convex hence the regression coefficients in this section are negative. Note, however, that the curves do not cross the horizontal axis, implying that the contributions of the two sections are adding up to the overall regression coefficients and one cannot use a transformation to change the (negative) signs of the regression coefficients. However, note that for the Gini regression we can evaluate, by adding and subtracting positive and negative areas, how many additional observations of the "wrong" sign we can add without changing the sign of the regression coefficient. We note that the regression coefficients obtained by the two methods are similar. However, using the Gini method and the figures which are derived from the Gini enable the user to make the partition into sections as shown above. To sum up: in this example one cannot use a monotonic transformation in order to change the sign of the regression coefficient, but throwing/adding observations may cause it to happen.

Figure 19.3 presents the NLMA of the regressions of monthly hours of work on age for men and women who work 30 hours or more per week. The curve for men can be divided into two regions: for the 45% of observations with younger ages (up to age 39) the regression coefficient is positive and every monotonic transformation of it will yield a positive coefficient, while for the 55% with older ages (ages from 39 up to 70) the regression coefficient for every monotonic transformation of it is negative. Table 19.2 confirms this result. Because the curve intersects the horizontal axis at

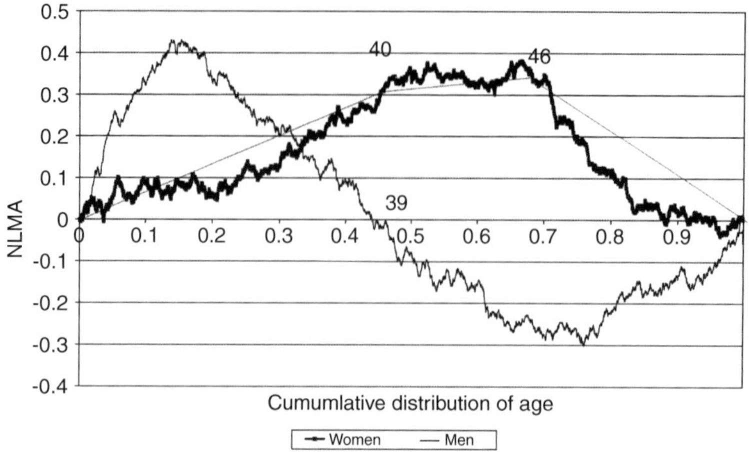

Fig. 19.3 NLMA curve for monthly hours of work and age by gender. (For individuals working 30 h and more. Reprinted with permission by Elsevier). Source: Yitzhaki and Golan (2010) and Yitzhaki and Schechtman (2012)

Table 19.2 Regression coefficients of working hours on age in the different sections by gender

	GINI				OLS			
	Men		Women		Men		Women	
Section	Weight	Beta	Weight	Beta	Weight	Beta	Weight	Beta
I	0.070	1.4731	0.097	−0.153	0.057	1.522	0.085	−0.166
II	0.206	−0.4515	0.007	−0.294	0.245	−0.439	0.007	−0.302
III	–	–	0.049	−0.568	–	–	0.066	−0.526
Between	0.724	−0.0009	0.848	0.246	0.698	−0.001	0.841	0.2480
Overall	1.000	0.0095	1.000	0.165	1.000	−0.022	1.000	0.158

Source: Yitzhaki and Golan (2010) and Economics Letters, Yitzhaki and Schechtman (2012) Reprinted with permission by Elsevier

this point, the Gini (and OLS) between-groups regression coefficient is zero[5] (it is close to zero due to the fact that the empirical distribution is discrete). A monotonic transformation that extends (shrinks) the range of ages at low values of age relative to high values of age will change the overall regression coefficient to be positive (negative) (e.g., using log(age) will increase the value of the regression coefficient, while using exp(age) will decrease it to be negative). The case of women is more interesting. Here we can divide the age into three age groups: up to 40, 40 to 46, and over 46. In each of those groups the regression coefficient is negative, but the overall coefficient is positive. Table 19.2 confirms this result. Note that the graph

[5] At the point of intersection with the horizontal line, the averages of the dependent variable for the above intersection and below intersection groups are equal. Thus the between-group regression coefficient is equal to zero.

intersects the horizontal axis only at the very end (right side). As can be seen from the graph and the table, the between-groups component is positive, leading to an overall positive regression coefficient. However a monotonic transformation of age may affect the sign of the regression coefficient. Finally, note that for women the OLS and Gini regressions result in an overall positive effect of age on working hours, while for men the sign depends on the regression method used (the difference may be statistically insignificant).

19.5 Data Manipulations

The following data manipulations can be handled by tools from the OLS and Gini methodologies and will be discussed below.

1. Omitting a group of observations.
2. Omitting irrelevant observations and using only a subgroup of the population—e.g., imposing bounds on observations that participate in the regression.
3. Substituting a continuous variable by a discrete one (given in intervals) with the data entries taken to be either midpoints or averages. Referred to as binning (Wainer, Gessaroli, & Verdi 2006).
4. Applying a monotonic nondecreasing transformation to the explanatory variable.

19.5.1 Omitting a Group of Observations

We start with the first two actions (actually the second one can be viewed as a special case of the first one)—omitting a group of observations. The decompositions of the OLS and Gini regression coefficients (19.2a) and (19.2b) allow us to investigate the effects of different groups of observations and of different actions on the regression coefficients in the simple regression framework.[6]

Generally, omitting a group of observations will affect the weighted sum of slopes (one slope will be missing and the remaining weights will be adjusted) and the between-groups component.

The effect on the weighting scheme depends on the variability measure used (variance in OLS or Gini in Gini regression), while the effect on the regression coefficient is determined by the slopes in the data as well as by the regression method (i.e., the variability measure) used. We discuss the effect on the OLS

[6] Applying a monotonic transformation requires a different approach and will be dealt with later.

regression coefficient via (19.2a). The discussion for the Gini case (via (19.2b)) is similar with one difference—the slopes and the weights are based on the Gini terminology.

In order to illustrate we simplify the problem and have only two groups of observations: the omitted group (denoted by O) and the remaining group (denoted by R). In this simplified case, (19.2a) becomes

$$\beta = \mathrm{w}_{\mathrm{R}}\beta_{\mathrm{R}} + \mathrm{w}_{\mathrm{O}}\beta_{\mathrm{O}} + \mathrm{w}_{\mathrm{B}}\beta_{\mathrm{B}}. \tag{19.3}$$

Omitting the group eliminates the second and third terms on the right hand side of (19.3) and set $\mathrm{w}_{\mathrm{R}} = 1$, so for this case the effect of the omission is that we set $\beta = \beta_{\mathrm{R}}$. The case of omitting one extreme observation can be handled as a special case of (19.3) by forming two groups: a group with only one observation and a group containing all other observations. In this case the middle term in (19.3) is zero (there is no slope with one observation) and (19.3) includes only two terms, because the between-groups component now disappears. By reporting the result of the decomposition proposed in (19.3) before and after the omission of a group of observations the reader can be better informed about the effect of the omitted group on the regression coefficient: is it the between-groups component or is it the intra-omitted group regression coefficient. The case of omitting just one extreme observation is a bit different: in this case the second term in (19.3) is zero anyway (there is no intra-omitted group coefficient), and the decomposition before elimination includes two terms—the first and the last. Note that the justification for omitting a group can be supported either by the wish to increase the robustness of the findings, or because the underlying economic model assumes a different behavior below or above a threshold.

The decomposition proposed here can be helpful because it enables the reader to see which effect has led to the result: is it the innocent need to make the results more robust or is it caused by an overzealous investigator who wants to prove his point.

19.5.2 Substituting a Continuous Variable by a Discrete One

This action is referred to by Wainer et al. (2006) as binning. Some econometricians tend to transform a continuous variable into a discrete one (binning) or even a binary one, indicating participating or nonparticipating in a program. See, among others, Chai and Moneta (2010) and the survey by Angrist and Krueger (1999) concerning random assignment. In this case, all intra-group components are omitted from (19.2a) and we are left with the between-groups component. If the sign of the between-groups component is different from the sign of the overall regression coefficient then this action causes a sign change. The same procedure can be applied to an instrumental variable (IV) estimator. In that case, it can totally change the direction in which one variable influences the other (Heckman, Stixrud, & Urzwa, 2006; Heckman & Urzua, 2009; Heckman, Urzua, & Vytlacil, 2006a,

2006b; Yitzhaki & Schechtman, 2004), which makes the effect on the sign of the regression coefficient even more difficult to analyze.

For example, consider the case of grouping of observations as in the case of Wald estimator. To overcome an error-in-variable problem the investigator uses only group averages (with or without omitting some observations). As can be seen from (19.2a) the Wald estimator is actually using the between-groups regression coefficient (Pakes, 1982; Wald, 1940) instead of the overall regression coefficient. One possibility is that the grouping is intended to reduce the variability in order to achieve a higher portion of explained variability.[7] Another possibility is that the grouping results in omitting some negating effects that happen to occur in the subgroup. Reporting the components of (19.2a) enables the reader to see which explanation is more reasonable.

19.5.3 The Effect of Transformations

Transformations can be applied to the dependent variable or to the explanatory variable. Concerning transformation of the dependent variable, it is easy to see that if there is a transformation of the dependent variable that can change the sign of the Gini regression coefficient, then there will also be one that can change the sign of the OLS regression coefficient. However, it need not be the same transformation. The explanation is the following: a transformation can change the sign of the Gini regression coefficient if the LMA intersects the horizontal axis. Note that the LMA that is relevant to Gini regression has the cumulative distribution F(X) on the horizontal axis, while in the case of the OLS we substitute F(X) by X on the horizontal axis. This horizontal movement does not affect the existence of an intersection.

In order to examine whether a transformation of the explanatory variable can change the sign of the OLS regression coefficient, all we have to do is to plot the LMA curve while switching the roles of the variables. However, note that the signs of the Gini and reverse Gini regressions need not be equal (Goldberger, 1984) so that the same procedure is not always applicable for the Gini regression.

19.6 Summary

Both OLS and Gini simple regression coefficients can be expressed as weighted averages of slopes. The weights assigned to the slopes are different. In some cases this difference may result in different signs of the coefficients. This fact is troublesome as it might affect (i.e., reverse) the conclusions drawn from the analyses.

[7] For a recent illustration, see Angrist and Pischke (2009, p. 129).

In addition both Gini and OLS coefficients can be decomposed into intra- and inter-groups components. This property may enable the user to affect the sign of the regression coefficient by manipulating the data by actions such as omitting a group of observations, grouping or applying a transformation. As was shown in earlier chapters, the Gini methodology reveals more and at the same time is harder to manipulate. It enables the user to uncover some of the implicit assumptions of the regression as well as the effects of some manipulations.

Among those assumptions that lie behind the OLS regression are the assumption that the model is linear and that the correlation coefficients between explanatory variables are symmetric. The extreme violation of the linearity assumption is having non-monotonic relationship between the variables.

In this chapter we illustrate (via a simulated example) a way one can use the EG regression in order to trace the curvature of a regression curve.

We note that the use of the EG to trace the curvature of a regression curve relies on aggregating slopes by a predetermined set of weights which are functions of the cumulative distribution of the explanatory variable. Therefore the outcome is one parameter, which may imply that we miss ranges with a change in the curvature of the curve.

For this purpose we introduce and illustrate the use of the LMA curve which is derived from the connection between the co-Gini and the ACC. The LMA curve is useful whenever one is interested in evaluating the effects of several common manipulations and the data such as omitting a group of observations, imposing bounds on observations that participate in the regression, binning and applying a monotonic nondecreasing transformation to the explanatory variable.

Our main focus is on the effects of the above-mentioned commonly used practices on the sign of the regression coefficient. The reason for stressing sign change is that it may reverse the conclusion reached. Instead of positive (negative) effect it may turn the effect into a negative (positive) one.

The application of the recommended method is simple and includes three steps: Plot the NLMA, look for convex and concave parts, and check whether the curve interests the horizontal axis or not. Based on the three steps one can learn whether there exists a sign change along the regression curve and whether one can find a monotonic transformation that can reverse the sign or not.

We illustrate the techniques using data taken from Income Survey, 2005, conducted by the Israeli Central Bureau of Statistics. The data consists of 12,685 observations for women and 11,213 for men and the variables of interest are participation and hours worked by age and gender.

Chapter 20
Gini's Multiple Regressions: Two Approaches and Their Interaction

Introduction

Our target in this chapter is to illustrate one of the major advantages of the GMD regressions: they offer a complete framework for checking and dealing with some of the assumptions imposed on the data in a multiple regression problem. There are two approaches that are related to the Gini—the semi-parametric approach and the minimization approach. The interaction between the two gives tools for assessing the adequacy of the model. In addition, there are two tools that enable the researcher to investigate the curvature of the regression curve: the extended Gini regression and the NLMA curve. The basic idea is the following: there is an unknown regression curve that relates the dependent variable Y and (all or some out of) a set of explanatory variables $X_1,\ldots,X_n$. The shape of the curve is not known. The curve is approximated by a linear model (which is then estimated from the data). However, each approach mentioned above leads to a (possibly different) linear model. The interaction between the two approaches can help to decide whether the original curve is linear (in each individual explanatory variable) or not. The suggested stages are the following: first one estimates the regression coefficients according to the semi-parametric approach without specifying a linear model. This means that at this stage the researcher decides only on the set of explanatory variables to be included in the regression model but not on the functional form.[1] Then one uses the residuals from the fitted curve and tests whether they fulfill the necessary conditions for the minimization approach (which were obtained assuming linearity) for each explanatory variable separately. If for any given explanatory variable the above conditions are fulfilled; that is, if the hypothesis that the two regression coefficients are equal is not rejected, then one concludes that the regression curve is linear in this variable. Otherwise it is not (see Chap. 7 for details or below for a brief review). This property is especially important in regressions with several explanatory variables. It enables the investigator to find a set of variables that allows linear predictions without having

[1] To be accurate, the investigator has also to decide whether the model is multiplicative or additive.

S. Yitzhaki and E. Schechtman, *The Gini Methodology: A Primer on a Statistical Methodology*, Springer Series in Statistics 272, DOI 10.1007/978-1-4614-4720-7_20,

to commit to the linearity of the model as a whole. Provided that the linearity hypothesis is not rejected for all explanatory variables one can examine the properties of the residuals such as their distribution, whether it is symmetric around the regression line or not, the serial correlation between them, etc., using the methodologies that will keep the analysis under the Gini framework.[2] Although each stage could be performed by alternative methods, we are not aware of any methodology that can offer a complete set of tests that is governed by a unified framework and therefore offers a method to test the assumptions behind the regression with an internal consistency. We note in passing that the suggested test for linearity does not require replications of observations, as is the case in the common tests for linearity.

Some of the stages can also be performed with the EG regression and some can be illustrated visually by the LMA and concentration curves. The LMA curve and the decomposition of the regression coefficient are useful whenever we are dealing with a simple regression case. We note that the same ideas and methodology can be used, without any adjustment, in the multiple regression case. However, its use in practice is much more complicated. The regression coefficient in the multiple regression setting is a function of the simple regression coefficients (as will be shown in Chap. 21). Therefore one needs to take into account combinations of multiple effects. That is, an action taken with respect to one variable may affect, through the effects on the covariances with other explanatory variables, several (or all) of the explanatory variables.

At this stage the application of those methodologies in a multiple regression context is cumbersome and therefore will not be pursued in this book. However, we do show some examples in specific simple cases.

The structure of the chapter is as follows: Sect. 20.1 is devoted to a brief review of Gini's multiple regressions. Section 20.2 concentrates on the relationship between the two types of regression methods, while Sect. 20.3 relies on the properties developed in Sect. 20.2 to assess the linearity of the model.

Section 20.4 introduces the LMA curve and shows the connection between the Gini regression coefficient and the curve, while Sect. 20.5 illustrates the methodology for the case of two explanatory variables. Section 20.6 applies the methodology to assess the linearity of consumption as a function of income and family size. Section 20.7 concludes and offers a direction for further research.

20.1 Gini's Multiple Regressions[3]

The aim of this section is to briefly review the results for Gini's multiple regressions. The full derivation can be found in Chaps. 7 and 8.

---

[2] See, for example, Frick et al. (2006) who developed ANOGI—the Gini equivalent of ANOVA, and Shalit (2010) for a test for normality.

[3] This section is based on Schechtman, Yitzhaki, and Pudalov (2011).

20.1.1 The Semi-Parametric Approach

Let $(Y, X_1,\ldots,X_K)$ be a $(K + 1)$-variate random variable with expected values $(\mu_Y, \mu_1,\ldots, \mu_K)$, respectively, and a finite variance–covariance matrix Σ. Assume that we have a general regression curve defined by

$$g(x_1,\ldots,x_K) = E\{Y|X_1 = x_1,\ldots,X_K = x_K\}.$$

The investigator is interested in estimating a linear approximation of the regression curve. That is, she needs to estimate a set of slopes (the constant term will be determined later) which are conditional slopes: the slope of Y on X_i is conditional on the other Xs in the model.

The steps taken are as follows. First, the linear approximation is defined. Then, the parameters (conditional slopes) are interpreted as the solutions of a set of linear equations which involve the (known) simple regression slopes, and the last step is the estimation procedure, based on the data.

The resulting vector of regression coefficients, β_N, is given by

$$\beta_N = [E(V'X)]^{-1}E(V'Y), \tag{20.1}$$

where $\beta_N = \{\beta_{N1},\ldots, \beta_{NK}\}$ is a $(K \times 1)$ column vector of the (conditional) regression coefficients, V is an $(n \times K)$ matrix of the cumulative distributions of $X_1,\ldots, X_K$ (in deviations from their expected values), Y is an $(n \times 1)$ vector of the dependent variable, and X is an $(n \times K)$ matrix of the deviations of the explanatory variables from their expected values. The elements of $E(V'Y)$ and $E(V'X)$ are $COV(Y, F(X_k))$ and $COV(X_j, F(X_k))$, respectively. It is assumed that the rank of $V'X$ equals K, the number of explanatory variables. This implies a restriction on the choice of the explanatory variables that does not exist in OLS: no explanatory variable can be a monotonic transformation of another explanatory variable because if it does it will imply identical rows in the matrix V'X (which depends on X_i via $F_i(X)$). The details of the derivation are given by Schechtman, Yitzhaki and Artsev (2008).[4]

The natural estimators of the regression coefficients are based on replacing the cumulative distributions by the empirical distributions (which are calculated using ranks):

$$b_N = [v'x]^{-1}(v'y), \tag{20.2}$$

where v is a matrix with elements $[n^{-1}(r(x_{ik})) - 1/2]$, and $r(x_{ik})$ is the rank of x_{ik} among $x_{1k},\ldots,x_{nk}$. Schechtman, Yitzhaki and Artsev (2008) prove that b_N is a consistent estimator of β_N and its limiting distribution is normal under regularity conditions.

Once the Gini regression coefficients are estimated, the constant term can be estimated by minimizing a function of the residuals. The exact function used

[4] If one is interested in overcoming the restriction, then one should use EG regression. See Chap. 21.

determines whether the regression passes through the mean, the median, or any other quantile. The multiple regression procedure, although it is not based on an optimization procedure, generates equivalents to the OLS's normal equations. This property plays an important role in this chapter, as will be shown in the next section. By defining the residual term and substituting for the multiple regression coefficients, it can be shown that

$$\mathrm{COV}(\varepsilon, \mathrm{F_k(X)}) = 0 \quad \text{for } \mathrm{k} = 1, \ldots, \mathrm{K}, \tag{20.3}$$

as stated in the following Lemma.

Lemma 20.1 *Define the vector* $\varepsilon = Y - X\,\beta_N$. *Then,* $E(V'\varepsilon) = 0$, *where 0 is a vector of zeros.*

This property holds in the sample as $v'e = 0$, where $e = y - x\, b_N$.

Finally, because each variance and covariance in OLS regression is substituted in Gini regression by GMD and co-Gini, respectively, it is easy to verify that other concepts used in the OLS such as partial correlation coefficients can be translated into the Gini regression. Among those concepts is R^2 of the regression, which can be considered as a measure to assess the share of the (square of the) GMD of the dependent variable which is explained by the model. That is, the R^2 for the Gini semi-parametric regression is defined (Olkin and Yitzhaki, 1992) as one minus the square of the GMD of the residual term *divided by* the square of the GMD of the dependent variable:[5]

$$\mathrm{GR}^2 = 1 - [\mathrm{cov}(\mathrm{e}, \mathrm{r}(\mathrm{e}))/\mathrm{cov}(\mathrm{y}, \mathrm{r}(\mathrm{y}))]^2. \tag{20.4}$$

However, because the decomposition of the GMD of a linear combination into the contributions of the different components is more complicated than the decomposition of the variance, the properties of GR^2 differ from the properties of the equivalent term, R^2 in OLS. For example, as will be seen in the next section, GR^2 will obtain its maximal value under the Gini minimization approach. Therefore, the GR^2 in the semi-parametric version would always be not greater than the GR^2 in the minimization approach. Equality holds when the model is linear in all the explanatory variables.

Additional measures of the quality of the fit of the model to the data in the GMD regression are the Gini correlations between the dependent variable and the predicted variable. Formally,

$$\Gamma_{\mathrm{Y\hat{Y}}} = \frac{\mathrm{cov}(\mathrm{Y}, \mathrm{F}(\hat{\mathrm{Y}}))}{\mathrm{cov}(\mathrm{Y}, \mathrm{F}(\mathrm{Y}))} \quad \text{and } \Gamma_{\mathrm{\hat{Y}Y}} = \frac{\mathrm{cov}(\hat{\mathrm{Y}}, \mathrm{F}(\mathrm{Y}))}{\mathrm{cov}(\hat{\mathrm{Y}}, \mathrm{F}(\hat{\mathrm{Y}}))}, \tag{20.5}$$

where $\hat{\mathrm{Y}}$ is the predicted variable. As a result of the differences between the properties of the decomposition of the variance and those of the decomposition of

---

[5] In the empirical application we use $\mathrm{GR}^* = 1 - \mathrm{cov}(\mathrm{e},\mathrm{r}(\mathrm{e}))/\mathrm{cov}(\mathrm{y},\mathrm{r}(\mathrm{y}))$.

the GMD, we substitute the R^2 of OLS by three measures: the one in (20.4) and the two in (20.5). Note, however, that in the OLS, the parallels to these three measures are numerically equal.

20.1.2 The Minimization Approach

This approach, which is based on minimization of the GMD of the residuals, has already been developed in the literature and it is referred to as R-regression (Jurečková (1969, 1971); Jaeckel (1972); McKean and Hettmansperger (1978); Hettmansperger (1984)). Therefore its properties will not be repeated here.[6] For our argument, only the orthogonality condition (also known as the normal equations) is needed. Note that this method requires the specification of a model.

Consider the following model:

$$Y = X\beta_M + \varepsilon, \tag{20.6}$$

with the usual assumptions on ε, that is, the ε's are independent and have mean zero and a constant variance, and the additional assumption that X_i and ε_j are independent for all i,j. (Note that this assumption follows automatically if the X_i's are considered deterministic.) The estimated equation is

$$y = xb_M + e_M, \tag{20.7}$$

where b_M is the estimator of the slope β using the minimization of GMD of the residual term e_M. Using the covariance presentation of GMD and imposing the restriction that the mean of the residuals is zero enables us to show that minimizing GMD of the residuals is equivalent to minimizing

$$\sum_{i=1}^{n} r(e_{Mi})\, e_{Mi}, \tag{20.8}$$

where $r(e_{M_i})$ is the vector of ranks of the residuals e_{Mi}. Because the semi-parametric approach for the Gini regression is used to estimate the regression coefficients, the only property required for our suggested approach for testing linearity is that minimizing the GMD of the residual term yields an orthogonality condition which is the equivalent of the OLS normal equation and is given by

$$x'r_M = 0, \tag{20.9}$$

[6] It is worth emphasizing that the connection between R-regression and GMD was not recognized in the literature mentioned above. Many of the properties of those regressions can be traced to the properties of GMD. Bowie and Bradfield (1998) compare the robustness of several alternative estimation methods in the simple regression case and find the minimization of the GMD of the residuals among the most robust methods.

where r_M is the vector of the ranks of e_M, rescaled and shifted to have a zero mean.[7] (The ith element of $x'r_M$ is $cov(x_i,r_M)$). Equation (20.9) says that the sample covariance between the rank of the residuals and the variate value of the explanatory variable is set to zero as a result of the minimization of GMD of the residuals.

20.2 The Relationship Between the Two Approaches

As seen in the previous section, each approach yields a set of estimators, a set of residuals, and a set of "normal equations." The aim of this section is to identify the conditions under which the two estimators actually estimate the same parameters in the population and to give necessary and sufficient conditions under which the two estimators are algebraically identical. Recall that b_M, the estimator obtained by the minimization of GMD of the residuals, estimates the vector of slopes β under the linearity assumption. On the other hand no model was required in order to derive b_N, which is based on weighted averages of slopes (see Chap. 7 for details). Therefore in general the two approaches may yield different estimators. However, when the model is linear the vector of (true) slopes of the regression curve under the semi-parametric approach, β_N, is equal to the vector of slopes which was obtained under the linear assumption, β_M. The reason is because when the model is linear, the slopes are all equal along the regression curve and the weighted average of them is that same constant, therefore $\beta_N = \beta_M = \beta$.

Hence, both b_N and b_M estimate the same vector of slopes, namely β, and the first-order conditions of both methods should hold with the same set of residuals. (The last fact will be our basic tool for assessing the linearity of the model, as will be discussed in the next section). To see that, let

$$Y = X\,\beta + \varepsilon, \tag{20.10}$$

Then, by (20.1)

$$\beta_N = [E(V'X)]^{-1}E(V'Y) = \beta + [E(V'X)]^{-1}E(V'\,\varepsilon). \tag{20.11}$$

The fact that $\beta_N = \beta$ implies that $E\{V'\varepsilon\} = 0$, which is the first-order condition for the semi-parametric approach (Lemma (20.1)).

By substituting $y = xb_M + e_M$ into (20.2) the following relationship holds between b_N and b_M in the sample:

$$b_N = b_M + (v'x)^{-1}v'e_M. \tag{20.12}$$

[7] Because (20.8), the GMD of the residuals, is a piecewise linear function, its partial derivative with respect to b_M may not exist because the derivative is a step function. In this case the solutions b_M to (20.9) form a segment on the real line and b_M is determined up to a range. The larger the sample the lower the probability that such an event occurs.

The following proposition gives necessary and sufficient conditions for b_N to be algebraically equal to b_M in the sample.

Proposition 20.1 *Let* $(y_i, x_{1i},\ldots, x_{Ki})$, $i = 1,\ldots,n$ *be a sample of size n from a continuous multivariate distribution with finite second moments. Then*

(a) $v'e_M = 0$ iff $b_M = b_N$.
(b) $x'r_N = 0$ iff $b_M = b_N$, where r_N is the vector of ranks of e_N.

Proof See Chap. 8.

The second term in (20.12) is equal to the semi-parametric estimator of a regression coefficient in which the dependent variable is the vector of the residuals of the minimization approach. Therefore one can view (20.12) as running the regression in two steps: in the first step the minimization approach is applied to the data. Then, the semi-parametric approach is applied to the residuals obtained by the minimization approach. If the regression curve is linear in an explanatory variable, then using the minimization approach's residuals as the dependent variable and running a regression with respect to the explanatory variable, using the semi-parametric approach, will yield a regression coefficient that is equal to zero. If, on the other hand, the regression is not linear in one of the explanatory variables, then the second-step regression coefficient for this explanatory variable will deviate from zero, and other coefficients may be affected as well due to possible correlations among the explanatory variables. We note that the order can be changed: one can run the semi-parametric regression first, and then use the residuals to test whether the first-order conditions of the minimization approach are fulfilled using the set of residuals of the semi-parametric approach. This order is suggested in the next section.

20.3 Assessing the Goodness of Fit of the Linear Model

In what follows, we treat each explanatory variable X_k separately. For simplicity, $F_k(X_k)$ will be denoted by $F(X_k)$. When the model is linear with respect to an explanatory variable X_k, the semi-parametric approach and the Gini minimization approach estimate the same parameter β_k. Lemma 20.1 and the assumptions of the linear model in the minimization approach imply that $COV(\varepsilon, F(X_k))$ and $COV(X_k, F(\varepsilon))$ are equal to zero for each X_k. That is, if the specification of the model is correct then the following relationships hold in the population:

$$COV(\varepsilon, F(X_k)) = 0 = COV(X_k, F(\varepsilon))\ (k = 1,\ldots,K). \qquad (20.13)$$

The left-hand side of (20.13) is the population version of the "normal equation" obtained by the semi-parametric approach, while the right-hand side is the population version of the "normal equation" obtained by the minimization. The proposed method will take advantage of the fact that two covariances are involved, which is special to the Gini regression approach.

In estimating a Gini regression coefficient, one sample covariance is *set to zero by construction*, according to the approach taken, but the other sample covariance can be used for the test. For example, by running the semi-parametric regression, the sample covariance of the left-hand side of (20.13) is set to zero by construction (with e_N as the residuals). Hence, one can test for linearity by testing (against a broad alternative) whether $COV(X_k, F(\varepsilon)) = 0$, using e_N. Note that this last covariance is set to equal zero under the minimization approach when using e_M as the residuals. Alternatively, one can run R-regression and reverse the procedure, as was mentioned at the end of the previous section. Starting with a semi-parametric regression to construct a linearity test has several advantages.

(1) The semi-parametric regression does not require specification of the model.
(2) Unlike the minimization approach, there is no problem of nonuniqueness of the estimated regression coefficient.
(3) The estimators of the semi-parametric approach can be written explicitly using OLS—like terminology.
(4) The point estimators of the semi-parametric approach can be calculated easily using the instrumental variable approach therefore standard regression software can be used.[8]

For these reasons the following procedure is suggested for assessing the linearity in X_k.

Step 1: Use the semi-parametric approach to estimate the Gini regression coefficients. Obtain the residuals e_N and the normal equation $cov(e_N,r_k) = 0$, where r_k is the vector of ranks of X_k.

Step 2: Use e_N to test H_0: $COV(X_k, F(\varepsilon)) = 0$. H_0 states that the normal equation of the minimization approach holds for the residuals of the semi-parametric approach. If H_0 is rejected, then one can conclude that the model is not linear in X_k. Recall that in the sample $cov(x_k, r(e_M)) = 0$ by construction (where $r(e_M)$ is the rank of the residuals according to the minimization approach).

A test of H_0 will be based on a U-statistic. Its consistency and asymptotic distribution under H_0 are given by Schechtman and Yitzhaki (1987). (For the general approach for testing see Chap. 10). Because (20.13) holds for each X_k separately, one can use the proposed test for each explanatory variable separately. However, if one wishes to test for linearity of several X's simultaneously, then one should run the regression twice—once for the full model and then for the reduced model (excluding the nonrelevant X's) and compare the Gini's of the residuals of the two models. The significance of the difference can be formally evaluated using the methods developed in Chap. 10. If one is interested in the model as a whole,

[8] The semi-parametric estimators can be viewed as OLS instrumental variable (IV) estimators, with the rank of each variable being used as an IV. However, note that the assumptions that are assumed here are entirely different (see Yitzhaki and Schechtman (2004)). Therefore the inference cannot be drawn from there.

then one would replace X_k by $\widehat{Y}$ in (20.13). That is, one would use $\widehat{Y}$ from the semi-parametric approach and test whether $COV(\widehat{Y}, F(\varepsilon)) = 0$, where $\widehat{Y}$ is the predicted value of Y. This test examines whether the same set of residuals and predicted values can serve as solutions to both methods.

Finally, it is worth mentioning that other assumptions imposed on the regression can be tested by using the GMD. For example, D'Agostino's (1971, 1972) test for normality of the residuals is based on the statistic $cov(e_N, r(e_N))/nS$ (where S is the sample standard deviation of the residuals and n is the sample size), whose numerator is the GMD of the residuals (see Chap. 7).

20.4 The LMA Curve

The Gini can be visually presented by using a Lorenz curve. The Gini coefficient is the area between the line of equality (i.e., the diagonal) and the Lorenz curve. Similar to the presentation of Gini by a Lorenz curve, one can also visually present the co-Gini (as well as the Gini correlation and the Gini regression coefficient) by using a curve, to be defined below.

Moreover, this curve is useful to identify sections of local regression coefficients with different signs if they exist, and/or detect whether a monotonic nondecreasing transformation of the dependent variable can change the sign of the Gini (or OLS) regression coefficients.

The curve is based on two curves: the line of independence (LOI) and the absolute concentration curve (ACC). The vertical difference between the two curves, which we call the LMA curve, is then normalized by dividing it by $cov(X,F(X))$. The resulting curve is denoted by NLMA and the area between it and the horizontal axis is equal to the Gini regression coefficient (see details in Chaps. 7 or 19).

In general the shape of the curve is as shown in Fig. 20.1

The relevant properties of the NLMA curve are

(a) The curve starts at (0,0) and ends at (1,0). It can take any shape depending on the data.
(b) The area between the curve and the horizontal axis is equal to the Gini regression coefficient. In the curve plotted in Fig. 20.1, the section OAB is contributing toward a positive value to the Gini regression coefficient, while BCD contributes toward a negative value.
(c) The same sections (in terms of the transformed values on the horizontal axis) are contributing toward the OLS regression coefficient. However, the magnitudes (but not the signs) may be different. Therefore, the overall sums of the areas may differ resulting in different signs of OLS and Gini regressions.
(d) If the curve changes from convex (concave) to concave (convex) then it is possible to change the sign of Gini and OLS regression coefficients by truncating the distribution of the explanatory variable.
(e) If the curve intersects the horizontal axis then there exist monotonic nondecreasing transformations of X that can change the sign of the OLS regression coefficient.

20.5 An Illustration: The Two Explanatory Variables Case

In order to be able to illustrate the properties of the Gini multiple regression in details and to investigate the roles of the different components, it is convenient to restrict ourselves to the two explanatory variables case. The objective of this section is to express the multiple regression coefficients in the two explanatory variables case as explicit functions of the simple regression coefficients so that it will be clear how each simple regression coefficient affects the multiple regression coefficients.

A detailed derivation of the general case is given in Chap. 8. Here we only give a brief review.

Restricting (20.1) to two explanatory variables, the matrix a = r'x (which is the equivalent of x'x in OLS) is given by

$$a = \begin{pmatrix} \mathrm{cov}(x_1, r_1) & \mathrm{cov}(x_2, r_1) \\ \mathrm{cov}(x_1, r_2) & \mathrm{cov}(x_2, r_2) \end{pmatrix},$$

where r_i is the vector of ranks of the explanatory variable X_i. Note that the matrix **a** is not necessarily symmetric. Dividing each row by the GMD of the diagonal element and solving the linear equations, we get an explicit presentation which is identical in structure to the OLS presentation. In order to show the similarity of b_N to OLS regression, let us rewrite (20.2) as follows

$$\begin{pmatrix} b_{N01.2} \\ b_{N02.1} \end{pmatrix} = \frac{1}{1 - \Gamma_{12}\,\Gamma_{21}} \begin{pmatrix} 1 & -b_{N21} \\ -b_{N12} & 1 \end{pmatrix} \begin{pmatrix} b_{N01} \\ b_{N02} \end{pmatrix},$$

where $\Gamma_{12}\Gamma_{21}$ is the symmetric version of the Gini correlation, b_{Nij} (i = 0,1,2; j = 1,2) indicates the simple Gini regression coefficient of variable i on j, with 0 denoting the dependent variable. The regression coefficients in the multiple regression are $b_{N0i.j}$ (i,j = 1,2). See Chap. 8 for details.

To make the analysis simpler, note that

$$b_{N12}b_{N21} = \frac{\mathrm{cov}(x_1,r_2)}{\mathrm{cov}(x_2,r_2)}\frac{\mathrm{cov}(x_2,r_1)}{\mathrm{cov}(x_1,r_1)} = \frac{\mathrm{cov}(x_1,r_2)}{\mathrm{cov}(x_1,r_1)}\frac{\mathrm{cov}(x_2,r_1)}{\mathrm{cov}(x_2,r_2)} = \Gamma_{12}\Gamma_{21}.$$

Using the above equation we rewrite the coefficients in the multiple regression case as functions of the simple regression coefficients as follows

$$\begin{pmatrix} b_{N01.2} \\ b_{N02.1} \end{pmatrix} = \frac{1}{1 - b_{N12}\,b_{N21}} \begin{pmatrix} b_{N01} - b_{N02}\,b_{N21} \\ b_{N02} - b_{N01}\,b_{N12} \end{pmatrix}.$$

Note that the denominator is always nonnegative because

$$b_{N12}b_{N21} = \Gamma_{12}\,\Gamma_{21} \le 1.$$

Therefore, the sign of $b_{N0i.j}$ is determined by the sign of ($b_{N0i} - b_{N0j}\,b_{Nji}$).

20.6 An Application: Assessing the Linearity of Consumption as a Function of Income and Family Size

In the previous section we expressed the multiple regression coefficients in the two explanatory variables case as explicit functions of the simple regression coefficients so that it will be clear how each simple regression coefficient affects the multiple regression coefficients. We now discuss the logical inconsistency between measurement and policy instruments used in order to deal with the effect of family size on economic well-being and in the next section we present the empirical analysis intended to shed some light on this internal inconsistency.

20.6.1 The Problem to be Solved

In this section we present an example to illustrate the properties of the Gini regressions, the specification tests, and the visual inspection of whether the association between random variables is monotonic over the entire range of the explanatory variable or not. In this example there is an incompatibility between the measurement of performance of a policy intended to improve the income distribution and the policy instruments that the government uses. An empirical examination can help to decide which way is the "right" way. The issue is the following: when the problem of interest is to measure inequality in economic well-being, differences in family sizes are commonly taken into account by looking at income per capita or some kind of an equivalence scale. Almost all equivalence scales in use are based on dividing the income by a number which is a function of the family size. When viewing consumption per adult equivalent as representing economic well-being we should expect consumption expenditures to be related to income and family size in a multiplicative relationship. On the other hand, most of the policy instruments in use in the income tax and benefits systems are based on an adjustment of the tax to family size by giving a tax relief which is based on decreasing the tax (or increasing the benefits) by an amount which is only a function of the family size. We argue that those instruments represent an additive relationship between consumption and family size. It should be emphasized that we are not dealing with the normative issue of how the tax system should treat families of different sizes. The normative issue needs further research. All we deal with is the issue—are the way in which performance of tax systems in the area of reducing inequality is measured and the policy instruments used compatible? In the rest of this section we present the distinction between a multiplicative relationship and an additive one in a formal way.

The first step in the measurement of inequality in economic well-being is to rank households according to economic well-being. One way of doing that is to use a multiplicative scale so that the ability to consume is defined as

$$e(Y, N) = E\left(\frac{Y}{a(N)}\right), \tag{20.14}$$

where Y is the net income of the household, N is the family size, a(N) is the adult equivalent scale, and e(Y,N) is the equalized income that represents economic well-being. For example, the European Union is using the scale of $a(N) = N^{0.5}$ as its official scale. Feldstein's (1976) principle of horizontal equity is that *"If two individuals would be equally well off (have the same utility) in the absence of taxation, they should also be equally well off if there is a tax"* (Italic at source, 1976, p. 83). Our interpretation of the principle is that the ranking of families according to before-tax economic well-being should be identical to the ranking of families according to after-tax economic well-being. Would we want the tax and benefit function to keep Feldstein's principle of horizontal equity, the structure of the tax and benefit function should have been

$$T(N, Y) = a(N)\, t\left(\frac{Y}{a(N)}\right), \tag{20.15}$$

where T(N,Y) is the total tax minus the benefits that the family receives and t( ) is the tax function defined over adult equivalent income.[9,10] On the other hand, when looking at tax and benefit systems, it turns out that most countries rely on an additive scale. This is the case whenever child allowances or exemptions are used to handle family size. In the case of exemptions, the structure of the tax function is

$$T(N, Y) = T(Y - EX(N)). \tag{20.16}$$

That is, to adjust the tax to family size, the amount of EX(N) is deducted from before-tax income, for all families of size N, as it is, for example, in the USA. In the case of tax-exempt allowances, the tax function is

$$T(N, Y) = T(Y) - AL(N), \tag{20.17}$$

where AL is the allowance, as is the case in Israel and Britain. In both cases (20.16) and (20.17) it is as if the state recognizes a given amount that should be added to the income to keep horizontal equity intact.

We argue that (20.14) on one hand and (20.16) and (20.17) on the other hand are incompatible because they violate Feldstein's principle of horizontal equity. To shed some empirical light on this issue it is worth to check whether the Engel curve, which relates consumption to income and family size, is additive or multiplicative.

---

[9] In some applications the model used is $T(N, Y) = Nt(\frac{Y}{a(N)})$, so that each member of the household is counted as one (see Ebert (2005, 2010) and Ben-Porath's comment by Bruno and Habib (1976)).

[10] The French tax system resembles this structure.

This means that we view consumption as representing economic well-being and household's size as representing needs. Assume that the appropriate specification of the Engel curve is linear. That is,

$$\hat{C} = \alpha + \beta Y + \gamma N, \tag{20.18}$$

where α, β, and γ are either chosen by the policy maker or estimated from the data. Then it is reasonable to argue that the effect of an additional family member on consumption is to increase consumption by a constant and then the appropriate tax adjustment should be of the exemption or allowance type. If, on the other hand, the appropriate specification of the Engel curve is multiplicative, that is, of the type

$$\ln(\hat{C}) = \alpha + \beta \ln(Y) + \gamma \ln(N) \tag{20.19}$$

then the effect of an additional family member on consumption is to increase consumption by a given percentage, which can be viewed as supporting a tax function of the multiplicative type as in (20.15). Our empirical research question is to decide which of the two alternatives is supported by the data and represents the appropriate specification of the Engel curve.

20.6.2 *Empirical Findings*

The alternative models discussed above were applied using Israeli Survey of households' Expenditures, 2008. (For a description of the sample see Central Bureau of Statistics (2009), S.P. 1363). The data consists of 5,971 observations. Each observation includes a weight which represents its weight in the population. Consumption includes the depreciation and value of forgone interest on capital invested in housing and vehicles. Income is after-tax overall income, which includes money income *plus* in-kind income *minus* income tax and social security taxes.

The structure of the empirical illustration is the following: we first present the simple regression coefficients that are the basic components of the multiple regression coefficients and only later we present the multiple regression coefficients. In our presentation, our main interest is to find out whether the relationships between the variables are monotonic. By "monotonic" it is meant that the sign of the regression coefficient does not change over the entire range of the explanatory variable. The final stage is to see which model fits the data better: the additive or the multiplicative?

We start with the relationship between the dependent variable and the explanatory variables.

Table 20.1 presents the simple regression coefficients between consumption expenditures and after-tax (net) income using linear and multiplicative specifications. For comparison we also present the OLS estimates.

Table 20.1 Simple OLS and Gini regression coefficients—consumption as a function of income[a]

	OLS			Gini					
Model	a	b	R^2	a (mean)	a (median)	b	$R(y,\hat{y})$	$R(\hat{y},y)$	GR
Linear	4,756	0.533 (0.000)	0.514	3,505	2,641	0.621 (0.011)	0.792	0.803	0.346
Multiplicative	4.216	0.538 (0.000)	0.489	2.735	2.707	0.698 (0.014)	0.811	0.791	0.372

[a]Standard errors in parentheses.
In Gini regression the standard errors were calculated using Jackknife fast method. [In using the jackknife method in a regression context there are two options: when dropping an observation from the sample, should one re-estimate the whole model again or is it permissible to drop an observation and to evaluate the effect on the regression coefficient. The former approach seems to be the appropriate one but it is time consuming and the difference between the two methods seems negligible. By fast method it is meant that the model was not re-estimated.]
Standard errors of the OLS are rounded to three decimal points
Source: Schechtman, Yitzhaki, and Pudalov (2011), Table 6.1, p. 86

The left-hand part of Table 20.1 presents the OLS simple regression coefficients of the additive and multiplicative specifications (20.18) and (20.19). The right-hand part presents the semi-parametric Gini regression equivalents. Because we are only interested in the components of the multiple regression the standard errors for the constant term are not estimated. As can be seen from Table 20.1, the marginal propensity to spend, that is, the simple regression coefficient, is smaller for the OLS than for the Gini in both specifications. This is a result of a combination of two factors: the Gini regression tends to give lower weights to extreme observations, and the marginal propensity to spend tends to decline with income. The constant terms of the Gini regressions are estimated in two ways depending on whether the regression passes through the median of the dependent variable or through the mean. The median constant is lower than the mean constant, especially for the linear regression. This indicates that the residuals tend to be larger for high income groups. It should be mentioned that the standard errors of the Gini and OLS regression coefficients are not comparable because for the Gini it is assumed that the explanatory variable is a random variable, while under regular OLS software it is assumed that only the residual term is random. Also note that the measures of goodness of fit are not automatically comparable. In order to make them comparable one should look at R for OLS, rather than at R^2. Figure 20.2 presents the LMA of consumption versus income (Line of Independence Minus Absolute Concentration Curve). The curve is concave and smooth and does not intersect the horizontal axis, implying that the relationship between consumption and income is monotonically increasing and there is no monotonic transformation of income that can change the sign of the OLS regression coefficient. Also the curve is increasing till about the 60th percentile, indicating that the consumption is lower than the average consumption, and then declines, indicating that consumption in above average consumption.

Table 20.2 is similar to Table 20.1, except that this time the regression is with respect to family size. Here, the results are a bit different from those in Table 20.1.

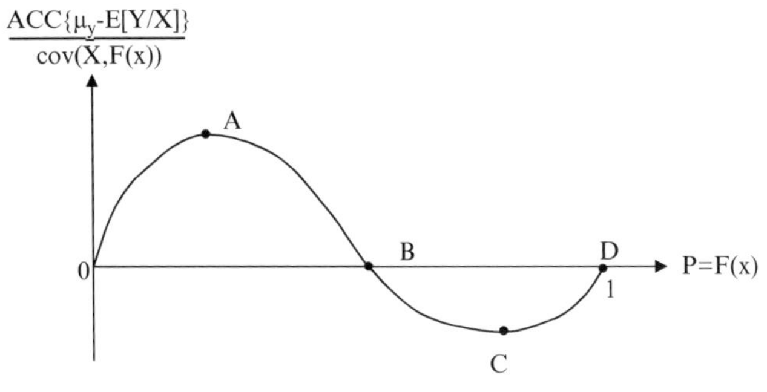

Fig. 20.1 The NLMA curve

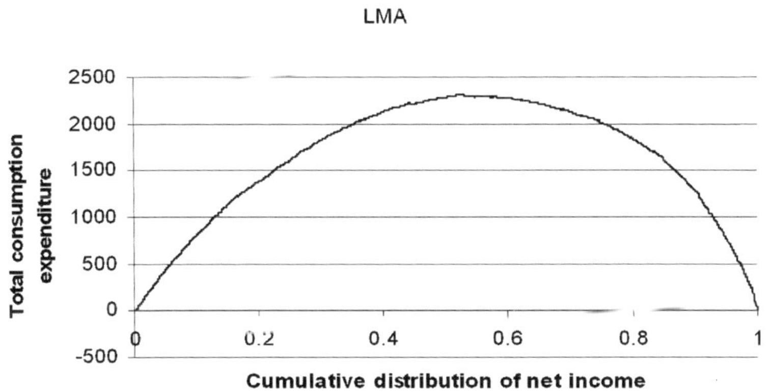

Fig. 20.2 LMA curve of consumption as a function of income. Source: Schechtman, Yitzhaki, and Pudalov (2011), Fig. 1, p. 85. Reprinted with permission by Metron International Journal of Statistics

Table 20.2 Simple OLS and Gini regressions—consumption as a function of household's size[a]

	OLS			Gini					
Model	a	b	R^2	a (mean)	A (median)	b	$R(y,\hat{y})$	$R(\hat{y},y)$	GR
Linear	8,191	1,251 (2.686)	0.093	7,195	5,674	1,551.6 (59.47)	0.399	0.399	0.056
Multiplicative	8.73	0.486 (0.001)	0.217	8.742	8.766	0.474 (0.015)	0.459	0.463	0.110

Source: Schechtman, Yitzhaki, and Pudalov (2011), Table 6.2, p. 86
Reprinted with permission by Metron International Journal of Statistics
[a]Standard errors in parentheses. In Gini regression standard errors were calculated using Jackknife fast method

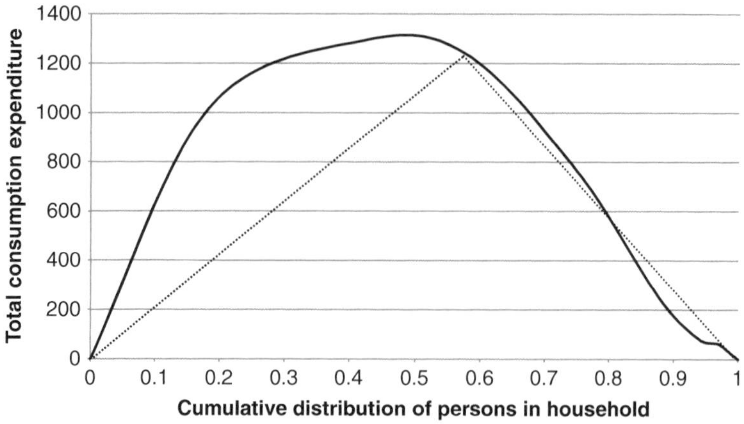

Fig. 20.3 LMA curve of consumption as a function of household's size. Source: Schechtman, Yitzhaki, and Pudalov (2011), Fig. 6.2, p. 86. Reprinted with permission by Metron International Journal of Statistics

Under the linear specification, having an additional person increases consumption by 1,552 NIS (New Israeli Shekel) according to the Gini regression, while under OLS the amount is smaller and is equal to 1,251 NIS. On the other hand, under the multiplicative specification, the OLS estimates of the percentage increase in consumption due to an increase in the household size is 0.49 which is lower than the Gini (0.47).

Figure 20.3 offers an insight to the results. Unlike the curve in Fig. 20.2 which is concave and smooth, Fig. 20.3 changes from a concave to a convex curve. This means that while the overall regression coefficient is positive (as seen in Table 20.2), it is positive among small households, but it is negative among the largest households as is illustrated next: using the Gini linear specification and estimating the regression coefficient for households of size 4 and above, which include 42% of the observations, we get $\beta = -173.5$, while restricting the regression to households of size 5 and larger, which amount to about 25% of the observations, we get an even smaller slope, $\beta = -827.3$. It is important to note that on the vertical axis we plotted overall consumption of the household, so that the decline in consumption with household size is of total consumption and not per-capita consumption.

Note, however, that the curve does not cross the horizontal axis, which implies that no monotonic transformation of household's size can change the sign of the regression coefficient. The explanation to this result lies in an additional decomposition that one can perform, which decomposes the regression coefficient into two components: a within-component (intra) and a between-components (inter) (see Chaps. 7 or 19). Assume that one divides the range of the explanatory variable into two sections. One section is composed of the sixty percent of the smallest households, while the other section is composed of the remaining 40% of the largest households. Then the overall regression coefficient can be expressed as a

Table 20.3 The simple regression coefficients between the explanatory variables[a]

	OLS			Gini					
Model	a	b	R^2	a (mean)	a (median)	b	R(y,ŷ)	R(ŷ,y)	GR
(a) Household's size as a function of income									
Linear	2.713	0.000042 (0.000)	0.055	2.393	1.96	0.000065 (0.000)	0.321	0.343	0.031
Multip.	−1.583	0.28 (0.000)	0.144	−2.113	−2.087	0.3366 (0.0117)	0.404	0.411	0.062
(b) Income as a function of household's size									
Linear	9,961.86	1,285.38 (3.687)	0.055	8,584	6,406	1,709 (82.4)	0.343	0.321	0.026
Multip.	8.781	0.514 (0.001)	0.144	8.802	8.863	0.4938 (0.024)	0.411	0.404	0.080

[a]Standard errors in parentheses.
In Gini regression standard errors were calculated using Jackknife fast method. See explanation to Table 20.1 and Yitzhaki (1991)
Source: as above, Table 6.3 p. 88

weighted sum of intra- and intersection regression coefficients weighted by the appropriate measure of variability used (variance of X for OLS; GMD of X for GMD regression). The between-groups component is reflected by the triangle which starts at the origin reaches the curve at the end of the first section and ends up on the horizontal axis at one. The intra-group components are reflected by the areas enclosed between the curve and the edges of the triangle. In our example (Fig. 20.3) the between-groups component contributes to the overall regression coefficient more than the intra-group components. Therefore it determines its sign. However, it should be clear that no linear model can explain such a pattern.

We now turn to describe the simple regression coefficients between the explanatory variables. Table 20.3 presents the last elements needed for the multiple regression coefficients.

It is worth mentioning that in OLS the ratio between a regression coefficient and the regression coefficient in a reversed regression is equal to the ratio of the variances. (i.e., in OLS $b_{yx}/b_{xy} = Var(y)/Var(x)$). In the Gini regression no such relationship has to hold. Moreover, at least in theory they can even have different signs! Figs. 20.4 and 20.5 present the LMA curves for the additive model. Figure 20.4 which portrays net income as a function of the size of the household indicates that while for small households the conditional expected value of income is increasing, this sign of the regression coefficient changes to a negative one among 45% of the largest households. Moreover, there is a small range (between 90th and 95th percentiles of household's size) in which the curve is below the horizontal axis, which means that a monotonic transformation that shrinks all the rest of the range of household's size will yield a negative regression coefficient in the OLS. Also, there is a transformation of income that can change both OLS and Gini regression coefficients. Obviously this will not be an acceptable treatment of the

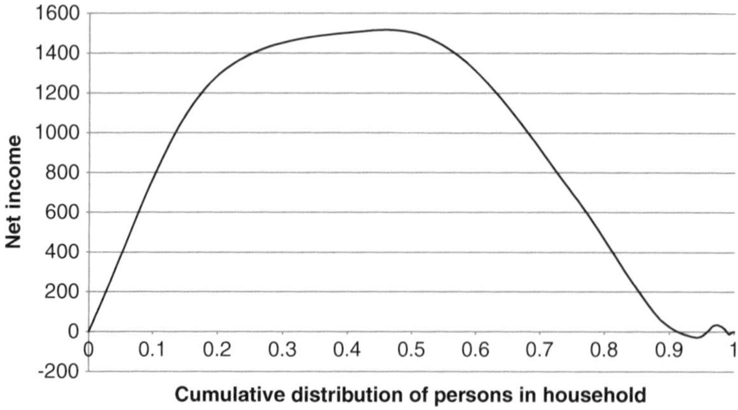

Fig. 20.4 LMA curve of income as a function of household's size

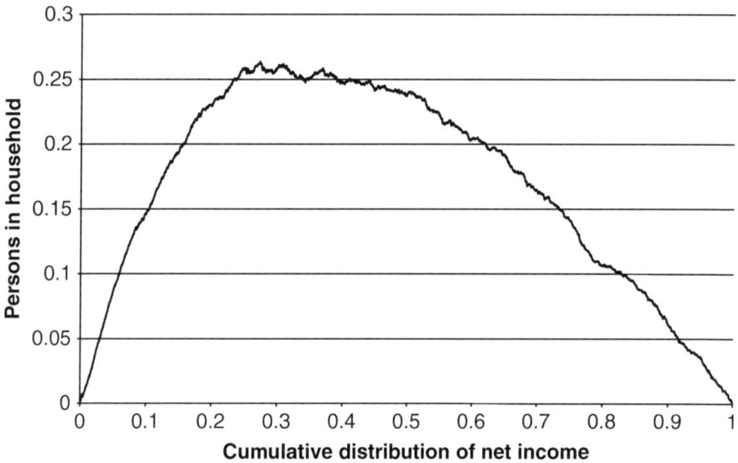

Fig. 20.5 LMA curve of household's size as a function of income. Source: same as above, Figs. 6.3 and 6.4 on p. 88 and 89

data. Figure 20.5, on the other hand, which presents household size as a function of income, is a concave and smooth curve.

As far as we can see, no simple model can explain such results. One possible explanation is that there are two models of behavior: the fertility in one group follows the regular pattern of bringing children to the world subject to having the ability to support them, while the fertility in the other group is not related to income. Our conclusion is that no simple model can explain such a curve.

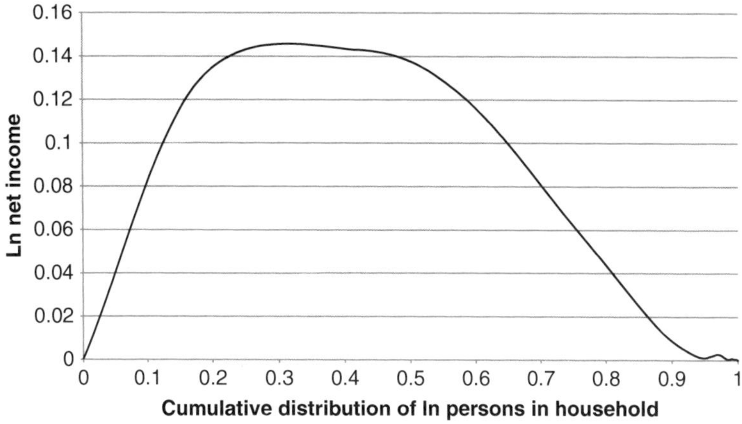

Fig. 20.6 LMA curve of ln(income) as a function of ln(household's size). Source: same as above, Fig. 6.5, p. 89

Figure 20.6 is added in order to explain the difference between the additive and the multiplicative models: because the horizontal axis portrays the cumulative distribution, it is the same as the horizontal axis of Fig. 20.4. Hence the only difference between the additive and the multiplicative models is that the latter shrinks the inconvenient deviations that existed in the additive model. As a result, in the latter model there is no monotonic transformation that can change the sign of the regression coefficient. However, the negative relationship in income as a function of the household's size for large households continues to hold. This is an indication that we should expect a better fit of the multiplicative model than the additive one.

Having described the components of the multiple regression coefficients, we now move to present the results of the multiple regressions. Table 20.4 presents the results.

Comparison of the regression coefficients between the OLS and the Gini in the additive and multiplicative models indicates that the marginal propensity to spend is larger under the Gini than under OLS regression, with the multiplicative model showing larger differences. On the other hand, both specifications indicate that the effect of an additional member in the household is larger under OLS than under the Gini method. The difference between the estimates of the simple regression coefficient and the parallel estimates in the multiple regression case indicates the effect of the association between the explanatory variables. The estimates are given in Table 20.5.

Would the explanatory variables be statistically independent, then there should have been no difference between the two. However, when the explanatory variables are correlated and the relationship is not linear then the effect of the correlation may be different under OLS and Gini regressions. The fact that the two methods yield regression coefficients that are different calls for further inspection of the way that the models fit the data. The range of the income variable is of a totally different magnitude than that of the household's size. This explains why the differences in

Table 20.4 The multiple regressions: consumption as a function of income and household's size[a]

	OLS				Gini						
Model	A	b (income)	b (size)	R^2	a (mean)	a (med.)	b (income)	b (size)	$R(y,\hat{y})$	$R(\hat{y},y)$	GR
Linear	3,132	0.508 (0.000)	598.6 (1.981)	0.534	2,173	1,346	0.585 (0.011)	556.4 (41.01)	0.801	0.813	0.363
Multip.	4.603	0.47 (0.000)	0.245 (0.001)	0.536	3.064	3.037	0.645 (0.013)	0.156 (0.012)	0.823	0.805	0.394

[a]Standard errors in parentheses.
In Gini regression standard errors were calculated using Jackknife slow method
Under the slow method, each time an observation was dropped from the sample, the model was re-estimated
Source: same as above, Table 6.4, p. 90

Table 20.5 The regression coefficients for the simple and multiple regression models[a]

	OLS				Gini			
	Income		Size		Income		Size	
Model	Simple	Multiple	Simple	Multiple	Simple	Multiple	Simple	Multiple
Linear	0.533	0.508	1,251	598.6	0.621	0.585	1,551.6	556.4
	(0.000)	(0.000)	(2.686)	(1.981)	(0.011)	(0.011)	(59.47)	(41.01)
Multip.	0.538	0.47	0.486	0.245	0.698	0.645	0.474	0.156
	(0.000)	(0.000)	(0.001)	(0.001)	(0.014)	(0.013)	(0.015)	(0.012)

[a]Standard errors in parentheses.
In Gini regression standard errors were calculated using Jackknife slow method.
Source: same as above

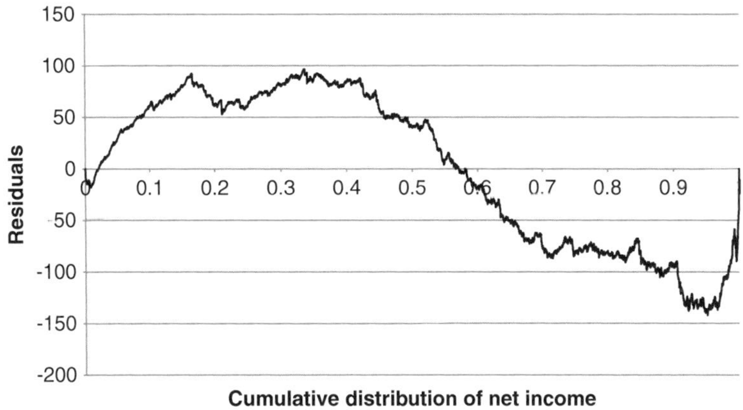

Fig. 20.7 LMA of residuals as a function of income: linear specification. Source: same as above, Fig. 6.6, p. 91

magnitude of the coefficients of income are smaller than the differences in the range of the coefficients for household's size.

Figure 20.7 presents the LMA curve of the residuals as a function of net income. By construction, the area between the curve and the horizontal axis is equal to zero (recall that $\text{cov}(e_N, F(X)) = 0$). A perfect fit of the model to the data will result in a curve which oscillates randomly around the horizontal axis. As can be seen, this is unlikely the case. For the lower 55% of observations of income (approximately) there is a positive (Gini and Pearson) correlation between the residuals and income, while the highest 45% of the observations on income reveal a negative correlation.

Note that by construction, the area enclosed between the curve and the horizontal axis equals to zero.

To check the quality of the specification Fig. 20.8 presents the LMA curve of income as a function of the residuals. Results show that Fig. 20.8 is almost a mirror image of Fig. 20.7. For small values of residuals the correlation is negative, while for large values of residuals the correlation is positive. Overall, $\text{cov}(x, r(e_N)) = -505.44$. To test whether the specification of the model is correct we estimated the (simple) Gini regression coefficient of income on the residuals and found that the regression

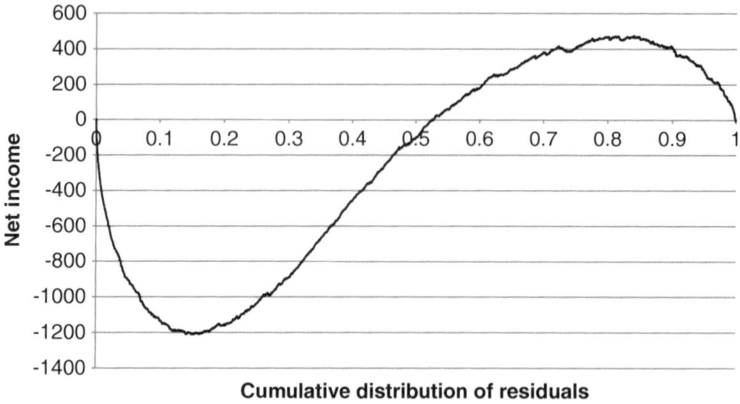

Fig. 20.8 LMA of income as a function of the residuals. Source: same as above. Figure 6.7 p. 91

coefficient $b_{x,r(e)} = -0.194$, $\hat{\sigma}_b = 0.042$, and because the estimator of the regression coefficient is approximately normally distributed, it turns out that the value of the test statistic is $Z = -4.609$ and the linearity of the model with respect to income is rejected.

We now inspect the quality of the specification of the model with respect to household's size. Figure 20.9 presents the LMA curve of the residuals as a function of household's size. Again, we remind the reader that the area enclosed between the curve and the horizontal axis is zero by construction. Similar to the case of income, there is a positive correlation between the residuals at low levels of household's size and negative correlation for large household's size. However, for about 25% of the observations of middle size households the curve is horizontal and close to the horizontal axis, indicating a good fit of the model to the observations.

Figure 20.10 presents the LMA curve of household's size as a function of the residuals. Again we get a mirror image of Fig. 20.9 although the quality of the "mirror" is worse than the one we got when we dealt with income. For small values of residuals we got a negative correlation between household's size and residuals, while for large values of residuals we got a positive correlation. To test for the quality of the linear specification we ran a (Gini simple) regression of household's size on the residuals. The estimated Gini regression coefficient is −0.0000247, its standard error is estimated to be 4.5197E−11, so that the test statistic indicates that the linear specification is rejected.[11]

[11] There are two problems with these results. The first problem is that household's size is a discrete variable. In this case there is a mismatch between the LMA curve and the definition of cumulative distribution, because the empirical cumulative distribution is defined as a step function, while in an LMA (and Lorenz) curve one connects different points of the curve by straight lines, which implies continuity (see Chap. 5). The other problem is the issue of rounding errors because of small numbers involved. Therefore one should be careful in interpreting this result. Further research is required to resolve this issue.

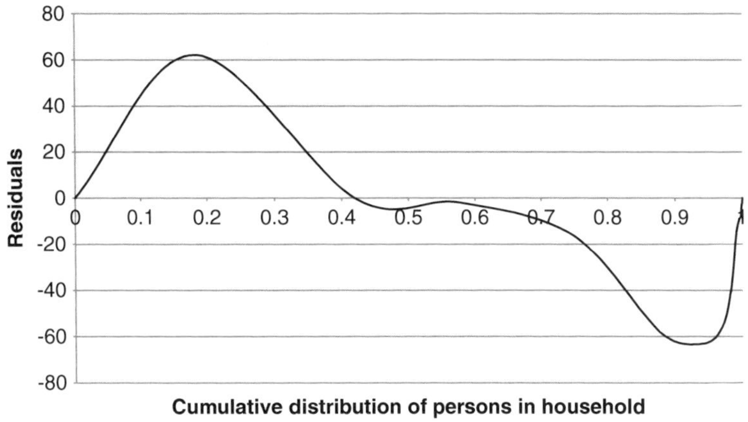

Fig. 20.9 LMA curve—linear specification: household's size

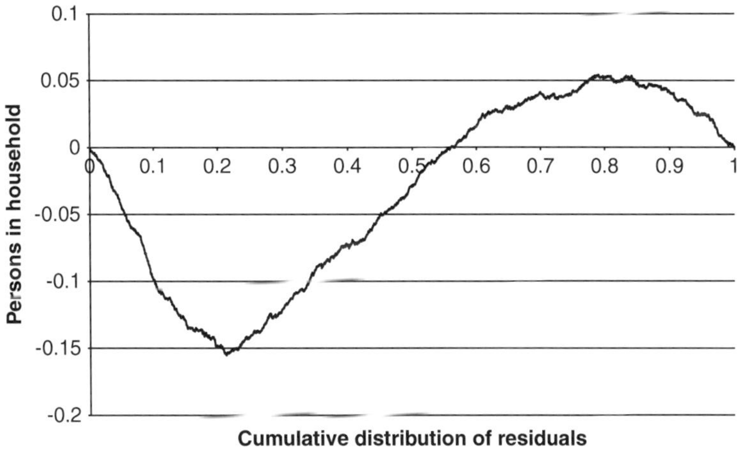

Fig. 20.10 LMA of household size as a function of residuals: linear specification. Source: same as above. Figure 6.7 p. 91

We now turn to inspect the multiplicative specification.

Figure 20.11 presents the LMA curve of the residuals as a function of ln(net income). Note that because the horizontal axis portrays the cumulative distribution of ln(net income) and the cumulative distribution of net income is not affected by monotonic increasing transformation, the horizontal axes in Figs. 20.11 and 20.7 are identical. The difference is only in the vertical axes. Comparisons of the two figures reveal that while in Fig. 20.7 low levels of income are associated with negative correlation with the residuals, the multiplicative specification has changed the sign and the order of correlations. Except for the lowest 20% of observations of income the curve seems flat indicating a low level of correlation between the

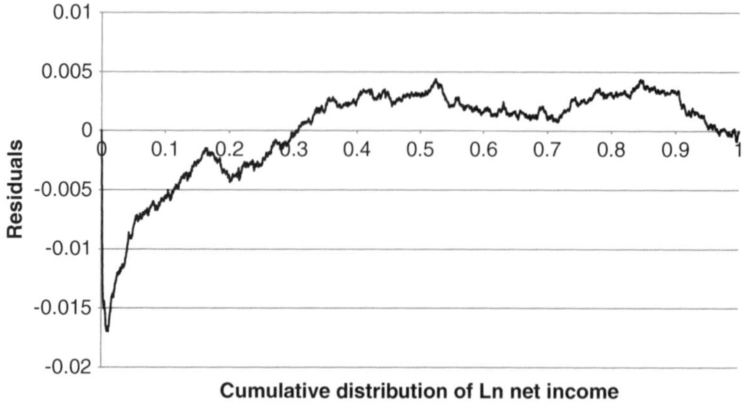

Fig. 20.11 Multiplicative specification: residuals as a function of ln(net income) Source: same as above, Fig. 6.10 p. 94

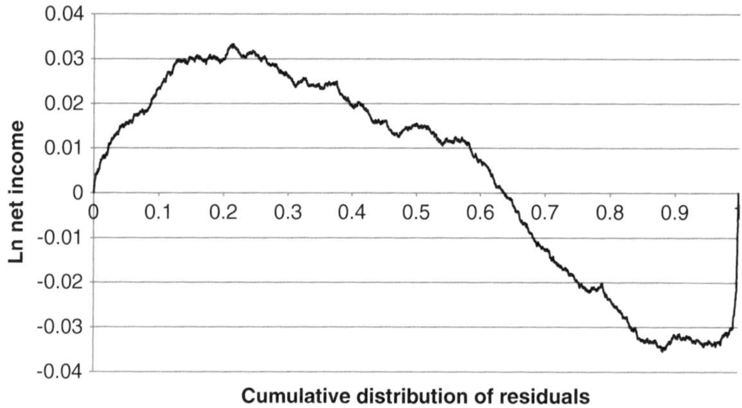

Fig. 20.12 LMA curve of ln(net income) as a function of the residuals. Source: same as above, Fig. 6.11 p. 94

residuals and income. Figure 20.12 presents the LMA curve of ln(net income) with respect to residuals. For low levels of residuals the correlation is positive, while for high levels of residuals the correlation is negative. The estimated cov(x,r(e)) $= -0.00803$, the regression coefficient of the simple (Gini) regression of ln(income) as a function of the residuals is $b_{x,r(e)} = -0.0365$, the estimated standard error is $\hat{\sigma}_b = 0.047$, the test statistic is $Z = -0.772$, so that we fail to reject the hypothesis that the model is linear with respect to income.[12]

[12] Standard errors were calculated using Jackknife fast method.

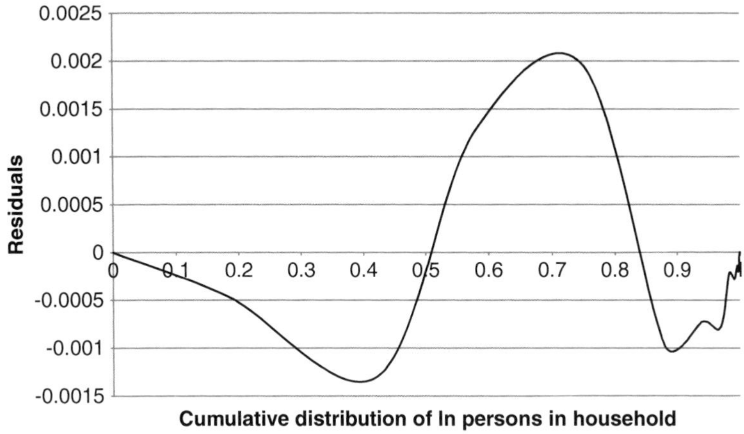

Fig. 20.13 Multiplicative specification: household's size. Source: same as above, Fig. 6.12 p. 95

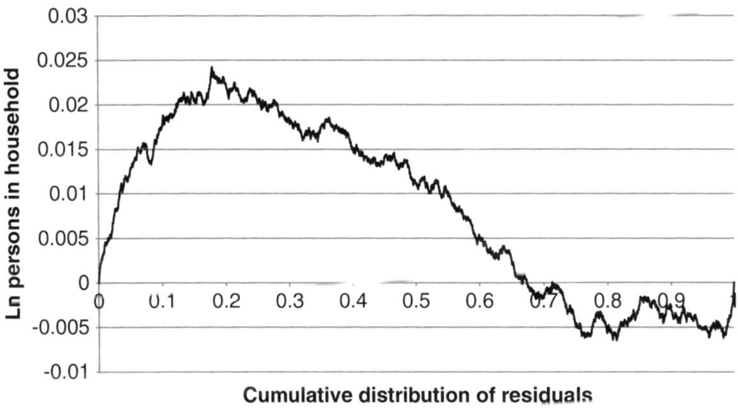

Fig. 20.14 LMA curve of household's size with respect to residuals. Source: same as above. Figure 6.13 p. 95

We turn now to the specification of the multiplicative model with respect to household's size. Figure 20.13 presents the LMA curve of the residuals as a function of the household's size. As can be seen, for low levels of household's size the correlation is negative, for middle-sized households it is positive and for large ones households it is again negative. Note that by construction, the overall area between the curve and the horizontal axis is equal to zero.

One can observe that the (Gini) correlation between household's size and the residuals is positive for small households and slightly negative for large ones (Fig. 20.14). The Gini covariance is cov(x,r(e)) = 0.01644, the (Gini simple) regression coefficient is $b_{x,r(e)} = 0.0748$, its standard error is $\hat{\sigma}_b = 0.026$, so that the test statistics is $Z = 2.907$. This means that although we have rejected the specification with respect to family size in the multiplicative model as well, the multiplicative model fits the data better than the additive one.

To conclude the empirical examination, we have found the multiplicative model to fit the data better than the additive one. If one has to choose between (20.8) and (20.9) then (20.9) is supported by the data better than (20.8).

20.7 Summary

In this chapter we have illustrated the use of the Gini multiple regression. Similar to the simple regression case, the Gini multiple regression offers two types of regressions. The first is a semi-parametric regression, which is an imitation of the OLS regression, and similar to it, the estimator can be explicitly written. The advantages are that no model has to be specified, and it is less sensitive to outliers than the OLS regression (because it is based on ranks). The other regression is based on minimization of the Gini of the residuals.

The combination of the two methods offers a built-in specification test. It is based on the fact that the Gini method has two covariances between each pair of random variables. In estimating a regression model one covariance between the residuals and each explanatory variable is set to zero, so that the other covariance can be used as a test for the specification of the model. One can start by estimating a linear approximation to the regression curve, and if one wants to use it for prediction, then the prediction will be restricted to the variables in which the model is linear.

The connection between the Gini parameters and concentration curves enables one to verify the monotonicity of the regression curve. The importance of verifying the monotonicity is stressed by Heckman, Urzua, and Vytlacil (2006a) in the context of an instrumental variable. In a Gini regression framework, monotonicity is not an assumption imposed on the data. It is a property that says that the regression coefficients on each section of the data have identical signs. It is up to the judgment of the reader to decide whether this requirement is violated and to what extent. The monotonicity can be graphically observed—if the LMA intersects the horizontal axis then the monotonicity is violated. We illustrate the methodology described above using data on consumption, income, and household's size from the Israeli households' expenditure survey. Schechtman, Shelef et al. (2008) suggest several statistical tests that will enable the user to formally test for the intersection of absolute concentration curves.

The basic OLS regression has many refinements. The similarity between the OLS and the semi-parametric Gini regression gives the hope that many of the refinements of the OLS can be developed for the Gini regression as well. A first step in this direction is offered by Yitzhaki and Schechtman (2004), where the instrumental variable approach is developed in a GMD framework. An additional tool is the decomposition of the Gini of a population by subpopulations (ANOGI) (Frick et al. 2006) which enables the user to imitate the decomposition of the variance (ANOVA) and to get an additional property—the stratification in a distribution.

ANOGI is the tool that enables the decomposition of the Gini regression coefficient into the contributions of different sections of the explanatory variable, as discussed by Yitzhaki and Schechtman (2012). Dividing the residuals into positive and negative groups and applying ANOGI enables the user to get further insights about the distribution of the negative and positive residuals along the range of the explanatory variable. Further research is needed to fully utilize the properties of the Gini in regression analysis.

Chapter 21
Mixed OLS, Gini, and Extended Gini Regressions

Introduction

The purpose of this chapter is to illustrate the use of the mixed regression technique. The meaning of mixed regression is that some of the explanatory variables are treated according to one regression method, while the others are treated according to another method. We extend this definition and include EG regressions for which different EG parameters may be attached to the different explanatory variables. Like any inbreeding the mixed regression does not have "pure" properties. Therefore the purpose for using it needs to be explained and justified.

We discuss two types of mixtures: mixing OLS and Gini regressions and mixing Gini and EG regressions. Mixing Gini and OLS is useful whenever estimating a model with OLS yields estimates that are different from the estimates obtained by the Gini method. Sometimes the estimates may even have different signs. In this case, we can move from one method to the other by switching one variable at a time to find the variable(s) that are responsible for the deviations of the estimates. For example, we can start by treating all the variables by OLS and then start treating them (adding one at a time) by Gini regression.

Mixed Gini and EG regression is mainly intended to trace the curvature of a conditional regression curve. That is, given a model that was estimated by the Gini regression (i.e., $\nu = 1$), we can trace the curvature of the regression curve with respect to each explanatory variable by changing its EG parameter and checking whether the conditional regression curve (given the model and its estimation methodology) is concave, linear, or convex.

Mixing Gini and EG regression may also be motivated by economic theory. Whenever economic theory calls for asymmetric treatment of one variable as is the case in welfare economics, and in finance the mixed Gini and EG regression is the preferred method because OLS and Gini treat the variables in a symmetric way. The Gini belongs to the EG family, therefore it is more convenient to mix it with EG than mixing OLS and EG. In welfare economics, theory calls for imposing the "social evaluation of the marginal utility of income" on the data, while in finance

S. Yitzhaki and E. Schechtman, *The Gini Methodology: A Primer on a Statistical Methodology*, Springer Series in Statistics 272, DOI 10.1007/978-1-4614-4720-7_21,

the same role is played by risk aversion. In those cases we may want to apply the EG regression to the income or wealth, while other explanatory variables can be handled by OLS or Gini.

It is worth noting that it is possible to develop a mixed regression method for other techniques such as quantile or Mean Absolute Deviation regressions. However, we will not discuss these issues in details, but only point out how to do them.

The structure of the chapter is as follows: Sect. 21.1 presents the methodology of mixed regression. In Sect. 21.2 we illustrate the use of mixed OLS and Gini regression and in Sect. 21.3 we illustrate the use of mixed Gini and EG regression. Section 21.4 concludes.

21.1 Mixing Gini, Extended Gini, and OLS in the Same Regression

In this section we present a new regression technique which is based on mixing ordinary least squares (OLS) and Gini or Gini and EG regressions. The basic idea is the following: it is shown in Chap. 7 that the regression coefficients in a simple OLS and in Gini and EG regressions can be interpreted as weighted averages of slopes defined between adjacent observations of the explanatory variable. The implication of this observation is that the OLS and Gini estimators of the regression coefficients do not rely on the linearity assumption of the regression curve. Schechtman, Yitzhaki, and Artsev (2008) used the concept of a linear approximation to a regression curve, that is, estimating a linear approximation to the model without assuming that the model is truly linear. The aim of this section is to briefly present the basic derivation of estimators within the framework of mixed OLS, Gini and EG regressions. We refer to those regressions as covariance-based regressions because the estimators of the regression coefficients in a multiple regression framework are derived by solving a set of linear equations that are composed of simple regression coefficients that play the role of the parameters in those equations (see Chap. 8 for a detailed discussion). The presentation is restricted to population parameters. All estimators are sample's analogues of the population parameters.

Let $(Y, X_1,\ldots,X_K)$ be continuous random variables that follow a multivariate distribution with finite second moments. For every choice of constants, $\alpha, \beta_1, \ldots, \beta_K$ define the random variable ε by the following identity

$$Y \equiv \alpha + \beta_1 X_1 + \ldots + \beta_K X_K + \varepsilon. \tag{21.1}$$

At this stage, $\alpha, \beta_1,\ldots, \beta_K$ are arbitrary constants ($\beta_1,\ldots,\beta_K$ will later stand for the multiple regression coefficients, while α will be a location parameter). The random variable ε is defined as a slack variable, intended to fulfill identity (21.1). The symbol $\equiv$ is used to indicate that at this stage there are no assumptions imposed on ε and all its properties are determined by the properties of the distribution of $(Y, X_1, \ldots, X_K)$. Identity (21.1) is a tautology, which means that no assumption has been imposed on the regression curve.

Let $T_1, \ldots, T_K$ be K random variables. The covariances between Y and these variables define a set of identities as follows

$$\mathrm{cov}(Y, T_1) \equiv \beta_1 \mathrm{cov}(X_1, T_1) + \ldots + \beta_K \mathrm{cov}(X_K, T_1) + \mathrm{cov}(\varepsilon, T_1)$$

$$\mathrm{cov}(Y, T_k) \equiv \beta_1 \mathrm{cov}(X_1, T_k) + \ldots + \beta_K \mathrm{cov}(X_K, T_k) + \mathrm{cov}(\varepsilon, T_k)$$

$$\mathrm{cov}(Y, T_K) \equiv \beta_1 \mathrm{cov}(X_1, T_K) + \ldots + \beta_K \mathrm{cov}(X_K, T_K) + \mathrm{cov}(\varepsilon, T_K). \quad (21.2)$$

Dividing each line by the appropriate covariance, subject to the assumption that $\mathrm{cov}(X_k, T_k) \neq 0$, $(k = 1, \ldots, K)$ we get

$$\beta_{01} \equiv \beta_1 1 + \ldots + \beta_K \beta_{K1} + \beta_{\varepsilon 1}.$$

$$\beta_{0k} \equiv \beta_1 \beta_{1k} + \ldots + \beta_K \beta_{Kk} + \beta_{\varepsilon k}.$$

$$\beta_{0K} \equiv \beta_1 \beta_{1K} + \ldots + \beta_K 1 + \beta_{\varepsilon K}, \quad (21.3)$$

where the index 0 indicates the dependent variable,

$\beta_{\varepsilon j} = \frac{\mathrm{cov}(\varepsilon, T_j)}{\mathrm{cov}(X_j, T_j)}$ and $\beta_{kj} = \frac{\mathrm{cov}(X_k, T_j)}{\mathrm{cov}(X_j, T_j)}$ are the general formulae for the regression coefficients in the simple regressions of X_k on T_j, $k,j = 1, \ldots, K$. Two special cases are the OLS (iff $T_j = X_j$), and the Gini (iff $T_j = F(X_j)$). Provided that the rank of the matrix of the coefficients composed of the β_{kj}'s is K we get the following "solution" of the identities in (21.3)

$$\begin{pmatrix} \beta_1 \\ \\ \beta_K \end{pmatrix} \equiv \begin{pmatrix} 1 & \beta_{21} & \beta_{K1} \\ & & \\ \beta_{1K} & \beta_{2K} & 1 \end{pmatrix}^{-1} \begin{pmatrix} \beta_{01} - \beta_{\varepsilon 1} \\ \\ \beta_{0K} - \beta_{\varepsilon K} \end{pmatrix} \equiv A^{-1}[\beta_0 - \beta_\varepsilon] \quad (21.4)$$

where A^{-1} is a K×K matrix and the **β**'s are K×1 vectors. The set of identities (21.4) is the basic structure of the identities that hold in an arbitrary model.

So far no assumption has actually been imposed, except that $\mathrm{cov}(X_k, T_k) \neq 0$, $k = 1, \ldots, K$, and that the rank of the matrix A is equal to K.

We now impose a set of restrictions. We impose them on the data in the sample. The restrictions hold in the sample by construction, and therefore cannot be verified nor tested without additional information.

The set of restrictions to be imposed, referred to as "orthogonality conditions" is given by

$$\beta_{\varepsilon k} = 0, \quad \text{for } k = 1, \ldots, K. \quad (21.5)$$

One possible interpretation of (21.5) can be that it represents first-order conditions for an optimization with respect to a target function. This is the case for a specific choice of the variables T_k. For example, if $T_k = X_k$ then we are in the OLS regression case. Alternatively, one can follow DeLaubenfels' (2006) geometric interpretation that the inner products of the vectors of explanatory variables and the residual are zero. That is, the explanatory vectors are orthogonal to the residual. In both cases it should be remembered that those conditions are *imposed* on the data and there is no a priori reason to believe that they exist in the population.

The consequence of imposing the orthogonality conditions is that (21.4) now turns from an identity to a solution of a set of linear equations, so that β_k ($k = 1,\ldots,K$) cease to be arbitrary constants but become the solutions of a set of linear equations.

Formally, using the restriction (21.5), the identities of (21.4) turn into (21.6)

$$\begin{pmatrix} \beta_1 \\ \\ \beta_K \end{pmatrix} = \begin{pmatrix} 1 & \beta_{21} & \beta_{K1} \\ & & \\ \beta_{1K} & & 1 \end{pmatrix}^{-1} \begin{pmatrix} \beta_{01} \\ \\ \beta_{0K} \end{pmatrix} = A^{-1}\beta_0. \tag{21.6}$$

The structure given in (21.6) is general, and it corresponds to all members of the covariance-based regressions, depending on the choice of T_k, $k = 1,\ldots, K$. Special cases include

(a) $T_k = X_k$ for all k, $k = 1,\ldots, K$. Then it is easy to see that (21.6) represents the OLS.
(b) $T_k = F(X_k)$ for all k, $k = 1,\ldots, K$. Then (21.6) represents the semi-parametric Gini regression.
(c) $T_k = -[1 - F(X_k)]^{\nu_k}$ for all k, $k = 1,\ldots, K$, and ν_k are given parameters supplied by the researcher. Then (21.6) represents the structure of the extended Gini regression.
(d) $T_k = Z_k$ for some k, $k = 1,\ldots, K$. Then (21.6) represents the structure of an OLS-IV regression.
(e) $T_k = F(Z_k)$ for some k, $k = 1,\ldots, K$. Then (21.6) represents the structure of a Gini-IV regression.

Several additional properties of (21.6) are worth mentioning.

By choosing T_k one is choosing the weighting scheme used in the regression, which is actually a choice of the variability measure used (variance in OLS (a), Gini or extended Gini in the regressions defined in (b) and (c), respectively). As a result, this choice determines the metric used (Euclidean in the case of OLS, city block in the case of Gini) and the "orthogonality conditions" applied. In the case of OLS the orthogonality conditions are $\text{cov}(X_k, \varepsilon) = 0$, under the Gini regression they are $\text{cov}(F(X_k),\varepsilon) = 0$, etc.

Each of the K equations in (21.4) can be defined with a different T_k so that one can have mixed regression methods: some equations can be defined as based on

GMD, others on OLS, etc. The advantage of a mixed method is that it enables the user to check the robustness of each imposed linear normal equation with respect to different regression methodologies, so that only the linear approximation of the regression curve that is not seriously affected by the choice of the methodology will be leading to a robust conclusion with respect to its sign and magnitude.

It is interesting to note that we could substitute one or several of the equations in (21.5) by orthogonality conditions (i.e., first-order conditions for optimization) taken from other regression techniques. However, in such a case one could not have linear orthogonal conditions. In this case, one would need to solve the first-order conditions numerically.

Having derived the regression coefficients, we turn to the constant term, α. To see whether the residuals are symmetrically distributed around the regression line, one can set the constant term so that the regression line passes either through the means or through the medians of the observations. Comparisons between the two estimates yield a quantitative evaluation on the quality of the fit of the regression line. To do that define a residual term, ε', as

$$\varepsilon'_i = y_i - \sum \beta_j x_{ji}. \tag{21.7}$$

Then if one wants the regression line to pass through the mean then one solves for α as

$$\alpha = E\{\varepsilon'\}. \tag{21.8}$$

On the other hand, if one wants the line to pass through the median, then one has to set α as the solution for

$$\underset{\alpha}{\text{Min}}\ E\{|\varepsilon' - \alpha|\}. \tag{21.9}$$

The estimators are sample's values of the population parameters, corrected for the degrees of freedom. Standard errors are calculated using the Jackknife method.

Having estimated the coefficients we turn to the quality of the fit of the linear approximation of the regression curve. Under OLS regime, the R^2 can be interpreted as indicating a measure of correlation between the fitted and the realization of the dependent variable, or, equivalently, as one minus the ratio of the variance of the residuals to the variance of the dependent variable. The Gini mean difference (hereafter Gini) method has two correlation coefficients between each two random variables, and the regression methodology used in this chapter does not minimize the Gini of the residuals (Olkin & Yitzhaki, 1992). Therefore, we substitute the R^2 by three indicators: the (Gini) correlations between the fitted and the realizations of the dependent variable, and one minus the ratio of the Gini of the residuals to the Gini of the dependent variable.

Formally

$$\Gamma_{\hat{y}y} = \frac{\text{cov}(y,F(\hat{y}))}{\text{cov}(y,F(y))} \quad \text{and} \quad \Gamma_{y\hat{y}} = \frac{\text{cov}(\hat{y},F(y))}{\text{cov}(\hat{y},F(\hat{y}))}, \tag{21.10}$$

where $\hat{y}$ is the linear approximation, while F() represents the cumulative distribution. The coefficient that is based on the ratio of the Ginis is

$$GR = 1 - \frac{\text{cov}(e,F(e))}{\text{cov}(y,F(y))}. \tag{21.11}$$

These concepts were introduced in Chap. 20. However, it is important to note that the Gini and the OLS are based on different metrics, and further research is needed in order to make the concepts of the quality of the fit comparable.

21.2 An Illustration of Mixed OLS and Gini Regression[1]

The purpose of this section is to illustrate the mixed OLS and Gini regression by investigating patterns of nonresponse in the social survey which is conducted by the Israeli Central Bureau of Statistics. The survey is conducted each year since 2002, and it covers the entire year. The sample is drawn from the population registrar. That is, the population registrar is the sampling framework. This is done several months prior to the interviewing stage, which is conducted by a face to face interview, using Computer Assisted Personal Interviews (CAPI).

The major statistical problem with nonresponse is that if the nonresponse is not random, then it may cause the estimates to be systematically biased. Other issues are concerned with increasing costs and frustration on behalf of interviewers.

Social surveys, which concentrate on subjective feelings, may seem more intrusive than surveys that are concerned with solid facts that seem objective and known not only to the interviewed.

In general, there are two ways of investigating patterns of nonresponse. One way is to analyze the characteristics of those who do not respond. We will refer to this method as the direct way of investigation. The alternative way is to rely on the process that is conducted by Statistical Bureaus in order to decrease random perturbations of the estimates and to correct for biases caused by nonresponse. This process is based of creating a weighting scheme attached to each observation so that each demographic group in the population is represented according to its weight in the population. By investigating the weighting scheme one can learn about nonresponse, because the bigger the weight attached to an observation, the less its characteristic is represented in the sample. We will refer to this way of

[1] This section is based on Golan and Yitzhaki (2010).

investigation as the indirect way, because one investigates nonresponse from the characteristics of those who responded.

Both methods are not perfect and each one has its own drawbacks. The direct way may suffer from errors in the sampling framework and in the classification of the reasons for nonresponse. For example, the population registrar includes individuals that may be outside the country. In the indirect way, there is no distinction between failing to contact a person that does not respond because he avoids any connection with the interviewer or because the person is outside the country for a long period of time and therefore should not be interviewed. The major advantage of the indirect way is that the sample size of the respondents is bigger than the sample size of the non-respondents, and it may include additional variables. Also, it is conducted after the interviewing stage is completed, so that it overcomes the lag in updating the population registrar. In this book we only present the indirect way of investigating patterns of nonresponse.

The Social Survey is conducted by the Israeli Central Bureau of Statistics (hereafter ICBS) since 2002. It comprises of a basic questionnaire that is administered every year, and an additional topic to be conducted in a sporadic way. The Statistical Ordinance makes the response to the questionnaire mandatory. However, no person was ever prosecuted if he or she refused to participate.[2] Because non-respondents make a small portion of the sample and the data on non-respondents is limited, the indirect way of using the sample to evaluate patterns of nonresponse is used.

The sample is drawn from the population registrar about 6 months prior to the year in which the survey is conducted. The population registrar includes the entire population of Israel. However, according to rough estimates about 10% of the population in the registrar do not live in the country. Based on other official records such as social security records, the population registrar is improved by the ICBS prior to the sampling but it is clear that the sampling framework is contaminated by records of individuals who do not belong to the target population of the survey. Hence, relying on the sampling framework may produce biased estimates of nonresponse. The population registrar includes demographic data only. For the purpose of this investigation, an additional variable is added to the registrar: the earned income reported to the tax authorities. The earned income added is the earned income of the individual and it does not include income from capital nor government transfers from the National Insurance Institute.

Table 21.1 describes the field reports accumulated over the period 2004–2008. Overall, about 22% of the individuals that were selected for the sample were not interviewed. However, one has to differentiate between those who were not supposed to be interviewed because of errors in the framework or administrative reasons and those that refused to be interviewed or the interview was not conducted because of other reasons. As can be seen from the table, the failure to interview is higher among

[2] Romanov and Nir (2010) present an excellent review of the considerations in handling nonresponse in the ICBS.

Table 21.1 The characteristics of respondents and non-respondents—2004–2008

		Respondents	Non-respondents
Total	Obs.	29,774	8,187
		78.4%	21.6%
Gender	Males	48.4%	52.0%
	Females	51.6%	48.0%
Age	20–24	11.9%	13.0%
	25–44	41.7%	39.7%
	45–64	30.2%	22.6%
	65+	16.2%	24.7%
	Average	45.1	47.9
Population group	Jews	81.9%	81.1%
	Others	18.1%	18.9%
Immigrants 1990+		14.2%	17.2%
% Employees		56.4%	35.2%
Average earned income (New Shekels, monthly)		7,290	5,953
% Self-employed		7.2%	3.6%
Average earned income (New Shekels, monthly)		5,623	5,857
% Not working		36.4%	61.2%

Source: Golan and Yitzhaki (2010)

the immigrants, the elderly, the nonworking population, and slightly higher among males and the young. Comparison with tax data enabled us to estimate the participation rates and average earned income according to the labor market type of employment. It can be seen that employees and self-employed are represented more among the participants than among the nonparticipants. However, the patterns are different: among the employees the participants have a higher average income, while among the self-employed we observe an opposite pattern. In general, it seems that the major difference between respondents and non-respondents is in participation in the labor market. It should be pointed out that the sample is a stratified sample, with higher probability of being drawn into the sample if the individual belongs to a group with a higher nonresponse rate. However, taking into account the stratification of the sample did not change the estimates of the regressions.

21.2.1 *The Indirect Way of Analyzing Nonresponse*

The indirect way of analyzing the effect of nonresponse is to use the sample of the respondents and the weighting scheme in order to analyze the effect of nonresponse. The advantages of this method over the direct way are the following: the weighting scheme is based on an updated framework. That is, while the sample is drawn about 6 months prior to the interviewing stage, the weights are derived after the interviewing stage is completed, and therefore the framework used is an updated one. The second advantage is that one can use both the variables in the framework and the responses of the respondents in the analysis. The third advantage is the

Table 21.2 Average satisfaction according to ethnic group[a]

	Weighted	Sample	Ratio
All			
2004	1.9426	1.9372	1.003
2005	1.9525	1.9522	1.001
2006	1.9225	1.9324	0.995
2007	1.8819	1.8867	0.997
All years	1.9251	1.9272	0.999
Jewish			
2004	1.8949	1.8957	1.000
2005	1.9083	1.9120	0.998
2006	1.8781	1.8938	0.992
2007	1.8388	1.8488	0.995
All years	1.8804	1.8878	0.996
Non-Jewish			
2004	2.1344	2.1354	1.000
2005	2.1273	2.1226	1.002
2006	2.0984	2.0949	1.002
2007	2.0530	2.0379	1.007
All years	2.1023	2.0965	1.003

Source: same as above

[a] The average satisfaction is based on individuals that belong to the same category who did respond

possibility of separating the contributions of different attributes. The disadvantage of the method is that we cannot classify nonresponse according to reasons and hence we cannot separate refusals from administrative errors. We start with simple tabulations and later we use multiple regression methods.

The simplest way to see the effect of nonresponse is to compare the means or the distributions of the variables using non-weighted versus weighted observations. This way we can learn about the quantitative effect of the weighting scheme on the expected value of a variable of interest.

Table 21.2 presents the average of satisfaction from life, weighted and non-weighted. Satisfaction is classified into four discrete categories: (1) very satisfied, (2) satisfied, (3) not so satisfied, and (4) not satisfied at all. As a result, the lower the value, the higher the satisfaction is. As can be seen, in most cases using the weights does not change the average in a noticeable way, implying that non-respondents tend to be, on average, equally satisfied with life as the respondents.

21.2.2 Empirical Results

In this analysis the dependent variable is the weight assigned to each observation which is derived in order to adjust the sample to the marginal distributions of key demographic properties of the population. The weights are produced by imposing several hundreds of linear constraints on the sample so that key demographic properties of the population are preserved.

The higher the weight assigned the higher the degree of nonparticipation in the survey. Nonparticipation can occur because the respondent was not located, he/she was not at home, or he/she refused to participate. For the issue of whether the sample is representative or not it does not matter what the reason for failing to participate was.

The explanatory variables include age, ethnic group, gender, household size, education level, and income. Some of the explanatory variables are categorical variables (education, gender, and ethnic group). The common practice is to represent a variable which has k possible categories by (k − 1) dummy variables (each one is binary). Note that for binary variables it does not matter whether one uses OLS or Gini regression.[3]

Two alternative ways to represent income are used in the regression. One is based on an administrative source and it is the before-tax earned income of the individual. This income is referred to as earned income. Note that it does not include income of other members of the household nor income from capital or transfers from the government. On the other hand, it includes the income of those who refused to answer the question about income. Earned income is measured in relative terms, that is, each income is divided by the average income in the sample for that year.

The other income used is the income reported by the individual in the survey about before-tax income of the whole household. The respondent was asked to choose among ten different ranges of income of the household. Then, the mid-range income was divided by the number of persons in the household and the results were grouped into three new discrete categories: (1) up to 2,000 NIS per person, (2) between 2,001 and 4,000 NIS per person, and (3) above 4,000 NIS per person. For our purpose we multiplied the income per capita by the number of persons in the household. This income is referred to as Household Income (HI). The difference between the two representations of income is stressed because it turned out that the way income is represented in the sample is crucial to the conclusions.

Table 21.3 presents the estimates of the mixed OLS and Gini regressions using the earned income: on the left-hand side are the OLS estimates, while on the extreme right-hand side are the estimates of the Gini regression. Columns 1–8 present the estimates of the mixed regressions, with the letter O representing an OLS weighting scheme, while G represents the Gini weighting scheme.

The basic regression is for the largest group which is composed of Jewish women with above secondary school education but without a B.A. degree.

Comparisons of the OLS regression coefficients with column (1) and the Gini regression with column (8) reveal that whenever the explanatory variable is binary

[3] A binary variable includes only two possibilities therefore it does not matter whether city-block or Euclidean metrics are used.

Table 21.3 Multiple Gini and OLS regressions

Regression coefficient	OLS	1		2		3		4		5		6		7		8		Gini
Age	−1.19	O	−1.19	G	−1.12	O	−1.12	O	−1.24	G	−1.05	G	−1.05	O	−1.14	G	−0.96	−0.96
	(0.06)																	(0.07)
Household size	11.19	O	11.19	O	11.37	G	13.29	O	12.59	G	13.47	O	13.03	G	15.38	G	15.87	15.87
	(0.52)																	(0.60)
Earned income	−4.32	O	−4.32	O	−4.31	O	−4.40	G	−27.13	O	−4.40	G	−26.58	G	−27.36	G	−26.86	−26.86
	(0.42)																	(0.87)
Elementary/middle school or other certification	23.32	G	23.32	O	22.73	O	22.45	O	10.44	G	21.94	G	9.19	G	9.21	O	8.02	8.02
	(3.29)																	(3.44)
Secondary school without matriculation	1.73	G	1.73	O	1.87	O	1.22	O	−6.01	G	1.33	G	−5.47	G	−6.73	O	−6.25	−6.25
	(3.10)																	(3.04)
Secondary school with matriculation	10.79	G	10.79	O	11.53	O	11.28	O	2.92	G	11.91	G	5.04	G	3.53	O	5.51	5.51
	(3.11)																	(3.28)
BA degree	−16.13	G	−15.13	O	−15.87	O	−15.68	O	−0.08	G	−15.44	G	0.25	G	0.62	O	0.94	0.94
	(3.29)																	(3.25)
MA+ degree	−4.51	G	−4.61	O	−5.00	O	−4.28	O	21.79	G	−4.60	G	20.16	G	22.38	O	20.89	20.89
	(3.57)																	(4.04)
Jewish male	−0.15	G	−0.15	O	−0.07	O	−0.23	O	16.03	G	−0.17	G	15.84	G	16.01	O	15.83	15.83
	(2.12)																	(2.12)
Non-Jewish male	13.97	G	13.97	O	14.37	O	11.95	O	18.42	G	12.25	G	19.36	G	15.76	O	16.54	16.54
	(3.72)																	(4.64)
Non-Jewish female	15.63	G	15.63	O	16.09	O	13.81	O	8.15	G	14.16	G	9.54	G	5.69	O	6.89	6.89
	(3.74)																	(4.83)
α (mean)	612.43		612.43		608.35		601.92		625.95		598.35		614.93		612.05		601.49	601.49
α (median)	593.67		593.67		589.82		583.36		608.54		579.88		597.48		594.97		584.41	584.41

The dependent variable is the weight attached to an observation. Income is taken from administrative files

$R^2 = 0.06$; $\Gamma_{y\hat{y}} = 0.29$; $\Gamma_{\hat{y}y} = 0.25$; GR = 0.01

Number of observations: 28,029. Source: same as above

it does not matter which regression method is used for that variable, as long as the continuous variables remain treated by the same regression method. Therefore the difference between the estimates produced by the two methods should be attributed to the three nonbinary variables: age, household's size, and earned income.

The regression coefficient of age is negative, indicating that for a linear approximation, the higher the age the lower the weight (i.e., the higher the response rate). However, the magnitude of its impact is about 20% higher under OLS regime than under Gini, which is a hint that it may be caused by extreme observations, either the young or the elderly. It seems that we can attribute roughly half of the difference to the direct impact of applying the Gini weighting scheme to age, and another half of the difference should be attributed to the covariance with earnings.[4] Comparison with Table 21.1 indicates that the relationship between nonresponse and age is not monotonic therefore any sign can be reasonable.

The impact of household size is positive which means that the larger the household's size the lower the participation. This finding negates the finding by Schechtman, Yitzhaki, and Artsev (2008) that the larger the household, the larger the participation rate. The latter was found in the Household's Incomes and Expenditures survey (hereafter HIES) and will be discussed in the next section. One possible explanation is that in the social survey the interviewer has to locate the individual, while in the Household's survey, the participation is of the household. The larger the household size, the higher the probability of establishing a contact with the household.

The impact of earned income on participation seems to be the most important factor in the regression. When the OLS weighting scheme is applied to this variable the estimate is around (−5), while when applying the Gini methodology the estimate is around (−26). This indicates that the higher the income the higher the participation. This also seems to be in agreement with the findings in the direct method reported in Table 21.1. It may also be the result of the tendency for higher nonparticipation among the ultra-religious, who also tend to have lower income. Also the effect of the correlations with the dependent and with other explanatory variables on the estimate of the coefficient of this variable is negligible. This finding is similar to the one found in Schechtman, Yitzhaki, and Artsev (2008) concerning participation in the HIES.

The rest of the variables are binary hence the estimates are not directly affected by the methodology applied to them, but they are affected by the covariation with other explanatory variables, especially of earned income.

The role of education on the participation rate seems to differ between the methodologies. According to OLS, the higher the degree held the higher the response rate, but in some cases that are closer to the base group the differences are not significant. On the other hand, under the Gini regime for earnings we get that high levels of education, holding a B.A. degree or M.A. degree worsen the response rate.

---

[4] Note that it is not meaningful to compare the standard errors of the Gini and OLS estimates because they are not statistically independent.

However, for low levels of education (elementary school) both methods agree that low level of education reduces the participation rate.

Being a male improves participation relative to the reference group in a nonsignificant way under OLS but reduces it significantly under Gini.

Being non-Jewish reduces participation rate under both methods. Again, this result is the opposite of the conclusion reached by Schechtman, Yitzhaki, and Artsev (2008) that participation rate of non-Jews is significantly higher than the participation rate of Jews. However, this result confirms (Feskens et al. 2007) that nonresponse may be more severe among minorities and excluded groups.

The constant term was estimated in two ways: one way is the usual way of imposing the restriction that the regression line passes through the means (21.8), and the other is to force the regression line to pass through the medians, as is the case in least absolute deviation (LAD) regression (21.9). In both methods the mean constant term is higher than the median constant term, indicating that the distribution of the residuals is skewed, having a larger tail of positive errors than negative ones. Moreover, the OLS constant term is higher than the Gini's counterpart, which is another indication that the distribution of the residuals is skewed, because the OLS is more sensitive to extreme observations than the Gini regression.

The quality of the fit of the regressions seems similar: while $R^2 = 0.06$, $\Gamma_{\hat{y}y} \cdot \Gamma_{y\hat{y}} = 0.29 \times 0.25 \approx 0.07$. However, the interpretation of comparison between concepts that are based on different metrics is not clear. All that one can say is that it seems that there is no significant gain in the explanatory power of the regressions when moving from one regime to the other.

A key variable for determining our conclusions is the treatment of the earned income variable. Hence, it is worth to dwell a bit on this variable.

Table 21.4 replicates Table 21.3 with one major difference. Instead of using the earned income that was taken from the administrative file, the income of the household reported in the survey is used. This difference is causing the following changes. (a) There are 4,093 observations with a missing response on income in the survey. Naturally, those observations did not participate in the regression. (b) The income reported in the survey includes all sources of income, in particular transfers from the government. (c)The income in the survey is a result of two stages of grouping, an issue that was discussed earlier. Comparison of the OLS column in Table 21.4 with the Gini column reveals that all the signs of the coefficients agree in the two columns. Therefore there is no qualitative difference between the results reported according to the methodologies, and even the magnitudes of the coefficients do not seem to deviate much from each other. It is interesting to note that the quality of the fit did not change.

Having found that the way income is included and the methodology of the regression may affect the conclusions with respect to participation of different groups deserves further investigation. This is the main advantage of the mixed regression over other methods because one can move from one method to the other in a gradual way. This is illustrated in Sect. 21.2.3.

Table 21.4 Multiple Gini and OLS regressions

Regression coefficient	OLS	1		2		3		4		5		6		7		8		Gini
Age	−1.15	O	−1.15	G	−1.11	O	−1.08	O	−1.14	O	−1.07	G	−1.11	G	−1.05	G	−1.05	−1.05
	(0.07)																	(0.10)
Household size	12.48	O	12.48	O	12.57	G	16.35	O	10.86	G	14.67	O	10.92	G	16.44	G	14.74	14.74
	(0.73)																	(0.92)
Survey's income	−1.56	O	−1.56	O	−1.56	O	−2.90	G	−0.15	G	−1.57	G	−0.14	O	−2.90	G	−1.57	−1.57
	(0.40)																	0.5
Elementary/middle school or other certification	24.49	G	24.49	O	24.18	O	22.22	O	26.27	G	23.92	G	26.02	G	21.97	O	23.72	23.72
	(3.56)																	(3.72)
Secondary school without matriculation	−0.34	G	−0.34	O	−0.29	O	−1.87	O	0.86	G	−0.73	G	0.90	G	−1.85	O	−0.70	−0.70
	(3.32)																	(3.26)
Secondary school with matriculation	11.17	G	11.17	O	11.50	O	11.66	O	11.34	G	11.78	G	11.62	G	11.91	O	11.99	11.99
	(3.35)																	(3.52)
BA degree	−17.00	G	−17.00	O	−16.87	O	−14.94	O	−18.72	G	−16.59	G	−18.62	G	−14.83	O	−16.50	−16.50
	(3.49)																	(3.33)
MA+ degree	−6.44	G	−6.44	O	−6.63	O	−4.55	O	−8.22	G	−6.23	G	−8.38	G	−4.68	O	−6.34	−6.34
	(3.85)																	(3.75)
Jewish male	−1.13	G	−1.13	O	−1.14	O	−0.30	O	−2.27	G	−1.35	G	−2.29	G	−0.30	O	−1.36	−1.36
	(2.20)																	(2.06)
Non-Jewish male	18.16	G	18.16	O	18.40	O	12.98	O	19.92	G	14.87	G	20.12	G	13.13	O	14.99	14.99
	(4.83)																	(6.55)
Non-Jewish female	34.28	G	34.28	O	34.58	O	28.62	O	36.53	G	30.98	G	36.80	G	28.81	O	31.14	31.14
	(5.08)																	(7.00)
α (mean)	594.10		594.10		592.07		585.33		591.15		583.36		589.46		583.73		582.05	582.05
α (median)	576.55		576.55		574.43		568.59		573.05		565.78		571.33		566.95		564.42	564.42

The dependent variable is the weight attached to an observation. Income is reported by the interviewed

$R^2 = 0.06$; $\Gamma\hat{y}y = 0.25$; $\Gamma y\hat{y} = 0.25$; GR = 0.03

Number of observations: 23,936

Source: same as above

21.2.3 A Search for an Explanation

We have seen in the last section that if one uses earned income from administrative sources then the signs of the regression coefficients of some explanatory variables obtained by the two methods may disagree, while if one uses the income reported in the survey then the two methods produce similar estimates. There are three major differences between the two incomes: the earned income variable includes 4,093 additional observations, of those with missing data on income in the survey; the earned income variable includes actual earned income while the income in the survey was grouped into rough categories. On the other hand the income variable in the survey includes income from all sources and not only earned income. In this section we investigate the effects of the differences between the variables.

It turns out that the density function of earned income is more skewed than that of the household's income reported in the survey. The grouping of observations makes it less asymmetric so that it is almost like a truncated normal. One possible conclusion is that decreasing the asymmetry of the distribution of income reduces the difference between the estimates derived by the two methodologies.

Next we checked whether three very extreme observations of earned income caused the difference between the estimates of the two methods. We omitted the three extreme observations of earned income. As can be seen from Table 21.5, the difference in the effect of earned income is still very big while the effects of having a B.A. degree are still with negating signs, although the differences between the estimates produced by the two methods were somewhat reduced. Other variables (such as having M.A. degree or being a Jewish male) show differences as well.

Table 21.6 replicates Table 21.5 with one major difference: all observations with no earned income were omitted from the regression. This means that nonparticipants in the labor market were omitted. A comparison of the two columns indicates that there is no disagreement with respect to the signs of the regression coefficients, although one can observe quantitatively large differences between some estimates. The impact of earned income is different: −3 in the OLS, −10 in the Gini. The effect of a B.A. degree is −6 and significant under the OLS, −0.07 and insignificant under the Gini.

Based on the comparison between Tables 21.5 and 21.6, it seems that the difference between the results produced by the two methodologies is affected by whether one includes observations of individuals with no earned income in the regression or not. If one omits those observations, then the two methods produce similar results. The major change that occurs is that the effect of education turned to be insignificant. An alternative way of getting similar results by both methods is by using the income definition as reported in the survey.

21.2.4 Summary of the Example

The advantage of using a mixed OLS and Gini regression is in avoiding conclusions that are due to the use of one methodology only. Using different methodologies can

Table 21.5 Multiple regressions: three extreme observations were omitted

Regression coefficient	OLS	Gini
Age	−1.23	−1.07
	(0.07)	(0.10)
Household size	11.59	16.28
	(0.58)	(0.64)
Earned income	−8.32	−27.38
	(0.61)	(0.85)
Elementary/middle school or other certification	20.17	7.81
	(3.60)	(3.80)
Secondary school without matriculation	−1.27	−7.97
	(3.37)	(3.29)
Secondary school with matriculation	10.01	5.52
	(3.41)	(3.59)
BA degree	−13.63	0.70
	(3.40)	(3.33)
MA + degree	0.98	21.17
	(3.94)	(3.82)
Jewish male	3.92	17.66
	(2.26)	(2.14)
Non-Jewish male	20.43	19.81
	(4.89)	(6.68)
Non-Jewish female	32.60	21.46
	(5.13)	(7.72)
α (mean)	614.58	604.29
α (median)	596.37	587.87

$R^2 = 0.07$; $\Gamma\hat{y}y = 0.22$; $\Gamma y\hat{y} = 0.21$; GR = 0.01

Number of observations: 23,933

Source: same as above

sometimes result in contradicting signs of the regression coefficients even if the model is identical. This phenomenon is bothersome because it means that changing the regression methodology used can reverse the conclusions. A higher reliance on mixed regression, so that only conclusions that are supported by both estimation techniques are considered meaningful will reduce the number of findings by omitting the conclusions that have a limited support. As a result it will contribute to the trust in empirical results. Unlike other regression techniques the mixed regression enables the identification of the variables that may cause the change in sign of the regression coefficients. In the illustration presented in this section it was possible to relate the cause for the change in sign to whether one includes participants and nonparticipants in the labor market in the sample. Our guess is that this result is due to the nonlinearity of the regression curve with respect to earned income when both participants and nonparticipants in the labor market are included in the regression. However, we cannot exclude other possible explanations such as grouping of the income variable. The advantage of the mixed regression methodology is that it enables us to identify the variable or the action that can change the sign of the regression coefficients and as a result can reverse the conclusions. Further research is needed to find out whether this fragility of the regression-based research is limited to extreme cases.

Table 21.6 Multiple regressions without observations with zero earned income

Regression coefficient	OLS	Gini
Age	−1.39	−1.18
	(0.10)	(0.10)
Household size	7.96	9.87
	(0.66)	(0.72)
Earned income	−2.93	−10.42
	(0.62)	(1.03)
Elementary/middle school or other certification	−1.16	−6.35
	(4.70)	(4.85)
Secondary school without matriculation	−5.77	−8.92
	(3.77)	(3.75)
Secondary school with matriculation	8.22	6.82
	(3.77)	(3.95)
BA degree	−6.22	−0.07
	(3.72)	(3.63)
MA + degree	1.54	11.38
	(4.29)	(4.36)
Jewish male	14.04	21.53
	(2.56)	(2.57)
Non-Jewish male	25.10	25.27
	(5.30)	(7.03)
Non-Jewish female	−49.47	−52.89
	(8.07)	(12.01)
A (mean)	606.74	598.87
A (median)	593.05	585.37

$R^2 = 0.04$; $\Gamma\hat{y}y = 0.17$; $\Gamma y\hat{y} = 0.17$; GR = 0.01

Number of observations: 15,135 (8,798 observations were omitted)

Source: same as above

21.3 An Illustration of Mixed Gini and EG Regression[5]

This section presents an illustration of a mixed Gini and EG regression. The idea is similar to the mixed OLS and Gini regression, but the motivation and the purpose of its use are different. While the OLS and Gini mixed regression is mainly motivated by the popularity of the OLS, the mixed Gini and EG regression can be motivated by all or some of the alternative motives listed below.

(a) Adjustment according to economic theory. In several fields such as income distribution and welfare economics, the economic theory may call for asymmetric treatment of the data (see Chaps. 14 and 17). In those cases, where the investigator is asked to present her social attitude or risk aversion and to impose them on the data, and the income (or wealth) is introduced as an explanatory variable, then the EG regression is called for. Alternatively, one

[5] This section is based on Schechtman, Yitzhaki, and Artsev (2008).

may want to use the EG to check whether the estimated coefficients are sensitive to the social or risk aversion considerations.

(b) Tracing the curvature of the conditional regression curve. The motive in this case is to trace the curvature of the regression curve with respect to one explanatory variable, conditional on the other explanatory variables in the model. If this is the case, then the motive is to improve the model so that it fits the data better. In some sense this is similar to applying a monotonic transformation to one of the explanatory variables, with one major and important difference. Because the data are not touched, the data continues to keep all the original properties. For example, if the sum of the expenditures is equal to the total expenditure then using the log transformation may violate the constraint that the sum of the expenditures on the item is equal to total expenditures (known as the adding up property), while using the EG will keep this property intact.

The methodology is illustrated by investigating the tendency not to respond to questionnaires on finances of the household in official surveys. The common wisdom with respect to this issue is that either or both rich and poor people tend to respond less than ordinary people. Because we have no firm priors with respect to the kind of relationship we expect to see, the need for a nonparametric method that can analyze the data arises.

The methodology is identical to the one presented in Sects. 21.1 and 21.2 except that each cumulative distribution F() is substituted by $-[1 - F()]^{v}$, where $v \in (-1, \infty)$ is chosen by the investigator. If $v = 1$ then the investigator measures variability according to Gini's mean difference, implying a symmetric weighting scheme around the median. If $v \to 0$, the investigator does not care about variability; the range $-1 \leq v < 0$ reflects giving higher weights to the upper part of the distribution of the explanatory variable, while $v \to -1$ implies an investigator whose attitude to variability follows the max–max strategy, that is, caring about variability around the highest part of the distribution only. It is worth noting that when $-1 \leq v < 0$, the index of variability is negative. In the extreme case ($v \to \infty$) the investigator cares only about the lowest part of the cumulative distribution, as if he is guided by the max–min criterion.[6]

[6] See Donaldson and Weymark (1983), Yitzhaki (1983), and Chakravarty (1988, Chap. 3, pp. 82–102) for descriptions of the properties of the extended Gini index. Garner (1993), Lerman and Yitzhaki (1994), and Wodon and Yitzhaki (2002) are examples of its decomposition and use in welfare economics; see Araar and Duclos (2003) for a possible extension; see Davidson and Duclos (1997) for statistical inference, and Millimet and Slottje (2002) for an application in environmental economics. Note that in the above-mentioned literature, the parameter is restricted to $v > 0$. Schechtman and Yitzhaki (1987, 1999, 2003) define and investigate the properties of the equivalents of the covariance and the correlation based on the EG. Yitzhaki and Schechtman (2005) offer a survey of the properties of the EG family and in particular they show the metric that leads to the EG. The decomposition of the extended Gini of a sum of random variables into the contributions of the extended Gini's of the individual random variables and the (equivalents of) correlations among them can also be found there. Serfling and Xiao (2007) define and investigate the properties of multivariate L-moments which include the EG measures as a special case.

21.3.1 Non-reporting in a Household Finances Survey

The framework from which the sample of Household Finances survey in Israel is drawn is the file of dwellings used for local tax payments in Israel. Refusal to respond is illegal, but so far nobody was prosecuted or even threatened to be prosecuted for noncooperating with the interviewer. Naturally, the survey suffers from refusals or administrative errors that affect the coverage of the target population. If non-responding is correlated with income, then the estimates of the mean income and the index of income inequality may be biased.

The purpose of this section is to describe non-responding as a function of several demographic variables (which can be used later in designing the sample) and one major variable, income. In general, the experience concerning nonresponse is that the propensity not to respond is a U-shaped function with respect to income. A plausible explanation can be that the rich tend not to participate, while the poor and the young are harder to be found at home. A study by Mistiaen and Ravallion (2003) presents a model in which compliance can either decrease or increase with income and also be of an inverted U-shape. Moreover, adding other arguments such as the ability to find the members of the households at home, finding the address, viewing participation as a democratic value, etc., can lead to almost all kinds of patterns. Mistiaen and Ravallion (2003) find that the nonresponse problem cannot be ignored and that there is a highly negative significant income effect on compliance. Deaton (2005) raises the plausible conjecture that richer households are less likely to participate in surveys, in order to explain the gap between growth estimates based on households' surveys and those that are based on national accounts. Comprehensive studies dealing with almost all aspects of nonresponse are detailed in Groves and Couper (1998) and Groves et al. (2002). The main conclusion from the existing literature is that nonresponse is a serious issue that may bias the estimates, but there is not enough knowledge to justify the assumptions needed for running OLS or other parametric regressions.

To overcome biases that are caused by the sample being a nonrepresentative one and to reduce standard errors, many statistical agencies adjust the distribution of the sample to fit known marginal distributions of current demographic estimates that are based on the census. The outcome of this adjustment is a weighting scheme: a weight is attached to each observation. A necessary condition to be able to perform such an adjustment is having a detailed census data. Also, there are other reasons for using those procedures, among them is to ensure that different samples, performed by different units of the agency, report the same demographic structure so that official statistics will not be blamed by the media for publishing contradicting estimates. This may explain why the adjustment to given margins is performed mainly by producers of official statistics. For a survey of the different methodologies used to construct weighting schemes see the survey by Kalton and Flores-Cervantes (2003). A detailed description of the method used in Israel is offered by Kantorowitz (2002). For the purpose of this section it is sufficient to say that the above-mentioned procedures change the weight of each observation so that they add up to given marginal demographic and geographic distributions.

The sample we are dealing with is a sample of dwellings. It is a stratified sample according to geographical areas and types of dwellings, but the probability of each dwelling to be included in the sample is the same. This implies that the expected value of the weight of each observation is equal to the ratio of the overall population to the sample size. When nonresponse occurs in a certain group, it will be under-represented in the sample, so that the weight that will be assigned to those who responded in that group will be higher than its expected value in case of equal tendency to respond.

The weighting scheme of the sample is produced by an algorithm for calibration, with several hundreds of constraints imposed, and is intended to make the sample representative (Kantorowitz, 2002). In particular, a constraint is imposed on the maximal weight assigned to each observation, so that standard errors do not increase unnecessarily. The constraints ensure that the reported age structure, geographic distributions, and household sizes will add up to given margins of the distributions of the population.

It is important to note that the income is not involved at all in the derivation of the weights. Hence there is no built-in correlation (i.e., spurious correlation) between the weight of each observation and its income. (For additional information on the sample and the way the weighting scheme was created, see http://www.cbs.gov.il/publications/expenditure_survey04/pdf/e_intro.pdf, p. XXIII.) Based on the discussion above, the idea behind the illustration is to build a regression model where the dependent variable is the weight given to each observation and the explanatory variables are income, household size, and ethnic grouping (to be detailed below).

21.3.2 The Data

The data consist of the observations of the surveys for the years 1997–2001. Table 21.7 presents the weights according to years and ethnic groups, where the groups are defined as the majority (which includes the Jewish population, except the ultra-religious group), the Arab population and the ultra-religious group.

Because the groups differ in household size, which may affect the probability of finding someone at home, Table 21.8 presents the average weights according to household size. It can be seen that for the majority, household of size 1 has the highest weight and the rest are similar (year 2001 is different). This may be a result of small households not being at home while the elderly, although being at home, do not have the patience to complete the questionnaire. (For Arabs there is no obvious pattern. Nothing can be said about the ultra-religious Jews because the sample sizes are quite small.)

To summarize: the dependent variable is the weight assigned to each observation by a calibration procedure, intended to represent the entire population. The sample is a stratified sample, but the probability of each dwelling and each person living in a dwelling to be included in the sample is equal. Hence, if the propensity not to be

Table 21.7 Descriptive statistics of household weights by ethnic grouping

			Weight			
Year	Ethnic group	N	Average	Max	Min	Std. Dev
	Majority	4,942	283	1,196	20	166
1997	Arabs	529	267	1,051	19	199
	Ultra-religious Jews	90	398	1,127	45	245
	Majority	5,068	286	1,196	18	169
1998	Arabs	606	256	1,049	24	176
	Ultra-religious Jews	98	321	766	26	166
	Majority	5,114	291	1,134	13	154
1999	Arabs	597	269	1,129	14	169
	Ultra-religious Jews	105	292	639	20	115
	Majority	5,146	301	1,195	22	170
2000	Arabs	629	260	959	43	142
	Ultra-religious Jews	89	310	1,017	33	148
	Majority	5,049	314	1,185	18	152
2001	Arabs	662	285	1,902	31	171
	Ultra-religious Jews	76	341	834	104	145

Source: HES 1997–2001, excluding the observations of East Jerusalem in 1997–1999
Source: Schechtman, Yitzhaki, and Artsev (2008), Table 1, p. 338

included in the sample is equal, either because of nonresponse or errors on behalf of the agency, the expected weight assigned to each observation should be equal. It may differ between years if the ratio of the population to the sample size changes between years. The weight is treated as an indicator of nonresponse. Having described the dependent variable, we now move to describe the results concerning the regression coefficients.

21.3.3 *Empirical Results*

There are two types of explanatory variables in the regression: numerical and categorical. The two numerical variables are income and household size. The categorical variable is ethnic group. It results in several binary variables. For a binary variable all EG regression coefficients are identical because only one slope is defined between two points. Therefore we start with the EG regression coefficient of income.

Table 21.9 presents the regression coefficients when regressing weight on household income for different values of v. The higher the parameter v, the more the regression stresses the slopes of the regression curve at the lower end of the income distribution. As can be seen, the regression coefficients are negative, which means that the higher the income—the lower the weight assigned to the observations, implying that nonresponse declines with increase in income.

Table 21.8 Means of household weights by ethnic grouping and household size

Year	Ethnic group		Household size				
			1	2	3	4	5+
1997	Majority	Mean	325	278	278	276	266
		N	840	1,199	807	910	1,186
	Arabs	Mean	288	246	255	236	281
		N	12	48	54	102	313
	Ultra-religious Jews	Mean	273	363	294	315	488
		N	2	18	16	13	41
1998	Majority	Mean	330	280	290	276	268
		N	869	1,281	788	964	1,166
	Arabs	Mean	316	338	241	252	247
		N	13	49	70	98	376
	Ultra-religious Jews	Mean	235	247	385	329	325
		N	3	12	13	12	58
1999	Majority	Mean	333	291	276	297	266
		N	892	1,258	878	919	1,167
	Arabs	Mean	230	248	321	302	256
		N	15	49	64	93	376
	Ultra-religious Jews	Mean	176	293	291	333	288
		N	3	20	11	12	59
2000	Majority	Mean	351	291	310	280	282
		N	875	1,296	867	965	1,143
	Arabs	Mean	323	223	272	381	237
		N	17	58	64	78	412
	Ultra-religious Jews	Mean	239	327	254	336	301
		N	3	21	1	13	51
2001	Majority	Mean	363	294	326	291	308
		N	886	1,325	838	949	1,051
	Arabs	Mean	521	261	290	304	271
		N	20	64	74	116	388
	Ultra-religious Jews	Mean	393	346	483	366	319
		N	3	16	4	8	45

Source: HES 1997–2001, excluding the observations of East Jerusalem in 1997–1999
Source: same as above, p. 339

For example, a value of (-0.0022) (obtained for $\nu = 2$) means that a unit increase in income will decrease the average of the dependent variable (weight) by 0.0022. The average weight is the ratio of the population to the sample size. Therefore, the value of (-0.0022) divided by the average weight gives the percentage rate of the increase in response rate with a unit increase in income. Even when high-income groups are stressed ($\nu = -0.5$) we still have a significant negative regression coefficient. The interpretation of this finding is that we have a monotonic relationship between nonresponse and income. We note that the pattern shown above repeats for the years 1997–2000 (not shown here).

Table 21.9 Regression coefficients of household weight on gross income per household, by extended Gini parameter (ν)

	ν for gross income			
Coefficient	3	2	1	−0.5
b	−0.0025[a]	−0.0022[a]	−0.0017[a]	−0.0007[a]
SE(b)	0.0003	0.0003	0.0002	0.0001
a (mean)	348.3	342.8	335.7	321.0
a (median)	320.7	314.9	307.8	293.2

Source: HES 2001. Units of income—New Israeli Shekel per month
Source: same as above, Table 3, p. 339
[a]Indicates a value significantly different than 0 (at $\alpha = 0.05$)

Table 21.10 Regression coefficients of household weight on household size, by extended Gini parameter (ν)

	ν for household size			
Coefficient	3.0	2.0	1.0	−0.5
B	−12.2[a]	−10.6[a]	−8.5[a]	−3.5[a]
SE(b)	1.4	1.3	1.1	1.2
a (mean)	354.1	347.5	339.3	321.4
a (median)	326.7	320.0	311.1	293.2

Source: HES 2001. As above, Table 4, p. 339
[a]Indicates a value significantly different than 0 (at $\alpha = 0.05$)

The rest of Table 21.9 presents the two alternative estimates of the constant term. One presents the constant term when the regression line passes through the median, while the other through the mean. It is interesting to note that the difference between the two is around 28 with a(mean) being higher than a(median) regardless of the value of ν. This is an indication that the error term tends to be asymmetric. The fact that the difference between the constant terms seems to be independent of the slopes requires more research.

Table 21.10 presents the simple regression coefficients of weight on household size. As in the regression on income, the larger the family size the higher the value of the regression coefficient (in absolute value), and in all cases the signs of the regression coefficients are negative. This means that nonresponse is higher among small households. Note that as before, the constant term of the regression passing through the mean is larger than the constant term of the regression passing through the median, but again, the difference between the two constants is around 28.

Tables 21.11 and 21.12 present the multiple regression coefficients with gross income, household size, and dummy variables for being a member of a minority group (Arabs, ultra-religious Jews) as the explanatory variables. In Table 21.11, the extended Gini regression method is used with a choice of values for ν. For completeness we analyzed the data using OLS regression as well.

The OLS method is used in Table 21.12 for the entire sample and then for each quartile (by income) separately.

Table 21.11 Gini multiple regression coefficients of household weight (0) on gross income per household (1), household size (2), and ethnic grouping dummy variables (3, 4) ([a]), by extended Gini parameter (ν)

	ν for gross income:			
Coefficient	3.0	2.0	1.0	−0.5
b_{01}	−0.0026[b]	−0.0021[b]	−0.0016[b]	−0.0006[b]
	(0.000)	(0.000)	(0.000)	(0.000)
b_{02}	−1.41	−2.54	−3.95[b]	−6.32[b]
	(1.44)	(1.39)	(1.35)	(1.31)
b_{03}	−40.8[b]	−36.0[b]	−30.1[b]	−20.1[b]
	(8.1)	(8.0)	(7.9)	(7.8)
b_{04}	18. 8	23.3	28.9	38.3[b]
	(17.4)	(17.2)	(17.1)	(16.9)
a (mean)	357.9	354.6	350.5	343.6
a (median)	330.2	326.6	322.6	316.1

Source: HES 2001. Units of income—New Israeli Shekel per month
Source: Same as above, Table 5a, p. 340
[a] The dummy variable no. 3 has the following values: 1 = "Arab", 0 = "Other"; The dummy variable no. 4 has the following values: 1 = "Ultra-religious Jew," 0 = "Other"
[b] Indicates a value significantly different than 0 ($\alpha = 0.05$)

Table 21.12 OLS multiple regression coefficients of household weight (0) on gross income per household (1), household size (2), and ethnic grouping dummy variables (3, 4) ([a])

	OLS:				
Coefficient	Entire population	First quartile	Second quartile	Third quartile	Forth quartile
b_{01}	−0.0008[b]	−0.0094[b]	−0.00447	−0.0003	−0.0006[b]
	(0.0001)	(0.003)	(0.0033)	(0.0016)	(0.0002)
b_{02}	−4.384[b]	−11.10[b]	−1.55	3.868[b]	−1.48
	(1.154)	(3.56)	(2.275)	(1.928)	(2.31)
b_{03}	−25.039[b]	16.856	−48.36[b]	−53.164[b]	−47.4[b]
	(6.83)	(14.972)	(12.05)	(12.73)	(20.68)
b_{04}	33.128	78.977[b]	−4.01	0.793	46.2
	(17.92)	(37.895)	(31.17)	(30.4)	(56.2)
a (mean)	340.615	404.02	363.2	290.2	322.4

Source: HES 2001. Units of income—New Israeli Shekel per month
Source: same as above, Table 5b, p. 340
[a] The dummy variable no. 3 has the following values: 1 = "Arab", 0 = "Other"; The dummy variable no. 4 has the following values: 1 = "Ultra-religious Jew," 0 = "Other"
[b] Indicates a value significantly different than 0 ($\alpha = 0.05$)

(a) The extended Gini regression. The parameter ν is set to 1 (symmetric around the median) for household size and minority groups (represented by dummy variables), and it varies for income only. As can be seen, the regression coefficients of weight on income decline in absolute value as ν declines (i.e., stressing higher incomes) by up to 0.001, so that the patterns detected in the simple regression continue to hold. However, the sign of the regression coefficients of household size remains the same (negative), indicating that larger households respond in greater proportion to the questionnaires. Because the only difference between the regressions is the change in the parameter of income, the decline in absolute value of the coefficient of household size should be attributed to a change in the pattern of association between income and household size. The higher the stress on high-income groups, the lower is the absolute value of the effect of household size on nonresponse. Also, the magnitude of the regression coefficient of household size has changed from (–8.5) in the simple regression case to (–4.0), which may be an indicator of the magnitude of association between income and household size. Given income and household size, Arabs tend to respond in higher proportion than the majority group, but the more we stress high income, the lower the effect (this may be due to small sample size of the Arab population in the upper range of incomes). One possible interpretation is that the higher the income, the lower the difference in response rates between the majority group and Arabs. On the other hand, the effect of stressing high-income range on ultra-religious Jews is the opposite. The more high incomes are stressed, the lower the response rate is. Because it is a group with a low response rate on average and seems to be motivated by an ideology, it is reasonable to conclude that the difference in response rate between this group and the rest of the population increases with income. However, the high standard errors show that only when high income is stressed, the dummy variable for ultra-religious Jews is significant. As before, the difference between the constant terms is approximately 28. The main conclusion of the empirical application is that nonresponse to the survey of household's expenditures in Israel is a decreasing convex function of income, and almost reaches a plateau when high-income groups are stressed. Nonresponse tends to be negatively related to household size. The nonresponse rate differs among ethnic groups: the Arab population shows below average nonresponse rate, while the ultra-religious Jewish group has above average nonresponse rate. This result holds even when the response rate is adjusted for income and household size.

(b) The OLS regression. In order to see the difference between Gini and OLS regressions, the estimation is replicated using the OLS. The OLS was used for the entire sample (n = 5,787), and then for each quartile separately (n = 1,446–1,448). The quartiles are Q_1 = 25th percentile = 6,599 New Israeli Shekels (NIS) per month, Q_2 = median = 10,927 NIS, and Q_3 = 75th percentile = 18,534 NIS. Note, however, that under the OLS, only one-fourth of the observations participate in the regression of each quartile, while in the Gini regression, all observations participate in the regressions. The pattern found in the Gini regression appears, in part, when using OLS. The regression coefficients of weight on income are all negative, but only two are significantly different than zero—the coefficients for the first and fourth quartiles. However, moving from the first quartile to the third, there seems to be a trend of decrease in absolute value. Concerning the effect of demographic characteristics, in the OLS the magnitude and the effect of the demographic characteristics change between quarters, while under the Gini they are forced to be the same. Looking carefully at the slopes of income (in absolute values), there is a slight indication of a U-shape pattern—going down from 0.0094 to 0.00447 and 0.0003, and then up to 0.0006.

21.4 Summary

In this chapter we have illustrated the use of the mixed OLS, Gini, and EG regressions. It is hard to justify the mixed regression on theoretical grounds because it does not have "optimal" properties of a reasonable target function, but on the other hand it seems very useful in investigating violations of the assumptions imposed on the data by the different methods. As far as we can see, the main purpose of the mixed OLS and Gini regression is to enable explaining the difference between the results produced by the different methods, while the Gini and EG mixed regression is mainly intended to impose considerations of economic theory on the regression or to investigate the curvature of the (conditional) regression curve.

Whenever it is not motivated by economic theory, the EG regression is a descriptive tool, enabling the researcher to improve the model by tracing the curvature of the (conditional) regression curve.

Although descriptive in nature, it can be turned into a standard analytical regression technique. By selecting the same weighting scheme for all explanatory variables, one can have the structure of the OLS with one simple modification: each variance is substituted by an extended Gini, and each covariance is substituted by the appropriate extended Gini covariance. The only difference is that the method offers an infinite number of alternative regression coefficients. Clearly, the method enables the investigator to verify whether the results are sensitive to the specific index of variability (weighting scheme) used.

Turning to our illustration of nonresponse, we have found that in the survey of household expenditure in Israel, nonresponse decreases with income, decreases with household size, and differs among ethnic groups. The Arab population tends to respond more than the majority, while the ultra-religious Jewish population tends to respond less than the majority group. These results are in contrast with Deaton's (2005) conjecture that high income groups tend to respond less to surveys. However, one should be aware that nonresponse is a survey-specific not to mention the possibility of a country-specific phenomenon.

As far as we can see, the presentation in this chapter covers only the top of the iceberg. In some sense the method offers an infinite number of estimation techniques based on the choices of the EG parameter ν. Adaptation to different fields will probably take years to accomplish.

Further research is needed to compare the EG regression approach vs. the decomposition approach of a regression coefficient. As shown by Yitzhaki (2002) and in Chap. 7 if one divides the range of an explanatory variable into two sections, then the EG regression coefficient (and OLS) can be presented as a weighted average of the two within-section regression coefficients and a between-sections regression coefficient. The weights are the relative contributions of each section to the intra- and inter-group Ginis (variance in OLS, EG in EG regression) of the explanatory variable. This decomposition can be easily expanded to an arbitrary number of sections. Further research is needed to apply this additional decomposition to get a piecewise linear approximation to the regression curve that is based on a between-sections component and within-section components of the approximation. The piecewise linear approximation will allow the estimate of the partial derivative to vary over sections of the explanatory variables.

Additional research is needed to compare the Gini IV method vs. OLS IV method. Both belong to the covariance-based family.

Chapter 22
An Application in Statistics: ANOGI

Introduction

This chapter deals with applications of the GMD and the Gini coefficient in statistics. It presents an application which replicates the ANalysis Of VAriance (ANOVA) and is referred to as ANalysis Of GIni (ANOGI).

The relationship between ANOGI and ANOVA resembles the relationship between the decomposition of the variance of a linear combination of random variables and its Gini analogue. In both cases the decomposition of Gini includes more parameters than the decomposition of the variance, and if certain properties hold in the underlying distributions then the decomposition of the Gini includes the decomposition of the variance as a special case (while replacing the Gini components by the respective variance counterparts). For this reason we refer to the Gini as revealing more. Whether those additional parameters are useful or not depends on the questions asked and the underlying distributions.

The idea behind ANOVA and ANOGI is to decompose the measure of variability (total variance in the ANOVA case and the Gini coefficient (or GMD) of the entire population in the ANOGI case) into "inter-groups" and "intra-groups" sources of variabilities. ANOVA and ANOGI are identical in structure if the distributions of the different subgroups do not overlap. That is, if each subgroup occupies a given range along the variable of interest and no member of the rest of the population is located in this range. However, if there is an overlapping between the subgroups then the Gini coefficient's (and the GMD) decomposition includes additional set of parameters that are referred to as the overlapping indices. In general the higher the overlapping the more of the variability is attributed to the intra-group component and the less to the inter-group component.

The inverse of overlapping is referred to as stratification (no overlapping means a perfect stratification). In some sense, we can claim that ANOGI also offers a quantitative measure of the quality of a classification. More specifically, if one compares two classifications then the ANOGI can provide a quantitative way to decide which way of classification separates the subgroups better. The empirical

S. Yitzhaki and E. Schechtman, *The Gini Methodology: A Primer on a Statistical Methodology*, Springer Series in Statistics 272, DOI 10.1007/978-1-4614-4720-7_22,

illustration that follows is based on applying the methodology to examine the success of the "melting pot" policy in Israel. The structure of the chapter is as follows: Sect. 22.1 offers a brief review of the methodology. The empirical illustration is detailed in Sect. 22.2, and Sect. 22.3 concludes.

22.1 A Brief Review of the Methodology

ANOGI—ANalysis Of GIni is based on decomposing the Gini coefficient of economic well-being according to population subgroups in a way which is similar to ANOVA—ANalysis Of VAriance (see Chap. 4 for a full description of the methodology and the derivation and the properties of the parameters. The estimators are presented in Chap. 9).[1] In this section we give a very brief review of the decomposition of the Gini coefficient of the entire population and explain how the decomposition enables us to answer the research question.

The Gini coefficient is the most popular measure of inequality. Naturally, one would wish to decompose the Gini of a population into the contributions of the subpopulations. It turns out that the Gini is not additively decomposable by population subgroups. As a result, many economists argue that it is not meaningful to decompose it (Cowell, 1980; Shorrocks, 1984). However, as shown in several papers (Frick et al., 2006; Lambert & Aronson, 1993; Lambert & Decoster, 2005; Milanovic & Yitzhaki, 2002; Yitzhaki 1994a, 1994b; Yitzhaki & Lerman, 1991) the decomposition of the Gini reveals more information about the distribution than the decomposition of alternative measures of inequality. In particular, it enables one to evaluate the quality of the classification by subgroups (Heller & Yitzhaki, 2006), a property that will be dealt with in depth following the description of the properties of the decomposition. As demonstrated by Frick et al. (2006), the Gini decomposition according to population subgroups offers a method which is on one hand similar to ANOVA, but on the other hand is superior to it because it can indicate the degree to which the population is stratified.

Let the population income distribution Y_u be composed of the income distributions Y_i, $i = 1,\ldots,n$, of the n subpopulations. The Gini coefficient of the entire population, denoted by G_u, can be decomposed into three components: a

[1] An alternative methodology for analyzing the melting pot policy is to compare the earnings of second generation of immigrants with the earnings of first generation (or the earnings of the natives) while controlling for other effects (Borjas (2006) and the references therein). However, this methodology requires longitudinal data and other characteristics of the population, while the methodology presented here can be applied to cross-sections. The price paid for the use of our methodology is that we end up with descriptive statistics, while regression-based methodologies offer a detailed analysis and the possibility to find causal relationship. For a regression-type analysis of discrimination and second generation analysis of the Israeli labor market see, among others, Semyonov and Cohen (1990) and Cohen and Haberfeld (1998).

Table 22.1 Comparing ANOGI and ANOVA

Components parallel to ANOVA	Formula	Range
Intra-group	$IG = \sum_{i=1}^{n} s_i G_i$	$0 \leq IG \leq G_u$
Between-groups-Pyatt	$BG_p = G_{bp}$	$0 \leq BG_p \leq G_u$
Additional information provided by ANOGI		
Overlapping effect on intra-group	$IGO = \sum_{i=1}^{n} s_i G_i (O_i - 1)$	
Overlapping effect on between-groups	$BGO = G_b - G_{bp}$	$- BG_p - IGO - IG \leq BGO \leq 0$

Source: Frick et al. (2006), p. 439, Table 3
Reprinted with permission by SAGE

"within" component (intra), a "between" component (inter), and a component that is a function of the amount of overlapping among the subpopulations.

The decomposition is best understood in comparison to ANOVA, as shown in Table 22.1.

Let

$$Y_u = Y_1 \bigcup Y_2 \bigcup \ldots \bigcup Y_n,$$

where Y_u is the income of the entire population and Y_i is the income of subpopulation i (i = 1,. . .,n).

The Gini coefficient of the entire population, denoted by G_u, can be presented as

$$\begin{aligned} G_u &= \sum_{i=1}^{n} s_i G_i O_i + G_b \\ &= \sum_{i=1}^{n} s_i G_i + \sum_{i=1}^{n} s_i G_i (O_i - 1) + G_{bp} + (G_b - G_{bp}), \end{aligned} \tag{22.1}$$

where s_i denotes the share of group i in the overall income, O_i is the overlapping index of subpopulation i with the entire population (explained below), G_b measures the between-groups inequality and G_{bp} is Pyatt's between-groups Gini (Pyatt, 1976).

The overlapping coefficient was introduced by Yitzhaki and Lerman (1991) and modified by Yitzhaki (1994a, 1994b). Intuitively, it measures to what extent one group is overlapped by the other. The extreme lower bound occurs when there is a complete stratification, i.e., when each group occupies a given range and the ranges do not intersect (no overlapping, perfect stratification). In this case the overlapping index between any two subgroups equals zero. The extreme upper bound for the overlapping of group A by group B occurs when group A is concentrated inside the range of B, around the mean of group B, with no member of group B lying inside the range of group A. In this case, group B cannot be considered as a group because the members of group A separate the members of B that are below the average of B from those that are above it. In this case the overlapping index of B by A is zero because no member of B lies inside the range of A, while the overlapping index of B in A is greater than one, its value depends on the distributions involved and is bounded by 2. Obviously, most cases are in between these two extremes. The measure is based on ranking the members of one group according to the ranking of the other. Its values range from 0 to 2, where 1 means that the distributions of the two groups are similar.

The overlapping coefficient can tell us how much the distributions are intertwined, or, in other words, tell us about the degree of assimilation. Details can be found in Chap. 4.

Formally, overlapping of the overall population by subpopulation i is defined as

$$O_i = O_{ui} = \frac{\mathrm{cov}_i(Y, F_u(Y))}{\mathrm{cov}_i(Y, F_i(Y))}. \tag{22.2}$$

The denominator is (one-fourth of) the Gini's mean difference of group i, while the numerator is the covariance between the same observations and their rankings in the overall distribution.

The other components of (22.1) that require an interpretation are G_b and G_{bp}. G_b is based on the covariance between the mean value of each subgroup and the average rank of its members in the overall distribution (this is not a Gini coefficient). On the other hand, G_{bp} is based on the covariance between the mean value of each subgroup and the ranking of the mean value in the distribution of mean values (this is a Gini coefficient). By construction $G_b \leq G_{bp}$. The role of the overlapping in (22.1) can be seen from the second and fourth terms on the right side of the equation. The terms G_u, G_i ($i = 1,\ldots,n$), and G_{bp} are not affected by the degree of overlapping. Therefore the higher the degree of overlapping between the subgroups the higher the second term on the right-hand side of (22.1) (intra-group component) and the lower the fourth term (between-groups component). The decomposition is best understood in comparison to ANOVA, as shown in Table 22.1.

As can be seen from Table 22.1, ANOGI offers an additional parameter to ANOVA—the parameter of overlapping, which can be interpreted as the inverse of stratification. The amount of overlapping affects both the intra- and the inter-group terms. Other parameters being equal, the higher the overlapping (i.e., the larger O_i is), the higher the intra-group component and the lower the between-groups component.

A further look at two parameters: O_i and G_b that are involved in the decomposition enable one to elaborate on which groups are contributing to the quality of the decomposition. Note that O_i as a weighted average of O_{ji}, where O_{ji} is the degree by which members of group j are included in the range of group i and p_j is the share of subpopulation j in the entire population

$$O_i = \sum\nolimits_{j=1}^{n} p_j O_{ji}, \tag{22.3}$$

where

$$O_{ji} = \frac{cov_i(Y,F_j(Y))}{cov_i(Y,F_i(Y))}.$$

The numerator of O_{ji} involves the covariance between an income of a member in group i with its rank, had it been ranked within the incomes of the members of group j. The other parameter, G_b, involves the covariance between the mean income of each group and the mean ranking of its members in the overall population. Note that

$$\mu_0 = \sum_{i=1}^{n} p_i \mu_i \tag{22.4}$$

$$0.5 = \sum_{i=1}^{n} p_i \bar{F}_i. \tag{22.5}$$

$\bar{F}_i$ is the average rank of the members of group i in the population. μ_i and $\bar{F}_i$ are the two components that represent group i in the between-groups component. In a perfectly stratified society, the ranking of μ_i and $\bar{F}_i$ are identical and all O_i and O_{ji} are equal to zero. If they are not, we get an indication about the groups that deteriorate the quality of the stratification.

The interpretation of the decomposition used in this chapter follows the one presented by Heller and Yitzhaki (2006) which deals with classification of snails, and the one presented by Frick and Goebel (2008) on the decomposition of well-being in Germany according to regions.

Assume we are given two alternative ways of classifications of the same entire population, into several subgroups. For example, the two alternative classifications can be classification according to gender or classification according to whether one is black or not. The variable we are interested in is income. Denote those ways of classification into subgroups by A, B,. . . The question we want to answer is which way of grouping is more stratified. Perfect stratification is defined as having the incomes of members of each subgroup confined to a given range, and no member of the other subgroups be located in this range. To see the meaning of perfect stratification consider the property "being black." If all blacks are poor[2] (rich) and all the poor are black (white) then we will say that we have a perfect stratification, and the quality of the classification (i.e., grouping) is perfect. The more whites (blacks) are poor (rich)—the lower the quality of the stratification. Consider now two classifications: one according to gender and the other according to being black or not. We will say that gender is a better classifier of the society if the inequality within women and within men is lower than the inequality within blacks and within

[2] By poor (rich) it is meant that the income is below (above) a certain level.

others (i.e. within a group the individuals are similar to each other) and the overlapping between women and men is lower than the overlapping between blacks and others (i.e., different from the other group). In other words, we will define one classification as better than the other if the members of each subgroup are similar to each other (low intra-group inequality) and different from members of other subgroups (low overlapping, high between-groups inequality). As can be seen, the higher the overlapping the lower the between-groups inequality and the higher the intra-group component. It is argued that given several classifications into subgroups of the same entire population, the grouping with the lowest overlapping (highest stratification) will be defined as the best grouping.[3]

22.2 An Illustration of ANOGI: The Melting Pot Policy[4]

Societies with large immigration tend to be sensitive to the assimilation of the immigrants into the society. Instead of a fragmented society, divided by ethnic rifts, the preference is for a society where origin ceases to be an identifying characteristic.

To describe the meaning of the ideal melting pot, we can't find a better description than the one presented by Zangwill (1914, p. 33), as quoted by Hirschman (1983, p. 397).

"America is God's Crucible, the Great Melting Pot where all races of Europe are melting and re-forming! Here you stand, good folk, think I, when I see them at Ellis Island, here you stand in your fifty groups, your fifty languages, and histories, and your fifty blood hatreds and rivalries. But you won't be long like that, brothers, for these are the fires of God you've come to—these are fires of God. A fig for your feuds and vendettas! Germans and Frenchmen, Irishmen and Englishmen, Jews and Russians—into the Crucible with you all! God is making the American."

Social integration includes many dimensions: cultural, language, common history, equal opportunities, to list a few. Also, it is not agreed by all that the melting pot policy, which destroys the diversity of cultural heritage, is really something that a society should aim for.[5] What seems to be noncontroversial is that society should not be stratified by ethnic grouping when restricting the attention to economic well-being. Unlike other dimensions of the melting pot policy—integration of ethnic groups into the society along the dimension of economic well-being is relatively

---

[3] An additional concept that is used extensively in the literature is polarization (see, e.g., Duclos, Esteban and Ray (2003) and the references there). However, further research is needed to establish the relationship between stratification and polarization.

[4] This section is based on Yitzhaki and Schechtman (2009).

[5] Glazer (1993) considers the decline in the positive attitude toward assimilation as an ideal for migrants in the USA.

easy to quantify and to agree upon.[6] It seems that it is agreed by all that a successful melting pot policy should abolish stratification of economic well-being according to ethnic groups.

The aim of this section is to apply ANOGI to compare two ways of classification and to see which one stratifies the society better according to ethnic groups. By comparing the change in the index of stratification over time we can evaluate the success of a melting pot policy according to this economic dimension. In other words—it enables us to see whether the background of origin plays an important role in stratification of a society according to ethnic groups and by tracing it over time we can learn about the achievements of a melting pot policy.

Israel is one example of a country where the melting pot policy was an officially declared policy (Lissak, 1999). We apply ANOGI to Israeli data in order to examine its success. Specifically we compare the stratification index (the inverse of overlapping) under two alternative definitions of ethnic groups. According to one definition—definition W (wide)—second generation Israelis, i.e., those who were born in Israel are defined as one group (regardless of their father's origin), while according to the other classification—definition N (narrow)—second generation Israelis are classified according to the ethnic group of the father.[7] A successful melting pot policy should have resulted in classification W revealing a more stratified society than classification N. The intuitive explanation of this kind of conclusion is that a successful melting pot policy should have resulted in a formation of a "new" group—second generation Israelis where (original) ethnic differences do not show up. On the other hand, if stratification is higher when one uses the N definition, i.e., when the second generation Israelis are classified according to the original (i.e. father's) ethnic group—then we conclude that the melting pot policy failed to create a new generation for which the (original) ethnic grouping ceases to be a stratifying variable. There are several reasons to suspect that definition W will be a better classifier than definition N, and they are detailed in Sect. 22.2.1.

It is worth emphasizing that the purpose of the application is descriptive. We are not trying to find out what causes success or failure of the melting pot policy nor whether there are other variables that may distinguish between the groups better than economic well-being. Therefore we do not use regression methods that can relate the difference in economic well-being to other attributes. Instead, we introduce a relatively new descriptive measure, the overlapping measure.

[6] Among the other aspects of assimilation that are not dealt within this chapter it is worth mentioning the acquisition of native language skills. See among others Chisweek (1978, 1998, 1999); Beenstock (1996) and the literature therein. Easterly and Levine (1997) relate ethnic diversity as impediment to growth.

[7] The classification of the ethnic group by the origin of the father is dictated by the available data.

22.2.1 *Definitions*

The variable of interest for classification is economic well-being, which is defined as after-tax income per equivalent adult, according to the official scale used in Israel.[8] To avoid the effect of different fertility rates we limit our population to adults only—age 30 and above. That is, although the sample is a sample of dwellings, our observations are adults of age 30+. We start by dividing the population according to the following distinctive groups that compose the entire population in Israel.

1. Jews born in Europe or America
2. Jews born in Asia or Africa
3. Jews born in Israel
4. Immigrants—those who migrated to Israel less than 10 years prior to the survey
5. Others—non-Jewish population.

To examine the success of the melting pot policy we compare two alternative definitions of the group "Jews born in Israel." By definition W (wide) "Jews born in Israel" are defined as those who were born in Israel, regardless of the father's origin. By the alternative definition N (narrow) "Jews born in Israel" are only those whose father was also born in Israel.[9] Otherwise those people are classified according to their father's origin. That is, the difference between the two alternative definitions is how the group of "Jews born in Israel but the father was born abroad" is classified. According to definition N this group is classified according to the place of birth of the father, while under definition W this group is classified as Israeli born. We apply this distinction only to those who were in the country for at least 10 years prior to the survey and are Jewish. We do not ignore the rest of the population. They are grouped as "immigrants" (those who migrated to Israel less than 10 years before the survey) and "others" (the non-Jewish population). The main point is that the definitions of those groups remain intact between the two alternative definitions.[10]

The comparison between the qualities of the decompositions one gets under definitions N and W is used to examine the success of the melting pot policy. If classification W shows higher stratification (lower overlapping) we will conclude that the melting pot policy was successful in creating a new group—those who were born in Israel are "similar within themselves and different from the other groups," where "other groups" include their parents. On the other hand, if classification N

[8] The equivalence scale used for comparison of economic well-being of households of different sizes is one-person: 1.25, two: 2.0, three: 2.65, four: 3.2, five: 3.75, six: 4.25, seven: 4.75, eight: 5.2 and 0.4 for each additional person. For additional explanations see Statistical Abstract of Israel, 2004, No. 55, p. 46.

[9] Data limitations do not allow us to refer to the place of birth of the mother.

[10] Note, however, the difference between "others" and "immigrants." Immigrants in an early survey may be defined as foreign born in a later period.

shows higher stratification, we conclude that the melting pot policy was not successful because the off-springs of the immigrants are similar to their parents,

Naturally, we would expect that the broader definition of Israelis (definition W) will create a separate group because

(a) They are expected to be younger and therefore they should be different from their parents.
(b) Although they are of mixed origins, they were raised in Israel. That is, they were educated in Israel so they should be different from their parents.
(c) Definition N suffers from misclassification and therefore the groups tend to be blurred over time because of mixed marriages of the parents of Israeli borns and because we are classifying the groups according to the place of birth of the father.

Therefore we would expect that the overlapping index for grouping N will be larger than the overlapping index for grouping W.

22.2.2 Data Description

The data consist of three Household expenditure surveys conducted in the years 1979/1980, 1992, and 2002 in Israel by the Central Bureau of Statistics and they are described in the publications of those surveys. There are several differences among the surveys that are important for the analysis carried out in this section.

(a) Coverage of the population: The survey in 1979/1980 includes only settlements with over 10,000 individuals, while the survey of 2002 includes settlements with population over 2,000. Because a large part of the Arab population live in rural areas, and because the population in many settlements has increased over time, the share of the Arab population that is covered has increased in a way that makes the comparison over the years seriously biased. Hence, we included the Arab population for completeness but one has to be careful in reaching conclusions because of sample selection bias.
(b) The accounting period has changed over time. In 1979/1980 the accounting period is 12 months. That is, the income reported is the income earned in the 12 months prior to the visit of the surveyors. However, in 1992 and 2002 the accounting period is composed of 3 months. The shorter the accounting period, the higher the inequality. Finkel, Artsev, and Yitzhaki (2006) estimated the bias to be of a magnitude of about 20% in the Gini of after-tax income of equivalent adult.

Having these limitations in mind—we can concentrate on the decomposition of the Gini according to ethnic groups.

To avoid the influence of different fertility rates, only adults of age 30 and above are considered in the analysis.

Table 22.2 The decomposition of the Gini coefficient of income among ages 30–65 according to ethnic groups[a]

Year	Definition	Overall Gini (1)	S*G*O (2)	G_B (3)	G_{BP} (4)	G_B/G_{BP} (5)
1979/1980	N	0.318	0.256	0.062	0.125	0.496
S. Error		(0.007)		(0.007)	(0.010)	
	W	0.318	0.257	0.061	0.1230	0.495
S. Error		(0.007)		(0.007)	(0.010)	
1992/1993	N	0.321	0.256	0.065	0.138	0.471
S. Error		(0.003)		(0.003)	(0.004)	
	W	0.321	0.270	0.051	0.120	0.427
S. Error		(0.003)		(0.003)	(0.004)	
2002	N	0.365	0.293	0.072	0.151	0.477
S. Error		(0.004)		(0.003)	(0.004)	
	W	0.365	0.305	0.060	0.122	0.495
S. Error		(0.004)		(0.003)	(0.004)	

[a]Definition N—Israeli born defined according to father's origin. Definition W—Israeli born defined as a separate group. Source: Yitzhaki and Schechtman (2009), Table 1, p. 143

22.2.3 *Results*

We analyze the data for the two alternative definitions in parallel. We present two tables of decompositions.

Table 22.2 presents the decomposition of the Gini coefficient of adult-equivalent after-tax income among prime-age Israelis according to ethnic groups. In the rows denoted by N, Israeli is defined as a person whose father was also born in Israel. Israeli born whose father was born abroad is defined as belonging to the original group (of the father). In the rows denoted by W, Israeli is defined as a person who was born in Israel. It can be seen that in both definitions, inequality has increased significantly between 1992 and 2002 from 0.321 to 0.365.

As can be seen, the Gini's between groups (G_B as well as G_{BP}) are bigger in definition N for the years 1992 and 2002. This means that when we classify second generation Israelis as belonging to the country of origin (of the father), we get a better stratified society. This conclusion is also supported by the overlapping term (column 2), where it is smaller by definition N than by definition W.

Table 22.3 presents the contributions of the components to the intra-ethnic group inequality, for different years. The ethnic groups used for definition N are as follows.

Eu-Am = born Europe-America, or born in Israel and father born in Europe-America
As-Af = born Asia-Africa, or born in Israel and father born in Asia-Africa
Israel = born in Israel and father born in Israel (or father's origin is unknown)
Immig = new immigrants, less than 10 years in Israel (migrated after 1970, 1982, 1992 accordingly).
Others = Non-Jewish or unknown origin.

Table 22.3 The contribution of the components to the intra-ethnic group inequality for different years, for definitions N and W

Category	Population share (1)	Relative mean income	Average rank (3)	Income share, S (4)	Gini, G (5) (SE)	Overlapping, O, (6) (SE)	SGO (7)
1979/1980							
Eu/Am N	0.429	1.243	0.629	0.533	0.282 (0.006)	0.868 (0.009)	0.131
W	0.324	1.232	0.615	0.399	0.296 (0.006)	0.910 (0.010)	0.108
As/Af N	0.362	0.786	0.430	0.285	0.266 (0.006)	0.899 (0.010)	0.068
W	0.332	0.771	0.390	0.256	0.265 (0.006)	0.893 (0.011)	0.061
Israel N	0.037	1.059	0.578	0.039	0.227 (0.093)	0.829 (0.136)	0.007
W	0.172	1.176	0.624	0.202	0.242 (0.006)	0.798 (0.011)	0.039
Immig N,W	0.079	1.049	0.552	0.083	0.279 (0.009)	0.971 (0.011)	0.022
Others N,W	0.093	0.645	0.217	0.060	0.394 (0.093)	1.160 (0.136)	0.027
N-SGO							0.255
W-SGO							0.257
1992/1993							
Eu/Am N	0.316	1.319	0.656	0.417	0.283 (0.004)	0.841 (0.006)	0.099
W	0.173	1.265	0.632	0.219	0.290 (0.003)	0.891 (0.006)	0.057
As/Af N	0.389	0.932	0.480	0.362	0.286 (0.004)	0.933 (0.005)	0.097
W	0.225	0.939	0.481	0.212	0.294 (0.005)	0.951 (0.006)	0.059
Israel N	0.048	1.152	0.577	0.056	0.310 (0.004)	0.951 (0.008)	0.016
W	0.355	1.140	0.576	0.405	0.298 (0.006)	0.913 (0.004)	0.110
Immig N,W	0.135	0.716	0.347	0.097	0.289 (0.003)	0.935 (0.006)	0.026
Others N,W	0.111	0.610	0.278	0.068	0.294 (0.004)	0.882 (0.008)	0.018
N-SGO							0.256
W-SGO							0.270
2002							
Eu/Am N	0.319	1.298	0.626	0.414	0.336 (0.006)	0.874 (0.006)	0.122
W	0.186	1.191	0.591	0.222	0.332 (0.004)	0.903 (0.006)	0.067

(continued)

Table 22.3 (continued)

Category	Population share (1)	Relative mean income	Average rank (3)	Income share, S (4)	Gini, G (5) (SE)	Overlapping, O, (6) (SE)	SGO (7)
As/Af N	0.343	0.943	0.501	0.324	0.324 (0.005)	0.931 (0.004)	0.097
W	0.133	0.955	0.506	0.127	0.347 (0.008)	0.953 (0.005)	0.042
Israel N	0.087	1.244	0.589	0.109	0.368 (0.003)	0.939 (0.005)	0.038
W	0.430	1.156	0.571	0.498	0.346 (0.005)	0.928 (0.005)	0.160
Immig N,W	0.105	0.788	0.418	0.083	0.284 (0.011)	0.870 (0.016)	0.020
Others N,W	0.145	0.486	0.228	0.070	0.307 (0.003)	0.749 (0.005)	0.016
N-SGO							0.293
W-SGO							0.305

Source: Yitzhaki and Schechtman, 2009, Table 2, p. 144

The ethnic groups used for definition W are as follows.
Eu-Am = born Europe-America
As-Af = born Asia-Africa
Israel = born in Israel
Immig = new immigrants, less than 10 years in Israel (migrated after 1970, 1982, 1992 accordingly)
Others = Non Jewish or unknown origin

As can be seen, the share of the Israelis is much bigger for definition W. (And obviously, the shares of Eu/Am and As/Af are declining). The overlapping index of the Israeli group is smaller in 1979 and gets larger and closer to 1 in 1992 and 2002. The two definitions show similar trends in the overlapping index. However, the overlapping of Israelis by definition N is bigger than by definition W in 1992, and the gap is much smaller in 2002. Also, the Gini coefficient of Israelis in 2002 is smaller by definition W than by definition N. (The two Ginis are similar in 1992.) The sum of the products SGO for Israelis is smaller for definition N than for definition W for the 3 years under study. This means that the intra-group inequality and overlapping between groups explain a greater portion of inequality leading us to conclude that definition N provides a more distinctive grouping of the socicty than definition W.

The decomposition of the Gini of the entire population by ANOGI provides additional information, given by the ranking of each group in terms of the others and by the overlapping of each group by the others. These measures are not reported here because they do not contribute to our discussion. However we report them, for completeness, in Appendix 22.1. In addition, we have performed ANOVA on the two different groupings. The results strengthen the conclusions of this chapter that definition N is a better stratifier and are reported in Appendix 22.2.

22.3 Summary

The objective of the chapter was to introduce an application of the decomposition of the Gini coefficient, called ANOGI and to illustrate its use. The purpose of the study was to evaluate the success of the melting pot policy in Israel. We concentrated on one aspect only—the melting pot in terms of economic well-being. We introduced a relatively new tool—the decomposition of the Gini coefficient of the entire population into the contributions of the individual Ginis of subpopulations (intra-group component), the between-groups inequality (inter-group component), and additional terms, defined as overlapping indices. The basic idea was to divide the entire population into subpopulations in two different ways (called here N and W), and check which one will stratify the population better. The conclusion from this study is that based on the between-groups Gini's, definition N stratifies better. That is, we can say that the melting pot did not succeed. The persons born in Israel are more similar to their parents than to each other.

Appendix 22.1

Tables 22.4 (N) and 22.5 (W) present the ranking of each group in terms of the other for the 3 years, for the two alternative definitions of the Israeli group.

Each entry in the tables presents the average rank of the members of the group indicated in the row, had they been ranked according to the ranking of the group indicated in the column. Looking at Table 22.4 (N) we see that the average rank of Jews born in Asia/Africa, had they been ranked according to Jews from Europe/America is 0.26 in 1979, 0.31 in 1992, and 0.37 in 2002. This is an indication that over time the relative status of Jews from Asia/Africa has improved. Looking at the column of Israeli born, the ranking in terms of Europe/America has slightly declined from 0.43 in 1979 to 0.42 in 1992, but has increased to 0.47 in 2002. On the other hand, the average ranking of the Arab population in terms of European/American born has increased from 0.12 in 1979 to 0.15 in 1992 but declined later (in 2002) to 0.13.[11]

Tables 22.6 (N) and 22.7 (W) present the overlapping index (and standard error) of each group in terms of the other for the 3 years, for the two alternative definitions

Table 22.4 (N): The ranking of each group in terms of the other, for the 3 years

	Eu/Am	As/Af	Israel	Immig	Other
1979/1980					
Eu/Am	0.5	0.74	0.57	0.58	0.88
As/Af	0.26	0.5	0.30	0.34	0.73
Israel	0.43	0.70	0.5	0.51	0.87
Immig	0.42	0.66	0.49	0.5	0.82
Other	0.12	0.27	0.13	0.18	0.5
1992/1993					
Eu/Am	0.5	0.69	0.58	0.80	0.85
As/Af	0.31	0.5	0.40	0.65	0.72
Israel	0.42	0.60	0.5	0.73	0.78
Immig	0.20	0.35	0.27	0.5	0.59
Other	0.15	0.28	0.22	0.41	0.5
2002					
Eu/Am	0.5	0.63	0.53	0.72	0.87
As/Af	0.37	0.5	0.40	0.59	0.79
Israel	0.47	0.60	0.5	0.68	0.83
Immig	0.28	0.41	0.32	0.5	0.75
Other	0.13	0.21	0.17	0.25	0.5

Source: Yitzhaki and Schechtman, 2009, Table 3, p. 146

[11] The disclaimer that the coverage of this population has changed over time, which may bias the results should be added.

Table 22.5 (W): The ranking of each group in terms of the other, for the 3 years

	Eu/Am	As/Af	Israel	Immig	Other
1979/1980					
Eu/Am	0.5	0.73	0.50	0.57	0.86
As/Af	0.27	0.5	0.25	0.33	0.72
Israel	0.50	0.75	0.5	0.57	0.89
Immig	0.43	0.67	0.43	0.5	0.82
Others	0.14	0.28	0.11	0.18	0.5
1992/1993					
Eu/Am	0.5	0.66	0.56	0.77	0.83
As/Af	0.34	0.5	0.40	0.65	0.72
Israel	0.44	0.60	0.5	0.73	0.79
Immig	0.23	0.35	0.27	0.5	0.59
Others	0.17	0.28	0.21	0.41	0.5
2002					
Eu/Am	0.5	0.59	0.52	0.69	0.85
As/Af	0.41	0.5	0.43	0.59	0.79
Israel	0.48	0.57	0.5	0.67	0.83
Immig	0.31	0.41	0.33	0.5	0.75
Others	0.15	0.21	0.17	0.25	0.5

Source: Yitzhaki and Schechtman, 2009, Table 4, p. 147

of the Israeli group. Each column represents the reference group (represented by the index j in the decomposition of O_{ji}), while the row represents i. Multiplying the elements of each row by the share in the population of the group and summing up yields the overlapping of the group with the entire population. That is, each row represents the overlapping of the group with other groups (and with itself. The overlapping of a group with itself is 1). The first line says that Europe/ America is a stratified group with respect to Asia/Africa (0.79), but it is less of a group when the reference group is Israeli born. It is definitely a group with respect to the "Others" group. In 1979 the group "Others" included several rich people so that it became a non-group with respect to all other groups.[12] However, in 1992 the "Others" became a distinct group relative to all others except immigrants, while in 2002 they were left behind by almost all other groups. Over time the groups Asia/Africa and Europe/America became less distinct from each other with the overlapping indices increasing from (0.79; 0.85) in 1979 to (0.92; 0.94) in 2002.

[12] We do not have a good explanation to this result. It may be caused by the members of the Christian-Arab population who were with relatively high income and emigrated from the country.

Table 22.6 (N): The overlapping index (and SE) of one group in terms of the other

	Eu/Am	As/Af	Israel	Immig	Others
1979/1980					
Eu/Am	1	0.79	1.06	0.95	0.43
		0.02	*0.03*	*0.03*	*0.02*
As/Af	0.85	1	0.97	0.87	0.75
	0.03		*0.04*	*0.03*	*0.03*
Israel	0.88	0.84	1	0.85	0.48
	0.04	*0.05*		*0.02*	*0.02*
Immig	1.00	0.99	1.14	1	0.66
	0.02	*0.03*	*0.01*		*0.03*
Others	1.17	1.19	1.20	1.15	1
	0.28	*0.09*	*0.18*	*0.16*	
1992/1993					
Eu/Am	1	0.86	0.92	0.65	0.54
		0.02	*0.01*	*0.01*	*0.01*
As/Af	0.94	1	0.94	0.87	0.75
	0.02		*0.02*	*0.01*	*0.02*
Israel	1.05	0.99	1	0.81	0.70
	0.02	*0.02*		*0.02*	*0.02*
Immig	0.82	1.02	0.88	1	0.91
	0.01	*0.03*	*0.02*		*0.01*
Others	0.69	0.96	0.78	1.05	1
	0.03	*0.02*	*0.02*	*0.01*	
2002					
Eu/Am	1	0.92	0.95	0.82	0.47
		0.01	*0.01*	*0.01*	*0.01*
As/Af	0.94	1	0.93	1.01	0.68
	0.01		*0.01*	*0.01*	*0.02*
Israel	1.04	0.98	1	0.91	0.59
	0.01	*0.01*		*0.01*	*0.02*
Immig	0.86	0.94	0.84	1	0.66
	0.03	*0.02*	*0.03*		*0.01*
Others	0.57	0.78	0.59	0.98	1
	0.02	*0.02*	*0.01*	*0.01*	

Source: Yitzhaki and Schechtman, 2009, Table 5, p. 148

Appendix 22.2: ANOVA

In addition to the decomposition of Gini, a decomposition of the variance was obtained by ANOVA. Note that there are only two components: between (intra) and within (inter). The results are given in Table 22.8. We note that the question asked by ANOVA is different—it is meant to compare the means of the subpopulations. As can be seen from the last column (the F ratio), the between MS is (relatively) larger for definition N for the 3 years under study, strengthening our conclusion that definition N is a better stratifier.

Table 22.7 (W): The overlapping index (and SE) of one group in terms of the other

	Eu/Am	As/Af	Israel	Immig	Others
1979/1980					
Eu/Am	1	0.82	1.11	0.99	0.47
		0.02	*0.02*	*0.02*	*0.02*
As/Af	0.82	1	0.90	0.86	0.77
	0.03		*0.03*	*0.03*	*0.03*
Israel	0.88	0.72	1	0.87	0.37
	0.02	*0.02*		*0.02*	*0.03*
Immig	0.97	0.98	1.11	1	0.66
	0.02	*0.03*	*0.02*		*0.03*
Others	1.17	1.19	1.19	1.15	1
	0.27	*0.09*	*0.20*	*0.16*	
1992/1993					
Eu/Am	1	0.91	0.99	0.72	0.60
		0.01	*0.01*	*0.02*	*0.01*
As/Af	0.96	1	1.00	0.88	0.77
	0.02		*0.01*	*0.01*	*0.01*
Israel	1.00	0.94	1	0.76	0.64
	0.01	*0.01*		*0.01*	*0.02*
Immig	0.84	1.00	0.92	1	0.91
	0.03	*0.02*	*0.03*		*0.01*
Others	0.72	0.95	0.82	1.05	1
	0.02	*0.02*	*0.03*	*0.01*	
2002					
Eu/Am	1	0.95	0.98	0.88	0.52
		0.01	*0.01*	*0.01*	*0.01*
As/Af	0.99	1	0.99	1.03	0.71
	0.01		*0.01*	*0.01*	*0.02*
Israel	1.02	0.97	1	0.92	0.57
	0.01	*0.01*		*0.01*	*0.02*
Immig	0.89	0.91	0.89	1	0.66
	0.03	*0.02*	*0.03*		*0.01*
Others	0.64	0.77	0.65	0.98	1
	0.02	*0.02*	*0.02*	*0.01*	

Source: Yitzhaki and Schechtman, 2009, Table 6, p. 149

Table 22.8 ANOVA for definitions N and W

	SS-within (df)	SS-between (df)	Total (df)	MS-within	MS-between	F = MSB/ MSW
N-1979	337,543 (2,868)	40,524 (4)	378,067 (2,872)	118	10,131	86
W-1979	338,961 (2,868)	39,106 (4)	378,067 (2,872)	118	9,776	83
N-1992	916,094 (7,039)	171,763 (4)	1,087,857 (7,043)	130	42,941	330
W-1992	957,762 (7,039)	130,095 (4)	1,087,857 (7,043)	136	32,524	239
N-2002	7,897,294 (8,149)	1,074,649 (4)	8,971,943 (8,153)	969	268,662	277
W-2002	8,132,176 (8,149)	839,767 (4)	8,971,943 (8,153)	998	209,942	210

Source: Yitzhaki and Schechtman, 2009, Table 7, p. 150

Chapter 23
Suggestions for Further Research

Introduction

Throughout the book we have stressed several properties that distinguish between the GMD and the variance, claiming that those properties give an advantage to using the GMD over the variance, in cases where the assumption of normality is not supported by the data. Among those properties are the following:

(a) The GMD enables one to impose economic theory on the statistical analysis. This property is relevant whenever economic theory calls for asymmetric treatment of the distribution.
(b) Decompositions: the decomposition of the GMD includes additional parameters. This property is relevant whenever: (1) stratification and quality of classification into groups play a role in the application or (2) one is dealing with a random variable which is composed of a linear combination of random variables.
(c) The Gini methodology enables one to see whether the association between variables is monotonic. This property enables one to identify cases where it is impossible to get conclusive evidence from the data concerning the association between random variables and as a result one has to restrict the analysis to regimes of the variables in order to assign a sign to the association.
(d) The GMD has two regression coefficients associated with it. This property is useful in cases that one wants to test for the linearity assumption, and as we argue later, it is also useful in time-series analysis.
(e) The GMD can be used simultaneously with the variance, a property that enables one to replicate the statistical analysis performed with the variance with the GMD in a stepwise manner so that one is able to find out which variable(s) is responsible for the non-robustness of the OLS model.

Those properties have some advantage in some fields more than in others. For example, property (a) is relevant in the areas of income distribution and finance because economic theory assumes an increasing concave utility function. The variance, on the other hand, is a convex function. This difference may cause

S. Yitzhaki and E. Schechtman, *The Gini Methodology: A Primer on a Statistical Methodology*, Springer Series in Statistics 272, DOI 10.1007/978-1-4614-4720-7_23,

conflict between the two approaches. As a result, to get compatibility, there is a need for imposing strong assumptions in order to eliminate the contradiction. The use of the GMD overcomes this problem. However, the implementation of the GMD comes with a price tag because it complicates the analysis. As a result it may be that it is not justified to use it to estimate production functions.

Properties (b) to (d) belong to the "Gini reveals more" argument and as such they are relevant for all areas in which the convenient world of normality does not hold. In this book we have concentrated on imitating the analysis performed by the variance. In this chapter we point out additional fields that may benefit from using the additional properties of the Gini methodology. Note, however, that our discussion in this chapter does not have the maturity of the rest of the book.

Section 23.1 suggests using the GMD for testing claims for convergence to the multivariate normal distribution. This section utilizes property (b.2) and is relevant for all procedures that rely on the central limit theorem. Section 23.2 suggests the use of the GMD methodology in the area of education. Our belief in the contribution of using the GMD in this area is based on the fact that ability is a latent variable while exams can be interpreted as applying monotonic nondecreasing transformations intended to reveal the latent ability. Another property of the GMD that is relevant in this area is the ability to identify a monotonic relationship, a property that is helpful in determining the classification of abilities into areas of ability. This issue is also discussed in Sect. 23.2. Section 23.3 comments on the use of the GMD in time-series analysis and diffusion processes. The advantage of the GMD in the former application is in having two regression coefficients associated with it that are not related to each other by a functional form as is the case in the OLS. Hence, we are able to check whether moving forward in time results in the same conclusions as moving backward in time. Also, the GMD may be useful in interpreting and estimating diffusion processes. Section 23.4 comments on an issue that troubles everyone who suggests a new methodology: is there an alternative methodology that outperforms the suggested one. In the case of the GMD the competition comes from other indices based on the L_1 metric. Section 23.5 comments on the need for software that can handle the method suggested in this book in a unified way.

The list above does not cover all possible areas that can benefit from implementing the GMD as the measure of variability. It only covers areas with low hanging fruits. Among the areas that we did not cover it is worth to mention the use of the Gini in classification and regression trees (Montanari & Monari, 2005 and the references cited there).

23.1 Convergence to the Normal Distribution

The central limit theorem says that by averaging a sufficient number of random variables, the distribution of the average converges to the normal distribution. This theorem is the base for many methodologies, including the well known nonparametric methods such as the bootstrap or the jackknife methods. The main question left is how to define the "sufficient number." An alternative way of presenting the same argument is how many moments of the distribution should we compare in order to claim "sufficiency" or convergence to the normal. Should we just rely on the mean and variance? Each additional required moment is going to increase the required sample size.

Here, one can use the properties of the GMD decomposition in order to find out what is the reasonable size of the sample. The basic idea is the following: a necessarycondition for the approximation to the normal distribution to be reasonable is that the distribution of the average of averages of observations will be of the same family. This kind of a test can rely on the decomposition of the GMD of a linear combination of random variables.

To see this assume that we have a sequence of averages of i.i.d. random variables and the question asked is the following: is the size of the sample used to create the averages sufficient in order to claim convergence to the normal distribution?

Let $X_1,\ldots,X_m$ be a random sample from an unknown distribution F, where $m = kn$ is large. Denote the average of the sample by $\overline{X}_m$ and let $\overline{X}_n^1,\ldots,\overline{X}_n^k$ be k averages of samples of k randomly selected disjoint subgroups of observations (each of size n). We use the decomposition of the Gini of $\overline{X}_m$ into the contributions of the individual averages and the following two facts: (1) For m large enough the sampling distribution of $\overline{X}_m$ belongs to the same family as the distributions of $\overline{X}_n$, and (2) a necessary condition for two random variables to be exchangeable up to a linear transformation is the symmetry of the Gini correlation between them. Using these two facts, let

$$\overline{X}_m = \frac{n}{m}\overline{X}_n^1 + \frac{n}{m}\overline{X}_n^2 + \ldots + \frac{n}{m}\overline{X}_n^k.$$

We now replicate the process so that we have a large number of observations of $(\overline{x}_m,\overline{x}_n^1,\ldots,\overline{x}_n^k)$. If averaging of n observations is sufficient in order to get that the sampling distribution of the averages is approximately normal, then one should expect that the two Gini correlations between $\overline{X}_m$ and each individual average will be equal. Note that because the $\overline{X}_n^k$ are drawn from independent samples, our interest is in the terms D in (4.6) because they indicate whether $\overline{X}_m$ and $\overline{X}_n^k$ belong to the same family of distributions. Since we know that the distribution of $\overline{X}_m$ converges to the normal it is sufficient to verify that the distributions converge to the same family.

To sum up: applying the test on equality of two Gini correlations enables us to determine whether n is a sufficient number of observations to claim normality of the distribution of the average of size n.

23.2 The Use of the Gini Method in the Area of Education

The area of education seems as an area that can benefit from using the Gini methodology because of the following properties that characterize the field:

(a) Ability, whether general ability exists or not, is a latent variable that can be discovered by asking questions and observing whether the examinee answered them correctly.
(b) The usual way to evaluate ability is by confronting the examinees with a set of questions and counting the number of questions that each examinee was able to answer correctly.
(c) It is not easy to observe which property (ies) of the examinees enable them to correctly answer the questions.

Those properties give an advantage to the GMD over alternative methods because the Gini "reveals more."

To see the possible contribution of the GMD we first argue that the robustness of ranking of groups according to average success in exams is related to first degree stochastic dominance. Then, we argue that whether a question belongs to the subject matter of the exam can be handled by examining the monotonicity of the relationship between the score in answering a question and the score in the exam.

23.2.1 Ranking Groups According to Average Success

To see the implications of those properties consider the following: assume that we can classify abilities to uni-dimensional abilities, and the distribution of the uni-dimensional abilities in the population is a continuous variable with an unknown density function f(a). The questionnaire is composed of n questions, with difficulties d_i ($i = 1,\ldots,n$). To simplify the arguments, assume that the probability to correctly answer a question is p(a − d), where d is the difficulty of the question. Note that both a and d are unobservable. The probability of successfully answering a question is

$$\mathrm{p(a-d)} \quad \text{with } \mathrm{p}'()>0. \tag{23.1}$$

The probability of success p is increasing in a and decreasing in d.[1] The expected score (and the probability of success) in a test with n questions, administered to an examinee with ability a, is:

$$S(a,d) = \frac{1}{n}\sum_{i=1}^{n} p(a,d_i), \tag{23.2}$$

[1] This assumption is known as the "monotonicity assumption." Additional assumptions that could be imposed on (23.1) below are local independence and local homogeneity (see Ellis & Wollenberg, 1993), but these additions are not relevant to our main argument.

(d is a vector whose components are d_i, i = 1,. . .,n). Equation (23.2) states that a subject's expected score is the average of the expected value of n binomial random variables. However, these random variables are not statistically independent—they are all affected by a, the subject's ability and d the difficulty distribution of the questions in the questionnaire. As we are interested mainly in the expected value of S(a, d), we ignore any random factors that might affect the results.

Finally, we assume that there exist x_{max} and x_{min} such that:[2]

$$p(a - d) = 0 \quad \text{for } a - d \leq x_{min} \quad \text{and} \quad p(a - d) = 1 \quad \text{for } a - d \geq x_{max}. \qquad (23.3)$$

Assumption (23.3) means that one can always compose a question that no one will ever answer correctly, and another that will always be answered correctly. This assumption eliminates the possibility of the probability of success being a constant that is independent of the task's difficulty.

Following are several results: (Yitzhaki & Eisenstaedt, 2003).

Proposition 23.1 *Individuals' ranking within a group cannot be altered by changing the difficulty distribution of the questions in the questionnaire.*

Proof Consider two individuals who have abilities $a_1 > a_2$. One has to prove that $S(a_1, d) \geq S(a_2, d)$ for all d. According to (23.2),

$$S(a_1,d) - S(a_2,d) = \frac{1}{n}\sum [p(a_1 - d_i) - p(a_2 - d_i)] \geq 0.$$

The non-negativity of the terms in the square brackets is caused by $p'(\,) \geq 0$.

We now move to evaluate ranking of groups, like schools, classes, ethnic groups etc. Groups' ranking, being more complex, needs an example. Take two groups of equal size, "blues" and "greens," where $a_1{}^b \leq a_2{}^b \leq \ldots \leq a_m{}^b$ and $a_1{}^g \leq a_2{}^g \leq \ldots \leq a_m{}^g$ denote blues' and greens' abilities, respectively. Denote by $F_b(a) = \frac{1}{m}\sum I(a_i^b)$ the cumulative distribution function, where I(X) equals 1 if X is true and zero otherwise. Assume that the ranking of the groups is determined by the difference in average scores achieved in the test, as follows:

$$\Delta R = \sum_{j=1}^{m} [S(a_j^b, d) - S(a_j^g, d)]. \qquad (23.4)$$

Can the sign of ΔR be changed by manipulating d?

Proposition 23.2 *Assuming that (23.4) is used to rank groups, and that (23.2) and (23.3) hold, then a necessary and sufficient condition for the impossibility of changing the sign of ΔR by manipulating d is that $F_b(a)$ and $F_g(a)$ do not intersect.*

---

[2] These are assumptions of convenience; the conclusions reached here are not affected by allowing a "guessing parameter" to affect the item response function (see Lord, 1980, p. 12).

If on the other hand, the cumulative distributions intersect, then one can always find two exams that will rank the groups differently.

Proof Begin with a test in which all questions are equally difficult, $d_1 = d_2 = \ldots = d_n = d_c$. In this case, it suffices to prove the proposition with a test composed of one question.

Suppose the distributions intersect only once, at a_3. That is, $F_g(a) > F_b(a)$ for $a \leq a_3$ and $F_g(a) < F_b(a)$ for $a > a_3$. If so, one can choose d_c such that $a_3 - d_c < x_{max}$. Now, all the subjects whose $a \geq x_{max} + d_c$ will score a hit with probability one, and since $1 - F_g(x_{min} + d_c) < 1 - F_b(x_{min} + d_c)$ there will be fewer greens than blues among them. As for the rest, since $F_g(a) > F_b(a)$, the blue with the poorest ability has better chances of scoring a hit than does the green with the poorest ability, the blue second in rank is more likely to score a hit than is the green second in rank, and so on. Blues will therefore perform better than greens in this test.

To change the groups' ranking it suffices to choose d_c such that $a_3 - d_c < x_{min}$. Now, only $1 - F_g(x_{min} + d_c) > 1 - F_b(x_{min} + d_c)$ will score a hit. We now scan from best to worst: the best green has a higher probability of scoring a hit than the best blue, the second-best green has a better chance than the second-best blue, and so on. This proves that if the distributions of two groups intersect, one can switch the rankings of average success. If the distributions do not intersect, then for any d_c chosen by the investigator, if the lowest ranking member of one group has a higher (lower) chance of scoring a hit than does the lowest ranking member of the other group, then the same can be said of the rest of the population.

The condition of nonintersecting distributions is identical to the condition that the groups can be stochastically ordered (Lehmann, 1955; Spencer, 1983a, 1983b), or, to use the term used in economics, that the distributions can be ranked according to First Degree Stochastic Dominance Criterion (FSD).

The proof of Proposition 23.2 relies on three assumptions: (1) That the groups are equal in size, (2) that the test consists of one question, and (3) that the distributions intersect only once. The proposition can be extended while relaxing these assumptions.

An important property of Proposition 23.2, which will come into play later, is that whether or not the distributions intersect does not depend on d. This is so because the subject's ranking is not sensitive to the test's difficulty distribution (see proof of Proposition 23.1). Since the relevant cumulative distributions are based only on individuals' abilities, changing the difficulty distribution cannot change the order in which the cumulative distributions are ranked. However, for a test to reveal more than gross clumping of performance levels, it is important that the empirical cumulative distributions be strictly increasing; a test that is too easy or too difficult may obscure finer degrees of differentiation in the subjects' abilities.

Note that the conditions for Proposition 23.2 are quite common. If the unobserved ability distributions are assumed to be normal, then it is sufficient for the variance of two groups to differ to cause an intersection of the ability distributions. This means that assuming normal distribution of abilities means that the examiner can cause rank reversal of average scores of groups simply by changing the difficulty distribution of the questions in the exam.

It should be pointed out that the use of the scores of exams in a regression should also be subject to the question of whether a monotonic transformation of the scores can change the sign of the regression coefficient. This is so because manipulating the average may also affect any aggregated number. Further discussion and illustrations based on empirical distributions can be found in Schechtman, Soffer, and Yitzhaki (2008) and Yitzhaki and Schechtman (2012).

23.2.2 A Gini Item Characteristic Curve[3]

Item Response Theory (hereafter IRT) has been developed between the fifties and the seventies of the last century. It became the main stream theory in the field of education with the publication of the book by Lord and Novick (1968). Since then there have been different versions and extensions of the theory. However, the basic framework remains the same. It imposes a given structure on the data, that is a given distribution of abilities among the examinees and a given structure of the relationship between the probability of success in answering a question and ability. This structure is imposed over the entire range of abilities. For an excellent review of the recent main developments in the literature see Lee, Wollack, and Douglas (2009).

The GMD enables to estimate a nonparametric version for IRT. The basic idea is to offer a nonparametric method for estimating the basic curve in the IRT model, the Item Characteristic Curve (ICC), which relates the probability of success in answering a question to the ability of the examinees and the difficulty of the questions. In this sense one can classify the Gini methodology as belonging to the literature presented by Sijtsma and Molenaar (1987), Ramsay (1995) and Bolt (2001). The advantage of the nonparametric version based on the GMD is that there is no assumption of monotonicity imposed on the data and it enables the decomposition of the estimate of the overall regression coefficient between the success in answering a question and ability in different ranges of ability, so that one can observe the contribution of different levels of ability to the overall regression coefficient. This decomposition enables one to see whether the relationship between the probability in answering a given question and the ability demonstrated in the exam is monotonic, that is whether the contribution of each section of ability to the regression coefficient is with the same sign. Would one find large ranges of abilities with negating contributions then it is an indication that either the question is not related to the subject matter of the exam although the overall relationship is positive and statistically significant, or that the "ability" cannot be viewed based on a single property. In other words, the conclusions derived from the nonparametric method are not based on averages or summary statistics, but on differentiating among different levels of abilities so that one can see how the average relationship is composed of and whether there are regions of abilities that do not behave

[3] This section is based on Yitzhaki, Itzhaki and Pudalov, (2011).

according to the overall structure. An additional advantage of the suggested method over alternative nonparametric approaches is that it does not require arbitrary window selection for smoothing the curve that can blur the graph and hide some non-monotonic sections. The method enables a better scrutiny of the characteristics of the question and its relationship with the particular ability the exam is intended to evaluate.

In particular the method enables one to check whether a question belongs to the subject matter (to be called "ability") of the exam or not. The basic idea is that if a question belongs to the subject matter of the exam then the scores on this question form a monotonic nondecreasing relationship with the scores of the exam (after excluding this question). Otherwise, it implies that the question does not belong to the subject matter of the exam or that the exam is not testing a single type of ability. We expect a good exam to include only questions for which the scores form a monotonic relationship with the grades in the exam. For an application of this methodology see Yitzhaki, Itzhaki, and Pudalov (2011). However, the paper deals with one dimensional ability. This idea can serve as a tool for classifying questions according to different abilities (exams). For simplicity assume that there are two exams. One can look for monotonic relationships between each question in one exam and the score on its own exam and the score of the other exam, and classify the question to the exam with which it has a monotonic relationship. A question that does not form a monotonic relationship with any exam is a question that is redundant and should be discarded, unless one declares that there are several abilities that are needed for success in the exam. Hopefully, this way will enable an improved classification of abilities. A possible direction of further research is extending the methodology to multivariate abilities by using multiple Gini regression techniques. A different direction of further research is needed to implement other properties of the Gini methodology into the IRT model. In particular, decomposing the success in answering a given question into the contributions of traits and background of the examinees via the Gini multiple regression (Schechtman, Yitzhaki, & Pudalov, 2011). Another direction to follow is the use of the GMD in classification (Calò, 2006). In other words, a promising direction of research is extending the Gini IRT model to multiple regression case.

23.3 The Use of the Gini Methodology in Time-Series

There are two major differences between time-series analysis and cross section regression: in time-series the observations are dependent, while in most cross-section analyses the basic assumption is that observations are randomly selected. Having a sample which is not i.i.d. opens the way to trends in the change of the distributions over time so that stationarity of the distribution over time cannot be taken for granted. However, assuming that the correlation between observations is less than one then it can be shown that weak forms of the central limit theorem hold for the sample (Serfling, 1968, 1980). Since the GMD is bounded from above by a linear function of the standard deviation it is clear that any proposition about convergence that holds for the variance also holds for the GMD. Hence, the basic requirement for application of the GMD to time-series is already proved.

The other difference between time series analysis and cross section is that economic theory is mainly relevant for the latter. For example, we have more confidence in the shape of the distribution of income than in the shape of mobility, which can be viewed as representing the changes in the income distribution.

The potential contribution of the GMD to this field seems to be based on the additional parameters of the Gini regression. Among those parameters are: (a) The GMD has two covariances defined between two variables. In the case of time series this property translates into four possible regression coefficients. To see this, assume that we have the series, x(1),. . ., x(t),. . .,x(T). Also assume that one is interested in the relationship between x(t) and x(t−1). Then one can define four Gini regression coefficients, two semi-parametric and two based on minimization of the GMD of the residuals. The semi-parametric ones are:

$$\beta_N^F = \frac{\mathrm{cov}(x(t), F(x(t-1))}{\mathrm{cov}(x(t-1), F(x(t-1))} \tag{23.5}$$

And

$$\beta_N^B = \frac{\mathrm{cov}(x(t-1), F(x(t))}{\mathrm{cov}(x(t), F(x(t))}. \tag{23.6}$$

For a sufficiently large sample, the difference between the denominators of (23.5) and (23.6) should be negligible because the difference is in changing only one observation. But there is no reason to expect the difference between the numerators to also be negligible because sometimes "what you see from here is not what you see from there." Only if the distribution of (x(t),x(t − 1)) is exchangeable up to a linear transformation, we should expect $\beta_N^F = \beta_N^B$.

One implication of a stationary series is that moving forward in time does not result in an estimate that is different from the one obtained when moving backward in time.[4] Hence, a test of whether $\beta_N^F = \beta_N^B$ can be used to test whether the time-series is stationary. This test augments the test on whether the residual and the explanatory variables are uncorrelated as explained in Chap. 8. Preliminary work in this direction can be found in Serfling (2010) and Shelef and Schechtman (2011).

An additional application of the GMD is in estimating diffusion processes. It can be found in Trajtenberg and Yitzhaki (1989). The application is based on (2.1) which can be rewritten as:

[4] See, for example, Wodon and Yitzhaki (2006) for a critique of the β convergence concept used in macro-economics in order to prove convergence in the growth rates of countries. Wodon and Yitzhaki found that this concept may lead both to convergence when moving forward and backward in time, which leads to a contradiction. See also O'Neill and Van Kerm (2008).

$$\Delta = E\{|t_2 - t_1|\} = 2 \int_{-\infty}^{\infty} \int_{t_1}^{\infty} (t_2 - t_1) f(t_2) f(t_1) dt_2 dt_1,$$

which can be interpreted as the expected time difference between any two adoptions over the whole diffusion process. Chandra and Singpurwalla (1981) also relate the adoption rate (or the hazard rate) to the Gini and the Lorenz curve.

23.4 The Relationship Between the GMD and Absolute Mean Deviation[5]

The main argument in this book is that the GMD is a better measure of variability than the variance whenever the normal distribution cannot be assumed. The only difference between the GMD and the variance is in the metric used: the GMD relies on city block metric (a.k.a. L_1), while the variance is based on the Euclidean metric. Therefore, one may suspect that other measures based on the city block metric are better than the GMD. The other measures include the Mean Absolute Deviation (MAD) and the Least Absolute Deviation (LAD).

The MAD is a measure of variability that is recommended because of its robustness. It is based on dividing the distribution of a random variable into two groups: above and below the mean and estimating the absolute deviations of the observations from the mean. Gorard (2005) presents an excellent review of the history of 90 years of debates on the properties of MAD versus the standard deviation, reaching the conclusion that MAD should be preferred over the standard deviation whenever the distribution differs from the normal. His conclusions, based on Barnett and Lewis (1978) and Huber (1981), are based on the argument that even a small deviation from the normal distribution should lead one to prefer the MAD over the standard deviation. A generalization of MAD can be referred to as QUAD, which is the absolute difference from a quantile of the distribution. An interesting and popular member of this family is the LAD, which turns out to be the absolute deviation from the median. The LAD, MAD, and QUAD are popular measures that are used in regression analysis (Bassett & Koenker, 1978; Koenker & Bassett, 1978, 1982; Koenker, 2005), in portfolio analysis (Konno & Yamazaki, 1991; Simaan, 1997) and in science. The references mentioned above present a small sample of the literature.

The aim of this section is to argue that MAD, QUAD, and LAD are actually special cases of the between-groups component of ANOGI. The difference between the different measures is in the definition of the range of the groups. Alternatively, one can view them as the GMD applied to specific distributions, the distributions of the between-groups component. As such they are actually conveying the same

[5] This section is based on Yitzhaki and Lambert (2011).

information as the between-groups component in ANOGI for a specific case: whenever the overall population is divided into two groups with no overlapping between members of the groups.

The proof of the argument is based on the following:

The expected absolute deviation from any quantile, q, is:

$$E\{|X-q|\} = \int_{-\infty}^{q} (q-x)dF(x) + \int_{q}^{\infty} (x-q)dF(x). \tag{23.7}$$

The left term on the right hand side is the low absolute deviation: it is equal to:

$$\int_{-\infty}^{q} (q-x)dF(x) = p\,q - \mu\,LC(p).$$

The right hand term on the right side of (23.7) is equal to:

$$\int_{q}^{\infty} (x-q)dF(x) = \mu\,[1-LC(p)] - q[1-p].$$

Inserting them into (23.7) we get:

$$E\{|X-q|\} = 2p\,q + \mu - q - 2\,\mu\,LC(p). \tag{23.8}$$

Figure 23.1 presents a typical LC. On the horizontal axis, p, is the value of the cumulative distribution. The vertical axis represents the cumulative value of the variable, divided by the mean. The curve starts at (0, 0) and ends up at (1, 1). Its derivative with respect to p is equal to $q^* = q/\mu$, which is the inverse of the cumulative distribution divided by the expected value. The second derivative is $1/f(q)$ which is positive. The curve is convex because the second derivative is always positive.

The following proposition states the relationship between the absolute deviation from a given quantile and Gini.

Proposition *Let q separate the two groups. Then the absolute deviation from a quantile is a function of the between-groups Gini.*

Proof It turned out to be convenient to prove the connection between the Gini coefficient and the absolute deviation from a quantile of the distribution by geometric arguments. The main property of the Lorenz curve used is its convexity. An additional simplifying assumption is that the expected value of the distribution, μ is positive. Figure 23.1 presents a Lorenz curve. The curve is 0JBKC. The section 0J is added in order to draw attention that the curve can also be defined for variables that include negative values. The line ABD is the line tangent to the curve at p, and its slope is q^*.

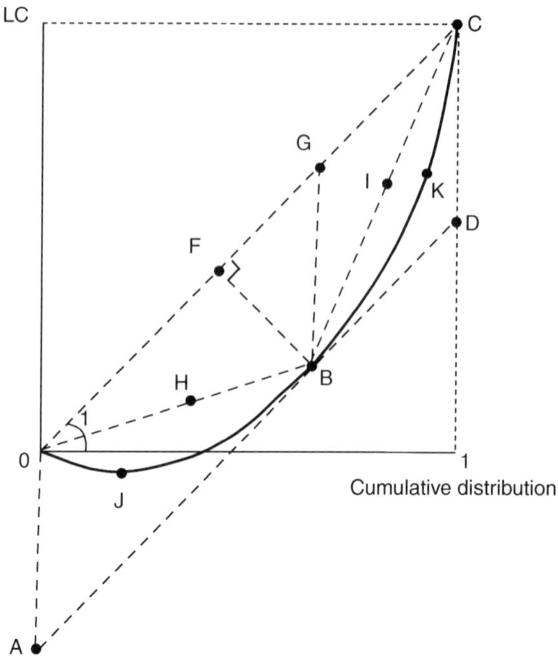

Fig. 23.1 The Lorenz curve. Source: Yitzhaki and Olkin (1988), Fig. 1

The left hand term in (23.7) is represented by $\overline{0A}$, while the right hand term is $\overline{CD}$. The absolute deviation from q (23.7) is $\overline{0A} + \overline{CD}$.

The Gini coefficient is equal to twice the area enclosed by 0JBKCGF. The between-group Gini is equal to twice the triangle defined by 0HBICGF. By Pythagoras theorem the length of the section OC is equal to $\sqrt{2}$, while $\overline{GB} = \sqrt{2}$ $\overline{FB}$. Therefore, the between-group Gini is a function of $\overline{GB}$. On the other hand, the left and right terms in (23.7) are also functions of $\bar{GB}$. To see this note that:

$$\overline{0A} = \overline{GB} + (q^* - 1)p$$

$$\overline{CD} = \overline{GB} - (q^* - 1)(1 - p)$$

So that:

$$E\{|X - q|\} = 2\overline{GB} + (q^* - 1)(2p - 1). \tag{23.9a}$$

q^* is a given constant while $p = F(q)$ is a function of the constant. Hence, the absolute deviation from q is a constant plus $\overline{GB}$. But the between-group Gini is also a function of $\overline{GB}$ so that both are determined by the same argument.

It is easy to see that we can omit the restriction $\mu > 0$. In this case, one has to use the Absolute Lorenz curve and follow the same line of the proof. We now deal with two special interesting cases: the MAD and the LAD.

(a) The relationship between MAD and Gini:
The MAD is the absolute deviation from the mean. In this case, the slope of ABD line is equal to 1. Hence, $\overline{0A} = \overline{GB} = \overline{CD}$. Therefore, the relationship between the between-groups Gini and MAD is relatively simple.

(b) The relationship between LAD and Gini:
The Least Absolute deviation is the deviation from the median.

Equating the derivative of (23.7) with respect to q to zero we get:

$$\frac{\partial E\{|X-q|\}}{\partial q} = 2p + 2qp'(q) - 1 - 2qp'(q) = 2p - 1 = 0$$

where $p = F(q)$. As is well known the minimum absolute deviation is from the median.

Then, (23.7) becomes:

$$E\{|X-q|\} = 2\overline{GB}. \tag{23.9b}$$

Having established that MAD and LAD and the absolute deviation from a quantile are equivalent to between-groups GMD implies that any property that the GMD possesses is also possessed by those measures. Alternatively, we can interpret them as the GMD for specific distributions, the distributions of the between-groups component. Therefore, we should expect future research to find out all the properties surveyed in this book applied and proved for those measures of variability. Moreover, since one limits the range of distributions to distributions of binary variables it may be that additional properties will be found. On the other hand, using only the between-groups GMD means that all the intra-group variability is ignored. As far as we can see users of MAD, LAD, and QUAD should justify dropping out this information, especially since by decomposing the GMD, one can find out what is lost by dropping the intra-groups variability. This brings us back to the debate raised by Grunfeld and Griliches (1960) whether aggregation prior to estimation is good? Alternatively, this brings us back as to whether Wainer, Gessaroli, and Verdi (2006) concept of binning is useful. In some sense the argument to prefer the MAD over the GMD is parallel to an argument suggesting the use of the between-groups variance over the variance as a way of increasing the robustness of the estimates. We are not aware of anyone suggesting the above argument. This topic is beyond the scope of this book.

23.5 A Comment on Required Software

A necessary condition for implementing a methodology is having a user-friendly software. The empirical part of this book is based on non user-friendly softwares that were developed specifically for one application. Examples of such softwares are DAD (Duclos, Araar, & Fortin, 2001; Duclos & Araar, 2006) for analyzing income distribution and poverty and Cheung et al. (2007) for portfolio construction. They are not useful for wide-spread use of the methodology. One reason for this unfortunate situation was the quick development in finding alternative formulas for the Gini, so that in many cases one has to abandon one formula for another approach. Following this book, we believe that the time is ripe for constructing a unified type of software that enables one to use all existing components of the OLS and Gini methodology in one package.

However, we believe that such a methodology will enables the zealous researcher who wants to prove his point almost anything that he wants to prove. To avoid such a situation it is worth to define standards of analysis of data so that only good quality research is published.

Following are several rules that should be investigated and agreed upon before the ground will be ripe for writing such software:

(a) The role of transformations: monotonic transformations of variables may affect the association between random variables through the change in the distribution of the variable. The OLS applies the transformations to the variables themselves, while the Gini methodology applies the transformations to the weighting schemes. The advantage of applying transformations only to the weighting schemes is that all properties of the data are kept untouched. For example, if the adding-up property is observed in the original data, it is automatically imposed on the estimation procedure in the Gini methodology. As an example, consider the estimation of a time-series of an aggregate variable and the components of the aggregate, as is the case in many cases where disaggregated time series are estimated. Since both methodologies are nonparametric in nature, we believe that almost all transformations should be banned, except a few that should be justified by the researcher.
(b) As a substitute to the flexibility offered by the use of transformations, one may consider extending the decompositions of the simple regression coefficients into the multiple regression framework. This procedure may be useful whenever the association between variables changes its sign, as may happen between the age and other variables. This procedure will allow "global" behavior of the regression curve to differ from the "sectional" behavior.
(c) However, a transformation that should be allowed is the decision of whether the model should be additive in the variables or additive in the logarithms of the variables. This is actually a decision of whether one uses an additive or multiplicative model. However, it should be kept in mind that in certain areas of applications the role of the estimated parameters can be reversed depending on whether one applies a multiplicative or an additive model. To see this let us present the following decision problem: you are given two investment opportunities. Assume that time is contiuous and one has to hold the investment for the whole period. The rate of return, r_i, is normally distributed with (μ_i, σ_i), $i = 1,2$. In an additive model, the expected return on the investment is μ_i and risk is represented by σ_i. In a multiplicative model the expected value of the rate

of return on the investment is μ_i+0.5 σ_i^2 while the risk is represented by σ_i^2. As a result one may prefer the investment with the higher variance because it yields a higher return. Yitzhaki (1987) used this difference in order to show that the rate of return of the rich on investments in the stock market is twice the rate of return of the poor. Hence, stochastic dominance rules for multiplicative models should be developed.

(d) As shown is Sect. 2.4 there are disagreements between the definitions of the GMD based on the covariance formula and those that are based on the ALC. This difference is accentuated when it is applied to relations between random variables such as Gini correlation or ANOGI. A decision on which way to follow is required.

(e) We suspect that there is a connection between the decomposition of the Gini of a linear combination of variables and ANOGI as hinted in Rao (1969). However, we did not find it yet. To see a possible connection define the linear combination of variables as:

$$Y = \sum_{i=1}^{n} I(i)\ \beta_i x_i,$$

where I(i) is the indicator function. Then the linear combination can be first decomposed according to ANOGI (using I(i)), and then decompose as a linear combination of random variables. Clearly, a combined version of decomposition will change the way decomposition is performed. Another possible direction is the one suggested in Cowell and Fiorio (2011).

23.6 Summary

In this chapter we have demonstrated the possible contributions of the GMD in different areas of research. It is clear to us that we have not exhausted all the areas that can benefit from using the properties of the GMD, nor that we have covered all the properties of the GMD. We believe that the GMD is a superior but unfortunate measure of variability that was ignored by the main stream of the literature. Its use may contribute to a better coordination between economic theory and econometric theory, and to improve the quality of statistical analysis of economic data. However, as far as we can see to fully use its properties several decades of research are needed. We hope that this book will convince other researchers to follow the Gini way.

References

Aaberge, R. (2000). Characterization of Lorenz curves and income distributions. *Social Choice and Welfare, 17*, 639–653.

Aaberge, R., Bjørklund, A., Jäntti, M., Palme, M., Pedersen, P. J., Smith, N., & Wennemo, T. (2002). Income inequality and income mobility in the Scandinavian countries compared to the United States. *Review of Income and Wealth, 48*(4), 443–469.

Achdut, L. (1996). Income inequality, income composition and macroeconomic trends: Israel, 1979–93. *Economica, 63*(250), S1–S27.

Aitchison, J., & Brown, J. A. C. (1963). *The lognormal distribution*. London: Cambridge University Press.

Alesina, A., & La Ferrara, E. (2005). Ethnic diversity and economic performance. *Journal of Economic Literature, 43*(3), 762–800.

Angrist, J. D. (1990). Life time earnings and the Vietnam era draft lottery: Evidence from social security administration records. *American Economic Review, 80*, 313–335.

Angrist, J. D., & Evans, W. G. (1998). Children and their parents' labor supply: Evidence from exogeneous variation in family size. *American Economic Review, 88*, 450–477.

Angrist, J. D., & Imbens, G. W. (1999). Comment on J.J. Heckman, Instrumental variables: A study of implicit behavioral assumptions used in making program evaluations. *Journal of Human Resources, 34*(4), 823–827.

Angrist, J. D., Imbens, G., & Rubin, D. (1996). Identification of casual effects using instrumental variables. *Journal of the American Statistical Association, 91*, 444–472.

Angrist, J. D., & Krueger, A. B. (1999). Empirical strategies in labor economics, Ch. 23. In O. Ashenfelter & D. Card (Eds.), *Handbook of labor economics* (1st ed.), Vol. 3, No. 3. Amsterdam: Elsevier.

Angrist, J. D., & Pischke, J. S. (2009). *Mostly harmless econometrics*. Princeton, NJ: Princeton University Press.

Araar, A., & Duclos, J. Y. (2003). An Atkinson-Gini type family of social evaluation functions. *Economics Bulletin, 3*(19), 1–16.

Arrow, K., & Debreu, G. (1954). Existence of equilibrium for a competitive economy. *Econometrica, 22*(3 (July)), 265–290.

Arvesen, J. N. (1969). Jackknifing U-statistics. *Annals of Mathematical Statistics, 40*, 2076–2100.

Atkinson, A. B. (1970). On the measurement of inequality. *Journal of Economic Theory, 2*, 244–263.

Atkinson, A. B. (1979). Horizontal equity and the distribution of the tax burden. In H. J. Aaron & M. J. Boskin (Eds.), *The economics of taxation*. Washington, DC: The Brookings Institution.

Atkinson, A. B. (1983). *The measurement of economic mobility*, in *social justice and public policy*. Sussex: Wheatsheaf Books Ltd.

Atkinson, A. B. (1987). On the measurement of poverty. *Econometrica, 55*, 747–764.

S. Yitzhaki and E. Schechtman, *The Gini Methodology: A Primer on a Statistical Methodology*, Springer Series in Statistics 272, DOI 10.1007/978-1-4614-4720-7,

Atkinson, A. B. (1993). *The institution of an official poverty line and economic policy*. Discussion paper no. WSP/98 (December). Suntory-Toyota International Centre for Economics and Related Disciplines, London.

Atkinson, A. B., & Bourguignon, F. (1992). *Empirical studies of earnings mobility*. Chur: Harwood.

Atkinson, A. B., & Stiglitz, J. (1980). *Lectures on public economics*. London: McGraw-Hill.

Bachi, R. (1956). A statistical analysis of the revival of Hebrew in Israel. *Scripta Hierosolymitana, 3*, 179–247.

Baker, R. D. (1997). A new perspective on the Cramer-Von Mises test. *Nonparametric Statistics, 7*, 255–277.

Barnett, V., Green, P. J., & Robinson, A. (1976). Concomitants and correlation estimates. *Biometrika, 63*, 323–328.

Barnett, V., & Lewis, T. (1978). *Outliers in statistical data*. Chinchester: John Wiley and Sons.

Barro, R. J. (1991). Economic growth in a cross-section of countries. *Quarterly Journal of Economics, CVI*, 407–444.

Barro, R. J., & Sala-i-Martin, X. (1992). Convergence. *Journal of Political Economy, 100*, 223–251.

Bartholomew, D. J. (1982). *Stochastic models for social processes* (3rd ed.). New York: John Wiley & Sons (First edition 1967).

Bassett, G., Jr., & Koenker, R. (1978). Asymptotic theory of least absolute error regression. *Journal of the American Statistical Association, 73*, 618–622.

Beach, C. M., & Davidson, R. (1983). Distribution-free statistical inference with Lorenz curves and income shares. *The Review of Economic Studies, 50*(4), 723–735.

Beach, C. M., & Richmond, J. (1985). Joint confidence intervals for income shares and Lorenz curves. *International Economic review, 26*(2), 439–450.

Beblo, M., & Knaus, T. (2001). Measuring Income Inequality in Euroland. *Review of Income and Wealth, 47*(3), 301–320.

Beenstock, M. (1996). The acquisition of language skills by immigrants: The case of Hebrew in Israel. *International Migration, 4*, 3–30.

Beenstock, M. (2002a). *Mobility and longitudinal inequality in earnings: Israel 1983–1995*. Falk Institute, Discussion paper no. 02.03.

Beenstock, M. (2002b). *Intergenerational mobility: Earnings and education in Israel*. Falk Institute, Discussion paper no. 02.05.

Beenstock, M. (2004). Rank and quantity mobility in the empirical dynamics of inequality. *Review of Income and Wealth, 50*(4), 519–541.

Beenstock, M., & Felsenstein, D. (2007). Mobility and mean reversion in the dynamics of regional inequality. *International Regional Science Review, 30*(4), 335–361.

Belsley, D. A., Kuh, E., & Welsch, R. E. (1980). *Regression diagnostics: Identifying influential data and sources of collinearity*. New York: John Wiley & Sons.

Ben Hur, D., Frantskevich, L., Schechtman, E., & Yitzhaki, S. (2010). Tracing the curvature of a regression curve. *Advances and Applications in Statistical Sciences*, 4(1), 31–56

Benabou, R. (2000). Unequal societies: Income distribution and the social contract. *American Economic Review, 90*(1), 96–129.

Ben-Porath, Y. (1980). The F-connection: Families, friends, and firms and the organization of exchange. *Population and Development Review, 6*(1 (March)), 19–39.

Bergson, A. (1938). A reformulation of certain aspects of welfare economics. *The Quarterly Journal of Economics, 52*(2 (February)), 310–334.

Berri, A., Lyk-Jensen, S. V., Mulalic, I., & Zachariadis, T. (2010). *Transport consumption inequalities and redistributive effects of taxes: A comparison of France, Denmark and Cyprus, ECINEQ WP 2010-159*. http://www.ecineq.org/milano/WP/ECINEQ2010-159.pdf.

Best, R. W., Hodges, C. W., & Yoder, J. A. (2006). Expected earnings growth and portfolio performance. *Review of Quantitative Finance and Accounting, 26*, 431–437.

Bibby, J. (1975). Methods of measuring mobility. *Quality and Quantity, 9*, 107–136.

Bibi, S., & Duclos, J.-Y. (2007). Poverty-decreasing indirect tax reforms: Evidence from Tunisia. *International Taxation and Public Finance, 14*, 165–190.

Bishop, J. A., Chakravarty, S., & Thistle, P. D. (1989). Asymptotically distribution-free statistical inference for generalized Lorenz curves. *The Review of Economics and Statistics, 71*, 725–727.

Bishop, J. A., Formby, J. P., & Smith, W. J. (1991). Lorenz dominance and welfare: Changes in the U.S. distribution of income 1967–1986. *The Review of Economics and Statistics, 73*(1), 134–139.

Bishop, J. A., Formby, J. P., & Zheng, B. (1997). Statistical inference and the Sen index of poverty. *International economic Review, 38*, 381–387.

Bliss, C. (1999). Galton's fallacy and economic convergence. *Oxford Economic Papers, 51*, 4–14.

Blitz, R. C., & Brittain, J. A. (1964). An extension of the Lorenz diagram to the correlation of two variables. *Metron, XXIII*(1–4), 137–143.

Bolt, D. M. (2001). Conditional covariance-based presentation of multidimensional test structure. *Applied Psychological Measurement, 25*, 244–257.

Boos, D. D. (1982). A test for asymmetry associated with the Hodges-Lehmann estimator. *Journal of the American Statistical Association, 77*(379 (September)), 647–651.

Borjas, G. J. (2006). *Making it in America: Social mobility in the immigrant population.* NBER WP No. 12088 (March).

Bossert, W., & D'Ambrosio, C. (2006). Reference groups and individual deprivation. *Economics Letters, 90*, 421–426.

Bossert, W., & D'Ambrosio, C. (2007). Dynamic measures of individual deprivation. *Social Choice and Welfare, 28*, 77–88.

Bossert, W., D'Ambrosio, C., & Peragine, V. (2007). Deprivation and Social Exclusion. *Economica, 74*(296), 777–803.

Boudon, R. (1973). *Mathematical structures of social mobility.* New York: Elsevier Scientific Publishing Company.

Bound, J., Jaeger, D. A., & Baker, R. M. (1995). Problems with instrumental variables estimation when the correlation between the instrument and the endogenous explanatory variable is weak. *Journal of the American Statistical Association, 90*(430 (June)), 443–450.

Bowie, C. D., & Bradfield, D. J. (1998). Robust estimation of beta coefficients: Evidence from small stock market. *Journal of Business Finance & Accounting, 25*(3/4 (April/May)), 439–454.

Breen, R., Garcia-Penalosa, C., & Orgiazzi, E. (2008). *Factor components of inequality: Cross-country differences and time changes.* LIS Working Paper Series #503.

Bruno, M., & Habib, J. (1976). Taxes, family grants and redistribution. *Journal of Public Economics, 5*, 57–79.

Budescu, D. V. (1980). Approximate confidence intervals for a robust scale parameters. *Psychometrika, 45*(3), 397–402.

Burkhauser, R. V., Frick, J. R., & Schwarze, J. (1997). A comparison of alternative measures of economic well-being for Germany and the United States. *Review of Income and Wealth, 43*(2), 153–171.

Burkhauser, R. V., & Poupore, J. G. (1997). A cross-national comparison of permanent inequality in the United States and Germany. *Review of Economics and Statistics, 79*, 10–17.

Burrell, Q. L. (2006). Measuring concentration within and co-concentration between informetric distributions: An empirical study. *Scientometrics, 68*(3), 441–456.

Calò, D. G. (2006). On a transvariation based measure of group separability. *Journal of Classification, 23*, 143–167.

Cambanis, S. (1991). On Eyraud–Farlie–Gumbel–Morgenstern random processes. In G. Dall'Aglio et al. (Eds.), *Advances in probability distributions with given marginals.* Dordrecht: Kluwer Academic Publishers.

Canner, N., Mankiw, N. G., & Weil, D. N. (1997). An asset allocation puzzle. *American Economic Review, 87*(1 (March)), 181–191.

Carty, K., Roshal, V., & Yitzhaki, S. (2009). *Inequality comparisons: The role of government and demographic structure.* Working paper. SSRN.COM.

Central Bureau of Statistics of Indonesia (1990). *National social Economic Survey.*

Central Bureau of Statistics of Israel (1989). *Family expenditure survey 1986/1987, part A: General summary*. Special Series No. 837, Jerusalem, Israel.

Central Bureau of Statistics of Israel (2009). *Family expenditure survey 2008*. Special Series No. 1363, Jerusalem, Israel.

Ceriani, L., & Verme, P. (2012). The origins of the Gini index: Extracts from Variabilità e Mutabilità (1912) by Corrado Gini. *Journal of Economic Inequality, 10*, 421–443.

Cerone, P. (2008). Bounding the Gini mean difference. *International Series of Numerical Mathematics, 157*, 77–89.

Cerone, P., & Dragomir, S. S. (2005). Bounds for the Gini mean difference via the Sonin identity. *Computers and Mathematics with Applications, 50*, 599–609.

Cerone, P., & Dragomir, S. S. (2006). Bounds for the Gini mean difference of an empirical distribution. *Applied Mathematics Letters, 19*, 283–293.

Chai, A., & Moneta, A. (2010). Engel curves. *Journal of Economic Perspectives, 24*(1), 225–240.

Chakravarty, S. R. (1983). Ethically flexible measures of poverty. *The Canadian Journal of Economics/Revue canadienne d'Economique, 16*(1 (February)), 74–85.

Chakravarty, S. R. (1988). Extended Gini indices of inequality. *International Economic Review, 29*, 147–156.

Chakravarty, S. R. (1990). *Ethical social index numbers*. Berlin: Springer-Verlag.

Chakravarty, S. R., & Mukherjee, D. (1999). Measures of deprivation and their meaning in terms of social satisfaction. *Theory and Decision, 47*, 89–100.

Chambers, R. G., & Quiggin, J. (2007). Dual approaches to the analysis of risk aversion. *Economica, 74*, 189–213.

Chand, S. K., & Parthasarathi, S. (1995). *Poverty alleviation in a financial programming framework: An integrated approach*. IMF working paper 95/29, Washington, D.C.

Chandra, M., & Singpurwalla, N. D. (1981). Relationships between some notions which are common to reliability theory and economics. *Mathematics of Operations Research, 6* (1 (February)), 113–121.

Chaudhury, M., & Ng, I. (1992). Absenteeism predictors: Least squares, rank regression, and model selection results. *The Canadian Journal of Economics/Revue canadienne d'Economique, 25*(3 (August)), 615–635.

Cheung, C. S., Kwan, C. C. & Miu, P. C. (2008), "A mean-Gini approach to asset allocation involving hedge funds", Andrew H. Chen, in (ed.) 24 (Research in Finance, Volume 24), Emerald Group Publishing Limited, pp. 197–212

Cheung, C., Sherman, C. C., Kwan, & YIP, P. C. Y. (1990). The hedging effectiveness of options and futures: A Mean-Gini approach. *Journal of Future Markets, 10*(1), 61–73.

Chotikapanich, D., & Griffiths, W. (2001). On Calculation of the extended Gini coefficient. *Review of Income and Wealth, 47*, 541–547.

Chow, K. V., Riley, W. B., & Formby, J. P. (1992). International portfolio selection and efficiency analysis. *Review of Quantitative Finance and Accounting, 2*, 47–67.

Clark, E., Jokung, O., & Kassimatis, K. (2011). Making inefficient market indices efficient. *European Journal of Operation Research, 209*, 83–93.

Clark, E. A., & Kassimatis, K. (2012). An empirical analysis of marginal conditional stochastic dominance. *Journal of Banking & Finance, 36*, 1144–1151.

Cohen, Y., & Haberfeld, Y. (1998). Second-generation Jewish immigrants in Israel: Have the Ethnic gaps in schooling and earnings declined? *Ethnic and Racial Studies, 21*(3 (May)), 507–528.

Conover, W. J. (1980). *Practical nonparametric statistics* (2nd ed.). New York: John Wiley & Sons.

Cord, L., & Wodon, Q. (2001). Do Mexico's agricultural programs alleviate poverty? Evidence from the ejido sector. *Cuadernos de Economia, 114*, 239–256.

Corlett, W. J., & Hague, D. C. (1953). Complementarity and the excess burden of taxation. *Review of Economic Studies, 21*, 21–30.

Coulter, F. A. E., Cowell, F. A., & Jenkins, S. P. (1992). Equivalence scale relativities and the extent of inequality and poverty. *The Economic Journal, 102*(414), 1067–1082.

Cowell, F. A. (1980). On the structure of additive inequality measures. *Review of Economic Studies, 47*, 521–531.

Cowell, F. A. (1984). The structure of American income inequality. *Review of Income and Wealth, 30*(3), 351–375.

Cowell, F. A. (1991). Grouping bounds for inequality measures under alternative informational assumptions. *Journal of Econometrics, 48*, 1–14.

Cowell, F. A., & Fiorio, C. V. (2012). Inequality decompositions - a reconciliation. *J Econ Inequality, 9*, 509–528.

Cramer, H. (1928). On the composition of elementary errors (Second paper: Statistical applications). *Skandinavisk Aktuareitidskrift, 11*, 141–180.

Creedy, J. (1979). The inequality of earning and the accounting period. *Scottish Journal of Political Economy, 26*, 89–96.

Creedy, J. (1991). Lifetime earning and inequality. *Economic Record, 67*, 46–58.

Crosby, F. (1979). Relative deprivation revisited: A response to Miller, Bolce, and Halligan. *The American Political Science Review, 73*(1 (March)), 103–112.

D'Agostino, R. B. (1971). An omnibus test of normality for moderate and large size samples. *Biometrika, 58*, 341–348.

D'Agostino, R. B. (1972). Small sample probability points for the D test of normality. *Biometrika, 59*, 219–221.

D'Ambrosio, C., & Frick, J. R. (2007). Income satisfaction and relative deprivation: An empirical link. *Social Indicators Research, 81*, 497–519.

Dagum, C. (1997). A new decomposition approach to the decomposition of the Gini income inequality ratio. *Empirical Economics, 22*, 515–531.

Dalton, H. (1920). The measurement of the inequality of income. *Economic Journal, 30*, 348–361 [Reprinted in the appendix to his the inequality of income. London: Routledge & Kegan Paul LTD.].

Daniels, H. E. (1944). The relation between measures of correlation in the universe of sample permutation. *Biometrika, 33*, 129–135.

Daniels, H. E. (1948). A property of rank correlation. *Biometrika, 35*, 416–417.

Daniels, H. E. (1950). Rank correlation and population models. *Journal of the Royal Statistical Society Series B (Methodological), 12*(2), 171–191.

Dardanoni, V. (1993). Measurement social mobility. *Journal of Economic Theory, 61*, 372–394.

Darling, D. A. (1957). The Kolmogorov-Smirnov, Cramer-Von Mises tests. *Annals of Mathematical Statistics, 28*, 823–838.

David, H. A. (1968). Gin'is mean difference rediscovered. *Biometrika, 55*, 573–575.

David, H. A. (1981). *Order statistics* (2nd ed.). New York: John Wiley & Sons.

David, H. A. (1998). Early sample measures of variability. *Statistical Science, 13*(4 (November)), 368–377.

Davidson, R. (2009). Reliable inference for the Gini index. *Journal of Econometrics, 150*, 30–40.

Davidson, R., & Duclos, J. Y. (1997). Statistical inference for the measurement of the incidence of taxes and transfers. *Econometrica, 65*(6), 1453–1465.

Davidson, R., & Duclos, J.-Y. (2000). Statistical inference for stochastic dominance and for the measurement of poverty and inequality. *Econometrica, 68*(6), 1435–1464.

Davydov, Y., & Egorov, V. (2000a). Functional limit theorems for induced order statistics. *Mathematical Methods of Statistics, 9*, 297–313.

Davydov, Y., & Egorov, V. (2000b). Functional limit theorems for induced order statistics of a sample from a domain of attraction of α-stable law, α ε (0, 2). In Puri, M. L. (Ed.) *Asymptotics in statistics and probability*. Papers in Honor of George Gregory Roussas, VSP, Netherland.

De Veaux, D. R. (1976). *Tight upper and lower bounds for correlation of bivariate distributions arising in air pollution modeling*. Technical Report No. 5, Department of Statistics, Stanford University.

De Vos, K., & Zaidi, M. A. (1997). Equivalence scale sensitivity of poverty statistics the members states of the European community. *Review of Income and Wealth, 43*(3), 319–333.

Deaton, A. (1979). The distance function in consumer behaviour with applications to index numbers and optimal taxation. *The Review of Economic Studies, 46*(3 (July)), 391–405.

Deaton, A. (2001). *Relative deprivation, inequality, and mortality*. NBER WP no. 8099 (January).

Deaton, A. (2005). Measuring poverty in a growing world (or measuring growth in a poor world). *Review of Economics and Statistics, 87*(1 (February)), 1–19.

Deaton, A. (2009). *Instruments of development: Randomization in the tropics, and the search for the elusive keys to economic development*. Research Program in Development Studies, Center for Health and Wellbeing, Princeton University.

DeLaubenfels, R. (2006). The victory of least squares and orthogonality in Statistics. *The American Statistician, 60*(4 (November)), 315–321.

Deltas, G. (2003). The small-sample bias of the Gini coefficient: Results and implications for empirical research. *The Review of Economics and Statistics, 85*(1), 226–234.

Dennenberg, D., & Leufer, N. (2008). Dual volatility and dependence parameters and the copula. *International Journal of Approximate Reasoning, 48*, 697–708.

Dennis, B., Patil, G. P., Rossi, O., Stehman, S., & Taille, C. (1979). A bibliography of literature on ecological diversity and related methodology. In J. F. Grassle, G. P. Patil, W. Smith, & C. Taillie (Eds.), *Ecological diversity in theory and practice, vol 1, CPH* (pp. 319–354). Fairland, MD: International Co-operative Publ. House.

Dentcheva, D., & Ruszczyński, A. (2006). Portfolio optimization with stochastic dominance constraints. *Journal of Banking and Finance, 30*, 433–451.

Denuit, M., & Dhaene, J. (2003). Simple characterizations of comonotonicity and countermonotonicity by extremal correlations. *Belgian Actuarial Bulletin, 3*, 22–27.

Deutsch, J., & Silber, J. (1999). Inequality decomposition by population subgroups and the analysis of interdistributional inequality. In J. Silber (Ed.), *Handbook of income inequality measurement* (pp. 363–397). Boston, MA: Kluwer Academic Publishers.

Diamond, P. A. (1975). A many-person Ramsey tax rule. *Journal of Public Economics, 4*, 335–342.

DiNardo, J., & Tobias, J. L. (2001). Nonparametric density and regression estimation. *Journal of Economic Perspectives, 15*(4), 11–28.

Donaldson, D., & Weymark, J. A. (1980). A single-parameter generalization of the Gini indices of inequality. *Journal of Economic Theory, 22*(1 (February)), 67–86.

Donaldson, D., & Weymark, J. A. (1983). Ethically flexible Gini indices for income distributions in the continuum. *Journal of Economic Theory, 29*, 353–358.

Dorfman, R. (1979). A formula for the Gini coefficient. *Review of Economics and Statistics, 61* (February), 146–149.

Dragomir, S. S. (2010). Weighted f-Gini mean difference for convex and symmetric functions in linear spaces. *Computers and Mathematics with Applications, 60*, 734–743.

Druckman, A., & Jackson, T. (2008). Measuring resource inequalities: The concepts and methodology for an area-based Gini coefficient. *Ecological Economics, 65*, 242–252.

Duclos, J.-Y., & Araar, A. (2006). *Poverty and equity, measurement, policy, and estimation with DAD*. Ottawa: Springer.

Duclos, J.-Y., Araar, A., & Fortin, C. (2001). *DAD: A software for distributive analysis/analyse distributive*. International Development Research Centre, Government of Canada, et CRÉFA, Laval University. http://www.mimap.ecn.ulaval.ca.

Duclus, J.-Y., Esteban, J., & Ray, D. (2004). Polarization: Concepts, measurement, estimation. *Econometrica, 72*(6), 1737–1772.

Duclos, J.-Y., & Makdissi, P. (2004). Restricted and unrestricted dominance for welfare, inequality and poverty ordering. *Journal of Public Economic Theory, 6*(1), 145–164.

Duclos, J.-Y., Makdissi, P., & Wodon, Q. (2005). Poverty-reducing tax reforms with heterogeneous agents. *Journal of Public Economic Theory, 7*(1), 107–116.

Duclos, J.-Y., Makdissi, P., & Wodon, Q. (2008). Socially improving tax-reforms. *International Economic Review, 49*(4 (November)), 1505–1537.

Durbin, J. (1954). Errors in Variables. *Review of International Statistical Institute, 22*, 23–32.

Easterly, W., & Levine, R. (1997). Africa growth tragedy: Policies and ethnic divisions. *Quarterly Journal of Economics, 112*, 1203–1250.

Ebert, U. (1997). Social Welfare when needs differ: An axiomatic approach. *Econometrica, 64*, 233–244.

Ebert, U. (1999). Using equivalence income of equivalent adults to rank income distributions when household types are different. *Social Choice and Welfare, 16*, 233–258.

Ebert, U. (2005). Optimal anti poverty programmes: Horizontal equity and the paradox of targeting. *Economica, 72*, 453–468.

Ebert, U. (2010). The decomposition of inequality reconsidered: Weakly decomposable measures. *Mathematical Social Sciences, 60*, 94–103.

Ebert, U., & Moyes, P. (2000). An axiomatic characterization of Yitzhaki's index of individual deprivation. *Economics Letters, 68*, 263–270.

Ellis, J. L., & van den Wollenberg, A. L. (1993). Local homogeneity in latent trait models: A characterization of the homogeneous monotone IRT model. *Psychometrika, 58*(3 September), 417–429.

Embrechts, P., Lindskog, F., & McNeil, A. (2003). Modelling dependence with copulas and applications to risk management. In T. R. Svetlozar (Ed.), *Handbook of heavy tailed distributions in finance* (pp. 331–384). New York: Elsevier.

Embrechts, P., McNeil, A., & Straumann, D. (2002). Correlation and dependence in risk management: Properties and pitfalls. In M. Alan & H. Dempster (Eds.), *Risk management: Value at risk and beyond* (pp. 176–223). London: Cambridge University Press.

Esteban, J., & Ray, D. (1994). On the measurement of polarization. *Econometrica, 62*, 819–852.

Esteban, J., & Ray, D. (1999). Conflict and distribution. *Journal of Economic Theory, 87*, 819–852.

Esteban, J., & Ray, D. (2001). Collective action and group size paradox. *American Political Science Review, 95*, 663–672.

Esteban, J., & Ray, D. (2005). *A model of ethnic conflict*. Working Paper, September.

Eubank, R., Schechtman, E., & Yitzhaki, S. (1993). A test for second order stochastic dominance. *Communications in Statistics, Theory and Methods, 22*(7), 1893–1905.

Fei, J. C. H., Ranis, G., & Kou, S. W. Y. (1978). Growth and the family distribution of income by factor components. *Quarterly Journal of Economics, 92*(1 (February)), 17–53.

Feldstein, M. S. (1972). Distributional equity and the optimal structure of public spending. *American Economic Review, 62*(1 (March)), 32–36.

Feldstein, M. S. (1976). On the theory of tax reform. *Journal of Public Economics, 6*, 77–104.

Feskens, R.C.W., Hox, J., Lensvelt-Mulders, G., & Schmeets, H. (2007). Nonreponse among ethnic minorities: a multivariate analysis. *Journal of Official Statistics, 23*(3), 387–408.

Fields, G. S., & Ok, E. A. (1996). The meaning and measurement of income mobility. *Journal of Economic Theory, 71*(2 (November)), 349–377.

Fields, G. S., & Ok, E. A. (1999). The measurement of income mobility. In J. Silber (Ed.), *Handbook of income inequality measurement* (pp. 557–598). Boston, MA. Kluwer Academic Publishers

Finkel, Y., Artsev, Y., & Yitzhaki, S. (2006). Inequality measurement and time structure of household income in Israel. *Journal of Economic Inequality, 4*(2), 153–179.

Fisher, G. M. (1992). The development and history of the poverty threshold. *Social Security Bulletin, 55*(4), 3–46.

Fisher, J. D., & Johnson, D. S. (2006). Consumption mobility in the United States: Evidence from two panel data sets. *The B. E. Journal of Economic Analysis & Policy, 6*(1). http://www.bepress.com/bejeap/topics/vol6/iss1/art16.

Fortin, N. M., & Lemieux, T. (1998). Rank regression, wage distribution, and the gender gap. *The Journal of Human Resources, 33*(3 (Summer)), 610–643.

Foster, J., Greer, J., & Thorbecke, E. (1984). A class of decomposable poverty measures. *Econometrica, 52*, 761–766.

Foster, J. E., & Shorrocks, A. F. (1988). Poverty ordering. *Econometrica, 56*, 173–177.

Fréchet, M. (1951). Sur les tableaux de correlation dont les marges sont données. *Annals Universite de Lyon, Sect. A. Ser., 3*(14), 53–77.

Frick, J., & Goebel, J. (2008). Regional income stratification in unified Germany using a Gini decomposition approach. *Regional Studies, 42*(4), 555–577.

Frick, R. J., Goebel, J., Schechtman, E., Wagner, G. G., & Yitzhaki, S. (2006). Using analysis of Gini (ANOGI) for detecting whether two sub-samples represent the same universe: The German Socio-Economic Panel study (SOEP) experience. *Sociological Methods and Research, 34*(4 (May)), 427–468.

Friedman, M. (1951). A method of comparing incomes of families differing in composition. *Studies in Income and Wealth*, vol. 15 (pp. 10–24). New York: National Bureau of Economic Research.

Friedman, M. (1992). Do old fallacies ever die? *Journal of Economic Literature, 30*, 2129–2132.

Fuller, M. (1979). The estimation of Gini coefficient from grouped data: Upper and lower bounds. *Economics Letters, 3*, 187–192.

Gail, M. H., & Gastwirth, J. L. (1978). A scale-free goodness of fit test for the exponential distribution based on the Gini statistic. *Journal of the Royal Statistical Society B, 40*(3), 350–357.

Galton, F. (1888). Co-relations and their measurement, chiefly from anthropometric data. *Proceedings of the Royal Society of London, 45*, 135–145.

Garner, T. I. (1993). Consumer expenditures and inequality: An analysis based on decomposition of the Gini coefficient. *Review of Economics and Statistics, 75*, 134–138.

Gastwirth, J. L. (1971). A general definition of the Lorenz curve. *Econometrica, 1971*, 1037–1039.

Gastwirth, J. L. (1972). The estimation of the Lorenz curve and Gini index. *Review of Economics and Statistics, 54*, 306–316.

Gastwirth, J. L., & Gail, M. H. (1985). Simple asymptotically distribution-free methods for comparing Lorenz curves and Gini indices obtained from complete data. In R. L. Basmann & G. Rhodes Jr. (Eds.), *Advances in econometrics, 4*. Greenwich, CT: JAI Press.

Gastwirth, J., & Glauberman, M. (1976). The interpolation of the Lorenz curve and Gini index from grouped data. *Econometrica, 44*(3), 479–483.

Gastwirth, J., Modarres, R., & Bura, E. (2005). The use of Lorenz curve, Gini index and related measures of relative inequality and uniformity in securities law. *Metron, LXIII*(3), 451–469.

Geweke, J., Marshall, R. C., & Zarkin, G. A. (1986). Mobility indices in continuous time Markov chains. *Econometrica, 54*(6), 1407–1423.

Gibbs, J. P., & Martin, W. A. (1962). Urbanization, technology and the division of Labor: International patterns. *American Sociological Review, 27*(October), 667–677.

Gibson, J., Huang, J., & Rozelle, S. (2001). Why is income inequality so low in China compared to other countries? The effect of household survey method. *Economics Letters, 71*, 329–333.

Giles, D. E. A. (2004). Calculating a standard error for the Gini coefficient: Some further results. *Oxford Bulletin of Economics and Statistics, 66*, 425–433.

Giles, D. E. A. (2006). A cautionary note on estimating the standard error of the Gini index of inequality: Comments. *Oxford Bulletin of Economics sand Statistics, 68*, 395–396.

Gili, A., & Bettuzzi, G. (1987). Concordance, ch. 6. In A. Naddeo (Ed.), *Italian contributions to the methodology of statistics* (pp. 231–243). Padova: Societá Italiana di Statistica.

Gini, C. (1914). Reprinted: On the measurement of concentration and variability of characters (2005). *Metron, LXIII*(1), 3–38.

Gini, C. (1921). Measurement of inequality of incomes. *Economic Journal, 30*(121 (March)), 124–126.

Gini, C. (1936). Sur le coefficient de correlation comme caracteristique du degree de la concordance. *Review of the International Statistical Institute, 4*(3), 355–366 (French).

Gini, C. (1965). On the characteristics of Italian statistics. *Journal of the Royal Statistical Society A (General), 128*(1), 89–109.

Gini, C. (1966). *Statistical methods*. Roma: Instituto Di Statistica E Ricerca Sociale "Corrado Gini", Universita Degli Studi di Roma.

Giorgi, G. M. (1990). Bibliographic portrait of the Gini concentration ratio. *Metron, 48*, 183–221.

Giorgi, G. M. (1993). A fresh look at the topical interest of the gini concentration ratio. *Metron, LI* (1–2), 83–98.

Giorgi, G. M., & Madarajah, S. (2010). Benferroni and Gini indices for various parametric families of distributions. *Metron, LXVIII*(1), 23–46.

Giorgi, G. M., & Pallini, A. (1987). About a general method for the lower and upper distribution-free bounds on Gini's concentration ratio from grouped data. *Statistica, XLVII*, 171–184.

Giorgi, G. M., Palmitesta, P., & Provasi, C. (2006). Asymptotic and bootstrap inference for the generalized Gini indices. *Metron, LXIV*(1), 107–124.

Glazer, N. (1993). Is assimilation dead? *Annals of the American Academy of Political Science, 530*, 122–136.

Golan, Y., & Yitzhaki S. (2010). *Who does not respond in the social survey: An exercise in OLS and Gini regressions*. Draft. Presented at the 31st IARIW, http://www.iariw.org/c2010.php.

Goldberger, A. S. (1984). Reverse regression and salary discrimination. *The Journal of Human Resources, 19*, 293–319.

Goldie, C. M. (1977). Convergence theorems for empirical Lorenz curves and their inverses. *Advances in Applied Probability, 9*, 765–791.

González-Abril, L., Valesco Morente, F., Gavilán Ruiz, J. M., & Sanchez-Reyes Fernández, L. M. (2010). The similarity between the square of the coefficient of variation and the Gini index of a general random variable. *Revista de Méthodos Cuantitativos para La Economia Y La Empresa* (December) (pp. 5–18). http://www.upo.es/RevMetCuant/art-40.pdf.

Gorard, S. (2005). Revisiting a 90-year-old debate: The advantages of the mean deviation. *British Journal of Educational Studies, 53*(4 (December)), 417–430.

Graves, S. B., & Ringuest, J. L. (2009). Probabilistic dominance criteria for comparing uncertain alternatives: A tutorial. *Omega, 37*, 346–357.

Gregory-Allen, R. B., & Shalit, H. (1999). The estimation of systematic risk under differentiated risk aversion: A mean-extended Gini approach. *Review of Quantitative Finance and Accounting, 12*, 135–157.

Grether, D. M. (1974). Correlation with ordinal data. *Journal of Econometrics, 2*, 241–246.

Groves, R. M., & Couper, M. P. (1998). *Non response in household interview surveys*. New York: John Wiley & Sons.

Groves, R. M., Dillman, D. A., Eltinge, J. L., & Little, J. A. (Eds.). (2002). *Survey non response*. New York: John Wiley & Sons.

Grunfeld, Y., & Griliches, Z. (1960). Is aggregation necessarily bad? *Review of Economics and Statistics, 42*(1 (February)), 1–13.

Gurin, P. (1985). Women's gender consciousness. *The Public Opinion Quarterly, 49*(2 (Summer)), 143–163.

Hadar, J., & Russell, W. R. (1969). Rules for ordering uncertain prospects. *American Economic Review, 59*(1 (March)), 25–34.

Hampel, F. R. (1974). The influence curve and its role in robust estimation. *Journal of the American Statistical Association, 69*(364), 383–393.

Hanoch, G., & Levy, H. (1969). The efficiency analysis of choices involving risk. *Review of Economic Studies, 36*, 335–346.

Härdle, W., & Stoker, T. M. (1989). Investigating smooth multiple regression by the method of average derivative. *Journal of the American Statistical Association, 84*(480), 986–995.

Harris, R. G. (1980). A general equilibrium analysis of the capital asset pricing model. *Journal of Financial and Quantitative Analysis, 15*(1 (March)), 99–122.

Hart, P. E. (1975). Moment distributions in economics: An exposition. *Journal of the Royal Statistical Society, A, 138*(Part 3), 423–434.

Harter, H. L. (1978). *A chronological annotated bibliography of order statistics, vol, 1: Pre-1950*. Washington, DC: U.S. Government Printing Office.

Haveman, R. H. (1987). *Poverty, policy and poverty research*. Madison, WI: University of Wisconsin Press.

Haveman, R. H. (1993). *Who are the nation's 'truly poor'? The Brookings Review*. Washington, DC: Brookings Institution.
Heckman, J. J. (1992). Randomization and social program evaluation. In C. Manski & I. Garfinkel (Eds.), *Evaluating welfare and training programs* (pp. 201–230). Cambridge, MA: Harvard.
Heckman, J. J. (1997). Instrumental variables: A study of implicit behavioral assumptions used in making program evaluations. *The Journal of Human Resources, 32*(3), 441–462.
Heckman, J. J. (1999). Instrumental variables: Response to Angrist and Imbens. *The Journal of Human Resources, 34*(4 (Autumn)), 828–837.
Heckman, J. J. (2000). Causal parameters and policy analysis in economics: A twentieth century retrospective. *Quarterly Journal of Economics, 115*, 45–97.
Heckman, J. J. (2001). Micro data, heterogeneity, and the evaluation of public policy: Nobel lecture. *Journal of Political Economy, 109*(4), 673–748.
Heckman, J. J. (2010). *Web Appendix for building bridges between structural and reduced form approaches to evaluation policy*. www.nber.org/data.../bui-bri-bet_web-app_2010-03-20a_jlt.pdf.
Heckman, J. J., & Smith, J. A. (1995). Assessing the case for social experiments. *Journal of Economic Perspectives, 9*(2), 85–115.
Heckman, J. J., Stixrud, J., & Urzwa, S. (2006). The effect of cognitive and noncognitive skills on labor market outcomes and social behavior. *Journal of Labor Economics, 24*(3), 411–481.
Heckman, J. J., & Urzua, S. (2009). *Comparing IV with structural models: What simple IV can or cannot identify*. IZA DP No. 3980.
Heckman, J. J., Urzua, S., & Vytlacil, E. (2006a). Understanding instrumental variables in models with essential heterogeneity. *Review of Economics and Statistics, LXXXVIII*(3 (August)), 389–432.
Heckman, J. J., Urzua, S., & Vytlacil, E. (2006b). *Supplement to "Understanding instrumental variables in models with essential heterogeneity"*. WEBAPPENDIX. March, http://jenni.uchicago.edu/underiv/webappendix.pdf.
Heckman, J. J., & Vytlacil, E. J. (1999). Local instrumental variables and latent variable models for identifying and bounding treatment effects. *Proceedings of the National Academy of Sciences, 96*(8), 4730–4734.
Heller, J., & Yitzhaki, S. (2006). Assigning fossil specimens to a given recent classification when the distribution of character variation is not normal. *Systematics and Biodiversity, 4*(2), 161–172.
Hettmansperger, T. P. (1984). *Statistical inference based on ranks*. New York: John Wiley and Sons.
Hey, J. D., & Lambert, P. J. (1980). Relative deprivation and the Gini coefficient: Comment. *Quarterly Journal of Economics, 95*(3), 567–573.
Hirschman, C. (1983). America's melting pot policy reconsidered. *Annual Review of Sociology, 9*, 397–423.
Hoeffding, W. (1948). A class of statistics with asymptotically normal distribution. *Annals of Mathematical statistics, 19*, 293–325.
Hopkins, E. (2008). Inequality, happiness and relative concerns: What actually is their relationship. *Journal of Economic Inequality, 6*, 351–372.
Huang, C., & Litzenberger, R. H. (1988). *Foundations for financial economics*. New York: North Holland.
Huber, P. (1981). *Robust statistics*. New York: John Wiley and Sons.
Iyengar, N. S. (1960). On a method of computing Engel elasticities from concentration curves. *Econometrica, 28*(4 (October)), 882–891.
Jaeckel, L. A. (1972). Estimating regression coefficients by minimizing the dispersion of the residuals. *Annals of Mathematical Statistics, 43*, 1449–1458.
Jäntti, M. (1997). Inequality in five countries in the 1980s: The role of demographic shifts, markets and government policies. *Economica, 64*, 415–440.
Jenkins, S. P. (1988). Calculating income distribution indices from micro data. *National Tax Journal, 41*, 139–142.

Jenkins, S. P. (1995). Accounting for inequality trends: Decomposition analyses for the UK, 1971–86. *Economica, 62*, 29–63.

Johnson, N. L., & Kotz, S. (1977). On some generalized Farlie-Gumbel-Morgenstern distributions–II: Regression, correlation, and further generalization. *Communication in Statistics, 6*, 485–496.

Jorgenson, D. W., & Slesnick, D. T. (1984). Aggregate consumer behaviour and the measurement of inequality. *The Review of Economic Studies, 51*(3 (July)), 369–392.

Juhn, C., Murphy, K., & Pierce, B. (1993). Wage inequality and the rise in returns to skill. *Journal of Political Economy, 101*(3), 410–442.

Jurečková, J. (1969). Asymptotic linearity of a rank statistic in regression parameter. *Annals of Mathematical Statistics, 40*, 1889–1900.

Jurečková, J. (1971). Nonparametric estimates of regression coefficients. *Annals of Mathematical Statistics, 42*, 1328–1338.

Kakwani, N. C. (1977). Applications of Lorenz curves in economic analysis. *Econometrica, 45*, 719–727.

Kakwani, N. C. (1980). On a class of poverty measures. *Econometrica, 48*(2 (March)), 437–446.

Kakwani, N. C. (1984). On the measurement of tax progressivity and redistributive effects of taxes with applications to horizontal and vertical equity. *Advances in Econometrics, 3*, 149–168.

Kakwani, N. C. (1995). *Income inequality, welfare and poverty: An illustration using Ukrainian data*. PRD working paper No. 1411, The World Bank. Washington, DC.

Kalton, G., & Flores-Cervantes, I. (2003). Weighting methods. *Journal of Official Statistics, 19*(2), 81–97.

Kantorowitz, M. (2002). *Reducing nonresponse bias of income estimates by integrating data from households' expenditure and income Surveys*. Paper presented at the International Conference on Improving Surveys (ICIS), Copenhagen, Denmark.

Karoly, L. (1994). Trends in income inequality: The impact of and implications for tax policy. In J. Slemrod (Ed.), *Tax progressivity and income inequality* (pp. 95–129). New York: Cambridge University Press.

Kaur, A., Rao, B. L. S. P., & Singh, H. (1994). Testing for second-order stochastic dominance of two distributions. *Econometric Theory, 10*, 849–866.

Keeney, M. (2000). The distributional impact of direct payments on Irish farm income. *Journal of Farm Economics, 51*(2 (May)), 252–263.

Kendall, M. G. (1938). A new measure of rank correlation. *Biometrika, 30*, 81–89.

Kendall, M. G. (1948). *Rank correlation methods*. London: Griffin& Co. (3rd ed., 1962).

Kendall, M. G. (1955). *Rank correlation methods* (2nd ed.). New York: Hafner Publishing Company.

Kendall, M. G., & Stuart, A. (1969). *The advanced theory of statistics*. London: Griffin.

King, M. A. (1983). An index of inequality: With application to horizontal equity and social mobility. *Econometrica, 51*, 99–115.

Klecan, L., McFadden, R., & McFadden, D. (1991). *A robust test for stochastic dominance*. Unpublished manuscript.

Kleiber, C., & Kotz, S. (2002). A characterization of income distributions in terms of generalized Gini coefficients. *Social Choice and Welfare, 19*, 789–794.

Koenker, R. (2005). *Quantile regression*. Cambridge: Cambridge University Press.

Koenker, R., & Bassett, G., Jr. (1978). Regression quantiles. *Econometrica, 46*, 33–50.

Koenker, R., & Bassett, G. (1982). Robust tests for heteroscedasticity based on regression quantiles. *Econometrica, 50*, 43–61.

Koenker, R., & Hallock, K. F. (2001). Quantile regression. *The Journal of Economic Perspectives, 15*(4), 143–156.

Kolm, S. C. (1976). Unequal inequalities. *Journal of Economic Theory, 16*, 416–442.

Konno, H., & Yamazaki, H. (1991). Mean-absolute deviation portfolio optimization model and its applications to Tokyo stock market. *Management Science, 37*(5 (May)), 519–531.

Kőszegi, B., & Rabin, M. (2007). Reference-dependent risk attitude. *American Economic Review, 97*(4 (September)), 1047–1073.

Kotz, S., & Seeger, J. P. (1991). A new approach to dependence in multivariate distributions. In G. Dall'Aglio et al. (Eds.), *Advances in probability distributions with given marginals*. Dordrecht: Kluwer Academic Publishers.

Kroll, Y., Levy, H., & Markowitz, H. M. (1984). Mean-variance versus direct utility maximization. *Journal of Finance, 39*, 47–61.

Kruskal, W. H. (1958). Ordinal measures of association. *Journal of the American Statistical Association, 53*(284 (December)), 814–861.

Lambert, P. J. (1985). Social welfare and the Gini coefficient revisited. *Mathematical Social Sciences, 9*, 19–26.

Lambert, P. J., & Decoster, A. (2005). The Gini coefficient reveals more. *Metron, LXIII*(3), 373–400.

Lambert, P. J. (2001). *The distribution and redistribution of income*. Manchester: Manchester University Press. Third edition.

Lambert, P. J. (2012). Lidia Ceriani and Paolo Verme's paper "The origins of the Gini index: Extracts from Variabilità eMutabilità (1912) by Corrado Gini". *J Econ Inequal*. 10:419–420.

Lambert, P. J., & Aronson, J. R. (1993). Inequality decomposition analysis and the Gini coefficient revisited. *Economic Journal, 103*, 1221–1227.

Lambert, P., & Yitzhaki, S. (1995). Equity, equality and welfare. *European Economic Review, 39*, 674–682.

Lasswell, T. (1965). *Class and stratum*. Boston, MA: Houghton Mifflin Company.

Layard, R. (1980). Human satisfaction and public policy. *The Economic Journal, 90*(360), 737–750.

Layard, R. (2006). Happiness and public policy: A challenge to the profession. *The Economic Journal, 116*(March), C24–C33.

Lazaridis, P. (2000). Decomposition of food expenditure inequality: An application of the extended Gini coefficient to Greek micro-data. *Social Indicators Research, 52*(2), 179–193.

Leamer, E. (1983). Let's take the con out of econometrics. *American Economic Review, 73*(1), 31–43.

Lee, A. J. (1990). *U-Statistics: Theory and practice*. New York: Marcel Dekker.

Lee, Y.-S., Wollack, J. A., & Douglas, J. (2009). On the use of nonparametric item characteristic curve estimation techniques for checking parametric model fit. *Educational and Psychological Measurement, 69*(2 (April)), 181–197.

Lehmann, E. L. (1955). Ordered families of distributions. *Annals of Mathematical Statistics, 26*, 399–419.

Lehmann, E. L. (1999). *Elements of large sample theory*. New York: Springer.

Lerman, R. I. (1999). How do income sources affect income inequality? In J. Silber (Ed.), *Handbook of income inequality measurement* (pp. 341–362). Boston, MA: Kluwer Academic Publishers.

Lerman, R. I., & Yitzhaki, S. (1984). A note on the calculation and interpretation of the Gini index. *Economics Letters, 15*, 363–368.

Lerman, R. I., & Yitzhaki, S. (1985). Income inequality effects by income source: A new approach and application to the U.S. *Review of Economics and Statistics, 67*, 151–156.

Lerman, R. I., & Yitzhaki, S. (1989). Improving the accuracy of estimates of Gini coefficients. *Journal of Econometrics, 42*(1), 43–47.

Lerman, R. I., & Yitzhaki, S. (1994). The effect of marginal changes in income sources on U.S. income Inequality. *Public Finance Quarterly, 22*(4), 403–417.

Lerman, R. I., & Yitzhaki, S. (1995). Changing ranks and inequality impacts of taxes and transfers. *National Tax Journal, 48*(1), 45–59.

Levy, H. (1989). Two-moment decision models and expected utility maximization: Comment. *American Economic Review, 79*(3 (June)), 597–600.

Levy, H. (1992). Stochastic dominance and expected utility: Survey and analysis. *Management Science, 38*, 555–593.

Levy, H. (2006). *Stochastic dominance: Investment decision making under uncertainty* (2nd ed.). New York: Springer Verlag.

Levy, H., Guttman, I., & Tkatch, I. (2001). Regression, correlation, and the time interval: Additive-multiplicative framework. *Management Science, 47*(8 (August)), 1150–1159.

Levy, H., & Markowitz, H. (1979). Approximating expected utility by a function of mean and variance. *American Economic Review, 69*(3 (June)), 308–317.

Levy, H., & Schwarz, G. (1997). Correlation and the time interval over which the variables are measured. *Journal of Econometrics, 76*, 341–350.

Lewis, J. D. (1993). *Energy pricing, tax policy, and the environment in Indonesia*. Development discussion paper no. 455. Harvard Institute for International Development. Cambridge, MA, June.

Lewis, R. J. (2000). *An introduction to classification and regression tree (CART) analysis, presented at the 2000 Annual Meeting of the Society for Academic Emergency Medicine in San Francesco*. http://www.citeseerx.ist.psu.edu.

Lieberson, S. (1969). Measuring population diversity. *American Sociological Review, 34* (December), 850–862.

Lien, D., & Tse, Y. K. (2002). Some future developments in future hedging. *Journal of Economic Surveys, 16*, 357–396.

Lissak, M. (1999). *The mass immigration in the fifties: The failure of the melting pot policy*. The Bialik Institute: Jerusalem (Hebrew).

Lord, F. M. (1970). Item characteristics curves estimated without knowledge of their mathematical form – A confrontation of Birnbaum's logistic model. *Psychometrika, 35*, 43–50.

Lord, F. M. (1980). Applications of Item Response Theory to Practical Testing Problems. Hillsdale, NJ: Lawrence Erlbaum Associates, Publishers.

Lord, F. M., & Novick, M. R. (1968). *Statistical theories of mental test scores*. Reading, MA: Addison-Wesley.

Lord, F. M. (1980). *Applications of item response theory to practical testing problems*. Hillsdale, NJ: Lawrence Erlbaum Associates, Publishers.

Lorenz, M. O. (1905). Methods for measuring concentration of wealth. *Journal of American Statistical Association, 9*(June), 209–219. New Series No. 70.

Lundin, D. (2001). Welfare-improving carbon dioxide tax reform taking externality and location into account. *International Tax and Public Finance, 8*, 815–835.

Luttmer, E. F. P. (2005). Neighbors as negatives: Relative earnings and well-being. *Quarterly Journal of Economics, 120*, 963–1002.

Mahalanobis, P. C. (1960). A method of fractile graphical analysis. *Econometrica, 28*(2), 325–351.

Makdissi, P., & Mussard, S. (2008). Analyzing the impact of indirect tax reforms on rank-dependent social welfare functions: A positional dominance approach. *Social Choice and Welfare, 30*, 385–399.

Markowitz, H. M. (1952). Portfolio selection. *Journal of Finance, 7*, 77–91.

Markowitz, H. M. (1970). *Portfolio selection: Efficient diversification of investments*. New Haven, CT: Yale University Press.

Marshall, A. W., & Olkin, I. (1979). *Inequalities: Theory of majorization and its application*. New York: Academic Press.

Mayshar, J., & Yitzhaki, S. (1995). Dalton-improving indirect tax reforms. *American Economic Review, 85*(4 (September)), 793–808.

Mayshar, J., & Yitzhaki, S. (1996). Dalton-improving indirect tax reform II—With differences in needs and ability. *Journal of Public Economics, 62*(3 (November)), 399–412.

McCall, J. J. (1991). Exchangeability and its economic applications. *Journal of Economic Dynamics and Control, 15*, 549–568.

McFadden, D. (1989). Testing for stochastic dominance. In T. Fombay & T. K. Seo (Eds.), *Studies in economics of uncertainty*. New York: Springer.

McKean, J. W., & Hettmansperger, T. P. (1978). A robust analysis of the general linear model based on one step R-estimates. *Biometrika, 65*, 571–579.

McKenzie, G. W., & Pearce, I. F. (1982). Welfare measurement—A synthesis. *American Economic Review, 72*(4 (September)), 669–682.

Mehran, F. (1975). Bounds on the Gini index based on observed points of the Lorenz curve. *Journal of the American Statistical Association, 70*, 64–66.

Merton, R. C. (1972). An analytic derivation of the efficient portfolio frontier. *Journal of Financial and Quantitative Analysis, 7*, 1851–1872.

Merton, R. K., & Kitt, A. (1950). Contributions to the theory of reference group behavior. In R. K. Merton & P. F. Lazarsfeld (Eds.), *Studies in the scope and method of "The American Soldier"* (pp. 42–53). Glencoe, IL: Free Press.

Meyer, J. (1987). Two-moment decision models and expected utility maximization. *American Economic Review, 77*(3 (June)), 421–430.

Milanovic, B. (1997). A simple way to calculate the Gini coefficient, and some implications. *Economics Letters, 56*(1), 45–49.

Milanovic, B., & Yitzhaki, S. (2002). Decomposing world income distribution: Does the world have a middle class? *Review of Income and Wealth, 48*(2 (June)), 155–178.

Millimet, D. L., & Slottje, D. (2002). Environmental compliance costs and the distribution of emissions in the U. S. *Journal of Regional Science, 42*(1), 87–105.

Millimet, D., Slottje, L. D., Yitzhaki, S., & Zandvakili, S. (2003). Return to schooling and rising inequality in the U. S. *Estadistica (Journal of the Inter-American Statistical Institute), 55*, 231–256.

Mirrlees, J. (1971). An exploration in the theory of optimum income taxation. *Review of Economic Studies, 38*, 175–208.

Mistiaen, J. A., & Ravallion, M. (2003). *Survey compliance and the distribution of income*. Washington, DC: The World Bank. June.

Modarres, R. (2008). Tests for bivariate exchangeability. *International Statistical Review, 76*(2), 203–213.

Modarres, R., & Gastwirth, J. (2006). A cautionary note on estimating the standard error of the Gini index of inequality. *Oxford Bulletin of Economics and Statistics, 68*, 385–390.

Montalvo, J. G., & Reynal-Querol, M. (2003). Religeous polarization and economic development. *Economic Letters, 80*, 201–210.

Montalvo, J. G., & Reynal-Querol, M. (2005). Ethnic polarization, potential conflict, and civil wars. *American Economic Review, 95*(3 (June)), 796–816.

Montanari, A., & Monari, P. (2005). *Gini's contribution to multivariate statistical analysis, Gini contribution to multivariate classification, regression trees.* http://www.unisi.it/eventi/GiniLorenz05/paper%2026%20may/PAPER_Monari_Montanari.pdf.

Monti, M., & Santoro, A. (2011). Stratification and between-group inequality: A new interpretation. *Review of Income and Wealth, 57*(3 (September)), 412–427.

Mookherjee, D., & Shorrocks, A. (1982). A decomposition analysis of the trend in UK income inequality. *The Economic Journal, 92*(368 (December)), 886–902.

Moothathu, T. S. K. (1985). Sampling distribution of Lorenz curve and Gini Index of the Pareto distribution. *Sankhya Series B, 47*(2), 247–258.

Moyes, P. (2007). An extended Gini approach to inequality measurement. *Journal of Economic Inequality, 5*, 279–303.

Murray, D. (1978). Extreme values for Gini coefficients calculated from grouped data. *Economics Letters, 1*, 389–393.

Mussard, S., & Richard, P. (2012). Linking Yitzhaki's and Dagum's decompositions. *Applied Economics, 44*(23), 2997–3010.

Newberry, D. (1970). A theorem on the measurement of inequality. *Journal of Economic Theory, 2*(September), 264–266.

Nielsen, L. T. (1990). Equilibrium in CAPM without a riskless asset. *Review of Economic Studies, 57*(2), 315–324.

Niewiadomska-Bugaj, M., & Kowalczyk, T. (2005). On grade transformation and its implications for copulas. *Brazilian Journal of Probability and Statistics, 19*, 125–137.

Nolan, B., & Maitre, B. (2000). A comparative perspective on trends in income inequality in Ireland. *Economic and Social Review, 31*(4), 329–350.

Nolan, B., & Smeeding, T. M. (2005). Ireland's income distribution in comparative perspective. *Review of Income and Wealth, 51*(4), 537–560.

Nygård, F., & Sandström, A. (1981). *Measuring Income Inequality*. Stockholm: Almqvist & Wiksell International.

O'Neill, D., & Van Kerm, P. (2008). An integrated framework for analysing income convergence. *The Manchester School, 76*(1 (January)), 1–20.

OECD. (2006). *What are equivalence scales, OECD Social Policy Division*. http://www.oecd.org/dataoecd/6152/35411111.pdf.

Ogryczak, W., & Ruszczyński, A. (2002). Dual stochastic dominance and related mean–risk models. *SIAM Journal on Optimization, 13*, 60–78.

Ogwang, T. (2000). A convenient method for computing the Gini index and its standard deviation. *Oxford Bulletin of Economics sand Statistics, 62*, 123–129.

Ogwang, T. (2003). Bounds of the Gini index using sparse information on mean incomes. *Review of Income and Wealth, 49*, 415–423.

Ogwang, T. (2004). Calculating a standard error of the Gini coefficient: Some further results: Reply. *Oxford Bulletin of Economics sand Statistics, 66*, 435–437.

Ogwang, T. (2006). A cautionary note on estimating the standard error of the Gini index of inequality: Comments. *Oxford Bulletin of Economics sand Statistics, 68*, 391–393.

Ogwang, T., & Rao, U. L. G. (1996). A new functional form for approximating the Lorenz curve. *Economics Letters, 52*, 21–29.

Ogwang, T., & Wang, B. (2004). A modification of Silber's algorithm to derive bounds on Gini's concentration ratio from grouped observations. *Statistica, LXIV*(4), 697–706.

Okunev, J. (1991). The generation of mean Gini efficient sets. *Journal of Business Finance and Accounting, 18*(2 (January)), 209–218.

Okunev, J., & Dillon, J. (1988). A linear programming algorithm for determining mean-Gini efficient farm plans. *Agricultural Economics, 2*, 273–285.

Olkin, I., & Yitzhaki, S. (1992). Gini regression analysis. *International Statistical Review, 60*(2), 185–196.

Pakes, A. (1982). On the asymptotic bias of Wald-type estimators of a straight line where both variables are subject to error. *International Economic Review, 23*(2 (June)), 491–497.

Pearson, K., & Lee, A. (1908). On the generalised probable error in multiple normal correlation. *Biometrika, 6*(1 (March)), 59–68.

Pedersen, A. W. (2004). Inequality as relative deprivation, a sociological approach to inequality measurement. *Acta Sociologica, 47*(1 (March)), 31–49.

Plotnick, R. (1981). A measure of horizontal inequity. *Review of Economics and Statistics, 63*, 283–287.

Podder, N. (1996). Relative deprivation, envy and economic inequality. *Kyklos, 49*, 353–376.

Portnov, B. A., & Felsenstein, D. (2010). On the stability of income inequality measures for regional analysis: Some evidence from simulation analysis and bootstrap tests. *Socio-Economic Planning Sciences, 44*, 212–219.

Post, T. (2003). Empirical tests for stochastic dominance efficiency. *Journal of Finance, 58*, 1905–1931.

Powell, J. L., Stock, J. H., & Stoker, T. M. (1989). Semiparametric estimation of index coefficients. *Econometrica, 57*(6 (November)), 1403–1430.

Prais, S. J. (1955). Measuring social mobility. *Journal of the Royal Statistical Society, A118*, 56–66.

Puyenbroeck, T. V. (2008). Proportional representation, Gini coefficients, and the principle of transfers. *Journal of Theoretical Polotics, 20*(4), 498–526.

Pyatt, G. (1976). On the interpretation and disaggregation of Gini coefficient. *The Economic Journal, 86*(342), 243–255.

Pyatt, G. (1987). Measuring welfare, poverty and inequality. *The Economic Journal, 97*, 459–467.

Pyatt, G., Chen, C., & Fei, J. (1980). The distribution of income by factor components. *Quarterly Journal of Economics, 95*(3), 451–473.

Quah, D. T. (1993). Galton's fallacy and tests of the convergence hypothesis. *Scandinavian Journal of Economics, 95*, 427–443.

Quah, D. T. (1996). Empirics for economic growth and convergence. *European Economic Review, 40*, 1353–1375.

Quenouille, M. H. (1949). Approximate tests of correlations in time-series. *Journal of Royal Statistical society Series B, 11*, 68–84.

Ramasubban, T. A. (1958). The mean differences and the mean deviations of some discontinuous distributions. *Biometrika, 45*, 549–556.

Ramasubban, T. A. (1959). The generalized mean differences of the binomial and Poisson distributions. *Biometrica, 46*, 223–229.

Ramasubban, T. A. (1960). Some distributions arising in the study of generalized mean differences. *Biometrika, 47*, 469–473.

Ramsay, J. O. (1995). A similarity-based smoothing approach to nondimensional item analysis. *Psychometrika, 60*, 323–339.

Randles, M., & Wolfe, D. A. (1979). *Introduction to the theory of nonparametric statistics*. New York: John Wiley and Sons.

Rao, R. C. (1982). Diversity: Its measurement, decomposition, apportionment and analysis, *Sankhya, Series A, 44*(Pt. 1), 1–22.

Rao, V. M. (1969). Two decompositions of concentration ratio. *Journal of the Royal Statistical Society Series A (General), 132*(3), 418–425.

Rashid, M., (1995). *Romania poverty assessment*. Washington, DC: World Bank. Unpublished.

Ravallion, M. (1993). *Poverty comparisons*. Switzerland/Philadelphia, PA: Chur/Harwood.

Ravallion, M. (1994a). Measuring social welfare with and without poverty lines. *American Economic Review, 84*, 359–364.

Ravallion, M. (1994b). *Poverty comparisons*. Switzerland: Harwood Academic Publishers.

Raveh, A. (1989). Gini correlation as a measure of monotonicity and two of its usages. *Communication in Statistics, Theory and Methods, 18*(4), 1415–1423.

Rilstone, P. (1991). Nonparametric hypothesis testing with parametric rates of convergence. *International Economic Review, 32*(1 (February)), 209–227.

Romanov, D., & Nir, M. (2010). Get it or drop it? Cost-benefit analysis of attempts to interview in household surveys. *Journal of Official Statistics, 26*(1), 165–191.

Rothschild, M., & Stiglitz, J. E. (1970). Increasing risk: I. a definition. *Journal of Economic Theory, 2*, 225–253.

Runciman, W. G. (1966). *Relative deprivation and social justice*. London: Routledge and Kegan Paul.

Ruszczyński, A., & Vanderbei, R. J. (2003). Frontier of stochastically nondominated portfolios. *Econometrica, 71*, 1287–1297.

Ryu, H. K., & Slottje, D. J. (1999). Parametric approximations of the Lorenz curve. In J. Silber (Ed.), *Handbook on income inequality measurement* (pp. 291–313). Boston, MA: Kluwer Academic Publishers.

Sala-i-Martin, X. (1996). Regional cohesion: Evidence and theories of regional growth and convergence. *European Economic Review, 40*, 1325–1352.

Samuelson, P. A. (1967). General proof that diversification pays. *Journal of Financial and Quantitative Analysis, 2*, 1–13.

Saunders, P. (2005). The $85 billion tax/welfare churn. *Issue analysis*, 57. Sydney: Centre for Independent Studies.

Schechtman, E. (2005). Stratification: Measuring and inference. *Communications in Statistics, Theory and Methods, 34*(11), 2133–2145.

Schechtman, E., Shelef, A., Yitzhaki, S., & Zitikis, R. (2008). Testing hypotheses about absolute concentration curves and marginal conditional stochastic dominance. *Econometric Theory, 24*, 1044–1062.

Schechtman, E., Soffer, E., & Yitzhaki, S. (2008). *The robustness of conclusions based on TIMSS mean grades, first draft.*

Schechtman, E., & Yitzhaki, S. (1987). A measure of association based on Gini's mean difference. *Communications in Statistics, Theory and Methods, 16*, 207–231.

Schechtman, E., & Yitzhaki, S. (1999). On the proper bounds of the Gini correlation. *Economics Letters, 63*(2), 133–138.

Schechtman, E., & Yitzhaki, S. (2003). A family of correlation coefficients based on the extended Gini index. *Journal of Economic Inequality, 1*(2), 129–146.

Schechtman, E., & Yitzhaki, S. (2008). Calculating the extended Gini coefficient from grouped data: A covariance presentation. *Bulletin of Statistics & Economics, 2*(Spring), 64–69.

Schechtman, E., & Yitzhaki, S. (2009). Ranking groups' abilities—Is it always reliable? *International Journal of Testing, 9*(3), 195–214.

Schechtman, E., Yitzhaki, S., & Artsev, Y. (2007). The similarity between mean-variance and mean – Gini: Testing for equality of Gini correlations. *Advances in Investment Analysis and Portfolio Management (AIAPM), 3*, 103–128.

Schechtman, E., Yitzhaki, S., & Artsev, Y. (2008). Who does not respond in the household expenditure survey: An exercise in extended Gini regressions. *Journal of Business & Economic Statistics, 26*(3), 329–344.

Schechtman, E., Yitzhaki, S., & Pudalov, T. (2011). Gini's multiple regressions: Two approaches and their interaction. *Metron, LXIX*(1), 65–97.

Schmid, F., & Trede,M. (1996). *Nonparametric inference for second order stochastic dominance.* (Unpublished manuscript).

Scholz, F.-W. (1978). Weighted median regression estimates. *The Annals of Statistics, 6*(3 (May)), 603–609.

Schweizer, B., & Wolff, E. F. (1981). On nonparametric measures of dependence for random variables. *The Annals of Statistics, 9*(4), 879–885.

Semyonov, M., & Cohen, Y. (1990). Ethnic discrimination and the income of majority-group workers. *American Sociological Review, 55*(1 (February)), 107–114.

Sen, A. K. (1973). *On economic inequality*. Oxford: Clarendon Press.

Sen, A. K. (1976a). Poverty: An ordinal approach to measurement. *Econometrica, 44*, 219–231.

Sen, A. K. (1976b). Real national income. *Review of Economic Studies, 43*(1 (February)), 19–39.

Sen, A. K. (1979). Issues in the measurement of poverty. *Scandinavian Journal of Economics, 79*, 285–307.

Sen, A. K. (1983). Poor, relatively speaking. *Oxford Economic Papers, 35*(2 (July)), 153–169.

Sen, P. K. (1986). The Gini coefficient and poverty indexes: Some reconciliations. *Journal of the American Statistical Association, 81*, 396. (Theory and Methods), 1050–1057.

Serfling, R. J. (1968). Contributions to central limit theory for dependent variables. *The Annals of Mathematical Statistics, 39*(8), 1158–1175.

Serfling, R. J. (1980). *Approximation theorems of mathematical statistics*. New York: John Wiley and Sons.

Serfling, R. (2010). *Autoregressive models via Yule-Walker equations allowing heavy tail innovations.* Working paper. http://www.utdallas.edu/~serfling/papers/Serfling2010_AR_MR.pdf.

Serfling, R., & Xiao, P. (2007). A contribution to multivariate L-moments: L-comoment matrices. *Journal of Multivariate Analysis, 98*, 1765–1781.

Shalit, H. (2010). *Using OLS to test for normality, DP No. 09-12*. Monaster Center for Economic Research, Ben-Gurion University of the Negev, Beer Sheva, Israel.

Shalit, H., & Yitzhaki, S. (1984). Mean-Gini, portfolio theory and the pricing of risky assets. *Journal of Finance, 39*(5 (December)), 1449–1468.

Shalit, H., & Yitzhaki, S. (1989). Evaluating the mean-Gini approach selection to portfolio selection. *International Journal of Finance Spring, 1*(2), 15–31.

Shalit, H., & Yitzhaki, S. (1994). Marginal conditional stochastic dominance. *Management Science, 40*, 670–684.

Shalit, H., & Yitzhaki, S. (2002). Estimating Beta. *Review of Quantitative Finance and Accounting, 18*(2 (March)), 95–118.

Shalit, H., & Yitzhaki, S. (2003). An asset allocation puzzle: Comment. *American Economic Review, 93*, 1002–1008.

Shalit, H., & Yitzhaki, S. (2005). The mean-Gini efficient portfolio frontier. *Journal of Financial Research, XXVIII*(1 (Spring)), 59–75.

Shalit, H., & Yitzhaki, S. (2009). Capital market equilibrium with heterogeneous investors. *Quantitative Finance, 9*, 757–766.

Shalit, H., & Yitzhaki, S. (2010). How does beta explain stochastic dominance efficiency? *Review of Quantitative Finance and Accounting, 35*, 431–444.

Shayo, M. (2007). *A theory of social identity with an application to redistribution*. mimeo.SSRN.com.

Shelef, A., & Schechtman, E. (2011). A Gini-based methodology for analyzing time series, draft. SSRN.com.

Sheshinski, E. (1972). Relation between a social welfare function and the gini index of income inequality. *Journal of Economic Theory, 4*, 98–100.

Shih, W. J., & Huang, W. (1992). Evaluating correlation with proper bounds. *Biometrics, 48*, 1027–1213.

Shorrocks, A. J. (1978a). Measurement of mobility. *Econometrica, 46*, 1013–1024.

Shorrocks, A. J. (1978b). Income inequality and income mobility. *Journal of Economic Theory, 19*, 376–393.

Shorrocks, A. F. (1983). Ranking income distributions. *Economica, 50*, 3–17.

Shorrocks, A. F. (1984). Inequality decomposition by population subgroups. *Econometrica, 52*(6), 1369–1385.

Shorrocks, A. J. (1993). On the hart measure of income mobility. In M. Casson & J. Creedy (Eds.), *Industrial concentrations and economic inequality*. Cambridge: Edward Elgar.

Sievers, G. L. (1978). Weighted rank statistics for simple linear regression. *Journal of the American Statistical Association, 73*, 628–631.

Sievers, G. L. (1983). A weighted dispersion function for estimation of linear models. *Communication in Statistics, Theory and Methods, 12*(10), 1161–1179.

Sijtsma, K., & Molenaar, I. W. (1987). Reliability of test scores in nonparametric item response theory. *Psychmetrika, 52*, 79–97.

Silber, J. (1989). Factor components, population groups, and the computation of the Gini index of inequality. *Review of Economics and Statistics, 71*(2), 107–115.

Silber, J. (1990). On a new algorithm to derive bounds of Gini concentration ratio from grouped observations. *Statistica, L*, 215–220.

Silber, J. (1995). Horizontal inequity, the Gini index, and the measurement of distributional change. *Research on Economic Inequality, 6*, 379–392.

Simaan, Y. (1997). Estimation risk in portfolio selection: The mean variance model versus the mean absolute deviation model. *Management Science, 43*(10 (October)), 1437–1446.

Simes, R. J. (1986). An improved Bonferroni procedure for multiple tests of significance. *Biometrika, 73*, 751–754.

Simpson, E. H. (1949). Measurement of diversity. *Nature, 163*, 688.

Slemrod, J. (1992). Taxation and inequality: A time-exposure perspective. In J. M. Poterba (Ed.), *Tax policy and the economy* (pp. 105–128). Cambridge, MA: MIT Press.

Slemrod, J., & Yitzhaki, S. (2001). Integrating expenditure and tax decisions: The marginal cost of funds and the marginal benefit of projects. *National Tax Journal, 65*(2 (June)), 189–201.

Slemrod, J., & Yitzhaki, S. (2002). Tax avoidance, evasion, and administration. In A. Auerbach & M. Feldstein (Eds.), *Handbook of public economics, 3, Ch. 22* (pp. 1423–1470). Amsterdam: Elsevier Science B.V.

Smirnov, N. V. (1937). On the Distribution of the w^2 criterion of von Mises. *Recreational Mathematics (NS), 2*, 973–993.

Spearman, C. (1904). The proof and measurement of association between two things. *American Journal of Psychology, 15*, 72–101.

Spencer, B. D. (1983a). On interpreting test scores as social indicators: Statistical considerations. *Journal of Educational Measurement, 20*(4 Winter), 317–333.

Spencer, B. D. (1983b). Test scores as social statistics: Comparing distributions. *Journal of Educational Statistics, 8*, 249–269.

Spencer, B. D., & Fisher, S. (1992). On comparing distributions of poverty gaps. *Sankhyā: The Indian Journal of Statistics Series B, 54*(1 (April)), 114–126.

Staiger, D., & Stock, J. H. (1997). Instrumental variables regression with weak instruments. *Econometrica, 65*, 557–586.

Stark, O., Taylor, J. E., & Yitzhaki, S. (1986). Remittances and inequality. *Economic Journal, 96*(September), 722–740.

Stark, O., & Yitzhaki, S. (1988). Labour migration as a response to relative deprivation. *Journal of Population Economics, 1*(1), 57–70. A similar version is reprinted in Maital S. (Ed.) (1989). Applied behavioral economics. Brighton Wheatsheaf Books NY and New York University Press.

Stigler, S. M. (1974). Linear functions of order statistics with smooth weight functions. *Annals of Statistics, 2*, 676–693.

Stigler, S. M. (1989). Francis Galton's account of the invention of correlation. *Statistical Science, 4*(2), 73–86.

Stouffer, S. A., Suchman, E. A., DeVinney, L. C., Star, S. A., & Williams, R. M. (1949). *The American soldier: Adjustment during army life*. Princeton, NJ: Princeton University Press.

Stuart, A. (1954). The correlation between variate-values and ranks in samples from a continuous distributions. *British Journal of Statistical Psychology, 7*, 37–44.

Suits, D. (1977). Measurement of tax progressivity. *American Economic Review, 67*, 747–752.

Taguchi, T. (1981). On a multiple Gini coefficient and some concentrative regressions. *Metron, XXXIX*(1–2), 39.

Taguchi, T. (1987). On the structure of multivariate concentration—some relationships among the concentration surface of two variate mean difference and regressions. *Computational Statistics & Data Analysis, 6*, 307–334.

Thistle, P. D. (1993). Negative moments, risk aversion, and stochastic dominance. *Journal of Financial and Quantitative Analysis, 28*(2), 301–311.

Tobin, J. (1970). On limiting the domain of inequality. *Journal of Law and Economics, 13*, 263–277. Reprinted in Edmund S. Phelps (ed.), Economic justice (pp. 447–463). Baltimore: Penguin Books Inc.

Trajtenberg, M., & Yitzhaki, S. (1989). The diffusion of innovations. A methodological reappraisal. *Journal of Business & Economic Statistics, 7*(1 (January)), 35–47.

Van Praag, B., & Ferrer-i-Carbonell, A. (2008). *Happiness quantified: A satisfaction calculus approach*. Oxford: Oxford University Press.

Von Mises, R. (1931). *Wahrscheinlichkeitsrechnung*, Leipzig-Wien.

Wainer, H., Gessaroli, M., & Verdi, M. (2006). Finding what is not there through the unfortunate binning of results: The Mendel effect. *Chance, 19*(1), 49–52.

Wald, A. (1940). The fitting of straight lines if both variables are subject to errors. *The Annals of Mathematical Statistics, 11*(3 (September)), 284–300.

Whiteford, P. (2006). The welfare expenditure debate: Economic myths of the left and the right revisited. *The Economic and Labour Relations Review, 17*(1), 33–78.

Wilks, S. S. (1962). *Mathematical statistics*. New York: John Wiley & Sons.

Wodon, Q. (1999). Between group inequality and targeted transfers. *Review of Income and Wealth, 45*(1 (March)), 21–39.

Wodon, Q. (2001). Income mobility and risk during the business cycle: Comparing adjustments in labour markets in two Latin-American countries. *Economics of Transition, 9*(2), 449–461.

Wodon, Q., & Yitzhaki, S. (2002a). Evaluating the impact of government programs on social welfare: The role of targeting and the allocation rules among program beneficiaries. *Public Finance Review, 30*(1 (January)), 100–121.

Wodon, Q., & Yitzhaki, S. (2002b). Inequality and social welfare. In Klugman, J. (Ed.), *Poverty reduction strategy papers sourcebook*. Washington, DC: The World Bank. http://povlibrary.worldbank.org/files/11044_chap2.pdf.

Wodon, Q., & Yitzhaki, S. (2003a). Inequality and the accounting period. *Economics Bulletin, 4*(36), 1–8.

Wodon, Q., & Yitzhaki, S. (2003b). The effect of using grouped data on the estimation of the Gini income elasticity. *Economics Letters, 78*, 153–159.

Wodon, Q., & Yitzhaki, S. (2006). Convergence forward and backward. *Economics Letters, 92*, 47–51.

Wodon, Q., & Yitzhaki, S. (2009). *Does growth lead to deprivation*. Mimeo. http://ssrn.com.

Wold, H. (1935). A study of the mean difference, concentration curves and concentration ratios. *Metron, 12*(2), 39–58.

Wolff, E. N. (1996). International comparisons of wealth inequality. *Review of Income and Wealth, 42*(4), 433–451.

Wolff, E. N., & Zacharias, A. (2007a). The distributional consequences of government spending and taxation in the US, 1989 and 2000. *Review of Income and Wealth, 53*(4), 692–715.

Wolff, E. N., & Zacharias, A. (2007b). *Class structure and economic inequality*. Working paper no. 487, Levy Economics Institute of Bard College, January.

Wong, W. K., & Au, T. K.-K. (2004). *On two-moment decision models and expected utility maximization*. Department of Economics, National University of Singapore. ecswwk@nus.edu.sg.

World Bank. (1991). *Poverty reduction handbook*. Washington, DC: The World Bank.

Xu, K. (2007). U-statistics and their asymptotic results for some inequality and poverty measures. *Econometric Reviews, 26*, 567–577.

Xu, K., Fisher, G., & Wilson, D. (1995). *New distribution-free tests for stochastic Dominance*. Working paper No. 95-02, February, Dept. of Economics, Dalhousie University, Halifax, Nova Scotia.

Yaari, M. E. (1987). The dual theory of choice under risk. *Econometrica, 55*(1 (January)), 95–115.

Yaari, M. E. (1988). A controversial proposal concerning inequality measurement. *Journal of Economic Theory, 44*(2), 381–397.

Yitzhaki, S. (1979). Relative deprivation and the Gini coefficient. *Quarterly Journal of Economics, 93*(2 (May)), 321–324.

Yitzhaki, S. (1982a). Stochastic dominance, mean variance, and the Gini's mean difference. *American Economic Review, 72*, 178–185.

Yitzhaki, S. (1982b). Relative deprivation and economic welfare. *European Economic Review, 17*, 99–113.

Yitzhaki, S. (1982c). A tax programming model. *Journal of Public Economics, 19*, 107–120.

Yitzhaki, S. (1983). On an extension of the Gini inequality index. *International Economic Review, 24*(3 (October)), 617–628.

Yitzhaki, S. (1987). On the Relation between return and income. *Quarterly Journal of Economics, 102*(February), 77–95.

Yitzhaki, S. (1990). On the sensitivity of a regression coefficient to monotonic transformations. *Econometric Theory, 6*(2), 165–169.

Yitzhaki, S. (1991). Calculating jackknife variance estimators for parameters of the Gini method. *Journal of Business & Economics Statistics, 9*(2), 235–239.

Yitzhaki, S. (1994a). Economic distance and overlapping of distributions. *Journal of Econometrics, 61*, 147–159.

Yitzhaki, S. (1994b). On the progressivity of commodity taxation. In W. Eichhorn (Ed.), *Models and measurement of welfare and inequality* (pp. 448–465). Heidelberg: Springer-Verlag.

Yitzhaki, S. (1996). On using linear regression in welfare economics. *Journal of Business & Economic Statistics, 14*(4 (October)), 478–486.

Yitzhaki, S. (1998). More than a dozen alternative ways of spelling Gini. *Research on Economic Inequality, 8*, 13–30.

Yitzhaki, S. (1999). Necessary and sufficient conditions for dominance using generalized Lorenz curves. In D. J. Slottje (Ed.), *Advances in econometrics, income distribution and scientific methodology: Essays in honor of Camilo Dagum*. Heidelberg: Physica-Verlag.

Yitzhaki, S. (2002). Do we need a separate poverty measurement. *European Journal of Political Economy, 18*, 61–85.

Yitzhaki, S. (2003). Gini's mean difference: A superior measure of variability for non-normal distributions. *Metron, LXI*(2), 285–316.

Yitzhaki, S. (2010). Is there room for polarization. *The Review of Income and Wealth., 56*(1 (March)), 7–22.

Yitzhaki, S., & Eisenstaedt, M. (2003). Ranking groups' versus individuals' ranking. In Y. Amiel & J. A. Bishop (Eds.), *Fiscal policy, inequality, and welfare, research on economic inequality, 10* (pp. 101–123). Amsterdam: JAI.

Yitzhaki, S., & Golan, Y. (2010). Monotonic and non-monotonic relationship in the labor market. *Economic Quarterly, 56*(2), 143–168 (Hebrew).

Yitzhaki, S., & P. Lambert. (2011). The relationship between Gini's mean difference and the absolute deviation from a quantile. Unpublished, http://ssrn.com

Yitzhaki, S., & Lerman, R. I. (1991). Income stratification and income inequality. *Review of Income and Wealth, 37*(3 (September)), 313–329.

Yitzhaki, S., & Lewis, J. (1996). Guidelines on searching for Dalton-improving tax reforms: Illustrations with data from Indonesia. *The World Bank Economic Review, 10*(3), 541–562.

Yitzhaki, S., & Mayshar, J. (2002). *Characterizing efficient portfolios.* Available at: http://papers.ssrn.com/sol3/papers.cfm?abstract_id=297899. Unpublished.

Yitzhaki, S., & Olkin, I. (1988). *Concentration curves.* WP no. 179, Dept. of Economics, The Hebrew University and technical report no.230, 1987, Dept. of Statistics, Stanford University. http://statistics.stanford.edu/~ckirby/techreports/NSF/OLK%20NSF%20230.pdf.

Yitzhaki, S., & Olkin, I. (1991). Concentration indices and concentration curves. In Karl Mosler and Marco Scarsini (Eds.), *Stochastic orders and decisions under risk, 19* (pp. 380–392). Institute Of Mathematical Statistics: Lecture-Notes Monograph Series.

Yitzhaki, S., & Schechtman, E. (2004). The Gini instrumental variable, or the "double instrumental variable" estimator. *Metron, LXII*(3), 287–313.

Yitzhaki, S., & Schechtman, E. (2005). The properties of the extended Gini measures of variability and inequality. *Metron, LXIII*(3), 401–443.

Yitzhaki, S., & Schechtman, E. (2009). The "Melting Pot": A success story? *Journal of Economic Inequality, 7*, 137–151.

Yitzhaki, S., Itzhaki, R., & Pudalov, T. (2011). A nonparametric ICC using the Gini's mean difference approach. Unpublished, http://ssrn.com

Yitzhaki, S., & Schechtman, E. (2012). Identifying monotonic and non monotonic relationships. *Economics Letters, 116*, 23–25.

Yitzhaki, S., & Semrod, J. (1991). Welfare dominance: An application to commodity taxation. *American Economic Review, 81*(3 (June)), 480–496.

Yitzhaki, S., & Wodon, Q. (2004). Inequality, mobility, and horizontal inequity. In Amiel, Y., & Bishop, J. A. (Eds.), *Research on economic inequality, studies on economic well-being. Essays in honor of John P. Formby*, 12 (pp. 177–198).

Yntema, D. B. (1933). Measures of the inequality in the personal distribution of wealth or income. *Journal of the American Statistical Association, 28*(184 (December)), 423–433.

Zangwill, I. (1914). *The melting pot: Drama in four acts.* New York: Macmillan.

Zenga, M. (1987). Concentration measures. In A. Naddeo (Ed.), *Italian contributions to the methodology of statistics* (pp. 42–82). Padua: Societa Italiana Di Statistica.

Zhang, D. (2009). A demonstration of the non-necessity of marginal conditional stochastic dominance for portfolio inefficiency. *The Quarterly Review of Economics and Finance, 49*, 417–423.

Zhang, X., & Kanbur, R. (2001). What difference do polarization measures make? An application to China. *Journal of Development Studies, 37*(3 (February)), 85–98.

Zheng, B. (1996). *Statistical inference for testing marginal changes in Lorenz and generalized Lorenz curves.* Working paper, University of Colorado at Denver.

Zheng, B. (2002). Testing Lorenz curves with non-simple random samples. *Econometrica, 70*(3), 1235–1243.

Zitikis, R. (2003). Asymptotic estimation of the E-Gini index. *Econometric Theory, 19*, 587–601.

Author Index

S. Yitzhaki and E. Schechtman, *The Gini Methodology: A Primer on a Statistical Methodology*, Springer Series in Statistics 272, DOI 10.1007/978-1-4614-4720-7,

Subject Index

S. Yitzhaki and E. Schechtman, *The Gini Methodology: A Primer on a Statistical Methodology*, Springer Series in Statistics 272, DOI 10.1007/978-1-4614-4720-7,

Printed by Printforce, the Netherlands